Energy Efficiency
in Household Appliances
and Lighting

Springer
Berlin
Heidelberg
New York
Barcelona
Hong Kong
London
Milan
Paris
Singapore
Tokyo

Paolo Bertoldi · Andrea Ricci
Anibal de Almeida
Editors

Energy Efficiency in Household Appliances and Lighting

With 280 Figures
and 155 Tables

 Springer

Dr. Paolo Bertoldi
European Commission, DG TREN
Rue de la Loi, 200
1049 Brussels
Belgium

Dr. Andrea Ricci
ISIS
Via Flaminia 21
00196 Rome
Italy

Prof. Anibal de Almeida
University de Coimbra
Dep. Eng. Electrotecnica
3030 Coimbra
Portugal

ISBN 3-540-41482-7 Springer-Verlag Berlin Heidelberg New York

Library of Congress Cataloging-in-Publication Data
Die Deutsche Bibliothek – CIP-Einheitsaufnahme
Energy efficiency in household appliances and lighting: with 155 tables / Paolo Bertoldi
... ed. – Berlin; Heidelberg; New York; Barcelona; Hong Kong; London; Milan; Paris;
Singapore; Tokyo: Springer, 2001
   ISBN 3-540-41482-7

Springer-Verlag Berlin Heidelberg New York
a member of BertelsmannSpringer Science+Business Media GmbH

© Springer-Verlag Berlin · Heidelberg 2001
Printed in Germany

Cover-Design: Erich Kirchner, Heidelberg
SPIN 10793833   42/2202-5 4 3 2 1 0 – Printed on acid-free paper

Preface

Household appliances encompass a large variety of equipment including the cold appliances (refrigerators and freezers), the wet appliances (washing machines, dishwashers and dryers), the space conditioning appliances (heaters, air-conditioners, heat pumps, fans, boilers), the water heaters, the cooking appliances, a wide array of consumer electronics (such as TVs, VCRs, HiFi systems) and miscellaneous small appliances (such as vacuum cleaners, irons, toasters, hairdryers and power tools). Household appliances save a large amount of domestic labour to perform the household tasks, as well as provide comfort conditions and convenience to the household occupants.

The European Community SAVE Programme has promoted the efficient use of energy, in particular in domestic appliances. SAVE has sponsored a variety of studies to characterise the use of the main household appliances and lighting and to identify cost-effective technical options to improve the energy efficiency, as well as to identify the strategies to promote the penetration of efficient equipment in the market place. National energy agencies, independent experts and appliance manufacturers have participated in the SAVE activities and have done a remarkable job. While the energy efficiency of the main household appliances has been improved, at the same time it was possible in most cases to improve the appliance performance, reliability and quality of service.

Household appliances use over 600 TWh of electricity and over 1600 TWh of gas per year in the European Union. Significant progress has been achieved in the past decade in the advancement of energy efficiency in the household, but with the savings potential still available, a large effort is required to continue the progress carried out so far. Considering only the electrical appliances, it is estimated that the use of the best available cost-effective technology could save over 180 TWh/year.

Domestic Lighting is responsible for about 5%, that is about 100 TWh/year, of the electricity consumption in the European Union. The development of compact fluorescent lamps was a milestone in residential lighting, by slashing the consumption of inefficient and low duration incandescent bulbs. The large scale application of compact fluorescent lamps can save about 25 TWh/year in the European Union. The continuous improvement of improved lamps, ballasts, fixtures and intelligent controls can contribute to further reduce the electricity consumption by over 30% of the present consumption.

The 1997 Kyoto Conference defined CO_2 emission targets for the developed regions of the world. The EU target of decreasing the emissions 8% below the 1990 level, by 2010, will require a very substantial effort covering basically all activities if such a target is to be reached. Energy-efficient household appliances and lighting can provide one of the most important opportunities to achieve electricity and gas savings in a cost effective way, avoiding at the same time the emission of hundreds of millions of tons of carbon.

The reduction of energy consumption through improvements in energy efficiency is one of the major instruments for developed and developing countries to meet the Kyoto commitments. Energy efficiency is also a key element of the European Union (EU) energy policy, since it improves the efficiency of the economy, increases energy supply security, and decreases harmful emissions due to electricity generation and gas combustion.

The 2nd International Conference on Energy Efficiency in Household Appliances and Lighting held in Naples, 27-29 September, 2000, follows on from the very successful First International Conference on Energy Efficiency in Household Appliances, held in Florence, Italy, 10-12 November, 1997. There is evidence that the discussions and networking resulting from these conferences play an important step forward to better define research, development, standardisation, policies and programmes to promote energy-efficient appliances and lighting around the world.

The 2nd International Conference on Energy Efficiency in Household Appliances and Lighting brought together over 200 prominent experts, representing a diversity of stakeholders from all over the world, providing a variety of experiences and perspectives, to discuss the latest developments covering:

- New technology developments, in different types of appliances and lighting, covering both research results as well as innovative applications
- Policies and policy instruments, including labelling and minimum performance standards
- International programmes for market transformation and energy efficiency improvements as well as novel alternatives for financing and promoting energy-efficient appliances and lighting projects

The conference main message was to substantially reinforce the policies and programmes for market transformation for domestic appliances and lighting, through the use of standards, labels, procurement and R&D. In particular international co-operation between countries and economic areas would enhance the efforts and results of individual countries, and meet the increasing globalisation of manufacturers of domestic appliances.

This book contains the keynote presentations made in the general sessions, as well as most the papers presented in the parallel sessions. It is hoped that the availability of this book will enable a large audience to benefit from the presentations made in this conference. Potential readers who may benefit from this book include researchers, engineers, policymakers, energy agencies, electric utilities, and all those who can influence the design, selection, application and operation of electric motor systems. The book is structured into the following chapters:

- General Sessions
- Washing Machines and Detergents Technology
- Refrigeration Appliances and Vacuum Panel Technology
- Gas and Electric Installed Appliances (Heating, Water Heating, Cooling)
- Motor and IT Technology for Appliances
- Consumer Electronics
- Domestic Lighting (Technology, DSM Programmes)
- Test Methods for Appliances
- End-use Monitoring (In-situ Monitoring, Smart Metering, Tariff)
- Policies and Governmental Programmes (Standards, Labelling)
- Policies and Governmental Programmes (DSM, Action Plan)

The 2nd International Conference on Energy Efficiency in Household Appliances and Lighting had as its main sponsor the SAVE II Programme (Specific Actions for Vigorous Energy Efficiency), which was adopted by the Council of the European Communities. We also wish to acknowledge the following institutions that helped by sponsoring the event: ANPA, ADEME, ASSIL, ASSOLUCE, CECED, ENEA, ENEL, FEDERELETTRICA, NAPOLETANAGAS and SAES GETTERS. The support of all the sponsors and their generous funding was very important in ensuring the success of this conference and is greatly appreciated.

The Editors

Paolo Bertoldi
Andrea Ricci
Anibal de Almeida

Contents

Opening Speech

Gunther Hanreich

Director for Energy Efficiency and Renewable Energy Sources
European Commission
Directorate Energy and Transport

Mr. Chairman, Ladies and Gentlemen,

First of all I would like to welcome you on behalf of Directorate-General Energy and Transport of the European Commission to the second International Conference on Energy Efficiency in Domestic Appliances and Lighting. We are proud to see so many participants from every corner around the word to this conference that we have organised and sponsored.

I believe that we are here to take stock of the recent progresses in technologies, policies and programmes to advance energy efficiency in this important sector of our economy, and most important to discuss the actions needed in this area to bring us to Kyoto and beyond.

When we met in Florence three years ago, we had great hopes for the Kyoto meeting. At the time the EU negotiating position was for a CO_2 emission reduction of minus 15%. We knew that this was what we needed to get a better world and more sustainable future for our children. We also believed at the time that this would have resulted in a much greater attention to energy efficiency and to the adoption of policies and programmes which would have speed up the huge economical and technical potential available in domestic appliances and lighting.

Whatever was the result of Kyoto and of the subsequent COP meetings, we have experienced during the last three years the same difficulties and barriers to energy efficiency, which have continued to hinder the adoption of the right technologies and policies.

Let me give you some facts and figures on the importance of this sector for the European Union.

The residential sector consumed 612.6 TWh of final electricity and 1619 TWh of final gas[1] in the EU in 1996. Of the electrical energy lighting accounted for about 85 TWh, space and water heating about 200 TWh and appliances for 350 TWh. Residential final electricity consumption grew by an average of 2.8% per annum

---

[1] For the residential and tertiary sectors combined

from 1990 to 1996. Residential and tertiary sector final gas consumption grew by an average of 5.4% per annum over the same period. In total EU residential electricity consumption will have given rise to about 276 million tonnes of CO_2 emissions in 1996. The total EU CO_2 emissions in 1996 of the residential sector were of about 700 million tonnes.

Recent SAVE sponsored studies give a rough indication that annual electricity savings of more than 180 TWh/year are potentially achievable across the EU by substituting appliances with efficiencies at the least life cycle cost for those in the current stock. As this coincides with the least life cycle product cost it would also save European consumers billions of Euro in life cycle costs for the appliances concerned. Annual running cost savings of about 18 billion Euro would be achieved through the increase in efficiency but this would be partially offset by the rise in purchase price of new products. In reality this substitution would occur over time as new appliances join the stock and old ones are retired.

What we have done in the last past three years to address this very large saving potential and what will do in the near future?

Let me start with the first question. We have continued our standard and labelling actions for domestic appliances and lighting. Minimum efficiency requirements together with labelling are the most important policy instruments in this area and are at the centre of our strategy. These instruments have proven to be very effective in transforming the market. We have implemented new labels for dishwashers and lamps, and now for the majority of domestic appliances on show in our shops are labelled.

A very important step was achieved with the entry into force of the minimum efficiency requirement for refrigerators and freezers in September 1999. After the implementation of the minimum efficiency standards Directive in September 1999 the average efficiency of cold appliances offered for sale is in the region of 74 to 76% of the 1992 average efficiency, suggesting that they would consume about 27% less energy than equivalent appliances sold in 1992. This represents an average annual energy efficiency improvement over the period of about 4.3% per year.

During the Florence conference we announced the conclusion of two negotiated agreements, one for TV and VCRs and the second for washing machines. I can report that the washing machines agreement resulted in the period 1994 to 1999 in average efficiency improvements of 15%, while for TV and VCRs in the period 1996 to 1999 the efficiency improvement in stand-by consumption has been of 55% and 36% respectively.

At the beginning of this year the Commission presented a comprehensive and ambitious Action Plan, which cover all the energy consuming sectors. The Action Plan aims at refocusing, strengthening and expanding existing measures and programmes. Moreover the Action Plan proposes the launch of new policies and measures.

The Action Plan includes new actions for the domestic and lighting sectors for which an average economic saving potential of 20% has been identified. I stress the fact that this is an economic potential which would benefit our citizens, our economies and in the end the whole society. And this without taking into account the large environmental benefits associated with the reduction CO_2 emissions and other pollutants due to the energy production.

What have we proposed as additional measures to move the market towards more efficient appliances and lighting?

First of all we will continue to work in close collaboration with manufacturers. The collaboration established in technical-economic analyses and in the subsequent policies has proven successful. However, we want to make sure that the negotiated agreement process is more transparent and that it really addresses the economic potential. The case of electric storage water heather shows that not all the agreement offered by manufacturers are ambitious and a large part of the saving potential is missed. We hope that manufacturers can deliver much more of what they have agreed to do.

To this end we plan to present in the coming months a Framework Directive which will raise the ambitions of negotiated agreements and, should it prove necessary, facilitate the adoption of mandatory minimum efficiency standards based on pre-established economic criteria, such as minimum life cycle cost. We will continue to give preference to negotiated agreement if they are ambitious. Future agreements will cover air conditioners, dishwashers, driers, electric motors, circulation pumps, and lamps.

It is important that labelling and standard are dynamic and continue to follow technological progress. A new SAVE study for refrigeration appliances has identified a new life cycle cost minimum corresponding to appliance with an energy efficiency index of about 50%. At the same time are appearing on the market appliances with energy efficiency index in the lower thirties. Both the new refrigeration appliance label and the soon to be discuss new efficiency standards shall move quickly the market to these levels and save society large amount of money. Efficiency improvements of the same order of magnitude exist for air conditioner, electric water heaters, and lamps.

Our policies are not only focused on phasing out from the market the lowest efficiency appliances, but also to stimulate the purchase and development of the most efficient models. An important SAVE project is testing at EU level the co-operative procurement instrument for a super efficient refrigerator. If successful, is shall considerably expand the market share of the most efficient refrigerators.

Another important area for action identified in the Action Plan are the so-called "installed appliances". Minimum efficiency requirements and the foreseen label for hot water boilers are not enough. There is the need to promote correct sizing of boilers according to the real need and good system design of the heating system. To this end we plan to amend the labelling Framework Directive 92/75 by adding measures to achieve efficient installed systems through best practice information, labelling and its extension into local information schemes. Additional supporting measures such as a public database to disseminate information on models, efficiency levels and prices will be implemented either by agreement with manufacturers or, failing this, by the above mentioned amendment to Directive 92/75 prescribing such databases.

Stand-by energy consumption is also a major source of energy waste in the domestic sector. A recent IEA study has identified that about 1% of global CO_2 emission are due to stand-by energy consumption, which is by definition wasted energy. The Commission has been very active on this topic and following our Communication to the European Parliament and Council of 1999, we have proposed to industry two voluntary Code of Conducts to reduce the stand-by power consumption of external power supplies and set top boxes. We are keen to continue our efforts on this topic and to join forces with other nations, which are also fighting stand-by consumption. I would like to acknowledge the positive response to our external power supply Code of Conduct by several companies, including several non-EU companies. I hope that our levels can become the de-facto world standard.

Lighting is also an important area for energy saving in the domestic sector. Here a well known technology has been available now for a number of years: the Compact Fluorescent Lamps. However too a high price together with the lack of suitable luminaires has prevented the widespread penetration of this efficient technology. Recently the price for CFLs has come down to a more realistic and sustainable level. Yet misinformation and lack of well design luminaires still impede their use in most households. We have tried to transform the market for CFLs. First, we have introduced the energy label for lamps, which shall improve the visibility of the label and draw consumers' attention to energy savings in lighting. Now we currently co-sponsoring a major campaign together with Eurelectric to promote CFLs to be carried out by the electricity supply companies. We hope that millions of EU citizens can receive correct information on the benefit of this light source. We are also trying to introduce well-designed luminaires on the market, which can take only CFL. The first European design competition has shown that this is possible, and if this action will be successful in the market we would soon be saying good-by to the old incandescent lamp.

The Commission has managed to create a partnership with domestic appliance manufacturers. I believe that it is of the utmost importance to continue and strengthen this partnership and extend it to all the other "stakeholders": builders, architects, designers, installers, retailers, regions, municipalities, consumers, because, in my view, this is essential to reach the consumption reduction objectives.

It is important that with electricity and gas utilities have a more pro-active role in promoting efficiency to household through active energy services. The progressive electricity and gas market liberalisation must translate in a bigger attention to the individual client and new types of energy services must be proposed to respond to the increasing demand of higher added value in the services provided. There is the risk that electricity and gas utilities initially will focus their attention on their most important client, such as the industrial user. Moreover they might neglect the need of the domestic customer, who individually consume little, but whom together are responsible for about one third of the demand. In the worst case it may be that some utilities will make the domestic clients pay for the services given to the industrial clients.

During the eighties and early nineties there has been a continuous development of demand side activities in the domestic sector, primarily targeting appliances and compact fluorescent lamps. This type of activities seems to have completed died due to the market liberalisation process and to increased competition.

A recent joint Commission/Eurelectric study has proposed an action plan to stimulate active energy services in the residential sector. We will do our best effort to help implement the action plan.

The Commission remains of the view that there is a need to increase the focus on the electricity and gas industry's role in promoting the development and use of energy services. Commitments will therefore be sought from utilities and service companies on a voluntary basis to include energy efficiency along with performance contracting and similar proven approaches to market energy efficiency as a part of their corporate goal, provided they meet normal criteria for cost-effectiveness. These efforts will be directed toward correcting the institutional barrier resulting from the continued practice of selling energy in the form of kWh instead of efficient heating and cooling, lighting and motive power, the services which the consumer actually wants. The use of information technology in providing energy services and technical assistance in reducing consumption should also help and accelerate this process.

Another important actors are the retailers. Retailers have a key role in the purchase choice made by customers. If the retailers do not mention the possible energy savings or if does not draw the client's attention to the energy label, this may not influence the purchase choice. Moreover it is important that retailers stock high efficiency appliances. Recent SAVE studies have identified their important role in co-operative procurement and market transformation especially for lighting.

I can assure you that the subject of this conference is very close to my heart and to that of my collaborators at the European Commission. I would like to conclude by launching an appeal: our activities to promote energy efficiency in domestic appliances and lighting must be continued and strengthened. And this is very urgent as well, if we want to bring a substantial contribution to the reduction of CO_2 emissions efforts and this at very low cost for our society. We, at the European Commission, are going to put all our efforts toward this goal, but we need two things:

First: a strong and constructive political support by the EU Institutions to make our proposal operational. I hope in a speedy implementation of the Action Plan and of the soon to be presented Framework Directive for efficiency requirements of electrical appliances.

Second: the full collaboration and participation of all the interested parties: manufacturers, utilities, retailers, installers, and consumer organisations,

Thank you for your attention and I wish you a very interesting and useful conference.

Speech by the Ministry

Angelo Rega

Italian Ministry of Industry and Trade

Before analysing the general theme of the Conference, some preliminary remarks should be made to focus on the objectives and viable solutions to be taken into account to reach the main common targets of the EU Fifteen Countries and the Commission.

The **macro-objectives** of both European and National actions to promote the rational use of energy, particularly in the field of domestic end-use are the following:

1. **the security of supply** – that is a constant variable of any energy policy - looking at the rational use of energy as a "virtual" source of primary energy;
2. **the environmental protection policy,** considering international agreements as the most effective tool to fulfil its goals.
3. last but not least, the **competition within the European industrial system** vs the US and Japanese markets, in which, even if we do not find juridical barriers for the entrance of EU products, the access and penetration in these markets are rather difficult due to the differences existing in product certification procedures. Moreover, in the Japanese case, it is also important to stress the great integration between their national distribution and production sectors.

Generally speaking, in the last few years European economies have been characterised by the **decreasing role of the State** in favour of private initiative. The most striking example is the creation of internal electrical and natural gas markets and the consequent ending of state monopolies. Moreover, the **administrative decentralization** in progress is diverting the implementation of energy programmes and actions, thus reducing the States' role to the definition and respect of the rules in force. We can therefore affirm that there is a consistency problem between the objectives of environmental policies and the energy and industrial market trends.

In fact, from one side governments are still involved in policies aiming at reducing climate changes with consequent effects on markets; on the other side industrial players feel this presence "intrusive" and an obstacle to market competition (subsidies without security, binding regulations greatly affecting industrial policies, actions aiming at forcing the relation between offer and demand, etc).

In my opinion, this Conference should address energy efficiency policies as follows: "Energy, environment and industry: how can they be combined to reach the macro-objectives of household appliances end-use, without jeopardising the general market trends?" If we say that it is not possible to find a right comprise between them, we give a too simple answer to this question as all the a.m. objectives are still absolutely necessary for Europe.

A sound opportunity has been offered by **voluntary agreements** between manufacturers, all notified and accepted by the EU DG Competition: they in fact represent a first example of exception of competition mechanisms in favour of a higher common interest, i.e. the environmental protection. If compared to directives, voluntary agreements can guarantee a higher level of the average energy efficiency of appliances in very short time and according to well-defined procedures, while the standards set forth by any directive do not provide an equivalent reliability.

However, for their own nature, voluntary agreements do not guarantee the 100% coverage of the whole offer, being manufacturers not obliged to sign them. If the offer is very diversified, it is also complex to handle such an agreement. On the European level, we can recall a successful example of voluntary agreement reached on dishwashing machines, notified and accepted by EU DG Competition.

In Italy, last year National Conference on Energy and Environment stressed the importance of adopting and promoting voluntary agreements.

A further step towards a better implementation of policies could be made by providing the industrial sector with an incentive mechanism for products that are energy efficient and competitive in the market. We can also refer to something similar to the mechanism adopted in Italy to promote electricity from renewable sources, known as the mechanism of "green paper". Therefore, a mechanism that – although having a distortion effect on the market - can ensure an economic incentive to manufacturers who are asked – on their turn – to produce with lower costs and in compliance with some well-defined rules (thus increasing energy efficiency).

In the household appliances sector, incentives (that could be also represented by "bonus") should be addressed to high energy efficient products able to access and compete the market for their low production costs. Even in this case, we will face a classical market distortion, but this will be more sustainable if we consider that such products will meet competition and environmental targets.

The a.m mechanisms should be guaranteed for all the time period necessary to increase the average energy efficiency and to promote an information public campaign among end-users who are currently well aware of the great importance of energy efficiency especially in this period of oil pre-shock.

Of course, it is absolutely worth considering the great role played by the Commission in the promotion and implementation of these mechanisms as it would give them the necessary transparency and increase competition among industrial players, as well.

I think this will be an interesting discussion topic of this international conference organised in the framework of the SAVE II Programme.

Thank you for your attention.

Energy Efficiency of our Appliances in the Household Industry

Mr. Hans-Peter Haase

President of CECED

Dear ladies and gentlemen,

First of all I would like to thank Mr. Bertoldi who has invited me here today, as President of CECED, to give you an update on the major achievements in our goal, to increase the Energy Efficiency of our Appliances in the Household Industry.

CECED stands for "European Committee of Manufacturers of Domestic Equipment". This European manufacturing sector is operating at the forefront of global markets, and maintains a significant contribution to the EU economy.

CECED is represented today by:
-13 National Associations of manufacturers and
-13 Direct Members, e.g. companies with own operations in European countries

CECED Membership

Direct Members	National Associations
Atag/ETNA	Austria
Bosch Siemens	Belgium
Brandt	Denmark
Candy	France
De' Longhi	Germany
Electrolux	Italy
Fagor	The Netherlands
Gorenje	Norway
Merloni	Spain
Miele	Sweden
Philips	Switzerland
Seb	Turkey
Whirlpool	United Kingdom

Naples, September 27 th 2000

ceced

Main Figures on the Industrial Sector:

- •200,000 employees, directly, and approximately 500,000 overall, considering upstream and downstream activities

- •280 companies, 90% of which are small to medium independent enterprises

- •Turn-over of approximately, 35 billion Euro
    - *63 % from large appliances*
    - *17 % from small appliances*
    - *20 % from HVAC appliances*

- •50 million large appliance units produced in Europe every year

The industrial main figures of our Association:

- The household appliance industry reports a turn-over of approximately 35 billion EURO, produced by more than 280 companies with 200.000 direct employees, and about 500.000 employees in total.

- Approximately 50 million large and 200 million small appliances, such as vacuum cleaners, irons or coffee makers, are sold in Europe every year. Thanks to the work of domestic appliances, 200 million hours are freed up from the consumers domestic hard work routine every day.

- Each single unit has a limited environmental impact, and even the cumulative effect represents a limited share of the total environmental issue. Nevertheless, our industry does not ignore the relationship between environment and use of domestic appliances.

With respect to this responsibility, the European Household Appliance Industry, organised within CECED, shares the targets declared in the Kyoto Protocol and agrees with the European Community's commitment to reduce CO_2 emissions.

It is our goal to supply durable and sustainable technology, conserving natural resources. This includes not only processes and materials used during production, but also includes the reduction of the environmental impact of the product during its use, and at the end of its life.

We are equally committed to the continuous research and development of appliances to improve the quality of life of our consumers.

The industry therefore commits itself specifically to the following priorities:

Firstly: delivering affordable, innovative, high-quality products to consumers to enhance their home lifestyles.

Secondly: providing a continued programme to improve energy efficiency and to reduce environmental impact during the whole lifecycle of the product, from the cradle to the grave!

The household appliance industry has worked in the past for the constant development of new technology. The remarkable results are:

Washing machines more than 20% reduction in 4 years achieved (equal to 7 Terawatthours per year)

Dishwashers 20% reduction in 6 years committed

El. water heaters ~15% reduction in 4 years committed

These are the results of our efforts to reduce energy consumption.

But, ladies and gentlemen, producing more efficient appliances by our industry and offering them to the market is not enough!

They must be chosen and installed in households to deploy their effect!

In addition, in order to achieve the Kyoto protocol targets as fast as possible, the existing appliances in the households must be changed to more efficient ones. That is what we call *early replacement.*

To illustrate the leverage effect of such a market transformation and early replacement, look at the following chart.

The potential of replacing products
with efficient ones

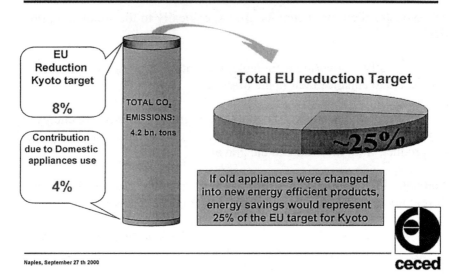

EU Reduction Kyoto target

8%

Contribution due to Domestic appliances use

4%

TOTAL CO$_2$ EMISSIONS: 4.2 bn. tons

Total EU reduction Target

~25%

If old appliances were changed into new energy efficient products, energy savings would represent 25% of the EU target for Kyoto

Naples, September 27 th 2000

ceced

Displayed on the left side is the total CO_2 equivalent emissions and the related CO_2 savings potential in the EU. The 4% indicates the total CO_2 emissions caused by electricity use in Domestic Appliances. Indeed, our sector is only one among other industries involved in the efforts of CO_2 reduction, and each industry deserves an appropriate level of intervention, depending on its contribution to the emission of CO_2.

The proposed savings at Kyoto are 8% for the whole EU. This 8% is the pie shown on the right side. 25% of this, which stands for ¼ of the total savings target, could be achieved by early replacement. Furthermore the energy efficient products to do so are already offered by the industry!

It is clear, that such a radical market transformation and early replacement is a long term strategic issue.

But the household appliance industry is convinced of its significance!

It is therefore important that there has to be an active, progressive commitment of all relevant bodies, with a clear knowledge about the costs of replacement and who is going to bear them.

This requires a long lasting policy to shift the demand of this consumer segment and to change their established habits.

This process should be complemented by the actions of all stakeholders and be focused on two goals:

No. 1: to increase consumer awareness towards an efficient pattern.

No. 2: to remove social barriers which limit the penetration and early replacement of energy efficient products.

The European household appliance industry is ready to be proactive within the European Climate Change Programme, in order to achieve the targeted reduction of emissions at the lowest cost for the society.

In order to take this process further, we have identified two strategic options:

Firstly: *Our technological options:*

European manufacturers have invested heavily over the last decades in efforts to reduce the environmental impact of household appliances. This is mostly driven by the strong competition between manufacturers on the market, but is also influenced by regulatory authorities setting thresholds for the energy consumption and other limits.

The development of production according to Energy Efficiency, shown with the product group refrigeration and washing as an example:

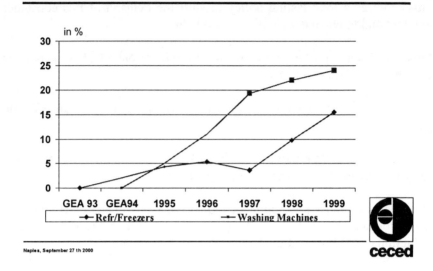

ceced

Experience has shown that voluntary agreements within the industry can be a bonus compared to less efficient governmental regulations. They are especially favourable in:

    - providing fast implementation,
    - finding the most effective approach in utilising the full knowledge of the industry, and
    - addressing the whole transformation of the market, not just the phase out of the least efficient products.

At present, the following voluntary agreements are in place:

Washing Machines	established in 1996
Dishwashers	established in 1999/2000
Electric Storage Water Heaters	established in 1999/2000

Let us make sure that we can continue in our successful co-operation with authorities, to contribute to the protection of our environment, in the spirit of voluntary agreements!

    The second strategic option: *the market transformation options:*

Technological development and innovations by the industry are only fruitful for the environment if we can implement a continuous process of market transformation. This requires a strong co-operation between all stakeholders.

Today, the household appliance industry has expressed its commitment to further contributions to market transformation. The emphasis must move from energy efficiency to the market aspects of Rational Use of Energy Programmes, anticipating EU top-down market transformation initiatives!

This is what the CECED strategy is focusing on: an interaction of interested parties via voluntary agreements, in order to reach the goal of low environmental impact with the highest level of transparency towards the consumer.

Two main areas are to be covered by market transformation:

- Further support of Energy Labelling activities by communication and incentives in order to raise consumers sensibility.
- Appropriate Social programmes to shift demand towards more efficient products, especially for the medium to low-income families. Traditional rebate programmes do not really address this area.

In this context I would like to make some remarks regarding the consistency of data, declared by individual manufacturers for the Energy Labels. There has been debate about the reliability and comparability of manufacturers data declarations.

Ladies and gentlemen, let me assure you that <u>none</u> of the manufacturers, will accept a distortion of the market, by allowing a competitor to cheat on a declaration, <u>nor will we accept</u> risking our image with our appliances as we have invested so much (about 200 Million Euro) to get our appliances fulfilling the requirements of the Energy label directives.

Manufacturers, consumers, and testing labs have conducted intensive investigations looking into the reasons for deviations. They have shown that
- different laboratory procedures and materials,
- different interpretation of standards,
- not considering necessary tolerances between measurements, and
- differences between hand-made samples and mass production
are mainly responsible for these deviations.

CECED has set-up an efficient system for the market surveillance of the Energy Label on a voluntary basis, to complement National Authorities enforcement activity. The main target of this verification procedure is to introduce a *self-control-system* within the competition of manufacturers.

In the case of wrong declaration, the challenged manufacturer has to inform his business partners, re-label the appliances, and renew the sales documents. The total costs of the verification procedure has to be paid by the "looser"!

It is our goal and responsibility: the consumer must receive an "A" appliance, if an "A" is declared by the manufacturer.
Let me summarise the achievements in protecting our environment. *The household appliance industry*

- *works proactively* within the European Climate Change Programme to achieve the targeted reduction of emissions,
- *is committed to improve the Energy Efficiency* of their products continuously!
- *agrees to reduce the environmental impact* during the whole life-cycle of the product, i.e. from the cradle to the grave,
- *delivers affordable, innovative, high quality products* to enhance consumers life-styles,
- *supports the European Labelling strategies* on a voluntary base,
- *take actions to avoid label declarations*, which can not be justified by test results from qualified labs,
- *contributes to market transformation* and early replacement actions to the Rational Use of Energy (RUE).

But we now need appropriate programmes to shift demand towards more efficient products, as it has been proposed by the representative of the Italian Ministry of Industry.

Thank you for your attention.

Domestic Lighting and Energy Efficiency: the Assoluce (the Italian Association of Lighting Manufacturers of Federlengo-Arredo) Position

Riccardo Sarfatti

Cahirman of AssoLuce, Milano, Italy

Good afternoon, ladies and gentlemen. Before going on with what I have to say, I would like to thank Mr. Bertoldi for having invited me here to speak to you.

My thanks are especially sincere because I particularly appreciate Mr. Bertoldi's invitation, as he well knows how my positions and those of the body I have the honour of chairing, ASSOLUCE, are quite different from the prevailing opinions in the majority of European bodies dealing with the subjects of this conference. At times, we have strongly disagreed and today I will illustrate the main reasons for this.

First of all, a few words about ASSOLUCE. It is one of the two Italian associations grouping together manufacturers of lighting appliances, in particular those manufacturing products considered "decorative", that is, those products which are more directly oriented for use in the home. There are just under a hundred member companies but they are all manufacturers of lighting appliances, whilst the other association (ASSIL) also has manufacturers of components for lighting and light bulbs (a few large multinationals) alongside manufacturers of lighting appliances.

The lighting sector in Italy is, by production value, second in Europe manufacturer (2.5 billion Euro), but the first in terms of exports. The Italian sector of decorative lighting products is the worldwide leader, with a share of 20% of world exports: I believe that this role as a leader comes from the fact that Italy is the only country in the world where all typologies of lighting appliances are produced. Manufacturers of all dimensions, from craftsmen to highly industrialized firms, produce every type of light possible: from wrought iron lamps and those with glass blown by the masters of the art in Murano or Tuscany, traditional, technical and decorative lamps up to those of the great Italian design.

This wealth and organization of production, in the modern world, can only be found in Italy; in particular, in no other country is there a similar wealth of firms with a high organizational and technical level that produce decorative and design lighting appliances. The members of ASSOLUCE also include the only two lighting appliance companies that have been awarded the prestigious ECDP (European Community Design Prize) by the European Community.

I mentioned earlier our disagreement on the currently prevailing opinions in Europe on energy saving in the home. I will now try and illustrate the reasons for this.

The first can be found in the very title of this conference, "Energy efficiency in household appliances and lighting": the distinction made between "lighting" and "appliances" clearly indicates the realization that "lighting" has different characteristics from all other "appliances" which use electricity in the home. Let us try and understand which elements lead us to consider "lighting" as something particular in the use of electricity in the home.

First of all, the characteristics of production.

The other household appliance manufacturing sectors (refrigerators, TVs, VCRs, ovens, cookers, water-heaters, air-conditions, stereos etc.) are characterized by a relatively small number of manufacturing firms, all with an industrial organization, of medium/large dimensions (over 300 employees), whilst the lighting sector in Europe has 8,000 manufacturers, with only about ten of these having more than 300 employees and, probably not more than thirty have 100 employees.
Today, many speakers have mentioned "voluntary agreements" as ideal ways and tools to reach sectorial agreements for the reduction of energy consumption and the consequent improvement in efficiency. How can we think of making "voluntary agreements" in a context with 8,000 subjects? Admitting this impossibility must not by any means lead to tools being imposed from above as suitable which, as well as being contrary to the logic of the free market, would never be accepted by the market itself in any case.

In the second place, the characteristics of the product. Unlike lighting appliances, other household appliances work from a limited power range. Televisions, refrigerators and ovens absorb powers that have by now been fairly well identified by technological development and consequently their performance standards can be defined and manufacturers can reach voluntary agreements for their improvement.
On the other hand, lighting appliances for the home use quite a wide wattage range (from 1 Watt up to 300 Watt). In addition, all the parts of the illuminating body, as a whole, are required to carry out its specific function. In the illuminating body, unlike other household appliances which have their "engine" and technology inside a casing, the external and visible part, contributes to the function of giving light or allowing its use. In this situation of differentiated power and a variety of solutions, the definition of standards of efficiency is rather more problematic.
For both these first two reasons (manufacturing characteristics and product characteristics) **we have been fighting for some time now against any hypothesis**, which is actually supported by many, **of classifying illuminating appliances, considering only the exclusive parameter of energy efficiency.**

But there are other reasons, perhaps even more decisive, for opposing classification using only the parameter of efficiency and they are all the more decisive because they deal directly with what **should always be the first parameter of reference for everything we do** (although unfortunately that is not always so): man, **the individual**. I would like each and every one of you to reflect in depth on what the consequences would be of a classification of illuminating bodies according to the exclusive parameter of energy efficiency. The first consequence would inevitably be to put at the top of the classification those appliances consisting merely of a low consumption energy source and a reflector to reflect the majority of the light emitted by the bulb-engine into the environment. Appliances like that, without any protective and diffusing element placed between the bulb and the environment would certainly be the most efficient. But we all know very well that our homes are not lit by appliances of this type: and I sincerely hope that our homes will not be lit by appliances of this type in the future either. As far as I am concerned, I am certain that they will not be, not even if some directive tries to impose them. I think that we all know well how our life is made up of both light or dark times or light and dark times together, and we also know that in our homes it is indispensable for us to enjoy lights, shadow, semi-darkness, the light and the dark. Diffusers of light, which man has always placed between light bulbs and the environment, whether made of glass, fabric, rice paper, perforated plating or plastics, are fundamental in determining the quality of light in any given environment. The diffusion of light and the reduction or elimination of glare produce different levels of visual comfort essential for the quality of human life. Well, friends, these indispensable diffusing elements reduce the light emitted by the bulb by between 40% and 50% and therefore all lights with diffusing elements would inevitably come at the bottom of a classification privileging illuminating bodies by the sole parameter of energy efficiency.

I would like to ask lovers of classifications, standards, labels and all those who believe, wrongly in my opinion, that these represent the scientific grounds for the solution of the problem of energy saving, if they have reflected well on what the light in our homes would be if, for lighting, the principle of efficiency were to prevail as a priority. I think that it would be a uniform, flat and wan light produced by anonymous appliances set in the ceilings and walls or, worse still, produced by glaring spotlights.

I think that all of us here, like the majority of people in the world, well realize how lights make up the environments we live in. They have always conveyed special values in addition to their elementary value of giving light: they convey values of culture, history, tradition, style, fashion and beauty. We need lights to be able to see and look but also because we like looking at them, using them and enjoying their presence. I think that there can be no doubt that the majority of products which give light in our homes have to be considered as products that go far beyond their lighting performance and therefore any attempt to classify them only according to their parameter of efficiency would be mistaken, as moreover, the market shows, and therefore doomed to failure.

And so does this position of mine mean that I do not consider the problem important? Obviously that is not the case. We are interested in the problem, we want to make our contribution but in such a way, correct as we see it, as to produce results that are far more significant than those which could be obtained from a forced classification and, I repeat, doomed to failure.

WHAT IS TO BE DONE?

Operate within a plurality of directions.

I will try and indicate some of these.
-        The first is that of increasing our investments in research to improve the technical characteristics of our products. In particular, research on materials to produce diffusers which allow a reduction in the absorption of the luminous flux emitted by the source, thereby improving efficiency without compromising levels of comfort.

-        The second is to produce more lights for Compact Fluorescent Lamps (CFL) with specific couplings (not E27 or E14) which do not allow the use of incandescent bulbs. To follow this direction, with increasing conviction, the major bulb manufacturers should continue their research and orient their investments in order to improve compact fluorescent lamps in reducing their dimensions, improving the quality of light, price levels, easiness of disposal at the end of their life cycle and certainty of ecological compatibility. It is equally obvious that we will be interested in and willing to support hypotheses of replacing all existing products by dedicated products, well aware that this hypothesis entails the production of hundreds of millions of new products for our homes. For all these reasons, we hesitated over contributing to the launch and progress of the "Lights of the Future" competition promoted by the EC for the design of new lighting products for the home dedicated to this type of light sources (CFL).

-        The third direction is that of making a greater commitment to campaigns for a correct use of energy in the lighting sector. We can do this both with direct contributions to such campaigns but, even more so, due to the capillary diffusion that such an action could have, by giving special messages to the consumer in each of our products.
In this regard, I would like to recall how appliances and systems that regulate the energy used for lighting in the home by computer, both in relation to external light conditions and in relation to the effective presence of people in the room are already technologically possible today.

-        The fourth and last direction that I feel I can indicate, and which probably is the most attractive even if, and I am a little surprised by this, no-one has made the slightest mention of it so far, is the attention that must be paid to all the experiments and research on the use of alternative and renewable sources of energy. Research and experimentation in this direction are much more advanced than we think. In Italy, the land of the sun, and in the southern part of the country where the sun shines longer and more brightly, public and private bodies are developing research and carrying out experiments with very significant results.

The underlying hypothesis is that of **conceiving a house which, by using new technologies capable of developing energy, both from natural external (the sun, water, wind) and internal sources (water circuits, use of waste, etc.), reverses the role of the house from that of a consumer of energy to that of consumer/producer of energy.** Thanks to photovoltaic and/or thermal facades, roofs with tiles and solar cells, co-generating systems (simultaneous generation of electricity and heat) or triple-generation (electricity, heat and cold), the correct use of domestic waste and water, new technologies for air-conditioning, intelligent and computerized management of loads and consumption, it is possible to think, as accurately observed by Prof. F.M. Butera, Chairman of AMG, the Palermo Gas board, of "new species" of homes and cities.

The evolution of the species is always a very long process, but it is always greatly accelerated by changes in the surrounding conditions. The possible worsening of the crisis of traditional energy sources, of which we are feeling the specific consequences at this very moment, and the worsening of environmental conditions (ozone hole, pollution, rising temperature) could make it indispensable in the next few years to proceed in this direction.

It is clear that for us it is important to support this research as from today and take an active part in all the experimentation on lighting homes and cities.

We have to give our constructive contribution to the consolidation of the hypothesis of the INTELLIGENT HOUSE and the SOLAR CITY. Of course it is not possible to proceed all over the world in this direction to the same extent but it is certainly possible in countries like ours, which is at least well endowed with these natural resources, just as it is in many of those countries which only very recently have started up significant processes of development.

This is the picture with all its components of what, for us, "is to be done".

Aware of what the effective consumption of electricity for domestic lighting is today (less than 6% of the total electricity produced), we refuse to take up exasperated positions that often tend, out of private industrial interests, to exaggerate problems and solve them with simple schematizations or classifications.

In the meantime we hope that as many people as possible can live in homes that are well lit by beautiful, well made and pleasant products, that give light according to the natural needs of the individual, without any waste of energy.

I have finished, dear friends, and I would like to conclude by telling you openly that it is always a great pleasure to be amongst a large number of "benefactors of humanity" as you certainly all are, due to your commitment to solving vital problems for the future of humanity. I fervently hope that after what I have said you will still wish to consider me by full right as one of your own.

Thank you.

The Road to Kyoto: What we Have Achieved, What Shall Be Done (Technologies)

Giannunzio Guzzini

ASSIL President

ASSIL – Associazione Nazionale Produttori Illuminazione, gathers 90 companies manufacturing luminaires, light sources and components.
Its main targets are the representation and the defence of the Member Companies. In order to achieve this scope, ASSIL's effort is devoted to the diffusion both on a national and international level of new products and new technologies according to comfort, efficient use of energy, respect for the environment in search of global quality.

On the whole, the sector represented by ASSIL can count on solid traditions and on a history full of significant successes coming from the quality of products and companies.
In particular, the Italian manufacturers show a marked tendency to export, thus offering an important active contribution to the balance of trade.
ASSIL is among the founders and an active member of CELMA, Federation of National Manufacturers' Associations for Luminaires and Electrotechnical Components for Luminaires.

As a mission, ASSIL's members are committed to deliver light sources, components and luminaries capable of providing end users with lighting solutions fulfilling high functional and esthetical requirements.

1   Slide 1. Trajectories of Technological Development in Lighting Solutions

It is a challenging mission and it requires mastering a wide range of technologies, from molecular chemistry to metallurgy, from optical engineering to electronics and lighting design.
Luminaries focus on design to integrate this wide spectrum of technologies in a package with the capability of delivering both functional and esthetical performances.
Besides "pure design", an important and distinctive element of Italian luminaries, research and development of new lighting solutions move along 5 key trajectories:

- System Efficiency
- Endurance
- System Miniaturization
- Light Comfort,
- Environmental friendliness

As outcome of continuous efforts, innovative technologies are bringing to lighting solutions relevant benefits, with ECG's and high frequency control at the heart of new high performance capabilities in term of light comfort, higher efficiency, longer life span, miniaturization and environmental friendliness resulting from above combined characteristics.

In this scenario, compact fluorescent lamps and electronic ballast's technologies play an increasingly vital role in delivering highly efficient lighting.

2 Slide 2. FLC's and Electronics Technologies: on the Main Road to Efficient Lighting

In order to achieve significant reductions in quantities of electricity consumption and CO_2 emissions related to lighting, a three-pronged approach can be pursued to leverage on benefits from above technologies:

A. To promote use of compact fluorescent lamps with integrated electronic ballast and E14/E27/ES sockets

B. To substitute conventional ballast's with electronic ballast's in fluorescent lighting systems

C. To develop luminaries based on fluorescent lamps (compacts, T5, and so on) and electronic ballast's.

Accordingly with our Conference theme, I would like to focus now on ASSIL's study on household lighting and on strong impact of programs aimed at reaching goal A.

In 1998, penetration of incandescent and fluorescent technologies in Italian households has been evaluated as follows:

Light source typology	Installed base	Typical lamp power	Nominal installed power	Switching on hours per day	Switching on days per year	Total electricity consumption per year
		W	GW	ore	giorni	GWh
Filament incandescent	350.000.000	60	21,00	1	330	6.930,00
Integrated fluorescent	7.500.000	15	0,11	1	330	37,13
Totals			21,11			6.967,13

(Source: ASSIL)

As key reasons for low CFL penetration our study pinpoints:

3   Slide 3. Key Factors Limiting CFL's Penetration into Households

. Low awareness of economic benefits

. Prejudices against fluorescent technology

. High price tags

4 Slide 4. FCL's Penetration into Households: the "Natural" Evolution

Based on these facts and market history, we projected following "natural" evolution for CFL:

YEAR	Market	Installed base
1996		6.000.000
2000	8.429.568	9.690.144
2004	13.264.088	15.384.651
2012	32.941.395	38.011.152

Accordingly with "natural" trend, without any specific incentive or promotion program, installed base will go from 10 million lamps to about 30 million lamps in 12 years.

5 Slide 5. FCL's Penetration into Households: Overcoming the Limits

Projections look different if we consider implementation of programs aimed at overcoming above limits:

. Institutional campaign, a mix of direct communication and advertising, in cooperation with Government agencies, aimed at raising awareness of economic and environmental benefits

. Sales-tax cuts and other Government incentives to be added to market prices reductions to lower end users' costs

6 Slide 6. FCL's Penetration into Households: the Stimulated Evolution

In this "what if" scenario following trends can be projected:

Year	Market	Installed Base
1996		6.000.000
2000	13.641.600	16.482.720
2008	47.491.719	49.630.412
2012	74.729.139	91.151.179

7 Slide 7. Saving from Stimulated Evolution of CFL's

As a result from above activities, figures for installed base, electricity consumption and energy savings can be projected as follows:
(upper table on page8)

Household /Residential – Stimulated evolution			
	Installed	Consumption	Reduction
1998			
Incandescent	350.000.000	6.930	
CFL	7.500.000	37	
		6.967	
2000			
Incandescent	341.000.000	6.752	
CFL	16.500.000	81	
		6.833	**134**
2005			
Incandescent	317.500.000	6.287	
CFL	40.000.000	197	
		6.484	**483**
2010			
Incandescent	287.500.000	5.693	
CFL	70.000.000	345	
		6.038	**929**

Accordingly with this projection, through an adequate program of communication and incentives, installed base can triple in just 4 years and go from 10 millions lamps to over 70 millions lamps by 2010. By that year, program can save 929 GWh, cutting by almost 500.000 tons CO_2 emissions into atmosphere

Strong back-up for projections comes from a 1996/97 campaign promoted by ASSIL and ACEA

On a four months period of time, a significant +30% increase over "natural" market growth was achieved thanks to a package of promotional activities including:

. **Billboard and radio advertising campaign**
. **Distribution of pamphlets to families through electricity bills**
. **Possibility of payment through bimonthly installments**

8   Slide 8. A Program Aimed at Stimulating CFL's Penetration into Households: Operazione Lampadina

After that, ASSIL defined an "OPERAZIONE LAMPADINA" program including activities such as:

-Distribution of informative pamphlets
-Demonstration points to compare the consumption of electricity of   incandescent
  lamps and CFLs
-Free sampling of CFL to end users
-Advertising Campaign

9   Slide 9. Operazione Lampadina: Past and Future

As you can see, ASSIL has carried out initiatives following the program "OPERAZIONE LAMPADINA" in co-operation with the municipality of Padua and Bologna and is preparing strong actions -in cooperation with FEDERELETTRICA and the Italian Ministries of Industry and Environment;
-with ENEL within the program sponsored by The European Commission and UNIPEDE/EURELECTRIC.

Energy Services in the Domestic Sector as a Result of the Liberalised Market

Guglielmo Gandino

It is a great pleasure for me to be here with you today, and have this unique opportunity to tell you, in the few minutes I have at my disposal, what we in Enel are planning to do, to improve our customer service versus the 30 million customer relations we manage on the Italian market today.

First of all, I have to underline that, out of 30 million, 24 are residential consumers (families), the rest being represented by operators in the so-called small and medium business segment areas.

So, after 300 days of hard project work, we are at last in a position to announce that we are starting the implementation phase of Enel's New Contact Management System, of which the very first centre will be in operation starting next January. After an adequate period of stress tests for all our processes and systems, we will then be able to roll out the whole organisation, during next year for completion within 2001.

Our National Contact Centre will consist of eight main sites, operating in a virtual environment, servicing customers 24 hours a day, 7 days per week.

This organisation will employ approx. 2,500 people, all of them coming from our present commercial network. For this purpose we will organise nearly 600 days of educational training, on customer care techniques and new applications.
Total investment is about 250 billion Lire or 130 million Euro.

Our Contact Centre will be supported by a National Scanning Centre, based in the south of Italy, where all customers' correspondence and returned contractual documentation will be received, scanned and filed. It will also be supported by a National Printing Centre, from where all answers and contracts will be duly printed and sent out to customers within agreed delays.

We are confident to be able to service customers in a qualified manner. Our target Key Performance Indicators have been set at a very aggressive level and compare quite favourably with best practices in Europe for similar operations.

We plan to achieve, in a reasonably short while, a rate of satisfied calls of 95% in 30 seconds with 90% of requests answered in "one stop solution". The balance is planned to be answered by the same operator in most cases ("one voice solution").

This to create a personal feeling between operator and customer, who should feel more and more at ease, talking to our customer service people.

Our Contact Centre will tend to be a highly dynamic centre of excellence, with professional teams of consultants supported and tutored by experts, all of them with a fair level of personal empowerment. It will really be the heart of our customer service organisation, focused on customer satisfaction, measured on customer loyalty, totally integrated with all other company functions. In a word, a Laboratory of Knowledge Enrichment, for motivated people that look for career improvement opportunities.

For all those customers instead that do not like to contact us by phone, e-mail or by a simple letter, but are familiar with a personal computer, a Self Service Area will be available on our Internet Portal. Contractual operations, meter reading, payments and pre-payments will be possible using a Personal Identification Number, that will be distributed to all our customers in time.

Another alternative will be constituted by 800 outlets of WIND (our TLC company), where a staffed Enel-in-WIND Corner will enable customers to operate through the Enel Portal, without having to necessarily do it themselves. We believe that this will dramatically improve our capillary presence in the Territory, with an increased degree of satisfaction for all those that prefer a face-to-face to a virtual contact.

Files with a high level of complexity, that forcibly require the physical presence of the customer, will continue to be handled through our directly managed Enel Points, existing in each province and in all cities of strategic importance.

With this totally renewed network, we will be ready, by the end of 2001, to face the challenge of further market liberalisation steps.

Both our Contact Centre and our Internet Portal will be a comprehensive interactive source of information for an intelligent and rational use of energy.

Let us make a few examples:
1. If you want to save energy by connecting your electric appliances during low-rate periods exclusively, you can call the Contact Centre and ask for a horizontal rate, choosing among many alternatives according to your needs. Your new rate will be immediately available after a simple remote command. Your appliances will start at a given time by means of a "smart socket".
2. If you want to have a complete information on electric reversible conditioners and the advantages they offer, Contact Centre or Internet Portal will offer you a vast range of available products supplying you with:
   • full specs and indicative prices,
   • approximate investment required,
   • cost savings and economical advantages.

If you prefer, a consultant from our Company will pay you a visit and give all necessary explanations.

3. If you want a remote controlled start of your electrical central heating in your holiday house, you ask Contact Centre for this service and you will be satisfied according to your requirements, at a reasonable price.

During 2001 then our services will expand from pure electrical business to other fields.

*   Other utilities is one example. In the areas where, through acquisitions and mergers, we will also sell water or gas, our network will be capable to offer a full multi-utility service.
*   Home services is another example. Our Contact Centre will be the ideal channel to receive, at any time, requests for maintenance contracts or emergency calls, that will then be handled by Enel.SI, the Enel specialised company in integrated services.

This ambitious plan, that will not exclude - in principle - any other type of additional service directed to families and people, is being developed in these months. Through a serious evaluation of pros and cons, we will finally understand which additional products are worthwhile to be concentrated on, both for energy and non energy supply markets.

This will allow us to build a true customer relationship, by better responding to their expectations, and to maximise in this way our customer retention rate, in a scenario of progressively free market.

Extra value for Enel, that will achieve operating synergies by using common structures, so reaching a distinctive competitive advantage.

Extra value for Enel's customers, that will soon have the possibility of acquiring bundled services with a consequent evident "plus", compared to purchasing single service components outside our network.

Our world of customers will be kept in constant touch with us, will be proactively informed about the "news of the market" through tele-selling focused campaigns, will receive direct mailing about new products, rates and services, and will be able to capitalise part of effected purchases, on "premium programmes" spendable as they like.

Ambitious programme? Maybe.
Realistic? For sure.

All of us are very determined and committed to these objectives. I am sure we will make it happen.

Thank you.

The Way Forward

Ruud Trines

Netherlands Agency for Energy and the Environment (NOVEM)

Since the first conference on energy efficiency of appliances in the domestic sector in Florence (1997) many international project were carried out on standards, labels and technical/economical developments and possibilities. Also (inter)national networks of professionals in this area are established. This is a good way and must be continued.

In the present second conference the European Commission has presented a framework directive for the coming years having the Minimum life cycle costs for consumers as basis for ambitions on energy efficiency. This is an important step forward.

Marketing on our mindset
Also I have noticed much more attention for marketing of efficient appliances and Market Transformation. We have had presentations of results on successful procurements, energy saving trusts, rebate systems en research on price elasticities. I think marketing has become part of the mindset of many people attending this conference.

Energy savings via consumer needs
However energy efficiency of the appliances that consume most energy has increased over the past years, energy consumption has increased as well. This can be explained by the fact that consumers buy more appliances and use them more frequent.
Apart from stand-by consumption appliances do not consume energy unless they are being used by consumers. With regard to the Kyoto targets it is necessary to save on consumption and this can only be done via the consumer and not only by improving the efficiency of appliances. To do this we need to understand the needs of the consumer. E.g.: a consumer does not need a washing machine or a freezer. However a consumer wants clean clothes or food. So in order to look for energy savings you have to consider the whole process and you may end up by fabric or food technology to save energy.
I also want to remind that the year 2010 is not that far away from now. Any washing machine sold today is most likely still in use in 2010. So improving energy efficiency will not influence energy consumption on a large scale, however daily consumer behaviour can contribute on a large scale.
It is my vision that the needs of the consumer and functions of appliances are starting point of considerations. In this respect I am happy with the overwhelming

number of participants during the parallel sessions on consumer behaviour. I think we need more socio economic research.

Interactive Policy planning

During this conference it became clear to me that several countries have a system for interactive policy planning. With interactive I mean discussing trends and interventions with the market parties. Not clear is whether these countries speak the same language in these models.

The market for appliances is becoming more and more international. The earlier mentioned European framework directive needs consensus on input on technical and economical data. Even better: there is a need for transparancy, interactiveness and sharing of views and reaching consensus as basis for policy planning on national and international scale.

I like to propose that either the EU (DG TREN) or IEA (within Implementing Agreement Demand Side Management/Annex 7 Market Transformation) will create a setting for a network and common language for international interactive policy planning.

Energy Efficiency Standards in the U.S.

Carl Adams

U.S. Department of Energy's Appliance Standards Program

Energy efficiency standards have played an important role in the U.S. as a successful way to reduce energy consumption. National energy efficiency standards started for some appliances in 1988 and now cover all major household appliances. Setting and improving these standards have been a high priority for the Department of Energy.

Appliance efficiency standards can save a significant amount of energy and result in substantial reductions in emissions. For example, appliance efficiency standards already in place have already saved 3.6 exajoules of energy. By the year 2020, the cumulative savings of these standards and the new and proposed standards planned standards being considered by the Department are expected to save an additional 22.7 exajoules. should be 2.8 quads. We believe these efficiency improvements and energy savings are a significant accomplishment.

If you were at the first conference in Florence, you heard a description of how we had changed and improved the process by which we set standards. These changes mostly involve increased dialogue with the manufacturers and other stakeholders at all stages of the process. U.S. law requires us to periodically set standards at the highest level which is technologically feasible and economically justified. As we go through the process of collecting data and conducting analyses, we make the results of our efforts available to the manufacturers and other stakeholders by methods such as posting them on our website and conducting workshops to discuss critical issues along the way. In this manner, we hope to get data and analyses that all sides can agree with. It is then up to the Secretary of Energy to decide what is the highest standard that is economicallycan be justified.

In a number of many cases, once we have reached this point, the manufacturers and other stakeholders have negotiated among themselves to propose to the Department where the standards should be set. Joint recommendations have been reached on fluorescent lamp ballasts and clothes washer standards. These joint recommendations have been based on the DOE analysis and, in all cases so far, the Department has been able to agree that the joint recommendations are the highest level that can be justified. Our last several standards have been set in this manner.

Next month new standards for room air conditioners, requiring a roughly 10 percent improvement in energy efficiency, go into effect. In July of 2001, new standards for refrigerators and freezers, requiring a roughly 30_ percent reduction in energy useimprovement go into effect. On September 19, just last week, we announced new standards for fluorescent lamp ballasts that will essentially require the use of electronic ballasts which will begin to go into effect in 2005. And in May of this year, Dan Reicher, an Assistant Secretary of Energy, joined industry representatives and other stakeholders, to announce proposed new standards for washingclothes washers machines that will save enough electricity to light 16 million U.S. homes for 25 years. By the year 2020, they would cut greenhouse gas emissions by an amount equal to that produced by three million cars every year.

Through our improved process, we strongly encourage manufacturers and others to negotiate standards to recommend to the Department. However, this doesn't always happen, but when it doesn't that doesn't stop the process. The process may take a little longer when there is no agreement, but it continues. For example, in April of this year, Vice President Gore announced proposed standards to improve the efficiency of residential water heaters, which would save American consumers more than $24 billion in energy costs over the next 20 years. Improving the efficiency of central air conditioners is also high on our priority list because it will significantly cut electricity consumption and lessen the strain on the electric grid during the summer, reducing the likelihood of brownouts and blackouts. We are working on both of these rules in the absence of any agreements so far and we are prepared to go forward to setting new standards.

While mandatory standards have been and are very important in improving the energy efficiency of household appliances, mandatory standards for the whole U.S. can only go so far. For example, in some areas of the country, where electricity prices might be higher than average, higher efficiency levels than those set for the entire country might be economical. Additionally, there are niche products like heat pump water heaters that are very efficient, but can't be used everywhere. To promote these even higher efficiency products, the Department of Energy and the Environmental Protection Agency (EPA) conduct the Energy Star Program, another high priority effort.

The Energy Star program–which illustrates how DOE, EPA, manufactures and retailers can cooperate for mutual benefit–has also had phenomenal success. EPA and DOE put together a program that makes money for its participants, doesn't cost a whole lot to run, and has had a significant impact on the entire marketplace for consumer products, computers, housing, and commercial buildings by promoting and giving recognition to higher efficiency products.

In 1999, sales of appliances with the Energy Star label doubled the sales gains of appliances without this designation. All five of the major U.S. appliance manufacturers (Amana, Frigidaire, GE, Maytag, and Whirlpool) label and promote Energy Star-qualifying models. Retailers such as Sears, The Home Depot, Circuit City and others enjoy increased sales from Energy Star-labeled products. According to a study DOE commissioned with the Gallup Organization in October 1999, over 80% percent of our industry partners agreed that Energy Star is an effective way to help them market energy efficient products.

The U.S. Department of Energy has worked hard to set the highest efficiency standards for appliances as our laws require. We and the EPA have worked hard to promote and recognize even higher efficiency products where those higher efficiencies make economic sense through the Energy Star program. With the cooperation of industry, these efforts have paid off handsomely in terms of reduced energy usage and environmental impacts. We hope to continue this effort into the future.

Thank you.

High Performance Clothes Washer in-Site Demonstration in a Multi-Housing Multi-User Environment

Graham Parker[1] and Greg Sullivan[2]

[1] Program Manager, Pacific Norwest National Laboratory
[2] Senior Research Engineer, Pacific Norwest National Laboratory

Abstract. The objective of the study was to measure, analyze and report on the efficiency of 4 high performance residential-style clothes washer brands compared to a conventional (baseline) clothes washer brand in multi-housing facilities at the Fort Hood Texas military base. The demonstration study also included a parallel study to ascertain the maintenance of these same clothes washer brands. This was the first independent in-situ evaluation of several brands of high performance clothes in a multi-housing environment in the U.S. This paper will focus on the energy savings for determining the cost-effectiveness of the machines for this multi-housing application.

The demonstration involved 6 conventional 6-year old washers manufactured by Roper (the baseline washers), and 6 new high performance washers from each of 4 manufacturers – Whirlpool, Inc., Maytag, Inc., Staber Industries, and Alliance Industries, Inc. (Speed Queen). Each of the 30 total individual washers in the study was metered in real-time for hot water use and temperature, cold water use and temperature, machine energy use, and the number of cycles completed. Data were collected from a central data logger and retrieved on a weekly basis over a phone line through the central polling computer over an 18 month period representing an average of over 350 uses (cycles) per machine.

The average machine electricity use of the baseline machines was 0.26 kWh/cycle and the machine energy use of 4 high performance brands averaged 0.20 kWh/cycle for a 23% reduction in machine energy use. The total average water use for the baseline machines was 35.4 gallons/cycle and the average for the 4 high performance brands was 18.8 gallons/cycle for a 47% reduction in water use. The baseline conventional machines used an average of 9.0 gallons hot water/cycle (5,610 Btu/cycle) whereas the 4 high performance brands used an average of 3.4 gallons hot water/cycle (2,120 Btu/cycle) for a 62% reduction in hot water use.

The average use of the washers in this study was 6.4 cycles/machine /day. Based on that average and extrapolated for an entire year (365 days), the total average water savings of the high performance machines compared to the baseline conventional machines is 38,780 gallons/year/machine. The machine energy savings is 140 kWh/year/machine and the hot water energy savings (at the clothes washer) for Fort Hood is 8.1×10^6 Btu/year/machine.

1 Washers and Demonstration Site Description

Fort Hood Army installation located near Killeen, Texas, was a site for a demonstration of high-performance commercial family-sized clothes washers. This demonstration was conducted by the Pacific Northwest National Laboratory for the U.S. Army Forces Command.

The objective of the study was to measure, analyze, and report on the efficiency of the high-performance clothes washers relative to the conventional (baseline) V-axis clothes washers in use at the installation. While the information reported here is believed to be accurate, it is not from a controlled experiment. All findings presented here are "average" consumption and use findings specific to the Fort Hood barracks setting and thus represent an accurate long-term "average" use profile of clothes washers at Fort Hood. The characteristics of the clothes washers evaluated in this study are shown in Table I.

Table I: Fort Hood clothes washer characteristics.

Clothes Washer Brand/Manufacturer (Model #)	Age of Equipment at Start of Study	Tub Volume[1] & Machine Weight	Axis of Rotation of Tub	Clothes Loading Location	Number of Access Doors for Loading
Roper/Whirlpool Corp. (AL6245VWO) **Baseline Clothes Washer**	6 years	2.50 cu.ft. ~170 lbs.	Vertical	Top	1
Maytag/Maytag Corp. (MAH14PNAWW)	New	2.86 cu.ft. 181 lbs.	Horizontal	Front	1
Speed Queen/Alliance Laundry Systems (SWF561)	New	2.80cu.ft. 240 lbs.	Horizontal	Front	1
Staber/Staber Industries, Inc. (2300)	New	1.93 cu.ft. 220 lbs.	Horizontal	Top	2
Whirlpool/Whirlpool Corp. (LSW9245)	New	3.0 cu. ft. ~175 lbs.	Vertical	Top	1

[1]Volume determined according to US DOE test procedure: Uniform Test Method for Measuring the Energy Consumption of Automatic and Semi-Automatic Clothes Washers," Code of Federal Regulations, Title 10, Part 430, Subpart B, Appendix J.

In comparing clothes washers it is important to note their tub volumes; smaller tub volume may result in more clothes washing cycles (thus more energy and water use) to wash a given volume of laundry.

Figures 1 below shows the 5 brands of clothes washers (baseline + 4 high performance washers) evaluated in this study.

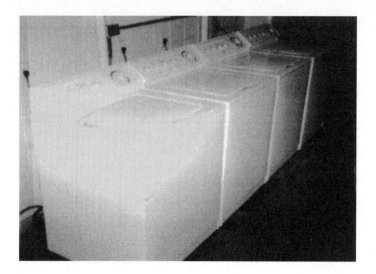

Figure 1: Whirlpool Model #LSW9245 V-axis clothes washers.

The demonstration involved three nearly identical barracks buildings of the same style, size, and occupancy levels (~140 troops/barracks). The barracks also housed soldiers from the same military assignment/training and thus had similar laundry use requirements. Each of the three barracks buildings has one central laundry room containing six clothes washers and six clothes dryers. Each of the three barracks buildings laundry rooms received identical end-use metering equipment. In each laundry room one central data logger was installed to record and store the relevant per-cycle energy and water data for each machine. A description of each monitored parameter is included below. The baseline V-axis and new high performance clothes washer monitoring strategy was identical. Figure 1 details the metering arrangement common to each clothes washer.

2 Metered Parameters

Clothes Washer Water Temperature: Water temperature, both hot and cold was monitored using resistance temperature detectors (RTDs). These are 1,000 ohm platinum RTDs (model S1764Pf) and were made by Minco Products, Inc. of Minneapolis, MN. The RTDs provided the temperature data to the central data logger where it was stored in a time-series format.

Clothes Washer Water Use: Water use was monitored by installed water flow meters on the hot and cold supply line to the machines. These water meters are conventional water utility nutating disk meters (model RCOL 25) made by Badger Meter, Inc. of Milwaukee, WI. To provide the appropriate output, the meters were modified with a reed switch, which opens and closes in proportion to the volume of water passing through. The output of these meters, conditioned to be a pulse output, provided per-cycle water use data to the central data logger where it was stored in a time-series format.

Clothes Washer Energy Use: Electrical Energy use was monitored by installed current transformers (CTs) on the power connections to the washers. The CTs provided per-cycle electricity use data to the central data logger where it was stored in a time-series format.

Clothes Washer Utilization: The total numbers of cycles per machine were captured by the CTs connected on the power line to the washer. The CTs provided the run-time data to the central data logger where it was stored in a time-series format.

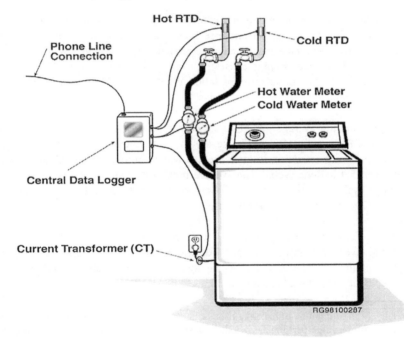

Figure 2: Clothes washer metering equipment and connections.

3 Metering Duration and Cycles

The metering of the six baseline conventional (Roper) clothes washers in one laundry room took place over a 2-month time period in late 1997 and included 1,050 wash cycles. The baseline clothes washers were then replaced by six high-performance clothes washers from a single manufacturer and these were likewise metered. High-performance clothes washers were also located in the other two identical-sized laundry rooms and were metered. Each metered laundry room was equipped with six high-performance clothes washers from the same manufacturer. Metering of the high-performance clothes washers took place over a 17-month time period from February 1998 through July 1999. During this metering period, the use of the high-performance washers ranged from 1,918 to 5,078 cycles/manufacturer, with an average of 3,026 cycles/manufacturer.[1]

4 Performance and Operations Results

Figure 3 presents the average motor and controls electricity (machine electricity) use in kWh/cycle. The four high-performance brands showed a reduction in machine electricity use over the baseline machine electricity use of 0.26 kWh/cycle. The average high-performance machine electricity use was 0.20 kWh/cycle. This resulted in an average electricity use reduction is 0.06 kWh/cycle (or 23%) for the four high-performance brands.

Figure 4 presents the average gallons/cycle with both the hot water and cold water components of the average total water use shown. The four high-performance brands showed a significant reduction in total average water use over the baseline machine water use of 35.4 gallons/cycle. The average high-performance total water use was 18.8 gallons/cycle, resulting in water savings of 16.6 gallons/cycle. These savings represent a 47% reduction in total water use.

The baseline conventional machines used an average of 9.0 gallons hot water/cycle whereas the average high-performance hot water use was 3.4 gallons/cycle. The average reduction in hot water use by the four high- performance brands was 5.6 gallons/cycle, or 62% of the baseline machine.

---

[1] The relatively short duration of metering the baseline clothes washers compared to duration of the metering of the high-performance clothes washers was due to the site scheduling the replacement of all their V-axis washers with new high-performance washers during the time of the baseline metering.

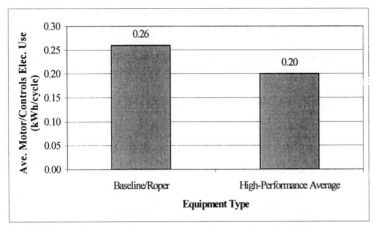

Figure 3: Average motor and controls electricity use (average kWh/cycle).

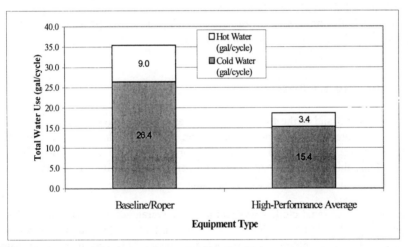

Figure 4: Average total water use (average gallons/cycle).

The average cycles/day for each machine in the study varied considerably ranging on average from 3.2 to 10.9 cycles/day for all 30 machines (6 conventional baseline machines + 24 high-performance machines) that were monitored. For the baseline (Roper) conventional machines, the average over the time period of monitoring for all six machines was 4.2 cycles/day. For the high-performance machines, the average was 7.0 cycles/day over the time period of the monitoring for all 24 machines.

In most cases, the variance in individual machine use was related to troop activity (i.e., variable occupancy levels due to field exercises). Other variables included the physical location of the machine relative to the laundry room door. The machines closest to the door received the greater use, which was expected. On average, the first two clothes washers nearest the door were used 55% more often than the two clothes washers farthest away from the door.[2]

It should be noted that in comparing between the high-performance clothes washers studied, consideration should be given to the clothes washer tub volume. Clothes washers of different tub volumes will be have an impact on the amount of clothes washed per cycle and therefore on the amount of annual energy and water use and savings; the relevance of this point is that three of the four washers studied here have significantly larger tub volumes than the fourth. In fact, while showing relatively similar energy and water use, the three larger machines theoretically would be capable of washing a load about 30-40% larger (based on their relative tub volumes) and thus have a higher efficiency per unit of laundry washed. This point is relevant in situations where full loads are commonly washed.

5   Economic Results

Based on Figures 3 and 4, the four high-performance brands saved an average of 5.6 gallons of hot water, 11.0 gallons of cold water, and 16.6 gallons of total water for each cycle of use compared to the average for the baseline conventional V-axis washers. Thus the savings by the four high-performance brands was 62% of hot water, 42% of cold water, and 47% of total water.

The baseline conventional clothes washers used an average of 5,610 Btu/cycle of hot water energy (at the clothes washer), and the average of the four high-performance brands used 2,120 Btu/cycle/machine in hot water energy. This is an average hot water energy savings of 3,490 Btu/cycle/machine and does not take into account hot water conversion inefficiencies. Given the average use of all five manufacturers' machines (baseline + high-performance) at Fort Hood over the time period of the study of 6.4

---

[2] In discussions with commercial clothes washer route operators, this same phenomenon necessitates these operators to rotate equipment so that equipment is used uniformly thus extending its life.

cycles/day/machine and extrapolated for an entire year (365 days), the total water savings of the high-performance machines compared to the baseline conventional machines at Fort Hood is 38,780 gallons/year/machine. The machine energy savings is 140 kWh/year/machine, and the hot water energy savings at the clothes washer is 8.1 million Btu/year/machine.

Based on Fort Hood utility rates,[3] the total water cost savings is $39/year/machine, the total machine electrical cost savings is $4/year/machine, and the hot water energy cost savings is $43/year/machine for the high-performance machines. This results in a *total cost savings of $86/year/machine* for the average of the four high- performance brands compared to the conventional baseline clothes washers.

Data are presented in Figures 5 and 6 showing expected lifetime water and energy cost savings of the high-performance clothes washers compared to conventional (baseline) V-axis clothes washers. The values used to develop the curves in Figures 5 and 6 are given in Table II below.

Figure 5 presents the present value of lifetime combined energy and water savings for the average of the four manufacturer's high-performance clothes washers (compared to the conventional baseline clothes washer) as a function of water/sewer price ($/1,000 gallons) electricity with a 100% conversion efficiency.

---

[3] Assuming 60% efficient hot water generation and distribution system, a 32 cents/therm natural gas cost; 3.2 cents/kWh electricity cost and $1.00/1000 gallons water/wastewater cost for Fort Hood.

Table II: Values used for clothes washer economic analysis.

Economic Analysis Metric	Value	Source/Notes
Baseline motor/controls electricity (kWh/cycle)	0.26	Average of the baseline (conventional) machines metered values
Baseline machines water consumption: hot/cold/total (gal/cycle)	9.0/26.4/35.4	Average of the baseline (conventional) machines metered values
High-performance machines motor/controls electricity Consumption (kWh/cycle)	0.20	Average of the 4 high-performance brands metered values
High-performance machines water consumption: hot/cold/total (gal/cycle)	3.35/15.35/18.7	Average of the 4 high-performance brands metered values
Clothes washer use (cycles/day/machine)	6.4	Average value of all machines metered in the study
Clothes washer life (years)	5	Typical commercial (OPL) washer life or lease term
Discount Rate (%)	3.1	Federal government discount rate for 1999

Figure 6 presents the value of lifetime combined energy and water savings for the average of the four manufacturer's high performance clothes washers (compared to the conventional/baseline clothes washers) as a function of water/sewer price ($/1,000 gallons) and natural gas price (cents/therm), assuming water is heated using natural gas with a 75% conversion efficiency. In Figure 6, the savings for the machine (motor and control) electrical energy is fixed at 6 cents/kWh and included in the analysis; in Figure 5, this savings is calculated based on the selected electricity rate.

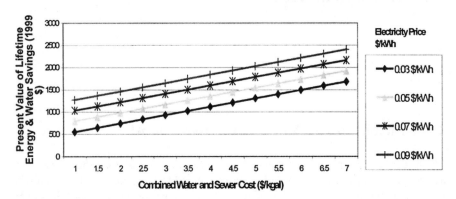

Figure 5: Average high-performance clothes washer lifetime savings – electric water heating.

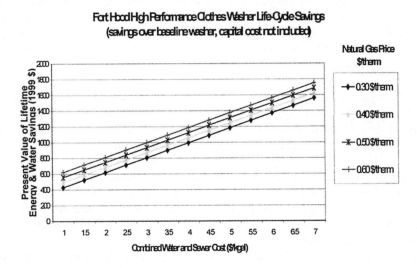

Figure 6: Average high-performance clothes washer lifetime energy savings-natural gas water heating

20% Less Energy on Washing Machines: How Were the Savings Achieved?

Rainer Stamminger

CECED

1   Content of the Voluntary Agreement

The threat of a climate change due to man-made emissions of green-house gases is an immense challenge to mankind. Politics seemed to have advanced only slowly for a long period of time. The climate summit conference in Kyoto however should turn out as a rather big step forward. The European Union for example promised a total 8% saving of CO_2-emmissions until the year 2012. A challenging task; for the politics of Member States, for industry and for all people whenever using energy. We have to admit that it is obviously rather difficult to change fundamentally the behaviour of the energy-users, the behaviour of people driving cars, heating houses or operating appliances. Maybe this is a deficit which should be more emphasised in the future. In any case, industry has to accept responsibility and the obligation to take all feasible measures to reduce energy consumption during the life cycle of their products.

For the household appliance industry, this implies to focus on the product's phase-of-use. Major household appliances have a rather long life: up to twenty years. So it is not surprising that around 90% of their environmental impact occur during the actual phase of use. Various studies supply evidence for that.

So the problems are obvious, the targets are well defined and the schedule to achieve these targets is urgent. The question is: how to achieve the targets efficiently and quickly ? The answer of the household appliance industry has been: Use Voluntary Agreements (or as it is named today: negotiated Agreements).
We were among the very first of all industry sectors to offer concrete and detailed proposals. First on a national basis in some countries, like Germany, and then on the European level.

The first Agreement we proposed was on domestic washing machines. There were good reasons to take this product group:
- there was no directive for efficiency limits as with refrigerators and freezers
- washing machines share a substantial part of electricity consumption in private households

- there were already basic studies defining the present situation, the so called basecase, and
- there was an energy label scheme, which could be taken as a tool for shaping the proposal.

It was not easy at all for us manufacturers to come to an internal agreement. Different products in terms of energy efficiency and price, different markets and different expectations of consumers all over Europe made it difficult to come to a common proposal for an Agreement.

I can tell from practical experience: The simple statement "industry agreed on a proposal" doesn't reflect at all the long discussions among manufacturers. A lot of ideas about targets and how to achieve them, were developed, were discussed and were dropped. Many times we had to try to balance the different interests and market strategies of our industry. But some ideas made it to the final proposal which was submitted to the European Commission.

In the following, I will go deeper in the content of the Commitment "to save 20% energy" and I will demonstrate how it was achieved by combining several means.

From the beginning it was clear, that this commitment should be driven by market forces, as no regulatory like system would be able to reach the same level of energy savings in that short period of time.

Therefore it was evident to base the Voluntary Agreement on the Energy Labelling scheme which was introduced for washing machines in 1996. This label is guiding the market towards more efficient machines by giving clear indications to the consumer and retailer on the efficiency of the washing machines offered on the market.

As any clear commitment should relate savings promised to a clear defined base-case, the outcome of the GEA-studies[i] was taken. Within this study a market survey on the offer of washing machines was done which resulted in an average specific energy consumption of the European market in 1993 of 0,26 kWh per kg of dry load. As these data were based on the measuring standard of this year (IEC 456:1989) this value has to be transformed to be comparable to values measured under the standard relevant for the Energy Labelling (EN 60456:1996). Work done within the GEA study have resulted in a need to add 0,04 kWh/kg to get values comparable to the new standard. Therefore as a base-case for the Voluntary Agreement a value of 0,30 kWh/kg was agreed to be correct!

The Voluntary Agreement itself consist of two kinds of targets, called 'soft' and 'hard' targets and some accompanying procedure for checking and scrutinising the correctness of the commitment.

The hard target itself consists firstly of the overall target to reach a saving on the specific energy consumption of 20% by the year 2000 and secondly by the elimination of the least efficient classes, namely classes D to G with only some few exceptions, realised in two steps.

Under the headline soft targets everything was subsumed which could not be expressed in numbers but which might help to support the saving of energy under household conditions and which could help to go forward for a possible next step of savings.
Special care was also taken to be able to verify if sufficient evidence is given that the target will be reached and can be checked by independent authorities.

When the Voluntary Agreement was finalised all major European manufacturers did sign it. Afterwards also some other manufacturers signed in like IAR Siltal and also from non EU countries manufacturers like Arcelik (Turkey) and Gorenje (Slovenia.) are now within the agreement. Today one can say that more of 95% of the European. offer of washing machines is governed by this agreement.

2 Achievements

2.1 Reducing Specific Energy Consumption

Concentrating on the energy reduction as the main target, the result is that the agreed target of an average energy consumption of 0,24 kWh/kg in the 60° cotton cycle for all machines produced for the European market was reached already in 1999 (Figure 1). This is confirmed by looking on the average consumption values as they are reported in the CECED database of all washing machine models updated once per year and also confirmed by the notary reporting system on the average of the production for each individual year. As both system refer to quite a different view on the market it is remarkable how well the correlation between these two reporting systems is (Figure 2).

How were these results achieved?
As discussed extensively during drafting the Voluntary Agreement, mainly two factors contribute to a saving in a voluntary schema: the elimination of the least efficient energy classes and the improvement of the remaining machines. Both effects have contributed to the savings for washing machines: Based on the market offer in year 1993 around 70% of the models had to be eliminated from the market and almost all of the remaining 30% had to be improved to reach the distribution of machines we see today (Figure 3).
The individual measures to improve the washing machine depended very much on the individual design and on how efficient the machine already was before the voluntary agreement. The following three main measures were applied by almost

all manufacturers to improve. But many other activities were undertaken on a more specific and individual level.

2.1.1 Reduction of Water During Main Wash

All the water needed during the main wash phase has to be heated up to somewhere close to 60°C. A considerable part of the total energy is spent for heating up this amount of water. Therefore the clear target was to reduce the amount of water needed for the main wash phase, while maintaining a good or even improved washing performance. To reach this target, a reduction of the tub/drum clearance was the simplest way to go, although this has caused quite big investments as either the tub or the drum of the washing machine had to be changed. Many manufacturers have taken this route and invested in new machinery equipment.

Additionally the level of water maintained during the washing phase had to be reduced. This could be realised either by reducing the levels and/or the tolerances of this level or by introducing some kind of spraying system to ensure a continues wetting of the load. Both improvements call for additional costs as reduced levels imply also reduced tolerances at higher component costs and an additional pump for spraying increases the costs for sure.

2.1.2 Reduced Temperature and Prolonged Time

Already investigated in the GEA studies is the effect of a 'time-temperature trade-off' meaning the possibility to reduce the actual washing temperature by some degrees but maintaining the washing performance by increasing the washing time. This trade-off was extensively used by many manufacturers as it does not cost any investment nor does it increase the production costs of the machine. Unfortunately this has led to washing times in the order of 3 hours which are hardly acceptable for any consumer for his daily wash. This is a clear negative effect of the energy label which should be corrected as soon as possible.

2.1.3 Use of Electronic Controls

As seen in other products as well, electronic controls are able to provide higher flexibility by having the same or even lower costs as the former electromechanical controls. This trend was therefore anticipated by the need to save energy and many 'intelligent' water and temperature control algorithms were introduced just to ensure the exact right amount of water and energy is used for the energy label program. If all of this is really for the sake of the consumer must be challenged and verified case by case. But electronic control have also improved the ease of use of washing machines considerable and therefore it is a must for today mid to high range machines anyhow.

2.2 Elimination of Least Efficient Energy Classes

As mentioned before, about 70% of the market offer of 1993 had to be eliminated as the energy consumption was ranked in the classes D to G which had to be eliminated almost completely by the Voluntary Agreement. Although this was agreed within industry in 1996 and negotiated with DG XVII in 1997 an exemption from §81 of the EC Treaty had to be asked for, as this common activity of industry is falling under the restriction of the European cartel laws. Finally these exemption was granted by decision of the European Commission dated 24th of January 2000.

Despite of this weak legal base, the elimination of classes D to G in energy has taken place. Today almost no washing machine in energy class worse the C should be found on European markets. And this was achieved by applying pure market forces driven by the energy label and supported by the voluntary agreement of European industry.

2.3 Maintaining Washing Performance

For all discussion about the reduction of energy and water: One must not forget that the task of a washing machine is firstly and mainly to clean clothes sufficiently! Therefore it is important to look at the washing performance, whenever energy savings are discussed. Looking at the development of the washing performance in the time between up to 1999 it is quite surprising to see, that no decrease of the washing performance seems to have appeared (fig 4). This is even more surprising as it would be easy to declare programs on the washing machine offering a very good rating on the optical dominating figure of the energy consumption but disregarding the 'less visible' washing performance rating. But perhaps the consumer is really more of a 'expert' about he really needs than many experts in Brussels and somewhere else? But especially when public discounts or rebates are offered, a good balance of the energy consumption values with performance values must be ensured. We, as European appliance industry would recommend not to accept a performance worse than class B for any publicly funded discount activity.

2.4 Influencing Consumers' Choice

As we all know, the energy label features only one program out of many programs offered by each washing machines. It is therefore not obvious on first sight, that this sole figure should represent the real life use of a washing machine. But, as shown earlier, most of the measures carried out to improve the energy label values do not influence only this one energy label program but influence and optimise the general energy efficiency of the washing machine. Therefore these improvements are relevant for many or even all programs the consumer may choose. The task of the declaration program (60° cotton cycle) is therefore not to represent the average program used by the consumer but to be represent all improvements in terms of energy efficiency of washing machines. And this job is done by the present program as it was proven by a study in France ("Ecodrôme"), where after one year of monitoring the real life use of appliances, these were exchanged by more efficient models (energy class 'A') and the effect on real energy used was studied. The observed savings of roughly 30%[ii] clearly confirm the general improvement of the energy efficiency on all programs for better rated machines.

Without doubt the consumer can contribute largely to energy and water saving simply by following some basic rules when operating a washing machine. Therefore CECED manufacturers have developed a standardised set of instructions to be inserted in manuals how to save energy and water by means of a proper use of the washing machine. It contains advices and tips concerning a correct loading, use of pre-wash, proper temperature setting and spin speed in case a tumble dryer is used. In the course of 1999, all participants to the Commitment engaged themselves to implement this uniform type of information in the operating instructions. The next step will be to introduce an information in sales brochures about the relation between energy consumption and the load. It is aimed at giving figures of energy consumption both for a cycle with full and with half load.

The washing temperature is a determinant factor for energy consumption. Therefore manufacturers are working intensively to improve the performance of the lower temperature cycles, in particular the 40°C cycle. Often a 40°C instead of 60°C programme is sufficient to get a sufficient washing result. However, it is important to ensure that the temperature is adjusted correctly, depending on the kind of laundry and its soiling. Manufacturers provide extensive information in the manuals how to set the correct temperature and washing programme.

When speaking about energy and water saving on washing machines, it is important to note that there is a strong interdependence between the technology of the washing process and the available detergents. Therefore, CECED suggested a close co-operation between manufacturers of washing machines and manufacturers of detergents. Consequently, a common working group with representatives of CECED and A.I.S.E. (European Association of Detergent Manufacturers) was set up to discuss technical and marketing related topics of common interest, to gather statistical data and to elaborate feasible strategies for further energy and water saving.

In 1998, CECED has developed a communication concept to support the visibility of the negotiated agreements and to spread the awareness of their results. The E&E (for Energy & Environment) White–star symbol, on the front cover, has been developed by CECED to mark the relevant steps in the process of market transformation, towards a better environment, achieved through voluntary actions of the European Industry and all the other actors involved in the process, which will accept to share our targets. 60.000 leaflets and 10.000 brochures have been printed and distributed during the Domotechnica trade exhibition and afterwards, by participating manufacturers.

2.5 Transfer to Washer-Dryer

Naturally, manufacturers are transferring new washing techniques, derived from washing machines, to washer-dryers as well. To monitor the resulting progress of energy and water saving, CECED committed to set up and to update regularly a database for washer-dryers. It can now be reported, that the average energy consumption of washer-dryers has been improved by 0,40 kWh/cycle comparing 1999 to 1997 and is now (1999 data) - in average - at 4,63 kWh/cycle. As most of the energy is consumed for drying the textiles, improvements were mainly reached by increasing the spinning efficiency of the washing cycle.

2.6 Work on 40°C Cycle

Manufacturers have also agreed to develop more efficient washing programs at lower temperatures, namely at 40° and to support standardisation to develop accurate measuring standard suitable for this temperature. While the improvement of the washing program is running, the development of an accurate standard for measuring at 40°C is delayed due to the enormous problems already appearing when measuring at 60°C. Neither there is a reference detergent suitable for 40°C (too much foam produced), nor are the present soils strips qualified (originally developed for 90°C wash), nor is their a chance to get a quantitative good differentiation of different washing technologies (relative tolerances of energy measurement are almost doubled compared to the 60° cycle). Therefore industry is quite reluctant to promote the use of a 40° cycle for the next step of a new designed energy label. This is also supported by the fact, that the program of most relevance in terms of energy consumption is still the 60° cotton cycle. But the running SAVE II project on washing machines is investigating this subject further and in more detail.

3 Outlook

The energy label of today has shown to have a high impact on the market, both on the retailer and the end consumer. By combining the effect of the energy label with the voluntary agreement a saving of 20% was already achieved in 1999, which has to be compared to a total saving potential (being technological proven and economically cost-effective) of 25% as found in the GEA studies of 1995[iii]. This shows impressively the advantage of a voluntary to a regulatory approach. But the consequence is also, that this process of improvement will slow down almost automatically as soon as this economic and technical reasonable potential is realised.?

We have (almost) reached what is the best technical and economical compromise. For a further substantial optimisation we would need a quite different approach in terms of technology or very different political and economical frame conditions. This can happen either by a break-through new technology (which is not visible today) or changed input parameters (e.g. much higher energy costs) or new political targets (e.g. highest priority for energy savings). As none of these seems to be likely to happen in the near future, an energy efficiency improvement of washing machines of about 25% (compared to the base-case of 0,30 kWh/kg for a cotton 60° program) is the maximum achievable target.
So the possibilities to reduce the specific energy consumption of new washing machines are limited. But are there means to reduce the total energy consumption for washing clothes in private households ? We think yes.

The first target for further actions must be to get the present offer of very efficient machines into the households. As the average life-time of a washing machine is in the order of 12 to 15 years, there are many machines in use, which are not at all as efficient as today's market offer. Therefore programs to support some early replacement of the present stock of machines are necessary. Why not - for example - giving the buyer of a very efficient washing machine the benefit of a lower VAT (value added tax) ?

The second target should be to continue to educate the consumer. To try to influence the consumers preference and habit in selecting the washing temperature and using the machine. Here some public advertising programs might help, targeting for teachers or other 'multiplicators'.

But we must be careful not to focus only on the environmental aspects. We must look at the performance issues as well. An increasing number of household use pre- and post-treatment processes to ensure a sufficient cleaning results (as reported by detergent manufacturers). And some newspaper articles on hygienic problems may indicate that we are on a shaky ground. Over all we have to respect the consumers request to get a well washed and refreshed laundry.

---

[i] published ....., *Group of Efficient Appliances, Study on Washing machines, Tumble Dryers and dishwashers,*

[ii] as reported *in SAVE II Study, Revision of Energy Labelling & Targets Washing Machines (Clothes), Interim Report-Final Draft*

[iii] See for example, Danish Energy Agency, June 1995, *Executive Summary, Washing Machines, Driers and Dishwashers, Group for Efficient Appliances,*

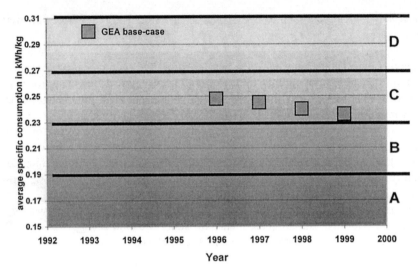

Figure 1: Average specific energy consumption of washing machines (in kWh/kg) from 1993 to 2000 (model average)

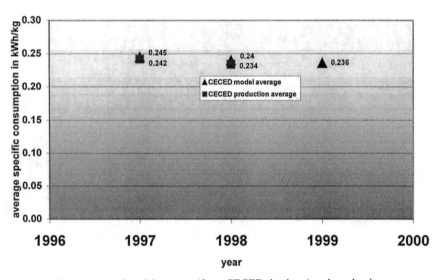

Figure 2: Comparison of model average (from CECED database) and production volume average (from notary reporting system) of specific energy consumption (in kWh/kg) for year 1997 to 1999

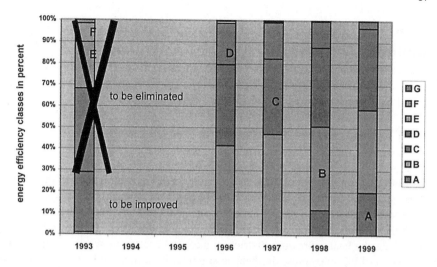

Figure 3: Representation (in %) of market offer of washing machines in EU regarding different energy classes in years 1993 to 1999

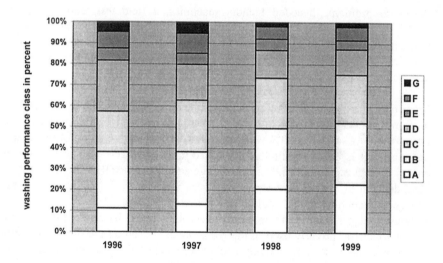

Figure 4: Representation (in %) of market offer of washing machines in EU regarding washing performance in years 1997 to 1999

Field Test of Heat-Fed Washing Machines and Laundry Dryers

ir. F.T.S. Zegers and dr. E.C. Molenbroek

ECOFYS energy and environment, Utrecht, The Netherlands

Summary. Application of heat-fed washing machines and tumble dryers is a promising fuel-switching option. It increases heat demand in district heating networks, it saves primary energy and it can save costs. In this way, it can benefit utilities, the environment and consumers at the same time. In heat-fed laundry machines, the heat source is a district- or a central-heating network. As opposed to hot-fill washing machines, the heat is supplied by a heat exchanger, rather than directly using the hot water for washing. In this way, the concept can also be used for drying and higher primary energy savings can be achieved.

Based on prototype heat-fed laundry machines, a field test, co-ordinated by Ecofys, was conducted. Seven utilities bought the machines to be placed in thirteen households for a one year -test. The aim of the test was to establish the technical performance in practice and to examine how users would appreciate the apparatus.

From the technical monitoring of the machines during the field-test, the washing machine turned out to function well. Averaged over all households a yearly primary energy saving of 43% was calculated. The average cooling of the district heating water is 26°C. The tumble dryer also functioned well. The yearly primary energy savings amounts to 52%. The average cooling of the DH-water is 16°C.
Heat demand in district heating networks can increase by 2.5 GJ per house. For energy efficient homes this would be an increase of 12%.
A household can save € 42 in yearly electricity costs, based on the day-time electricity tariff currently used in the Netherlands. Based on the night-time tariff this would be € 9. For actively stimulating heat-fed apparatus it could be considered to change tariffs to make heat consumption more interesting for consumers.

The heat-fed apparatus has shown to perform as well when connected to central heating systems. However, the primary energy savings are lower. From the field-test data savings of 18 and 23% were calculated for the washing machine and the tumble dryer, respectively. On the other hand, the cost savings are higher: € 54 for the day-time tariff and € 21 for the night-time tariff, respectively.

1 The Field Test

1.1 Introduction

In previous projects prototype heat-fed washing machines and tumble dryers were developed by adapting existing electric machines [VHK96a] [VHK96b]. The results of tests done with the machines in the lab were very satisfactory and it was decided to continue the project with a field test.

Seven utilities were part of the project. They recruited a total of thirteen households for participation in the field test. Out of these households, 11 homes were heated by district heating (DH) and two by central heating.

The financing of the project was done by EnergieNed, Novem and Gasunie. Seven utilities bought sets of the heat-fed laundry apparatus and recruited households. Ecofys Energy and Environment was main-contracter and project co-ordinator.

1.2 Development of Heat Fed Laundry Apparatus

The design of the prototypes and redesign for the field-test in case of the dryer was done by Van Holsteijn & Kemna.

The heat-fed washing machine[VHK98a]

The heat-fed washing machines were made by modifying the Miele comfortline W959. To be able to use hot water as a heating source an extra heat exchanger was placed at the bottom of the machine between the drum and the bottom of the tub. For the heat exchanger to fit the tub had to be modified. The electronic circuitry of the machine has undergone some changes as well. For example: if there is insufficient power from the heat exchanger to achieve the desired washing temperature, the electrical heating element is turned on. This makes it possible to do laundry at 95 °C if the DH-water is only 70 °C.

Table I: Technical specifications heat-fed laundry machine.

Technical specifications heat-fed laundry machine [VHK98a]		
Energy consumption[1] electricity	Wh	203
Energy consumption[1] heat	MJ	3,33
Consumption DH-water[2]	liter	42
Flow DH- water	liter/min	2
ΔT DH-water[2]	°C	19
Power[2/3]	kW	0,2 (2,7)
Water consumption	liter	61
Cycle time	min	108
Size	mm	900 x 595 x 600

1   60°C cotton without prewash

2   at a supply temperature of 70°C

3   in between brackets: the power of the electrical element

Heat-fed tumble dryer [VHK98a]

The heat-fed laundry dryer was made by modifying the Miele comfortline T559C. Just as for the washing machine, the main change was the addition of the heat exchanger for the DH-water. The redesign of the prototype resulted in a larger condensor and higher power ventilators for the drying air as well as the cooling air. In the redesigned machine the program time was reduced from 140 to 120 minutes. This is still longer than the time it takes for its electric counterpart (80-100 minutes). However, in practice, less laundry is put in the machine, which will also result in cycle times that are less than 100 minutes.

Table II: Technical specifications heat-fed tumble dryer.

Technical specifications heat-fed tumble dryer [VHK98a]		
Energy consumption[1] electricity	Wh	770
Energy consumption[1] heat	MJ	15,1
Consumption DH-water[2]	liter	180
Flow DH- water	liter/min	1,5
ΔT DH-water[2]	°C	19
Power[2/3]	kW	0,46 (3,6)
Cycle time	min	120
Size	mm	900 x 595 x 600

1   cotton cupboard dry cycle

2   at a supply temperature of 70°C

3   in between brackets: the input power of the electrical element

2 Field-Test Results

2.1 Results Washing Machine

Results from the lab-test preceding the field-test show that the washing machine performs as well as the prototype from the previous project (phase 3b).
In Table III the parameters of a standard cycle (60 °C cotton) is given, for the lab-test preceding the field-test and the average for the field-test itself.

Table III: Consumption of energy and water, average of lab-test and field-test for the standard program (60 °C cotton cycle).

Parameter	unit	Lab-test	Field-test	Conventional
Energy consumption electricity	Wh	216	230	1050
Energy consumption heat	MJ	3.9	5.0	-
Total water consumption	l		64.7	
Water consumption main wash cycle	l		16.3	15
Flow DH- water	liter/min	1.9	1.7	
ΔT DH-water	°C	17.6	19.3	
Laundry weight	kg	5.0	4.5-5.5	5.0
Cycle time	min	117	121	

The table shows that the lab-test and field-test are comparable as well. The heat consumption is higher in the field-test. This partly caused by the higher water usage in the main wash cycle in the field-test. The heat consumption also varies with parameters like DH-flow, supply-temperature and ambient temperature. These variations cause the difference between the lab-test and field-test.

2.2 Results Condensing Dryer

The redesigned dryer has a significantly lower energy consumption compared the the prototype from phase 3b. Consumption data and other parameters for the lab-test preceding the field-test and for the field-test itself are given in Table IV for the standard drying program. In the field-test hardly any drying at the standard load of 5.0 kg was done. Therefore, energy consumption and cycle time data for lower weight categories were calculated for 5.0 kg dryweight using the weight-dependency found earlier in lab-tests.
Comparison of the lab-test and field-test show that electricity consumption is somewhat higher in the field-test while the heat consumption is somewhat lower.

Table IV: Consumption of energy and water, average of lab-test and field-test for the standard drying program (cupborad dry).

Parameter	unit	Lab-test	Field-test	Conventional
Energy consumption electricity	Wh	747	845	3500
Energy consumption heat	MJ	13	11	-
Moisture content before drying		70	70	
Spinning speed	rpm	900	900	
Flow DH- water	liter/min	1.9	2.0	
ΔT DH-water	°C	17.3	16.4	
Dryweight laundry	kg	5.0	5.0	5.0
Cycle time	min	115	115	88

2.3 Cooling of District Heating Network

Cooling of district heating water by the washing machine
The cooling of the DH-water in the field-test for the 60°C cotton cycle of the washing machine is 20°C. For all programs together this turns out to be 26°C. The larger cooling for the average program compared to the standard program is caused by the fact that the average program has a lower washing temperature than the standard program, and therefore needs less DH-water.

Cooling of district heating water by tumble dryer
The cooling in the field-test by using the standard cupboard dry cycle at average load (this was 2.0 kg instead of the standard 5.0 kg) was 17.4°C. Averaged over all programs used in the field-test the cooling was 14.6°C. However, just like for the washing machine, a supply temperature below 70 °C negatively affects the energy performance of the dryer. If all cycles that took place at a supply temperature below 70 °C are filtered out, a cooling of 18.1°C results for the standard cupboard cycle and a cooling of 16.1°C for the field-test as a whole.

2.4 Transport Losses

Heat loss per cycle outside the heating season
Calculations show that heating losses through piping can vary widely, depending on pipe length, pipe diameter and whether they are insulated or not. For example, for a maximal pipe length of 2x12 m (supply and return) the losses in uninsulated pipes can be up to 52% of the useful heat consumption in the washing machine and up to 16% for the tumble dryer (insolating would bring this back to 13 and 4%, respectively). However, even if heat loss through the pipe *per cycle* is large, the heat loss averaged over a year that should be contributed to the laundry machines is small. The reasons for this are:
•   In half of the total number of cycles both machines are used at the same time.

- If the space heating is turned on heat loss through the pipe should not be attributed to the heat-fed laundry apparatus.

Heat loss on a yearly basis
For a loss per cycle of 52% for the washing machine the loss averaged over a year will not exceed 13%. For a loss per cycle of 16% for the tumble dryer the loss averaged over a year will not exceed 4%. This includes the assumption that in 50% of the cycles the washing machine and tumble dryer are used simultaneously, and that the space heating season is eight months long.

2.5 Increased Heat Demand

The total heat demand in district heating networks could increase through the use of heat-fed laundry apparatus. The yearly increase of heat consumption per household would be approximately 2.5 GJ. The total heat consumption of an energy efficient house (heat consumption for central heating 14 GJ, for tapwater 7.3 GJ) could increase by 12% through the use of heat-fed apparatus.

3   Energy and Cost Savings

3.1 Efficiencies and Costs

For calculating primary energy savings one has to decide how much primary energy is 'assigned' to the production of heat and to the production of electricity in a power plant. In the table below, input data for energy savings and cost saving calculations are summarized.

Table V:

Efficiency electricity generation	49% (steam and gas) [CO2-98]
Primary energy use district heating	500 MJ$_{primair}$/GJ end user [CO2-98]
Efficiency heat from central heating	90-100% on lower heating value
Running cost electricity for households	€ 0,145/kWh daytime and € 0,082/kWh night-time tariff (VAT included; more than 3000 kWh consumed) [ENE99]
Running cost heat for households	€ 13.28/GJ (VAT included) [ENE99]
Running cost natural gas for households	€ 0,2936/m³ (VAT included; more than 2100 m³/year consumed) [ENE99]

All tariffs include an energy tax (REB, 'Regulerende EnergieBelasting'). If this tax would not be channeled to the consumer in the case of heat production, the tariff would be € 9,69/GJ (VAT included). To indicate the sensitivity of the cost savings for the cost of heat, the cost savings are also calculated using this tariff.

3.2 Washing Machine

Energy- and cost saving in practice
The average primary energy saving in practice, averaged over all different programs and temperatures and loads used in practice amounts to 43%. This primary energy saving can be considered representative for an average dutch household.
In Table VI, the most relevant results on energy consumption, energy savings and running cost savings for the washer are given. The first column gives the electricity consumption and primary energy consumption in case these 350 cycles per year would have been done using the conventional fully electric machine. In the second column, energy use, energy savings and cost savings are given for the heat-fed machine. Naturally the cost savings depend on whether the washing is done in the hours where the night-tariff for electricity is in place or in the hours when the day-tariff is in place. Both options are presented here. The relative savings of running cost are less than the primary energy savings, because district heat is relatively expensive.

Table VI: Energy consumption, energy saving and cost saving on a yearly basis for the washing machine, based on the the households in the field test (350 cycles per year)

Energy consumption, energy saving and cost saving on a yearly basis	Conven-tional	Heat-fed	Heat fed, green tariff and 333 MJ/GJ	Heat-fed for central heating
Primary energy (GJ)	1,60	0,92	0,78	1,54-1,67
Heat consumption (GJ)	-	0,83	0,83	0,83
Electricity consumption (kWh)	218	68	68	68
Primary energy saving with respect to conventional (%)	-	43%	51%	4-18%*
Electricity saving (%)	-	69%	69%	69%
Running cost saving with respect to conventional (%)- day-tariff	-	34%	43%	47%
Running cost saving with respect to conventional (%)- night tariff	-	7%	24%	30%
Absolute running cost saving / yr – day tariff	-	€ 11	€ 14	€ 15
Absolute running cost saving / yr – night tariff	-	€ 1	€ 4	€ 5

*Spread caused by variation in efficiency heat production central heating (90-100% of lower heat value).

In the third column, the sensitivity for two parameters is shown. At first, a efficiency of 333 $MJ_{primary}$/GJ is assumed instead of 500. This positively affects the primary energy and the primary energy savings. Second, it is shown how changing the tariff for heat can positively affect the running cost savings.

In the last column the primary energy savings and running cost savings are calcuated in case heat from a central heating system in a home is used. This results in lower primary energy savings but higher running cost savings.

3.3 Condensing Dryer

Energy- and cost saving in practice

The average primary energy saving for the drying behavior found in the field-test (with respect to programs used, average weight, moisture content, cycles per year) is 52%. In Table VII the most relevant results on energy consumption, energy savings and running cost savings for the dryer are given. Because the fractional primary energy savings is hardly dependent on the type of program used, the determination of the absolute energy savings is based on the energy savings of the standard cupboard program. The first column gives the electricity consumption and primary energy consumption in case these 250 cycles per year would have been done using the conventional fully electric machine. In the second column, energy use, energy savings and cost savings are given for the heat-fed machine, using the assumptions of the previous paragraph. In the third column, the sensitivity for two parameters is shown. At first, a efficiency of 333 $MJ_{primary}$/GJ is assumed instead of 500. This positively affects the primary energy and the primary energy savings. Second, it is shown how changing the tariff for heat can positively affect the running cost savings.

In the last column the primary energy savings and running cost savings are calcuated in case heat from a central heating system in a home is used. Again, this results in lower primary energy savings but higher running cost savings.

Table VII: Energy consumption, energy saving and cost saving on a yearly basis for the tumble dryer, based on the households in the field test (250 cycles per year)

Energy consumption, energy saving and cost saving on a yearly basis	Conven-tional	Heat-fed	Heat fed, green tariff and 333 MJ/GJ	Heat-fed for central heating
Primary energy (GJ)	3,54	1,69	1,41	2,95
Heat consumption (GJ)	-	1,68	1,68	1,68
Electricity consumption (kWh)	481	116	116	116
Primary energy saving with respect to conventional (%)	-	52%	60%	17-23%[*]
Electricity saving (%)	-	76%	76%	76%
Running cost saving with respect to conventional (%)- day-tariff	-	44%	53%	56%
Running cost saving with respect to conventional (%)- night tariff	-	19%	35%	40%
Absolute running cost saving / yr – day tariff	-	€ 31	€ 37	€ 39
Absolute running cost saving / yr – night tariff	-	€ 8	€ 14	€ 16

* Variation caused by variation in efficiency central heating (90-100% of lower heatint content).

3.4 Heat-Fed Against Other Fuel Switching Options

In Table VIII an overview is given of energy savings and estimated time for return on investment for the heat-fed apparatus in comparison with competing concepts for washing and drying. The additional cost for the heat-fed apparatus includes additional cost for the machines as well as for installation of the machines.

For the washing machine, it can be concluded that
- Heat-fed washing machines connected to district heating results in the highest energy savings, much more than hot-fill machines. However, the extra investment cost for the hot-fill machine is paid back more easily.
- The energy savings of heat-fed washing machines connected to central heating are comparable to hot-fill machines

The time for return-on-investment of a heat-fed washing machine is less than five years (assuming the day-time electricity tariff).

For the tumble dryers the conclusions are as follows:

- Heat-fed dryers connected to district heating save similar amounts of energy as the heat-pump dryer. However, the time for return-on-investment is half that of the heat-pum dryer.
- The heat-fed dryer connected to district heating has higher energy savings than the gas-fed dryer, with a return-on-investment time that is slightly longer.
- The time for return-on-investment of the heat-fed tumble dryer is less than four years (assuming the day-time electricity tariff).

4   Experiences Users and Utilities from Field Test

The heat-fed washing machine performed very well. Most households had a positive opinion on the machine as it comes to wash performance, spinning, noise, stain removal, technical reliability, user friendliness, avoidance of wrinkles.

The opinion of the households with respect to the performance of the tumble dryer varies widely from one household to another. This may be related to the amount of technical difficulties each household has experienced with the dryer, but also with other factors, such as the type of laundry used and program choice. It is noticed that almost all households were disturbed by the noise the heat-fed dryer makes. Noise production should be a point of attention for a next series.

Table VIII: Energy savings and estimated time for return on investment of heat-fed apparatus and competing concepts for washing and drying (on the basis of labtests with standard program).

	electricity consum-ption	gas consumption	heat consum-ption	primary energy consumption	primary energy savings
	kWh/cycle	m³ gas/cycle	MJ$_{DH}$/cycle	MJ$_{prim}$/cycle	%
Washing machines (60° cotton, 5 kg)					
Conventional washing machine, av.)	1.10			8.1	
Heat-fed washing machine - DH	0.22		3.9	3.5	56%
Heat-fed washing machine - CH	0.22	0.12		5.5	32%
Heat-fed washing machine - CH + solar	0.22	0.09		4.5	44%
Hot-fill washing machine	0.50	0.06		5.5	32%
Tumble dryer (cupboard dry, 5kg)					
Conventional condensing dryer (av.)	3.49			25.6	
Heat-fed tumble dryer - DH	0.75		13.0	12.0	53%
Heat-fed tumble dryer - CH	0.75	0.41		18.5	28%
Heat-fed tumble dryer - CH + solar	0.75	0.31		15.2	41%
Heat pump tumble dryer	1.75			12.9	50%
Conventional air dryer (av.)	3.27			24.0	
Heat-fed tumble dryer - DH	0.45		12.6	9.6	60%
Gas-fired tumble dryer	0.25	0.41		14.8	39%

	additional cost	running cost	running cost savings	running cost savings	return-on-investment time
	Euro/machine	Euro/cycle	Euro/cycle	Euro/year	year
Washing machines (60° cotton, 5 kg)					
Conventional washing machine, av.)		0.16			
Heat-fed washing machine - DH	111	0.08	0.08	27	4.1
Heat-fed washing machine - CH	157	0.07	0.09	32	4.9
Heat-fed washing machine - CH + solar	157	0.06	0.10	,5	4.4
Hot-fill washing machine	66	0.09	0.07	25	2.7
Tumble dryer (cupboard dry, 5kg)					
Conventional condensing dryer (av.)		0.51			
Heat-fed tumble dryer - DH	211	0.28	0.23	56	3.7
Heat-fed tumble dryer - CH	211	0.23	0.28	69	3.0
Heat-fed tumble dryer - CH + solar	211	0.20	0.31	77	2.7
Heat pump tumble dryer	454	0.25	0.25	63	7.2
Conventional air dryer (av.)		0.48			
Heat-fed tumble dryer - DH	204	0.23	0.24	60	3.4
Gas-fired tumble dryer	238	0.16	0.32	79	3.0

Assumptions

1 primary energy use district heating	500 MJprim/GJfinal
2 Efficiency central heating	100% on lower heating value
3 Efficiency tap water heating	65% on lower heating value
4 Efficiency electricity consumption	49%
5 Solar contribution (4 m² collector, 45° South)	25%
6 Water use main washing cycle hot-fill	11.4 liter
7 Heating main cycle hot-fill	25 °C (15 -> 40°C)
8 gas price	0.29 Euro/m³
9 electricity price	0.15 Euro/kWh
10 heat price	13.28 Euro/GJ
11 number of washing cycles per year	350
12 number of drying ycles per year	250

It should be noted that the data in the table is based on energy consumption figures from lab-tests at a standard program and load [VHK94], [VHK98d], [VHK99], for the heat-fed apparatus as well as the competing concepts.

5 Conclusions

The field-test
The purpose of the field-test was to establish the technical performance in practice of newly developed heat-fed washing machines and tumble dryers and to examine how users would appreciate the apparatus. In both aspects, the field-test was succesful. Product development preceding the field-test has resulted in a reduction of the drying time of 20 minutes compared to the earlier prototype.

Energy and cost savings for district heating applications
From the technical monitoring of the machines during the field-test, the washing machine turned out to function well. Averaged over all households a yearly primary energy saving of 43% was calculated. The average cooling of the district heating water is 26°C. The tumble dryer also functioned well. The yearly primary energy savings amounts to 52%. The average cooling of the DH-water is 16°C.
Heat demand in district heating networks can increase by 2.5 GJ per house. For energy efficient homes this would be an increase of 12%.
A household can save € 42 in yearly electricity costs, based on the day-time electricity tariff currently used in the Netherlands. Based on the night-time tariff this would be € 9. For actively stimulating heat-fed apparatus it could be considered to change tariffs to make heat consumption more interesting for consumers.

Performance of heat-fed apparatus in central heating systems
The heat-fed apparatus has shown to perform as well when connected to central heating systems. However, the primary energy savings are lower. From the field-test data savings of 18 and 23% were calculated for the washing machine and the tumble dryer, respectively. On the other hand, the cost savings are higher: € 54 for the day-time tariff and € 21 for the night-time tariff, respectively.

Heat loss
The heat loss through the pipes averaged over a year is calculated to be at most 13% for the washing machine and 4% for the tumble dryer (based on 2x12 m pipe length). This has little influence on the calculated primary energy savings.

Supply temperature
It is important that the supply temperature to the heat-fed apparatus is at least 70°C. If this is not the case, the washing machine will turn on its electric heating element at a 60°C wash cycle and the tumbly drying cycles will take longer.

User appreciation
From the inquiries it is concluded that the participating households were very content with the washing machine. Appreciation of the dryer varies significantly from one household to another. Noise production seems to be the biggest factor

causing mixed feelings about the dryer. Some households find the drying cycles take too long. Some households find the dryer dries better than their previous machine, whereas others think its performance is inferior to their previous dryer.

Further developments
During the field-test technical problems have occurred every now and then. None of these were of a fundamental nature and all problems have been solved. From a technical viewpoint the field-test has shown there are no barriers for further development and large scale application of the heat-fed apparatus. In addition, users have indicated their satisfaction in using the heat-fed apparatus.
A brief comparison of the heat-fed concept with other energy saving concepts (hot-fill washing machines, gas-fired dryers, heat-pump dryers) show that the heat-fed apparatus combine large energy savings with moderate cost, and therefore are promising options for fuel switching. A brief inquiry also showed interest by utilities Europe-wide for application of heat-fed apparatus [ECO99b].

6   References

[CO2-98]   Uitvoeringsregeling subsidies CO_2-reductieplan' d.d. 30 juni 1998.
[ENE99]   Energietarieven warmte per 1/1/99, EnergieNed, feb. '99.
[ECO99]   Veldexperiment Warm Witgoed – Resultaten technische meting, Ecofys, Utrecht (1999).
[VHK94]   Stadsverwarmingskeuken: technische en economische haalbaarheidsstudie, Van Holsteijn en Kemna, fase 1 en 2, Delft (1994).
[VHK96a]   Stadsverwarmingskeuken technische haalbaarheidsstudie fase 3a: meetrapporten wasmachine Miele W957 en wasdroger Miele T557C, Van Holsteijn en Kemna, Delft (1996).
[VHK96b]   Stadsverwarmings-warmtewisselaar voor de wasmachine en de condenserende wasdroger, fase 3b 'Test en optimalisatie', Van Holsteijn en Kemna, Delft (1996).
[VHK98a]   Warm witgoed, onderzoek aansluitmogelijkheden, Van Holsteijn en Kemna, Delft (1998).
[VHK98b]   Warm witgoed, installatieinstructie, Van Holsteijn en Kemna, Delft (1998).
[VHK98c]   Warm witgoed, meetrapport 0-serie, c
[VHK98d]   Haalbaarheidsstudie warmtepompdrooghok, Van Holsteijn en Kemna, Delft (1998).
[VHK99]   Gasconcepten, Van Holsteijn en Kemna/Ecofys, Delft 1999.
[ECO99b]   Haalbaarheidsonderzoek EU-project Warm Witgoed, Utrecht 1999.

Heat Pump or Gas-Heating: New Approaches to Save Energy for Drying of Laundry

Hans-Joachim Klug

Electrolux Home Products

Abstract. Electric clothes dryers are often accused of producing too much hot air and wasting energy. While the ecological benefit of washing machines and dishwashers is beyond any questions, the youngest member of the so-called "White Goods Family" is under discussion from the beginning because of its high energy consumption. On the other hand, in spite of this bad reputation the saturation of electrical operated tumble dryers on the European Market is increasing continuously, showing that there is a clear request from the consumer. Within this background it is necessary to start activities which are suitable to improve the energy efficiency but without reducing the well known positive advantages.

Most of the electric tumble dryers available on the European Market are well optimised. Because of this high standard it is difficult to improve the energy efficiency with an acceptable effort by using the current dryer technique. More energy reduction needs new technical solutions. One possibility is the Heat Pump Dryer - a combination of condensation dryer and heat pump. With the introduction of this technology it is possible to reduce the energy consumption in comparison to traditional electrical drying processes by roughly 50 %.
A comparison, based on primary energy consumption shows that also the tumble dryer with Gas Heating is a very interesting alternative in relation to the traditional electrical tumble dryer. By using this technology, it is also possible to save approximately 50 % of primary energy due to the missing conversion losses from primary energy to electricity.

1 The Electric Tumble Dryer – A Useful Domestic Help

Electrical operated tumble dryers in European households have been in use for more than 25 years. Although many improvements have been introduced during this long period, the bad image because of its relative high energy consumption still remains. Even the best appliances on the marked need two to three times more energy for the drying process than most of the modern energy optimised washing machines for washing the same amount of load.

But this is only one side of the medal. The other side shows a steady increase. The overall European saturation in 1998 was 27 % (Figure 1).

Tumble Dryer: Saturation By Country

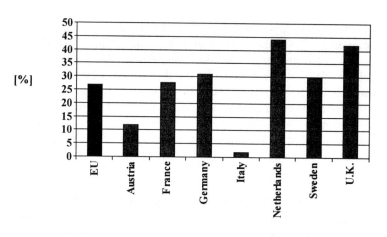

Source: APPLIANCE Magazine

Figure 1

There are some convincing reasons for such a positive development. Our modern way of life very often leaves no room for extensive housework, and besides washing, drying or ironing are not very attractive. Additionally the requirements regarding cleanliness and quick availability of the preferred articles become more and more important and this for sure is the point where the electrical tumble dryer shows its biggest advantage.

In front of this background there is the only possible prognosis for the future: In the European Community the electrical operated tumble dryer will extend its place because of its obvious advantages. As a consequence, without improvements concerning the energy efficiency, the energy consumption for drying of laundry will become the most electrical consumer in the household. Therefore there is a huge challenge for all the manufacturers of household appliances to develop energy efficient alternatives to the existing drying technique.

2 The Energy Consuming Factors of a Conventional Dryer

If we split the energy consumption of a current condensation dryer we can see that already 60% is needed for evaporating the water in the textiles. The remaining 40% of electrical energy is required for all the other functions like driving the drum and the air flow, for heating up the appliance, the load, the condensed water and for the heat losses (Fig. 2). The energy required for evaporating the water is a physical constant and because of that not practicable for measures to reduce the energy consumption. A realistic judgement based on the conventional drying process shows an energy reduction potential of maximum 5%. More energy savings need totally different technical solutions. One of these solutions may be the application of the Heat Pump , another one the use of gas to heat the drying air.

Energy Distribution
Conventional Condensation Dryer

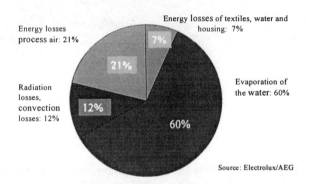

Energy losses of textiles, water and housing: 7%

Energy losses process air: 21%

Radiation losses, convection losses: 12%

Evaporation of the water: 60%

Source: Electrolux/AEG

Figure 2

3 The Condensation Dryer with Heat Pump – An Excellent Example for Applied Heat Recovery

To combine a conventional condensation dryer with a heat pump system is not a real new idea. Industrial solutions based on this technical principle were already known from the past. And for this reason technicians especially in the USA and Europe tried to find a solution in order to use this concept also for household application. As a first result of this effort Electrolux launched in 1997 the first Tumble Dryer with Heat Pump - the AEG ÖKO Lavatherm WP.

In a Heat Pump Dryer a complete heat pump system including evaporator, condenser, compressor, expansion valve etc. takes over the work of the traditional electrical heating unit and the condensation system . In detail the electrical heating must be replaced by the condenser and the air-air-heat exchanger must be replaced by the evaporator of the heat pump system (Figure 3).

The heat used for the drying process is absorbed by the refrigerant from the drying air leaving the drum and given back to the air entering the drum but on a higher temperature level due to the compressor and condenser of the heat pump system. This is an excellent application of heat recovery (Figure 4).

Figure 3

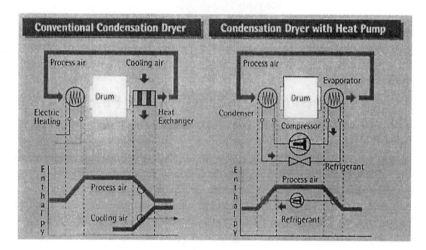

Figure 4

The utilisation of a heat pump system in combination with an condensation dryer offers several benefits. First of all a tremendous 50 % reduction in energy consumption. And because of this it is possible to reach the Energy Class A of the European Energy Label for a dryer.

4   The Gas Heated Dryer – Advantages in Saving Primary Energy

In the viewpoint of primary energy consumption all electrical operated devices suffer from energy losses during transportation and in particular from transformation losses when changed from primary energy to electricity (power plant efficiency). To avoid this high transformation losses of around 60 % it is necessary to use one form of primary energy like natural gas, oil etc.. This is the field were the Gas Dryer derives its advantage from. In a Gas Heated Dryer a gas burner heats up the drying air. Electricity is only used for the drum drive system and electronic control. In this way the energy for heating up and evaporating the water, which is roughly 80% to 90% of the complete consumption, is covered by primary energy.

Gas Heated Dryers are normally designed as air vented versions. Industrial applications are well known and especially in American households gas heated dryers are very common. But also some household appliances manufacturers in Europe offer gas heated tumble dryers as an alternative to the electrical operated tumblers. In 1997, the German manufacturer MIELE launched an electronic controlled gas heated Vented Air Dryer on the European Market. This tumble dryer offers all features of a high level electronic controlled version but with reduced consumption in primary energy and a lower CO_2–emission. Furthermore the consumer gets lower operating expenses due to the better cost situation with gas and a very short drying duration in comparison to conventional Vented Air Dryers due to the high heating power which is feasible with a gas burner.

System Advantages / System Disadvantages	
Heat Pump Dryer	**Gas Heated Dryer**
• 50% reduced electric energy consumption	• 45 % reduced primary energy consumption
• Reduced CO_2 emission	• Reduced CO_2 emission
• Lower operating noise due to elimination of the cooling air cycle	• Lower textile abrasion due to 40% shorter operating duration
• Gentle treatment of laundry due to lower drying temperatures	• Very low power input due to gas heating
• Lower power input due to elimination of the electric heating	• Reduced Operating expenses
• Reduced operating expenses	• Additional installation expenses due to gas supply
• High production costs	• Service expenses for gas operated device

Figure 5

5 Heat Pump or Gas Heating – Comparison of the Systems

For a comparison on primary energy basis the electrical energy consumption has to be corrected by a factor when converting into primary energy. The primary energy factor considers transformation losses as well as transportation losses for electricity. For the calculation in this report a primary energy factor of 2.5 is used. The Gas Heated Dryer from MIELE needs for drying a 5 kg load according to the European standard EN61121 only 14% electrical energy and for the rest (86%) primary energy in the form of natural gas. The comparison of the different drying systems shows for both, the Heat Pump Dryer and the Gas Heated Dryer, an improvement in primary energy consumption of approximately 50% with small advantages for the Heat Pump Dryer. Concerning the drying efficiency, which is the energy consumption calculated without correction via a primary energy factor, the Gas Heated Dryer takes 123% of the energy consumption of the conventional condensation dryer and is the worst one of this comparison. That means, while the Heat Pump Dryers saves primary energy by means of heat recovery the Gas Heated Dryer saves primary energy by avoiding the use of electrical energy.

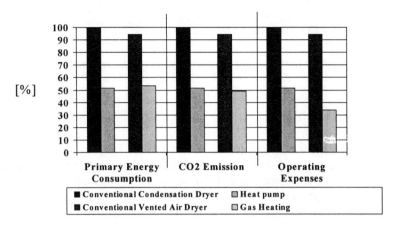

System comparison
Energy, CO$_2$ Emission and Operating Expenses

Figure 6

Basis: *Primary energy factor: 2,5*
CO$_2$ – emission factor for electricity: 0,55 kg/kWh
CO$_2$ – emission factor for gas combustion: 0,20 kg/kWh (methane)
Costs: 0,27 DM/kWh for electricity, 0,056 DM/kWh for gas

The CO$_2$ emission is directly connected to the energy consumption for a given energy production scenario. Because of that fact also the evaluation of the different drying systems leads to a similar result. That means both alternatives, the Heat Pump Dryer and the Gas Heated Dryer, show a approximately 50% reduction of CO$_2$ emission compared to the conventional counterparts but with small advantages for the Gas Dryer.

The operating expenses show clear advantages for the Gas Heated Dryer because of the lower gas price. In comparison to the conventional condensation dryer the operating expenses are only 32% but also the Heat Pump Dryer gives a reduction of the running costs of approximately 50%.

Figure 1 - Tumble Dryer: Saturation by Country. *Source: Appliance Magazine: November 1999 Appliance –European Edition*
Figure 2 - Energy Distribution, Conventional Condensation Dryer. *Source: ELUX*
Figures 3, 4, 5 - *Source: ELUX*
Figure 6 - System comparison - Energy, CO$_2$ Emission and Operating Expenses. *Source: Miele puplication, November 1997, Gastrockner*
   BGW: May 1997, Gas-Wäschetrockner im häuslichen Bereich

New Washing Technologies for the New Millennium

Prof. Dr.-Ing. Hans G. Hloch and Dr. H. Kruessmann

W- Research Institute for Cleaning Technology, Krefeld, Germany

1   Introduction

About 120 million washing machines nowadays are installed in the European Union according to Ceced. More than 11 million washing machines are sold annually in the Community Market, mainly for replacement. Most of the machines are constructed as horizontal axis drum washing machines, which were introduced at the beginning of the 50s in Western Europe. The Asian countries still prefer pulsator machines and in Northern America agitator machines are common. As shown in many studies, up till now the lowest water and energy consumption can only be achieved by drum washers. Modern "high-tech" machines especially which feature electronics and partly smart sensors meet most of the requirements needed by consumers and governmental regulations.

Two of the reasons for the achievement of low consumption values of washing machines are the introduction of Energy Labelling[1] and the Voluntary Agreement for clothes-washers between the European machine manufacturers association Ceced and the European Commission. As seen during Confortec 2000 trade fair already at least 8 clothes-washers show triple-A rating (energy consumption, washing performance, spinning efficiency) and as demonstrated in Figure 1 some 88 % of 165 clothes-washers from 17 brands have an energy-efficiency rating of A and B, and no appliance falls below a C-classification[2].

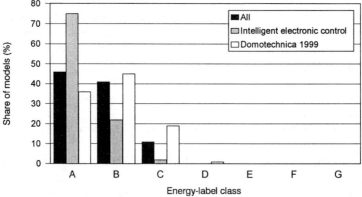

Figure 1: Energy efficiency of 165 clothes –washers promoted at Confortec 2000 according to [2]

These low consumption values with at the same time sufficient washing perform-ance are reached by introduction of electronic components, electronic controlling, application of new drum construction and drives, general reduction of water levels during wash and rinse phase, lowering of washing temperature and introduction of detergents with less environmental impact and especially good performance at lower temperature (surfactants, enzymes, bleaching systems, etc.)[3-6].

Independent of these developments we carry out the washing procedure in an inefficient system which still needs high amounts of water, electricity and deter-gents. Therefore the following paper tries to point out new washing technologies based on the basics of soil removal. There will be presented no mature technolo-gies which can be immediately applied in current machine constructions without any further R&D investigations. It is attempted to show general possibilities to get to new washing technologies, whereby "new millennium" not only means the next 10 years, but may be the next 100 years.

2 General Aspects of Soil Removal

The mechanisms of soil removal can be seen as sum of different phenomena[7]. There exist sorption effects of water and detergent, like roll up of oily soil, pene-tration of soil and solubilization and emulsification. These effects are mainly de-termined by type and amount of soil and by type and amount of detergent. The mechanical action (mechanical work) of the machine causes hydrodynamic flow, fibre flexing, abrasion and swelling of the fibre or finish.

The basic forces in the wash liquor, which are strongly influenced by the above mentioned effects, are interfacial tension forces, electrostatic effects, steric effects and ionic effects. According to SHORT[8], these forces involved in soil removal can be considered under three arbitrary size classifications. Large scale effects – mainly influenced by machine parameters - are those which involve the directed motion of the surface to be cleaned and of the main fluid body. Object of this directed motion is the mixing of textiles and free liquid. Turbulence favours mass transport perpendicular to flow direction and boundary layer (shear plane) may become very small.

Small scale effects are due to mass transfer processes (detergent to cloth - soil from cloth), the purely mechanical removal of microscopic dirt trapped between fibres, and the dissolution of solid materials near the cloth surface. Soil removal determining effects for this transfer are flow effects with random velocity fluctua-tions partly parallel/perpendicular to surface, the diffusion of chemicals and soil within boundary layer. The velocity is determined by fluid drag forces in the vicinity of the textile surface. Besides the characteristics of the soil, these effects are mainly influenced by the machine parameters time and temperature and by mainly influenced by the machine parameters time and temperature and by chemistry.

Molecular scale effects have to do with the forces, which attach soil material to a surface. At low distances from the fibre surface the liquid velocities become negligible, and the intensity of mechanical agitation has little influence on the breaking of soil retention forces, particularly compared to the influence of the mechanical action. Dominating forces are the interaction between born repulsive and van der Waals attractive forces, electrical double layer forces and effects of surface layer adsorption.

Before discussing new washing technologies a general consideration of the above mentioned effects with respect to the machine is given and compared to the current state of washing technology under the aspect of the washing technology.

3 Determination of Mechanical Action

By comparison of the theoretically needed mechanical input for soil removal with the current energy consumption it should be possible to give a rough estimate of the possible potential of energy reduction.

Because of lack of suitable measurement systems, procedures or calculations for determination of the mechanical action during washing, the following assessment is based on calculations of the "residual work of detergency" according to the papers[7, 9-10]. Based on the removal of oily soil the work of detergency is determined by interpretation from the point of interfacial phenomena. According to literature, these forces are dominant and significantly higher than electrical, ion and mass adhesion forces.

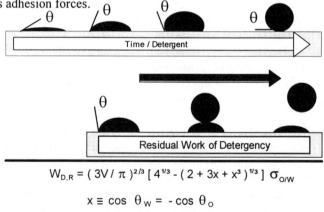

$$W_{D,R} = (3V / \pi)^{2/3} [4^{1/3} - (2 + 3x + x^3)^{1/3}] \, \sigma_{O/W}$$

$$x \equiv \cos \theta_w = - \cos \theta_o$$

$$W_{D,R} \approx \Delta F \, \sigma_{O/W}$$

Figure 2: Displacement of oily soil by surfactants and residual work of detergency

Basis of the evaluation of new technologies must be an evaluation of washing performance which covers consumers requirements: for instance stain removal. The assessment therefore shall not base on washing performance values, which show very little correlation to practical washing conditions. This means results on energy label measurements or evaluations cannot be used for decisions concerning the importance of a new washing technology.

Energy label testing is only for comparison of different machines concerning consumption values under specific testing conditions. It does not include the total spectrum of washing performance criteria. The given washing performance index covers only a very small range of soil removal effects, rinsing efficiency is not evaluated. In addition, especially effects correlating to textile effects (abrasion, pilling, creasing, dye transfer, etc.) and hygiene aspects are not considered.

The following discussion is based on the partners of washing.

5 Partners of Washing Process

Figure 4 shows the partners dominating clothes washing. These mainly are the solvent (water, including the detergent), the clothes, the soil (pigments, fat etc.) and the processing. The latter comprises machine construction features and the washing parameters according to Sinners' circle, like temperature, mechanical action, time and detergents.

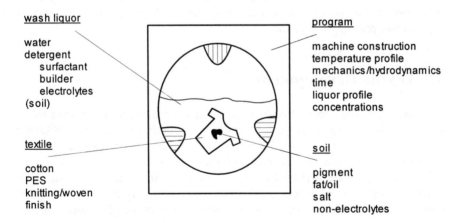

Figure 4: Partners of clothes washing

6 Washing Technologies for the Next Years

Within the next decades the common drum type washing machines will be the dominant machine construction. **Figure 5** gives a survey of possible features of the future washing machine[4].

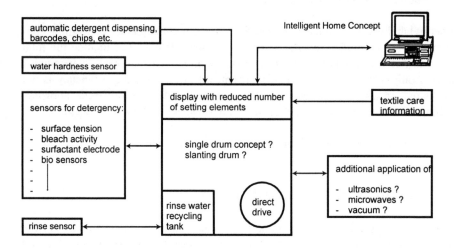

Figure 5. Features of Future Washing Machines

Concerning the technology of washing the following discussions show possible developments and their consequences for washing technology under the aspect of the different partners of washing.

7 Textile Developments

Washing technology (mechanical action and temperature) is determined by the textile and its soiling. Only the application of the detergent has a low influence. The current trend to less soiled and low load size of laundry (washed in 5 kg-machines) needs fewer requirements to both wash factors. *Short wash programs* with low temperature will be the future, whereby stain removal is achieved by separate application of special stain removers (spotting agents). The increasing wear of synthetic fibres, which can be more easily cleaned has similar consequences.

The development of "easy care textiles" including easy soil removal will be probably the future. This will be achieved by application of fibre/clothes coatings (finish, membranes etc.) in a conventional way or by finishing the clothes according to the Lotus effect. Whilst the above trends lead to easier cleaning, the increasing trend to more coating (UV-protection) can have the consequence of more gentle program design.

New "functional" textiles showing special effects like improved wearing comfort, cooling/heating properties and the ability to transfer medicine to the body will probably need a special program design. This only can be designed, when these textiles are available on the market. Probably partly professional cleaning will be necessary.

The application of textile identification systems (barcode labels, radio frequency identification) as already applied in professional laundries, sewed into the textile, yarns etc. will open the possibility for *automatic adaptation of the program design* to the care requirements.

8 Detergent Developments

The influence of further developments in the field of detergents is very important. New enzymes for removal of all types of stains and new bleach systems (which are already in pre-testing phase) will give the ability for *cold wash* and at the same time sufficient wash performance and hygiene standard. Probably liquids will be applied more intensively for cold wash. Lowering of temperature leads to less importance of water hardening which improves the dosage of detergents. The application of different detergent components will only succeed to a noticeable extent when automatic dosage systems are developed and the handling of the components is more consumer friendly, as discussed above. None of the current component systems on the market fulfils this criteria.

The introduction of devices for substitution of detergent components into the washing machine as *ozone application, electrochemical systems, activating of water* etc. will probably not succeed because of low efficiency compared to the single detergent component[14]. Further reductions of the detergent dosage will require more exactly *dosage* which can only be fulfilled by application of automatic systems. With regard to this liquids will show advantages.

9 Washing Technology Developments

The current washing technology is caused by the type of washing machine. Further reductions of energy demand will be the application of *low temperature* which is facilitated by application of new detergent systems and *lowered water levels* during main wash phase.

In the conventional machine system the *substitution of thermal by mechanical energy* is limited by textile damage (prolongation of wash duration) and the requirement of minimum water levels which hinder high mechanical action by flow effects. Therefore technologies causing intensive liquor exchange by *forced flow* (centrifugal wash, spray wash) have been developed and will be improved. Probably the application of *direct drive* with the possibility of exact and fast changing of the drum movement can produce higher mechanical input.

Reductions of water consumption during rinse with the higher part of the total water consumption can be reached by similar processing. In addition, *water recycling* (last rinses are used for wash phase and first rinse) will be applied more intensively, as it is already common in professional textile care. Arising hygiene problems can be solved. The most important prerequisite is the acceptance by the consumer.

Further extensive water reduction will only be possible by application of technologies with forced exchange of bound and free water. Besides high intermediate spin speed improvements technologies with *intensive flow effects* will result in further water saving efficiency. In addition, these efforts will be supported by textiles with low water (and detergent) absorbency.

Significant improvements in energy and water reduction can be attained by application of higher *efficient heat and detergent transfer* to the cleansing surface (soil/textile). In discussion are the application of *ultrasonic waves* for faster distribution of detergent and support of mechanical action[15], more intensive wetting the textile with detergent solution by *vacuum systems*[14], improved bleaching efficiency (including disinfection) and faster heat up (even under conditions with low bound water) with *microwaves*[16]. All these technologies nowadays are in the stage of pre-testing, their realisation needs extended further research and may be unsuccessful.

Discussions about new technologies should include general aspects of the application of *water for washing*. Water as solvent is used (up till now) because of low costs, no environmental impact and high availability. In theory clothes cleaning without any solvent would be the ideal system. Therefore for instance *foam washing* (as applied in hard surface cleaning) should be considered as one of the new technologies to be transferred and adapted to textile cleaning. In addition, the application of new solvents – like supercritical CO_2 as dry cleaning solvent – should be considered in the ongoing technology discussions, even the transfer into domestic appliances at this time seems to be impossible.

At the same time new processing should be seen under the aspect of *flat processing*. Advantage of this technology is the possibility of *directed treatment* of the textiles with low amounts of water, energy and detergents by spraying, sucking, extraction and foam application. Transfer of heat and detergent by more efficient systems and for instance *laser cleaning* would be much easier.

Figure 6 shows a textile care processing system where clothing is conveyed through wash, rinse and drying areas[17]. Smart sensors recognise each garment, and variable agitation washes all types of fabric with efficiency and speed.

Proposed *Fast Laundry* Wash-Process
[Pat. Pending]

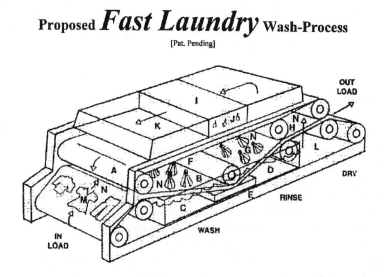

Figure 6: Continuous Clothes Laundry System according to [17]

New washing techniques are definitely more intensively determined by application of *electronics and computer controlling*. Most of the R&D work nowadays is invested in the field of development of electronic components, control systems and smart sensors. The consequence is the introduction of wash and rinse sensors which give the opportunity for textile care processing corresponding to the type and amount of the laundry in the machine and according to type and intensity of soiling. These developments would ease the handling of laundry and the operation of the washing machine, probably including drying and finishing of the clothing. Therefore the most important requirements of the consumers would be fulfilled.

The discussed possible changes in washing technology are extensively influenced by the development of *machine construction*. On the other hand the development of machine construction can be determined vice versa. In addition, *governmental regulations* determine further developments.

10 Summary

Besides changes in consumer habits, the introduction of new washing technologies is determined by developments in the field of textiles, soiling, detergents, machine construction and governmental regulations. Basis of new technologies should be the consumers requirements for lower consumption values and ease of use. Prerequisite of all technologies is the guaranteed cleaning of the clothes without textile damage and sufficient hygiene level. Therefore decisions about applicability of new technologies should base not on energy label testing, but on consumers requirements. The discussed technologies already are partly in development stage, in fundamental R&D phase or just fiction.

11 References

[1] Hloch, H.G., Kesel, W., Kruessmann, H., *Proceedings of the 51st Annual International Appliance Technical Conference 2000, Lexington/KY, USA, May 8-12*, p. 429

[2] N.N., *Appliance Efficiency*, 4 (2000) 1, 1

[3] Hloch, H.G., Kruessmann, H., *Proceedings of the 48th Annual International Appliance Technical Conference 1997, Columbus/OH, USA, May 14*, p. 1

[4] Hloch, H.G., *Proceedings of the 39th* wfk-*International Detergency Conference, Luxemburg, 6 - 8. September 1999, p. 31*

[5] Krings, P., Vogt, G., *Proceedings of the 39th* wfk-*International Detergency Conference, Luxemburg, 6 - 8. September 1999, p. 1*

[6] Bohnen, J., Foellner, B., Hloch, H.G., Kruessmann, H., *Proceedings of the Symposium on Technology and Culture of Washing and Laundry, Kobe, Japan, March 16-18, 1999*

[7] Kissa, E., *Textile Research Journal* (1975), 736

[8] Short, B.A., *Surfactant Science Series, Vol. 5, Marcel Dekker, New York 1972*, p. 237

[9] Kling, W., Koppe, H., *Melliand Textilberichte* 30 (1949), 23

[10] Saito,M. et.al., *Textile Research Journal* (1985), 157

[11] Kling, W., *Melliand Textilberichte* 46 (1965),957

[12] Kling, W., *Melliand Textilberichte* 29 (1948),427

[13] Beer, W.L., *51st Annual International Appliance Technical Conference 2000, Lexington/KY, USA, May 8-12*

[14] Hloch, H.G., Kruessmann, H., *Hauswirtschaft und Wissenschaft* (1989), 75

[15] Hloch, H.G., *Tenside Detergents* 19 (1982), 30

[16] Kruessmann, H., Kesel, W., *AiF-report* 11084N, 1998

[17] Porter, D., *www.fastlaundry.com/wash.htm*, 1999

How Precise Labelling Can Be Done?
Experience with Energy-Label Verification Procedures for Washing Machines

Dr. Ulrich Sommer

Federal Institute for Materials Research and Testing, Berlin, Germany

Summary. The intention of the Energy-Label for washing machines is to inform the consumers about energy consumption and related performance criteria prior to buying such an appliance. At the moment a 60°C Cotton programme has to be labelled for consumers and competitors interest.

The washing performance is a basic criterion of such an appliance and is more or less directly related to the main energy consumption of a washing programme. In contrary to that the spin drying efficiency is no energy related criterion for this appliance itself. But it is an important energy concerning data, because the generally high energy consumption of electric clothes dryers, very often used after a washing programme, directly depends on the water extraction efficiency of the washing machine.

The Energy Label for washing machines seems to have high importance for the competitive industry, an accurate labelling therefore is requested. The labelling system is based on EN 60456 „**Clothes washing machines for household use. Methods for measuring the performance"** This standard does not only contain the description of a method with very specified conditions for the determination of different performance criteria but also the tolerances and control procedures for the declared values of the Energy Label.

This evaluation method is very difficult to perform with sufficient accuracy as the high ongoing activities in international and European standardisation work on this reference standard show. Additionally the experience of the last years with several verification procedures indicates, that there is a visible need for better definitions of several parameters of this method in order to obtain reproducible results between different testing laboratories. The current discrepancies or deviation problems of this method will be shown in this paper with recent results of verification procedures.

The Energy-Label has an significant influence to the share of the market. Washing machines are mainly sold referring to the declared performance criteria of the Label followed by additional mentioned other features of the machine. Accuracy of labelling therefore is essential. In 1999 BAM has done a high number of so called verification procedures. Some important discrepancies caused by inadequate formulations and specifications in the standard EN 60456 were observed.

The Energy-Label shows the total Energy consumption of a 60°C Cotton cycle, the specific values in kWh/kg load classified in a 7 steps scale, the washing performance and spin drying efficiency in an analogous scaling system and the water consumption without an corresponding declaration of the rinsing efficiency, because an reproducible method for this criterion is not yet available. Water consumption therefore should not be stressed from the market as it is done at the moment.

The Energy-Label data are based on measurements according to the EUROPEAN STANDARD EN 60 456, June 1999; Clothes washing machines for household use, Methods for measuring the performance, (IEC 60 456 : 1998, modified).

Most important and most critical materials in this standard are the cotton base load, the soiled strips IEC 105 and the reference detergent IEC A. Other important conditions are better specified e.g. water hardness (2.5 +/- 0.2 mmol/l), water temperature (15 +/- 2°C), ambient temperature (20 +/- 5°C), Wascator FOM 71 LAB (Reference machine), Spectrophotometer and other instruments

The classification of the results has to be realised in a 7 steps scale with very small ranges, where normal standard deviation of results often covers two third of a class and more.

Classification of Energy Consumption

Energy Efficiency Class	Energy Consumption ‚C' in kWh per kg
A	$C \leq 0.19$
B	$0.19 < C \leq 0.23$
C	$0.23 < C \leq 0.27$
D	$0.27 < C \leq 0.31$
E	$0.31 < C \leq 0.35$
F	$0.35 < C \leq 0.39$
G	$0.39 < C$

Classification of Washing Performance

Washing Performance Class	Washing Performance index ‚P'
A	$P > 1.03$
B	$1.03 \geq P > 1.00$
C	$1.00 \geq P > 0.97$
D	$0.97 \geq P > 0.94$
E	$0.94 \geq P > 0.91$
F	$0.91 \geq P > 0.88$
G	$0.88 \geq P$

Tolerances and control procedures are defined in this standard as well. There are foreseen two verification steps:

	1st Step (1 machine)	2nd Step (3 machines)
Energy consumption	+ 15 %	$X_3 + 10 \%$
Washing performance	- 0,03	$X_3 - 0,02$
Spin extraction	+ 15 %	$X_3 + 10 \%$
Spin speed (max. spin speed during 60 s)	10 % or 100 rpm (lower number)	each machine as step 1
Water consumption	+ 15 %	$X_3 + 10 \%$
Programme duration	+ 15 %	$X_3 + 10 \%$

A verification procedure is established as follows:

Competitor A tests a competitor B machine
Results show, that Energy-Label declaration is not in line with test results
Competitor A repeats tests on more machines : same results

Competitor A contacts Competitor B, offers his results and asks him for test reports or other results or to change Labelling

Competitor B does not agree

Discussion on higher level leads to the decision to perform a verification procedure . Looser has to pay the test.

Test lab buys one machine, performs the test according the standard and confirms the declaration or starts second step of the verification procedure with lower tolerances for 3 machines

The test results are discussed between Competitor A and Competitor B and the consequences concluded

Next table shows that load characteristics have high influence on the results. The loads H, J, K were prepared artificial in the reference machine but according to the standard and were used in this example for the first time for a performance test. Load A is a load used according to the standard since longer time. The machines 68, 69, 70 and 88 are all of the same type. Washing performance results are different, depending on the load. Performance class C was proved as the right declaration with other loads too.

Influence of load on washing performance

Machine	68	69	70	88	88
Base Load	H	K	J	A	K
Water consumption During main wash l	17,1	15,3	15,8	17,2	16,5
Carbon black / oil (EMPA 106) %	34,6	33,2	32,7	42,1	38,4
Blood (EMPA 111) %	75,3	78,6	76,8	82,9	81,2
Cacao (EMPA 112) %	37,4	37,0	36,8	48,9	42,7
Red wine (EMPA 114)%	75,1	72,9	73,1	76,6	75,0
Sum of remission values (machine) %	222,4 ± 7,4	221,6 ± 5,0	219,4 ± 6,5	250,0 ± 1,8	237,3 ± 6,1
Sum of remission values (Wascator) %	243,0 ± 2,8	243,0 ± 2,8	243,0 ± 2,8	253,0 ± 2,7	253,0 ± 2,7
Ratio	0,915	0,912	0,903	0,990	0,938
Classification	E	E	F	C	E

The same influence of the „critical" loads H and K could be seen in the machine 60 which was tested with two „normal" loads D and C, comparable with above mentioned load A.

Influence of load on washing performance

Machine	60	60	60	60
Base Load	D	H	C	K
Water consumption during main wash l	14,9	14,8	14,3	15,2
Carbon black / oil (EMPA 106) %	46,0	45,3	48,0	44,1
Blood (EMPA 111) %	83,0	84,9	84,7	83,0
Cacao (EMPA 112) %	49,0	50,8	51,6	48,6
Red wine (EMPA 114)%	80,0	76,3	78,6	77,4
Sum of remission values (machine) %	258,0 ± 2,4	257,3 ± 3,9	262,8 ± 3,0	253,1 ± 3,0
Sum of remission values (Wascator) %	243,7 ± 2,2	252,1 ± 2,0	253,0 ± 2,7	253,0 ± 2,7
Ratio	1,059	1,020	1,039	1,004
Classification	A	B	A	B

Another reason for such discrepancies is that at the moment no load is available that really is according to the standard, although the specification level is low:

There are for example „IEC"-towels available in the market with very different shrinkage rates

Fabric	EMPA	WFK	Greve
Shrinkage warp / weft After 1 cycle	5,9 / 7,4	13,2 / 9,5	18,1 / 16,3
Shrinkage warp / weft After 3 cycles	8,0 / 7,6	19,3 / 11,9	20,0 / 18,1
Shrinkage warp / weft After 9 cycles	9,1 / 8,5	21,4 / 12,6	22,6 / 18,9
Shrinkage warp / weft After 12 cycles	9,2 / 8,7	21,7 / 13,0	22,7 / 19,1
Stable after 10 cycles			

The influence of different batches of the soils on washing performance is significant too.

Batch number 5 classifies the washing performance of the machines 60 and 67 lower than batch number 3

Machine	67	67	60	60
Strips batch	3	5	5	3
Carbon black / oil (EMPA 106) %	47,5 ± 3,7 / 3,0	45,4 ± 4,2 / 3,4	45,3 ± 3,2 / 2,2	46,0 ± 2,4 / 1,9
Blood (EMPA 111) %	82,1 ± 0,5 / 0,4	84,2 ± 0,7 / 0,6	84,9 ± 1,5 / 1,2	83,0 ± 0,5 / 0,4
Cacao (EMPA 112) %	47,8 ± 2,7 / 2,2	50,4 ± 3,0 / 2,5	50,8 ± 2,7 / 2,2	49,0 ± 1,5 / 1,2
Red wine (EMPA 114)%	77,0 ± 2,5 / 2,0	75,9 ± 1,4 / 1,1	76,3 ± 2,5 / 2,0	80,0 ± 0,9 / 0,8
Sum of remission values (machine) %	254,4 ± 4,2	255,9 ± 4,4	257,3 ± 3,9	258,0 ± 2,4
Sum of remission values (Wascator) %	244,7 ± 1,5	252,1 ± 2,0	252,1 ± 2,0	243,7 ± 2,2
Ratio of washing performance	1,040	1,015	1,020	1,059
Classification	A	B	B	A

Finally the effect of the ageing of the detergent (foam inhibitor) can be shown. Foam decreases washing performance because mechanical action is lower then. This effect of lower performance with higher foam level is more evident in the Wascator reference washing machine than under the conditions of household machines. This influences the washing performance ratio "machine under test/Wascator reference" in that way, that with lower sum of remission values in the Wascator reference machine the washing performance ratio for the household machine increases with the ageing of reference detergent. Refreshing of foam inhibitor leads back to higher performance in the reference machine and lower performance ratio for the machine under test in this example.

Strips batch	3					
Programme No.	99					
Date	23.03.99	15.06.99	12.07.99	13.07.99	19.07.99	20.08.99
Water consumption during main wash l	26,1	25,8	26,3	26,4	26,5	26,3
Carbon black / oil (EMPA 106) %	44,8	42,3	41,7	42,0	43,6	44,5
Blood (EMPA 111) %	80,3	81,0	78,4	78,6	82,2	80,7
Cacao (EMPA 112) %	50,8	47,9	47,4	47,6	48,6	50,1
Red wine (EMPA 114)%	76,5	75,9	76,2	76,2	75,5	74,7
Sum of remission values %	252,3 ± 1,7	247,1 ± 1,5	243,7 ± 2,2	244,4 ± 1,9	250,0 ± 2,2	250,0 ± 2,0

Conclusion

How precise labelling can be done at the moment? Not precise enough, because the standard EN 60456 is not precise enough and there is actual need for improvements!

Assessment of the Cleaning Efficiency of Domestic Washing Machines with Artificially Soiled Test Cloth

P. M. J. Terpstra

Wageningen University, Consumer Technology & Product Use

Abstract. The aim of a laundering process is the elimination of visible soil and stains from textile articles, the cleaning performance. A detailed test method for the assessment of the cleaning performance has been published by the International Electrotechnical Committee. In this test method two measurement principles for the assessment of the cleaning performance are elaborated. One is based on the visual assessment of washed naturally soiled household laundry. The second is based on the removal of visible soil from artificially soiled test fabrics (soil tracers). The soil removal is then measured by optical measurement of the test materials. The validity of the method with soil tracers, however, has not been reassessed since its first adoption in 1972.

The primary aim of the research is to evaluate whether the cleaning performance of domestic washing machines can be assessed with test soils. The secondary aim of the research is to investigate to what extent the IEC-60456 test method is suitable for the evaluation of the cleaning performance of modern washing machines.

In the experimental part of the research, wash tests in domestic washing machines were executed. The cleaning performance was visually assessed and the soil removal was established by optical remission measurement of soil tracers. Then it was investigated if the soil tracers data could predict the cleaning performance.

The results of the research show that the IEC-60456 test method can be used for the assessment of the cleaning index of wash processes but also that the methods that have been developed in the research are substantially more accurate.

1 Object of the Research

In our society cleaning plays an important role. A substantial part of our resources and labour capacity is used to clean our habitat and our working environment. Because on the one hand cleaning is essential for human well-being and on the other hand cleaning has a significant environmental impact, it is necessary to clean in the most effective way. This implies the use of effective and environment-friendly appliances and cleaning agents, an efficient laundering practice and an appropriate lifestyle. For that purpose efficient technology has to be produced and consumers have to be informed about the efficiency of products. This consumer information is often supplied by consumer associations. To perform this task they have a continuous need for up-to-date, reliable and cost- and time-effective product evaluation methods.

On March 1 1996 an energy label for household washing machines was introduced Europe-wide 1996. The intention of the European Community was to inform consumers better about the environmental effects of consumer goods. This energy label has to provide information on energy and water consumption and on some functional properties. Manufacturers are obliged to prepare technical documentation with a report of the relevant measurements as evidence.

The two examples above stress the need for reliable and practicable test methods for the evaluation of the performance properties of domestic textile cleaning.

The international test method for the evaluation of the washing performance of washing machines is specified in IEC-publication 60456; "Clothes washing machines for household use - Methods for measuring the performance" (SC59D 1998). The measurement of the cleaning effect is based on four test soils on cotton fabric. The test soils are washed in the washing machines under household conditions and with a representative detergent. Then the cleaning effect of the washing machine is deduced from the removal of the four test soils. This deduction is partly based on the results of an international round robin test that was executed before the adoption of the test method in 1972 (Nieuwenhuis and Spiegel 1970; IEC-59D 1971). A comprehensive systematic selection of test soils or an assessment of the optimal transformation equation has not been done.

Though washing behaviour, washing machines and detergents have changed significantly in the past few years (Puchta and Krings 1994), the measurement principle of the IEC test method for the evaluation of the washing performance has not been revalidated since their first adoption in 1972. This implies that it is not known to what extent the results obtained with these methods reflect the cleaning performance as perceived by consumers in normal practice. This research has been initiated to investigate if the cleaning performance of washing machines can be assessed with test soils in general and to investigate if the IEC method is still appropriate for today's washing practice. The previous considerations have led to the following two research objectives.

- The primary objective is to investigate if the evaluation of the cleaning effect of a wash process as perceived by consumers can be deduced from artificially soiled test material.

- The secondary objective of the research is to investigate if the test method for the evaluation of the cleaning effect of washing machines as described in IEC-publication 60456, provides a test result that correlates with the cleaning performance as perceived by consumers. Which implies that the validity of the method for today's laundering practice will be evaluated.

The research is primarily focused on washing machines, but could serve as an initial basis for future test methods for fabric detergents. The project is supported by the STISAM.

2 Set-Up of the Experiments

To evaluate the suitability of soil tracers for the assessment of the cleaning performance it has to be tested if there exists a quantitative relation between the cleaning index and the soil removal from one or more soil tracers.

There is substantial knowledge on individual soil removal mechanisms of textile laundering but understanding of the behaviour of the integral process is still limited (Löhr and Jacobi 1987). Therefore it is not feasible to deduce such relation on the basis of the present laundering theory. The relation, a model equation, therefore has to be developed on the basis of empirical data from wash tests. The approach for the development of the model equation is as follows. Of a series of wash processes the cleaning performance (cleaning indices) is measured. Of the same series of wash processes the soil removal (soil removal indices) of a number of soil tracers is measured. In a third step it is investigated if and how (model equation) the soil tracers can be used to predict the cleaning indices.

The application of the derived model equation can lead to a new measurement method for the cleaning index. In this new measurement method for the cleaning performance, soil tracers are used and the test result is calculated with the aid of the model equation.

For the purpose of the second research objective the IEC soil tracer set is included in the set of soil tracers used. After the experiments it is tested to what extend the IEC soil tracers correlate with the cleaning index.

2.1 Test Load and Fabric Samples

All machines were loaded with a load conform the rated capacity as specified by the manufacturer. This test load consisted of soiled household laundry, a set of soil tracers, a clean dummy load and additional control cloths. The soiled load used consisted of 6 kitchen towels and 3 pieces of underwear that had been soiled in household conditions. In addition two specimens of the 18 different standard soils were added to each load. Each load was filled to the rated capacity with clean laundry articles. The specifications of the soil tracers are given in Table I. Part of the soil tracers are prescribed in the IEC test method, the other soils are commonly used specific soil types meant for the evaluation of the cleaning effect of a laundering process. The soil tracers are classified in five stain categories, based on their behaviour and typical sensitivity.

2.2 Test Machines

A series of twelve domestic household washing machines together with the International reference machine were involved. All the machines were of the horizontal axis type. Two machines were top-loading machines, the other ones were so-called front loaders. At the start of the research all the machines were new, before the start of the test the suppliers were supposed to have checked the machines for proper functioning and compliance with the factory specifications.

2.3 Wash Processes and Repetitions

In each of the 13 machines two wash programmes are executed; the 40°C-programme and 60°C-programme for cotton textiles. In the 40°C test series, two programmes with the reference machine are run: a 40°C-process and a 60°C-process. This approach allows the comparison of the results from both series. For each wash temperature all wash programmes in one repetition are run on one and the same day. Each separate wash programme is repeated twelve times.

Table I: Soil tracers

Stain category	Soil code	Type	Carrier cloth	Composition
Bleach	EMPA-114	IEC	Cotton	Red wine
Bleach	TNO-SUNAK		Cotton	Black currant juice
Bleach	CFT-BC1		Cotton	Tea
Blood	EMPA-111	IEC	Cotton	Pig's blood (with potassium citrate)
Blood	CFT-CS1	IEC	Cotton	Pig's blood
Detergency	TNO-KWYOVE		Cotton	Iron oxide, mineral oil, arachide oil and SiO_2
Detergency	CFT-AS9		Cotton	Pigment with arachide oil
Detergency	EMPA-106	IEC	Cotton	Mineral oil and carbon black
Detergency	EMPA-101		Cotton	Arachide oil[1] and carbon black
Enzymatic	CFT-CS2	IEC	Cotton	Chocolate milk
Enzymatic	TNO-VEKOPROP		Cotton	Chocolate milk
Enzymatic	CFT-AS5		Cotton	Blood, milk and ink
Enzymatic	CFT-AS3		Cotton	Chocolate milk and carbon black
Enzymatic	EMPA-112	IEC	Cotton	Chocolate milk
General	EMPA-104		PE/Cotton	Olive oil and carbon black
General	WFK-10C		Cotton	Lanolin[2]
General	WFK-20C		PE/Cotton	Lanolin
General	EMPA-116		Cotton	Blood, milk and ink

[1] Peanut oil.

[2] A soft pale yellow wax. Lanolin is a purified form of wool "grease". Chemically, it is a mixture of cholesterol esters. Its resistance to rancidity, together with its emulsifying properties, slightly antiseptic effect, and capability of forming a stable emulsion with water, permit it to be widely used as a basis for ointments, emollients, salves, cosmetics, soaps, and shampoos.

2.4 Detergent

All tests were run with a standard detergent as specified in IEC-publication 60456. This detergent doesn't contain tripolyphosphate but instead polycarboxylates and zeolithes as a builder system. The detergent dosage used was also in accordance with IEC-publication 60456 and is specified in Table II.

Because the reference machine is derived from a professional washing machine, it has a higher suds level. Therefore the detergent dosage had to be higher too, 180 grams.

Table II: Detergent dosage

	Wash load in kg					
	1.6	2.0	2.5	4.5	4.7	5.0
Detergent dosage in grams	80	86	94	126	130	134

2.5 Water Used

Conditioned tap water was used for the wash tests.

Temperature:	$15°C \pm 1°K$
Hardness:	1.6 ± 0.1 mmol Ca/Mg per litre ($9.0 \pm 1.0 °$ Dh)
Alkalinity (M-value):	1.95 ± 0.10 mmol NaOH per litre

2.6 Assessment of the Cleaning Performance and Soil Removal

The cleaning index, by definition, is the measure for the cleaning performance as perceived by consumers in the real life situation. It should be emphasised here that the cleaning index, also by definition, refers to the single wash cleaning effect. For its assessment a panel of two judges was instructed to give each piece of the washed laundry a score ranging from 1 to 8 observer units depending on its cleanness. On this measurement scale a higher figure means a cleaner article. The judgement of one series of wash cycles, one run in all the washing machines, was always performed on one and the same day. The methodology of the assessment is described in the Dutch standard test method NEN-8284 (NEC59D 1989) and in IEC-publication 60456. The cleaning index of one single wash cycle of a wash process is calculated as the linear addition of the judgement scores of the individual washed articles in that cycle. The cleaning index of a wash process is calculated as the mean value of cleaning indices of the individual wash cycles. The unit is by definition the observer unit (OU).

The soil removal is assessed by means of remission measurement (59D/WG10 1998). For this purpose the light remission of the washed soil tracers was measured. The remission measurements were carried out with a Minolta Cr210 chromameter with diffuse illumination and measurement with a 0° viewing angle. For the remission measurements the tristimulus Y-values on two spots on each washed soil tracer were recorded.

3   Data Treatment and Test Results

In a first step the raw judgement data of both judges were studied Outliers and extremes were compared with other observations by the same judge and similar observations by the other judge. If this led to the assumption that these observations were judgement errors, they were excluded from the data set. After that the mean cleaning index of both judges for each wash cycle was calculated with the new data set. Again outliers and extremes were detected and in case of doubts about their validity excluded from the data set. The final data set, from which outlying judgements and cycles were excluded, is assumed to be representative for the cleaning indices. The cleaning indices, expressed in observer units, for the different wash processes at both temperatures are shown in figure 1.

After the wash treatment, the tristimulus Y-remission of the 18 different soil tracers was assessed. The remission values of each type of tracer were averaged per wash cycle, thus giving one test result (remission value) for that particular soil type. As for the cleaning index data, the initial analysis of the remission data implies a basic screening of the measurement values. Outliers and extremes were detected and excluded in the case of doubts about their validity. The final data set, from which outlying soil tracers and wash cycles are excluded, is assumed to be representative for the remission values of the wash processes investigated and is used for further analysis and for the model equation.

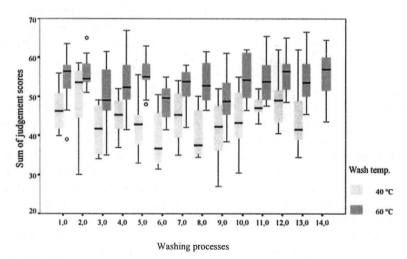

Figure 1: Cleaning indices (OU) clustered by temperature

4 Construction of the Models

Once the cleaning performance (cleaning index) and the soil removal for all processes were available, the analysis related to the first research aim was initiated. It had to be investigated if the soil tracers data could predict the cleaning indices. For this purpose two questions had to be answered: what soils have to be used and how can the relation between the soil removal and the visual judgement data be deduced. The combination of a soil tracer set and the mathematical expression that represents the relation with the cleaning index (model equation) is called a model. For the selection of the soil tracers two fundamentally different selection procedures were applied.

In first procedure, strategic regression, the set of soil tracers of which the linear combination describes the cleaning performance best, was retrieved. Multiple linear regression also provided the mathematical relation between the cleaning performance and the soil removal of the test soils.

The second procedure is based on a data reduction technique known as factor analysis (Norusis 1994). With this technique the variation in the soil removal of all test soils was analysed. The technique resulted in sets of test soils that explain a substantial and predefined part (e.g. 95%) of the variation in the whole data set. These subsets were substantially smaller than the complete data set. Subsequent multiple linear regression was used to find the (linear) relation between the soil removal from the subsets of tracers and the cleaning performance.

The quality of a model is determined by the adjusted multiple R^2 (Devore and Peck 1993) (Norusis 1994) value, the standard error and the residuals of the regression. Apart from this the regression coefficients should differ from zero with statistical significance.

Both mentioned approaches resulted in a number of suitable models. Three to five soil tracers appeared sufficient to build an accurate model. The three most accurate models that are derived from the data set are shown in Table III.

Table III: Details of the most accurate models

Model	Number of soils	Adj. R^2	Standard error in OU	Maximum residual in OU	Minimum residual in OU
1	5[1]	0.99	1.14	1.74	-2.85
2	3[2]	0.99	1.36	2.99	-2.35
3	5[3]	0.99	1.18	2.10	-2.0

[1] EMPA-101, TNO-kwyove, CFT-CS1, WFK-10C,WFK-20C

[2] EMPA-106, CFT-AS5, TNO-vekoprop

[3] TNO-kwyove, CFT-AS5, WFK-10C, WFK-20C, EMPA-104

102

The residuals of these models are small compared to the accuracy of the cleaning performance data (visual judgement). In Figure 2 the relative residuals, the residual as a percentage of the related cleaning index, of the three models are shown for all wash processes. The wash processes are shown in order of the magnitude of the cleaning index.

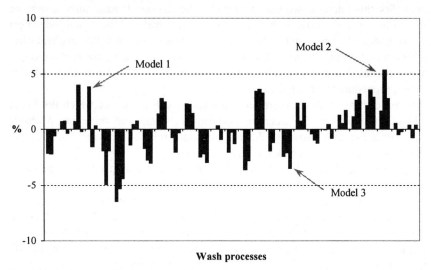

Figure 2: Relative residuals of model 1, 2 and 3

After the development of linear models it was attempted to find if non-linear factors could lead to more accurate models. None of these attempts led to substantially better models.

To investigate the validity of the IEC test method, a regression analyses with the remission data of the IEC soil tracer set was executed. Though the regression coefficients were significant, the multiple R^2 was high and the residuals were small, the fit was worse than that of the deduced models. The relative residuals of the IEC model in comparison with model 3 are shown in Figure 3.

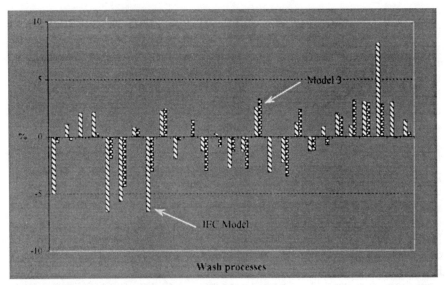

Figure 3: Relative residuals of the IEC model and model 3

Regression analysis with the separate IEC soil tracers resulted in a model with non-significant coefficients. The latter model as well was inferior to the models deduced in the research.

5   Conclusions and Discussion

The primary aim of the research was to test the hypothesis that there is a mathematical relation between the cleaning performance, as perceived by consumers, and the soil removal of a set of soil types. In the research it has been demonstrated that the developed models with soil tracers show a good correlation with the cleaning performance as perceived by consumers. It may therefore be concluded that the related model equations are useful descriptions of the mathematical relation between the soil removal of these tracers and the cleaning performance as perceived by consumers and that the hypothesis can be adopted.

The secondary aim was to investigate if the IEC test method is still an appropriate method for the evaluation of the cleaning performance. For this purpose the visual judgement method is again taken as the reference assessment method. The IEC test method may be considered appropriate if the cleaning index that it predicts demonstrates a substantial correlation with the cleaning index and if the standard error of the residuals is comparable with the error of the assessed cleaning index.
With an adjusted multiple R^2 of more than 0.9 and a standard error that is comparable with the experimental error of the assessed cleaning index, the previous conditions are considered fulfilled. This implies that the IEC test method

104

as it is, may be considered as an acceptable method for the evaluation of the cleaning performance of washing machines. .

It should also be noted that in the applied test set-up a naturally soiled load has been used. As was shown earlier (Besselink 1998) the cleaning performance may be influenced if a clean wash load is applied. This means that the applicability of the assessment method cannot be extended to testing set-ups in which a clean wash load is used. For this purpose additional research is needed.

The models in this research are based on the results of visual judgement of washed laundry by a panel. They are therefore linked to the judgement criteria of this panel. If the research would be repeated it is likely that the new panel will come up with different judgements, even in the hypothetical situation that the test specimens would be identical. Therefore the use of one specific panel implicates the adoption of a specific measurement scale for the cleaning index.

In this research the cleaning indices have been assessed on cotton laundry. As different substrates show different adhesion and soil release properties, it should be noted that the models are not applicable for other than cotton wash processes. Further research should be executed to test the applicability of the present models for other substrates.

The present models are deduced from a series of wash processes in different washing machines. The detergent was the same for all wash tests. Therefore it has not been tested to what extent the models are suitable for the evaluation of fabric detergents. Artificial test soils are frequently used for the evaluation of detergents. Therefore it may be rewarding to start a research into the establishment of models for the evaluation of fabric detergents using the methodology of this research.

Finally it should be noted that the models deduced in this research are substantially more accurate then the IEC model. As there is a general need for accurate assessment methods for washing machines, this implies that it is desirable to revise the present IEC test method.

6 Literature

[1] 59D/WG10, I. (1998). *Amendment to IEC 60456, Ed. 3 - Reflectance measurements for test samples*. Rajamaki (SF), International Electrotechnical Commission.

[2] Besselink, R. (1998). *Een onderzoek naar de invloed van vuilbelasting op de waswerking van huishoudelijke wasmachines*. Consumer Technology and Product usage. Wageningen, Wageningen Agricultural University: 44.

[3] Devore, J. and R. Peck (1993). *Statistics. The exploration and analysis of data*. Belmont, Wadsworth Publishing Company.

[4] IEC-59D (1971). *Comments of the Netherlands on: Proposal by Preparatory Working Group 1*. Standard artificial dirt. Geneve, International Electrotechnical Commission.

[5] Löhr, A. and G. Jacobi (1987). *Detergents and Textie Washing*. Weinheim, Duitsland, VCH Verlagsgesellschaft.

[6] NEC59D (1989). *Elektrische wasbehandelingstoestellen voor huishoudelijk gebruik*. Delft, Nederlands Normalisatie-Instituut.

[7] Nieuwenhuis, K. J. and W. Spiegel (1970). *An attempt to establish the correlation between the reflectance values of artificially soiled fabrics and the stain removal assessed on laundry articles*. Delft, Institute for Cleaning technology TNO.

[8] Norusis, M. J. (1994). *Advanced Statistics*. Chicago.

[9] Puchta, R. and P. Krings (1994). "Waschmittel, Waschhilfsmittel, Reinigungsmittel - Heute." Hauswirtschaft und Wissenschaft 1994(5): 209-221.

[10] SC59D, I. (1998). Clothes washing machines for household use - Methods for measuring the performance. Geneva, International Electrotechical Commission.

Development of a New Energy Efficient Combined Refrigerator/Freezer

Per Henrik Pedersen[1] and Eivind Sallo[2]

[1] Danish Technological Institute
[2] Vestfrost A/S

Abstract. A development work was commenced in 1998 with the aim to develop an energy efficient, combined refrigerator/freezer optimised in relation to Energy Class A. The work is a co-operation between Vestfrost, the Danish Technical University and the Danish Technological Institute. The work is supported by the Danish Electricity Saving Trust.

In phase 1 of the project simulation models and several prototypes have been tested including a concept based on an idea from the Vestfrost company. In this concept the appliance has one refrigeration system with a variable speed compressor. The temperature in the freezer compartment controls the speed of the compressor. The temperature in the fresh food compartment is controlled by a small fan blowing cold air intermittently from a remote part of the evaporator in the fresh food compartment.

Two different appliances according to this principle were tested by the National Consumer Agency of Denmark showing excellent results. The energy efficiency indices were 40 and 46 %, well below the 55 % index limit for energy efficiency class A.

Results from the lab tests are in fine agreement with simulation results. An economic calculation shows that the cost of the appliance carries a modest 5 % penalty only, and the simple pay back time will be 1 – 2 years.

Phase 2 of the project is now progressing and 60 units will be tested in typical apartments over the next 12 months.

1 Introduction

An investigation done by the Danish Technological Institute for the Danish Electricity Saving Trust in the spring of 1998 showed, that a major part of cold appliances sold in Denmark are combined refrigerators and freezers. The figure is about 45 % of the market. The investigation also showed that the introduction of new technology might reduce the energy consumption of the appliances.

This is the background for the Danish Electricity Saving Trust to support a project started in the summer of 1998. The goal for this project is to develop a new type of combined refrigerator and freezer with the following characteristics:

- Energy Efficiency Index less than 50 % ("super A-class")
- No use of harmful greenhouse gases
- About 200 liter refrigerator and 90 – 140 liter freezer
- Width: 60 cm.
- Economical realistic and competitive

A project group was created with representative from the Vestfrost Company, the Danish Technical University (Institute for Energy Technology and Institute for Building and Energy) and the Danish Technological Institute.
The Danish Technological Institute is the project manager.

Two different technologies were tested in the project. Both could fulfil the goals for the project, but the technology described in this paper was chosen because the components involved are commercial available and the principle showed to be the most efficient and economically realistic solution.

2 Baseline

The project started by testing two existing appliances BKF-354 and BKF-375 from Vestfrost. The cabinets from these appliances were later modified to create the basis for the new appliances.

Table I: Data for the two standard appliances

	BKF-354	BKF-375
Energy label	B	A
Volume (ref/freezer) netto, liter	184 / 125	200 / 96
Insulation thickness, freezer mm	72	80
Insulation thickness, refrigerator, mm	46	50
Compressor, freezer	HV70AH	HV57AH
Compressor, refrigerator	HV44AH	HV44AH
Thermostat	Mechanical	Mechanical
Energy consumption, measured by DTI, kWh/day	1,29	1,03

3   The Vestfrost Concept

The idea is to use one efficient compressor in a combined refrigerator/freezer. The concept is especially suited for compressors with variable speed.

So far it has been an advantage to use two compressors for combined refrigerators and freezers, because the thermodynamic process for cooling the refrigerator is better when the evaporator temperature is higher (eg. $- 10\ ^0C$) compared to the one compressor system, where the evaporator temperature is about $- 25\ ^0C$. Theoretical calculation shows that the saving in energy consumption is of the order 20 % for a two door combined refrigerator/freezers.

However, increased insulation in current appliances results in a decreased demand for cooling capacity, especially in the refrigerator compartment. This requires either a very small compressor, which will be less efficient or a low relative duty time for a normal size compressor, which also will result in a poor efficiency.

Therefore, a single compressor system with only one high efficient compressor will be superior.

In traditional single compressor appliances one important disadvantage often appears when the appliance is placed in a room with a low temperature (like in the winter time). If the compressor is controlled by the temperature in the fresh food compartment this might result in spoilage of the food in the freezer compartment.

This disadvantage is not present in the new appliance described in this paper, since the temperature in the freezer and the fresh food compartments can be controlled independently.

The compressor is controlled mainly by the temperature in the freezer compartment.

The temperature in the fresh food compartment is controlled by a thermostat controlled fan, which circulates fresh food air over a small part of an evaporator placed in a small compartment behind the fresh food compartment.

The small and efficient fan with less than 1 Watt power consumption ensures a constant temperature in the fresh food compartment. The temperature difference is less than one degree, and the vertical temperature gradient almost non-existant, similar to a laboratory appliance.

The electronic control of the refrigeration system has a build-in defrosting mode, which is activated every 9 hours. When the evaporator in the fresh food compartment is defrosting, the compressor is stopped until the temperature of the evaporator has reached a certain level above the freezing point.

The compressor is a Danfoss TLV6K equipped with adaptive energy optimizer (AEO). The electronic thermostat has a small temperature difference, which during normal conditions will ensure a nearly continuous duty of the compressor.

The AEO adaptive control will find the right speed of the compressor. If the freezer load is high, the compressor revolutions will ramp up after 60 minutes of continuous duty. This will continue until the maximum revolutions of 4000 rpm are reached and/or the electronic thermostat stops the compressor.

If the freezer load is low and the compressor stops after less than 60 minutes, the revolutions will ramp down until the lower limit (2000 rpm) is reached. The efficiency of the compressor is highest at the lower speed limit.

By limiting the compressor off cycles to less than one minute, the losses otherwise encountered due to pressure equalization in the refrigeration system is minimized. This is possible because the variable speed compressor is able to start against pressure.

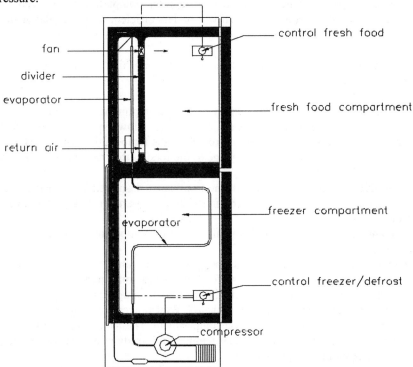

Figure 1: The principle in the new single compressor system. The compressor is controlled by the temperature in the freezer compartment. The temperature in the fresh food compartment is controlled by an efficient fan, controlled by a thermostat.

Data for the refrigeration system:
- Compressor: Danfoss TLV6K
- Refrigerant: 80 g R600a
- Electronic thermostat Danfoss ETC
- Condenser: integrated in cabinet

- Evaporator in freezer compartment: wire on tube
- Evaporator in fresh food compartment: serpentine type
- Fan in fresh food compartment: Sunon
- Thermostat in fresh food compartment: Danfoss 077B20
- Expansion: Capillary tube

4 Test of Prototypes

Two prototypes based on BKF354 and BKF375 were built and tested in climate chamber. The new appliances are now named BKF854 and BKF875. The first measurements were used for finding the correct size of the fresh food evaporator and location of the defrost sensor.

After the prototypes were tuned, they were shipped to the National Consumer Agency of Denmark.

Figure 2: X-ray picture of the combined refrigerator/freezer. The air flow in the fresh food compartment is illustrated by arrows.

The Lab at the National Consumer Agency is certified to measure in accordance to EN-153.

Table II: Results from measurements at the National Consumer Agency of Denmark

	BKF854	BKF875
Energy consumption, EN-153, kWh/day	0.861	0.707
Reduction in energy consumption, compared to standard units	33 %	31 %

By definition the fresh food compartment is of the no-frost type and the energy efficiency indices for the two models become 46 and 40 % respectively.

EI (energy efficient index) $= E / E_{reference}$

$E_{reference} = 0.777 * V_{corr} + 303$

$V_{corr} = 1.2 * V_{refrigerator} + 2,15 * V_{freezer}$

Where V is net volume in liters, E is the actual energy consumption in kWh/year and EI is the energy efficiency index.

5   LCA (Life Cycle Assessment)

A simple life cycle assessment was made in the project. The assessment concluded that the new types of appliances have less impact on the environment than the standard units.
The biggest impact on the environment is related to the energy consumption during the life time of the appliance.

Table III: The climate impact from the appliances during 12 years lifetime. Only impact from energy consumption is shown. The conversion factor for Denmark (in the year 2000) is 0.78 kg CO_2/kWh.

[tons CO_2]	BKF354/BKF854	BKF375/BKF875
The standard unit	4.407	3.519
The new appliance	2.942	2.415
Difference	1.465	1.104

6   Field Test

60 units of the new appliances will be tested in typical family homes from June 2000. The test will be conducted for one year. The aim is to obtain feed back from the consumers and measure the energy consumption at normal usage conditions. The Danish Technological Institute conducts the tests, which will take place in the suburbs of Copenhagen.

It is planned to measure:
- Energy consumption
- Temperature in freezer compartment
- Temperature in fresh food compartment
- Ambient temperature
- Accumulated door opening time, fresh food compartment
- Accumulated door opening time, freezer compartment

The meters will be read every month during the first three month, followed by once every three months. A report will be produced after every reading.

7   Cost

The cost of the new appliance is expected to carry a 5 % penalty compared to the standard units. This makes the new appliance very competitive on the marked, since the additional cost of the appliance will be paid back in one or two years, depending on the local electricity price.

The variable speed compressor is the most expensive component in the appliance. Typical current cost is two to two and a half times the cost of a standard compressor. The variable speed compressors are still produced in limited numbers. Increased demand for this type of compressors may eventually reduce the cost.

In the appliance one (1) variable speed compressor substitutes two (2) standard compressors, minimizing the additional production costs.

The new appliances are expected to go in regular production ultimo 2000.

The Effect of Thermal Aging Polyurethane to Increasing the Energy Consumption of Refrigerator and Freezer

Fatih Ozkadi

Arcelik A.S. R&D Center, Turkey

Abstract. The role of the polyurethane is very important for energy consumption of refrigerator and freezers. The thermal conductivity of the polyurethane has been taken into account when designing the refrigerator and freezers (R/Fs). Since the diffusion process has occurred inside the cells of the polyurethane the thermal conductivity has been increased by the time. This results in increasing heat gain of the cabinet and increasing energy consumption of R/Fs.

1 Introduction

As known energy consumption of refrigerator and freezers has been labeled in Europe and US. When labeling energy consumption value, initial energy consumption test results have been used. These products are being stored as a certain of time after the production. Because of the thermal aging of the polyurethane, energy consumption of those has been increased in this storage period. Sometimes R/Fs has been taken from market, then their energy consumption value has been checked by some consumer organizations. Increase in energy consumption can be important in this manner.

Information about the thermal aging of polyurethane foam has been given in some papers. A good prediction model was explained as follows : The k-value for 50% c-Pentane/CO_2 gas mixture increases from 13 mW/m.K to 20 mW/m.K after 20 years of aging, or an increase of 53% [1]. The 25-year aged thermal conductivity values were found to correspond to the formula for HCFC-141b blown products of $(k_1+6.7)$ mW/m.K (where k_1 = initial thermal conductivity) and for n-Pentane blown products of $(k_1+5.4)$ mW/m.K; the comparable formula for CFC blown products was found to be $(k_1+6.5)$ mW/m.K [2]. The mixture of some blowing agents was investigated on thermal aging [3]. Polyurethane supplier companies have started to present new foam systems, which have good thermal aging characteristic [4]. Nevertheless there is limited information and test results on heat gain aging and increasing energy consumption for R/Fs in the literature.

The following parameters should be taken care for increasing energy consumption of R/Fs by the time :

- Increasing thermal conductivity of polyurethane
- Decreasing the sealing properties of gaskets
- Decreasing heat transfer coefficient of condenser because of dusts occurred
- Increasing or decreasing COP (Coefficient of Performance) value of compressor because of changing the mechanical efficiency by running parts inside the compressor

Since among these effecting parameters dominant one is mostly increasing thermal conductivity of polyurethane, this topic is discussed as detailed in this paper.

In this study a number of polyurethane, which were blown by c-Pentane from different suppliers, were examined and the thermal aging test data by measuring the thermal conductivity has been obtained. The heat gain of several cabinets by performing reverse heat leakage test has been measured by the time and the results of these tests have been given in this paper. In addition the relation between change of the thermal conductivity and change of the energy consumption by means of heat gain have been explained.

2   Thermal Aging of Polyurethane

Four different polyurethane systems blown c-Pentane were evaluated as follows :

- Normal thermal aging at 23°C ambient temperature with 10°C of mean temperature for thermal conductivity measurement, Figure 1.
- Normal thermal aging at 23°C ambient temperature with 24°C of mean temperature for thermal conductivity measurement, Figure 2.
- Accelerated thermal aging at 70°C temperature with 24°C of mean temperature for thermal conductivity measurement, Figure 3.

116

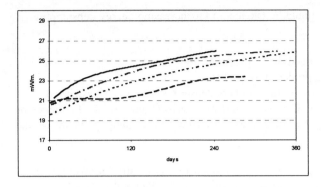

Figure 1: Thermal conductivity aging, k @ 10°C

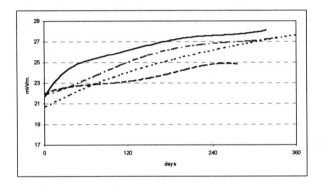

Figure 2: Thermal conductivity aging, k @ 24°C

As seen from Figure 1 and Figure 2, there is nearly 1mW/m.K difference between thermal conductivity data at 10°C and 24°C of mean temperature. Four different polyurethane foam show interesting aging characteristics. While polyurethane marked _ _ _ (blue line) has not low initial thermal conductivity, it has very good thermal aging curve with respect to others. In addition while polyurethane marked . . . (red line) has good initial thermal conductivity, its thermal aging curve worse than polyurethane marked _ _ _ (blue line).

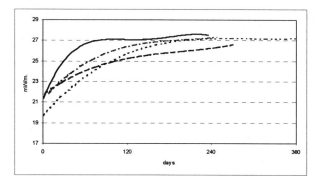

Figure 3: Accelerated thermal aging, k @ 24°C

Figure 3 presents thermally aged polyurethane as worst case scenario. To accelerate diffusion of blowing agents from the cell, foam samples were stored at 70°C of an oven. Thermal conductivity measurements were performed at 24°C of mean temperature. All measurements explained in Figure 1, 2 and 3 were done with Holometrix Rapid K 30 according to ASTM C518.

When comparing normal aging and accelerated aging data for one polyurethane system it is clear to see aging characteristics are the similar for all tests. Due to fact that performing accelerated aging can be useful to understand normal aging of a polyurethane system. Regarding the duration of such a test 120 days seem good enough to get an indication about aging characteristics.

3   Thermal Conductivity & Temperature

In general thermal conductivity has been measured and declared at 10°C and/or 24°C of mean temperatures. The cold plate and hot plate temperatures are 0°C and 20°C respectively for 10°C of mean temperature. The cold plate and hot plate temperatures are 10°C and 38°C respectively for 24°of mean temperature. Actually when measuring the energy consumption of R/Fs according to ISO norms ambient temperature is 25°C. The freezer compartment is –18°C and fresh food temperature is 5°C. Polyurethane is subject to average temperature of 3.5°C at freezer region and to average temperature of 15°C at fresh food region. As going down temperature of polyurethane the thermal conductivity is decreased. In Figure 4 this relation can be seen for c-Pentane based Polyurethane and c-Pentane/iso-Pentane based Polyurethane

Arcelik R/F plant since 1996. These were tested with LaserComp Fox 600 in Arcelik R&D lab.

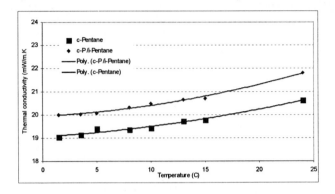

Figure 4: Changing thermal conductivity with temperature for two fresh different foams (tested by LaserComp Fox 600 and ASTM C518)

From Figure 4 it can be said there is 1.1 to 1.8 mW/m.K of difference between thermal conductivity at 24°C and thermal conductivity at range of 3.5°C to 15°C.

4 Heat Gain Aging

Reverse heat leakage tests were performed for three larder type cabinets (fully fresh food) by the time to understand heat gain aging. Inside temperatures of test cabinets and climatized test room were measured during tests. A fan and a resistance heater group were inserted into the cabinet door. The fan was then used over the resistance heater for getting of fine temperature distribution in the cabinet. To avoid gasket effects all gasket and flange region were removed from the cabinet. After the test set-up was completed, the door of cabinet was closed tightly. According to the test procedure, the room was cooled down to 5 C and cabinet was heated up to 25 C.

Figure 5 shows heat gain aging of cabinets. As seen, heat gain increase and also energy consumption is nearly 5%, 7% and 8% after 1 year, 2 year and 3 year of aging respectively.

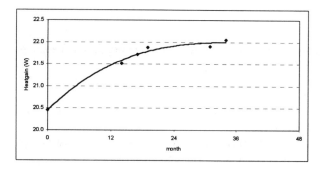

Figure 5: Heat gain aging and increase in energy consumption

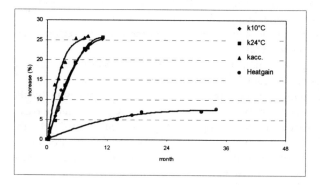

Figure 6: Comparison of increase in thermal conductivity and heat gain

In Figure 6 it can be seen increasing thermal conductivity for measurement of 10°C and 24°C and accelerated thermal aging data and increasing heat gain. All polyurethane system used is same in this graph. As seen increase in thermal conductivity at 10°C and 24°C are almost similar to each other. Data for accelerated thermal conductivity aging show a sharp increase with respect to data for normal aging as expected. Increase in heat gain aging is very low when comparing change in thermal conductivity by the time.

5 Conclusions

- Polyurethane is subject to 3.5°C to 15°C of mean temperature during the energy consumption tests at 25°C of ambient temperature according to the ISO norms, thermal conductivity of polyurethane should be measured at 3.5°C and 15°C instead of 10°C and 24°C.

- There is 1.1 to 1.8 mW/m.K of difference between thermal conductivity at 24°C and thermal conductivity at range of 3.5°C to 15°C.

- As polyurethane has low initial thermal conductivity it can has very fast thermal aging characteristic which is not desirable. The important one for a polyurethane system is has to have good thermal aging for low energy consumption of a R/F during its whole lifetime. Low initial thermal conductivity as alone should not be very important for decision of new polyurethane system.

- Since gas diffusion phenomena inside the cell of polyurethane foam during the normal and accelerated aging test is different and faster than real life of polyurethane between plastic liner and metal face in the R/Fs, to perform heat gain tests by the time is very useful.

- Heat gain and energy consumption increase of a cabinet is nearly 5%, 7% and 8% after 1 year, 2 year and 3 year of aging respectively.

- Increase in heat gain aging is slower than rise in thermal conductivity by the time. If initial thermal conductivity can be said 21 mW/m.K for c-Pentane based polyurethane, after 1 year of normal aging test it is obtained 25 mW/m.K, the real thermal conductivity of poyurethane inside the cabinet is around 22 mW/m.K. After 3 year of aging of polyurethane inside the cabinet the real thermal conductivity is nearly 23 mW/m.K.

Acknowledgements

The author extends his appreciation to Arcelik A.S for permission to publish this paper. The author also acknowledges all of the help and suggestions that were given by those involved in this work, especially for Mr. Refik Üreyen, Technology Coordinator in Arcelik AS, Mr. F. Cavusoglu, Research Technician and Mr. N. Kandemir, Research Technician.

6 References

[1] Biesmann, De Vos, Rosbotham 1994, *Comparison of Experimental and Predicted Ageing of PU Foams Blown with CFC Alternatives, Utech, 1994, 32-1*

[2] Ball 1997, *Thermal Conductivity of Rigid Cellular Polymers : Predicting 25-year Values, Urethane Technologies, August/September 1997, 50*

[3] Fleurent, Thijs 1995, *The Use of Pentanes as Blowing Agent in Rigid Polyurethane Foam, Journal of Cellular Plastics, November 1995, 580-598.*

[4] Seifert 2000, *Long-term Energy Efficiency of PU-insulation for Refrigeration, Utech 2000, 47*

Vacuum Insulation Panels (VIPs) Technology : a Viable Route to Improve Energy Efficiency in Domestic Refrigerators and Freezers

Paolo Manini

Rizzi Enea, SAES Getters SpA, Italy

The growing attention to the environmental issues is leading policy makers to take actions to reduce CO_2 emission and limit as much as possible the inefficient use of non renewable energy resources.

Improving the efficiency of thermal insulation is a key option to increase energy savings in refrigerators and freezers, which account for about 20% of the total electricity consumption in household appliances in Europe.

Vacuum insulation panels (VIPs), having thermal conductivity 3 to 7 times lower than conventional closed cell polyurethane foams can realize more efficient thermal barriers, thus providing significant energy savings without the need to increase the insulation thickness. Alternatively, VIPs can be used to reduce the total thickness of the insulation without penalizing performances. Thanks to this, vacuum insulation panels are finding increasing acceptance in a variety of domestic and industrial applications encompassing refrigerators and freezers, vending machine, shipping containers or insulated transportation.

In the present paper, some recent developments in VIP technology are reviewed, focusing the attention on the selection and processing of the components and the VIP design. Examples of application of vacuum panels and their potentials to save energy in appliances, as well as the main open issues to be addressed to ease the diffusion of this promising technology, are also presented and discussed

1 Vacuum Panels Design

Vacuum insulation panels (VIPs) have thermal conductivity 3 to 7 times lower than conventional insulating materials such as closed cell polyurethane and polystyrene foams or fiber glass, as showed in the figure. They can therefore be used to significantly reduce the energy consumption in a variety of domestic, industrial and civil applications. VIPs are manufactured by inserting a micro-porous low thermal conductivity filler material in a highly impermeable gas barrier which is evacuated and sealed.

Glass wool and compressed powders, such as precipitated silica and perlite, have been proposed in the past as core materials for VIPs [1]. To overcome some of their limitations, open cell polyurethane [2][3] and polystyrene foams [4] and improved fumed silica powders [5], have been recently developed and put on the market.

Each of these insulating materials presents specific advantages and disadvantages in term of cost, handling and pre-treatment, weight and thermal conductivity, which may render it particularly suitable for some applications and less for others.

All these fillers offer improved insulation performances, provided a suitable vacuum level is drawn and maintained during the VIP projected service life.

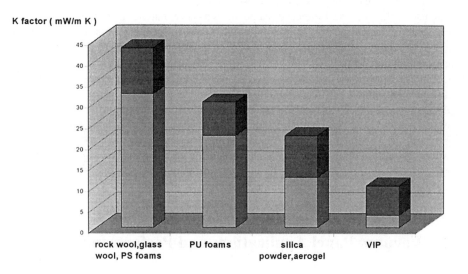

In the case of the open cell foams, having cell diameter typically comprised between 30 and 150 micron, this vacuum level ranges between 0,1 and 1 mbar, which requires the use of very good quality gas barrier envelopes to limit air and moisture infiltration during time. Multilayered vacuum-metallized barriers as well as laminates incorporating a continuous aluminium foil (usually 5-7 micron thick) have been proposed to this purpose [6][7]. Metallized films have the great advantage of minimal heat flow across the panel flanges (edge effect) which ensures full exploitation of the super insulation properties of the core material. Laminates incorporating a continuous aluminium foil provide better gas barrier properties and are therefore ideal for long term applications (e.g. 15-20 years). However, to reduce the impact of the edge effect on the thermal performances, the use of large size panels (e.g. >0,5 m^2) is here recommended.

Even using a high quality barrier envelope, pressure build-up in the panel will take place over time, due to outgassing from the barrier and core material and gas permeation through the barrier itself. A gas adsorbent or getter device is generally necessary to remove gases and stabilize the pressure over lifetime. A specific type of getter device , the COMBOGETTER has been purposely developed by SAES Getters after extensive studies addressed to identify the main gas sources deteriorating the vacuum in a panel [8]. This getter can also compensate for gas burst generated in the panel by fluctuations in the quality of the components or by panel processing (e.g. foaming in place of the panel in the household appliances).

The use of the improved fumed silica powders as core material increases by more than one order of magnitude the internal tolerable pressure and poses less requirements on the gas tightness of the barrier film. For silica filled vacuum panels, the getter is also generally not necessary.

Compared to the open cell foams, silica powders are denser, more expensive and more difficult to handle.

Another panel design makes use of stainless steel to envelope the core material, which may be glass wool, compressed powder or open cell foams. This panel design is generally very robust and well suited for high temperature working conditions or where good structural properties are mandatory (e.g. building or logistics). Being generally heavy and expensive, the adoption of stainless steel panels in refrigerator and freezers has been so far quite limited.

2   Vacuum Panel Applications and Performances

Thanks to the recent progresses in core, films and getters, VIPs are now finding increasing acceptance in a variety of domestic and industrial applications encompassing refrigerators and freezers, vending machine , shipping containers or insulated transportation.

The main reason supporting the use of VIPs in household appliances is the need to reduce energy consumption and comply with the forthcoming more demanding energy regulations.

A second driving force for the appliance industry is the possibility to increase the internal volume of refrigerators and freezers without increasing the outer dimensions. This aspect is particularly important in Europe and Japan, where the built-in appliance market is an important segment and space constraints play a role. For household appliances, depending on the surface coverage of the cabinet and the panels thickness, energy savings from 15% to 30% have been reported [3][9][10][11] using open cell foam-filled panels packaged in a 6 micron aluminium foil-based barrier.

Further improvements can be obtained decreasing the aluminium foil thickness or using a metal foil-free barrier and/or decreasing the thermal conductivity of the core material.

Open cell polyutrethane foam-filled panels are being extensively used in vending machines, which are especially popular in Japan, Korea and the Far East. Most Japanese vending machine manufacturers use one or two panels to separate the hot and cold beverage compartments. In this case, the main driving force for VIP adoption has been the possibility to increase the internal volume for the storage of beverages, without reducing the insulation efficiency.

Vending machine is quite a demanding application for VIPs, since the panel temperature is cycled between room temperature and 70°C and the lifetime is more than 5 years. Ageing tests have been carried out by V/M manufacturers to assess VIP usability and reliability in this application.

The deterioration of the thermal conductivity, due to the temperature-enhanced outgassing and permeation, is quite rapid and must be compensated with the getter.

Another application for VIPs is in low temperature freezers for biomedical and laboratory applications. These are very special equipment which are designed to operate at very low temperature (e.g. from –30 °C to –86°C) to age samples or to store valuable and perishable goods, like organs and tissues, biological and medical samples or vaccines. Vacuum panels are used mainly to increase the internal storage volume without increasing the energy consumption. Since the conventional insulation must be very thick to ensure the achievement of such low temperatures, a partial replacement of the conventional insulator with VIPs contributes significantly to increase the internal volume, from 20% up to 40% or more in specific models.

As a second advantage, panels provide an overall superior passive insulation, which means that the temperature rise in the case of power failure is less steep and more time is necessary before a given critical temperature is reached. This provides extra-safety for delicate articles which may rapidly deteriorate when their temperature exceeds a given value.

Ultra-low temperature freezer models using foam-filled vacuum panels have been successfully placed on the market by leading companies since a few years and other companies are expected to follow soon.

Large vacuum insulated containers, ships and trucks , as well as small commercial shipping containers to store and deliver pharmaceutical, food and valuable products, are also under evaluation by various companies world wide.

A variety of other potential applications exist for VIPs in the construction industry, from cold stores to insulation in industrial plants (e.g. industrial reactors or liquid natural gas tanks), water heaters or heat pipes. Key issue to the adoption of VIPs in this last family of applications is the possibility to bend the vacuum panel into a round or cylindrical shape.

Insulation in buildings, especially in prefabricated roofs, wall constructions, doors and facades, is another area where vacuum panels have the potential to provide interesting contributions both in term of energy and space savings. Specific and challenging requirements are posed by this application, since mechanical stability, very long lifetime (> 50 years), thermal conductivity and cost issues have to be properly addressed.

3 Open Issues

To find larger acceptance in Household Appliances , further improvements in VIP technology are necessary both in term of performances and cost. The refrigerator/freezer is in fact a very cost-sensitive and competitive market where a favourable cost to performance ratio is key for the adoption of a technology

Better performances can be obtained through a careful selection and processing of the components and by improving the panel design. Attention is now particularly paid at reducing the edge effect which still prevent VIPs from achieving their full potentials. To this purpose, new families of improved metallized products, as well as laminated barriers with thinner aluminium foil have been recently developed and are under testing in a variety of applications.

The actual evaluation of the heat flow through a panel and in the refrigerator cabinet is also key to optimize the VIP design and its integration in the appliance. For this reason SAES Getters has developed specific expertize in the mathematical modelling of insulating structures of complex geometry and can provide technical support for the optimal use of panels.

To ease the process of technology assessment, SAES Getters recently made a major investment in a VIP pilot manufacturing line capable to produce panels in a variety of designs and dimensions. This line, fully operative since late 1999 can handle different types of filler and film components.

An ugrading of the production capacity is under study with the aim to deliver the appliance market mass production volumes of high quality cost-effective vacuum panels.

The recent breakthroughs in film, core, getter design , as well as in their mass production manufacturing processes, is also bringing results in term of cost reduction. It is the author's opinion that this ongoing activity should be effective in reducing the cost per unit performance of VIPs to a level where this technology becomes a viable route to improve energy efficiency not only in high end domestic refrigerators and freezers but also in a significant portion of main stream models.

4 References

[1] Fine."Advanced Evacuated Thermal Insulations : the State of the Art", Journal of Thermal Insulation, **12**, 1989, 183.

[2] De Vos and Rosbotham, "Polyurethane Foam Based Vacuum Panel Technology", Cellular Polymers, **13**, N° 2, 1994,147.

[3] Tao, Sung and Lin. " Development of Vacuum Insulation Panels Systems", Journal of Cellular Plastics, **33**, 1997, 545

[4] Pendergast and B. Malone. "Characterization and Commercialization of INSTILL Vacuum Insulation Core, a Unique Polystyrene Vacuum Insulation Filler from the Dow Chemical Company", Vuoto, **XXVIII**, n° 1-2, Jan/Jun 1999, 27.

[5] Boes, Kelly, Roderick, Brungardt, Braun and D. Smith " Vacuum Panels based on nanoporous Fillers ", presented at the VIP Symposium,Baltimore, 1999, SAES Getters Web site, http://www.saesgetters.com

[6] Sugiyama, Tada and Yoshimoto."Gas Permeation Through the Pinholes of Plastic Film Laminated with Aluminium Foil", Vuoto, **XXVIII**, n° 1-2, Jan/Jun 1999, 51.

[7] Lamb and Zeiler. "Designing Non-Foil Containing Skins for VIP Applications", presented at the VIP Symposium, Baltimore, May 199, SAES Getters web site, http://www.saesgetters.com.

[8] Manini," The Combogetter as a Key Component in the Vacuum Insulated Panels (VIPs) Technology", Vuoto, Apr./Jun 1997, **XXVI**, n°2, 45

[9] Wacker, Christfreund, Randall and Keane. "Developments of Vacuum Panel Technology Based on Open Celled Polyurethane Foam", Proceeding of the Polyurethanes Expo '96, 1996, 35

[10] Hamilton,"An Evaluation of the Practical Application and use of VACPAC panel technology, Vuoto, XXVIII, n°1-2, Jan/Jun 1999, 27.

[11] Zoughaib and Clodic " Technical and economical Evaluation of Vacuum Insulated Panels (VIPs) for a European Freezer "Proceeding This Conference.

Technical and Economical Evaluation of Vacuum Insulated Panels for a European Freezer

Assaad Zoughaib and Denis Clodic

Ecole des Mines de Paris, Centre d'Energétique

Abstract. Vacuum insulated panels (VIP) is an emerging technology that may constitute an alternative to rigid polyurethane (RPU) foam as insulation in refrigerators and freezers. The VIP "skin effect" caused by the aluminum layer in the envelope complicates the evaluation of the global thermal resistance. This paper presents a new method of VIP characterization. Values of the global thermal resistance obtained are used in a dynamic simulation tool ENEREF® to perform predictive calculations. Direct experimental measurements permit to validate calculations. Finally, a life cycle cost analysis is performed and payback period of the technical option is calculated.

Nomenclature

k	Thermal conductivity , W/m.K
P,W	Electrical power, heat flux, W
e	Thickness, m
A	Area, m^2
ΔT	Temperature difference, K
G	Specific heat loss coefficient, $W/m^3.K$

1   State of Art

The Montreal Protocol phased out CFCs refrigerants. R-11 (CCL_3F) being used as an expanding gas in the rigid polyurethane foams (PUR) is replaced by alternative fluids such as R-141b, cyclopentane and CO_2. As a consequence, when using cyclopentane, the thermal resistance of PUR foams is reduced by about 33%. In parallel, international and particularly U.S. and European efforts [EurDir96] are focused on the reduction of appliance energy consumptions. The contradiction between lower thermal resistance of insulation and lower energy consumption of refrigerating appliances pushed the insulation manufacturers to find alternative solutions such as VIPs.

2   VIP Structure

Vacuum panels are composed of three main components: a core material, an envelope and a getter. Many materials can be used as a core material in VIP such as open cell polyurethane, expanded polystyrene (XPS), silica powder, and perlite... Therefore, for technico-economical considerations, the open cell PU and the XPS are the most usual materials [Dow99].

The internal absolute pressure of a VIP is about 50 Pa abs. The thermal resistance of the panel is directly related to this pressure level; i.e. for open cell PU, at a pressure of 10 Pa abs the thermal conductivity in the center of the panel is about 6 mW/m.K, which increases to 25 mW/m.K when the pressure reaches 1000 Pa abs [Man97]. The envelop and getter roles are to maintain this pressure during the panel lifetime.

The getter is a chemical adsorbent composed of barium and lithium alloy that adsorbs nitrogen in large quantities. Additions of calcium and cobalt oxides permit the adsorption of humidity, hydrogen, R-141b...

The envelope is a multilayer polymer (*Figure 1*). The first layer is in nylon that stops water molecules. The polyethylene teraphtalate (PET) and aluminum layer prevent the passage of smaller molecules, and finally, the polyethylene (PE) layer permits the welding [Toyo99]. This envelope guarantees the leak tightness. Therefore, the aluminum layer creates a "skin effect", which is the heat transfer along the envelope.

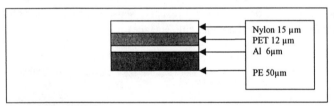

Figure 1: Envelop composition

Considering the high thermal resistance of the panel in the normal direction, the heat transferred through the aluminum layer is not negligible and the equivalent thermal resistance is dimension dependent.

3   VIP Characterization

3.1 Standard Methods For Thermal Conductivity Characterization

The standard methods of thermal conductivity characterization are described in ISO 8301 and 8302 [Pon91] and called respectively the "heat flow meter apparatus method" and "the guarded hot plate". Both methods are not convenient for the VIP characterization because of the apparatus size constraint. Only small sample can be measured. Because of the "skin effect", the VIP characterization shall be performed in real dimensions.

The "guarded hot box" method described in ASTM [ASTM99] is used in general to characterize building assemblies. The method uses a measurement cell in which the temperature is maintained and a test cell where a thermal flux is generated. The tested material constitutes the partition wall between the two cells. As it is described, this method cannot be used to characterize super insulation since the estimated heat loss through the test cell walls are not negligible compared to the heat flow going through the VIP. An adaptation of the method is necessary to permit the exact measurement of the VIP equivalent thermal resistance.

3.2 Adaptation Of "The Guarded Hot Box Method"

The test bench used for the VIP characterization consists of two identical cells in XPS separated by the tested panel (*Figures 2a and 2b*). Each cell is equiped with an electrical heating device. PT100 resistive sensors measure the cell temperatures and the temperatures of the panel side surfaces

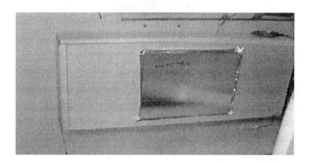

Figure 2a : Testing cell showing the VIP

The test is performed in two steps and the ambiance temperature is kept constant.

- A first step is needed to characterize the cell heat losses. Both cells are heated to have the same mean temperature on both surfaces of the tested panel. Since the separation wall is adiabatic, the heating power of each cell (P_{adiab}) corresponds to the cell heat loss.

Figure 2b : Both cells closed

- In the second step, one cell is removed and the remaining cell is heated up to the same temperature reached in the first step. In this case, the heating power (P_{spec}) is equal to the heat loss measured in the first step and to the heat flow passing through the tested panel.

The difference between the two electrical powers measured in steps 1 and 2 is the heat flux that went through the tested panel. Both VIP side mean surface temperatures are measured and the geometric characteristics are known. The VIP equivalent thermal conductivity can be calculated using equation (1).

$$k = \frac{\left(P_{spec} - P_{adiab}\right)e}{A\,\Delta T} \tag{1}$$

When the tested VIP does not fit exactly in between cells, a previously tested material is used as a frame, as shown in *Figure 2a*. The heat flux (P_{frame}) passing through the frame can be calculated and equation (2) is used to calculate the VIP equivalent thermal conductivity.

$$k = \frac{\left(P_{spec} - P_{adiab} - P_{frame}\right)e}{A\,\Delta T} \tag{2}$$

The novelty of the method consists in the measurement of the cell heat loss rather than trying to estimate or to calculate them; the first step permits this by keeping the separation wall adiabatic.

3.3 Results

Measuring the thermal conductivity of a known material validates the method. A test is performed on RPU. Comparison between the given and measured values of the thermal conductivity is presented in the *Table I*.

Table I: Compared values of thermal conductivity.

Material	Measured thermal conductivity (mW/m.K)	Given thermal conductivity (mW/m.K)	Difference (%)
Usual PU	20	21	5

Measurements are performed on two types of VIP using different core material. The first is an open cell PU VIP and the second is using XPS. Temperature distribution on the VIP surface for both panels shows the existence of "the skin effect". On the VIP cold side, the lowest temperature is in the center and the highest one is located on the borders (*Figure 3*).

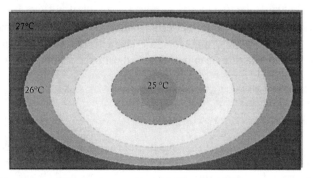

Figure 3: Temperature distribution on the cold side of the VIP

The equivalent thermal conductivity of VIP panels and the conductivity in the center (the one measured in the center by the mean of a heat flow meter apparatus) are given in the *Table II*.

Table II: Equivalent thermal conductivity of the tested VIP

Core material	Dimensions (mm)	Equivalent thermal conductivity (mW/m.K)	Conductivity in the center (mW/m.K)
Open cell PU	450x450x20	12.7	9
XPS	450x450x20	9	5.5

4 Insulation Design Options for Reducing Energy Consumption of Refrigerators/Freezers

Improving the appliance insulation is one of the main design options for reducing energy consumption. Therefore, different possibilities exist to improve the insulation.

Basically, four possibilities are the most common.

- Increasing the insulation thickness by reducing the internal volume: this option is very penalizing since the consumer is interested by the storage volume of the appliance. Thus, with this option the new appliance can not be compared to the base line because they will not have the same internal volume.
- Increasing the insulation thickness by increasing the external volume: this option is more interesting than the first one since the useful volume is the same. The drawback of this option is that the appliance external dimension become larger and the appliance does not fit in the kitchens. This is, till now, not quantified as an economic penalty. Thus in life cycle cost studies this option seems to be the most viable. Without economic penalty evaluation, manufacturers will not be capable to compare exactly this option with the others.
- Increasing the insulation thickness by increasing the height of the appliance: this option permits to obtain the same internal volume and the same width and depth. The energy saving for the same thickness increase is less than the second option because of the ratio area/volume. The height increase is limited by the usability.
- VIP integration: the insulation thickness is kept constant. The cost of this option is not defined exactly and thus, by being conservative or not, the viability of this option can be discussed.

In the next sections, the last option is evaluated for different appliances.

5 VIP Integration in Appliances

This option has the advantage of improving the insulation without changing the appliance geometry. Therefore, an exhaustive investigation on the different types of appliances permits a better and optimized use of this technology. The appliances can be divided into two groups: the single volume appliances and the multiple volume appliances.

5.1 The Multiple Volume Appliance

These types of appliances are generally multi temperature appliances. For the two volume appliances, which is the most common, one volume is a refrigerator and the other one is a freezer. The temperature difference between the two volumes constitutes the design complexity of these appliances. Two main subcategories are observed: the single system, and the double system.

- The single system appliances: the cooling capacity necessary for the two volume is produced by one system; the evaporators are in series and the control system is in one of the volumes generally in the refrigerator. The temperature is guaranteed by the evaporator exchange area ratio that corresponds to the heat loss ratio. Thus, when modifying the insulation of such appliance the heat loss ratio must be conserved otherwise the evaporators must be modified as well. These constraints are to consider when analyzing the VIP integration.

- The double system appliances: this subcategory includes the double compressor appliances and the single compressor with independent cooling capacity distribution and double control system. The insulation modification of this type of appliances does not have any constraints besides the optimal VIP panel dimensions (to avoid an important skin effect). Thus the designer can either super insulate the freezer where the insulation gain is the most important, or super insulate the entire appliance.

5.2 The Single Volume Appliances

This type of appliances can be either a simple refrigerator, or a freezer. For this type of appliances there are no design constraints. The energy saving potential for this category is a function of the operating temperature; thus it is obvious that for a freezer compartment, insulation improvement can reduce energy consumption more than in a refrigerator.

6 Evaluation of VIP Integration in a European Upright Freezer

By using VIP in the insulation of a refrigerator/freezer, the energy consumption can be reduced without changing the thickness of walls. The freezer is the most advantageous model when using super insulation since the temperature difference is greater than in a refrigerator.

6.1 Simulations of Different Integration Scenarios

A European upright freezer has been chosen for evaluation. A dynamic simulation software ENEREF® [EN99] (*Figure 4*) is used to calculate the appliance dynamic behavior when adding VIP to the insulation. The reference is simulated and results are compared to the measured annual consumption measured and the specific cabinet heat loss (*Table III*).

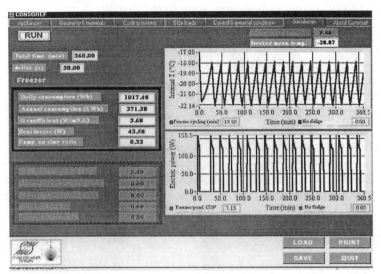

Figure 4: ENEREF® output screen

Table III: Reference simulation.

	Measured	Calculated	Difference
Energy consumption (kWh/an)	535	502	6 %
G (W/m³.K)	4.73	4.82	2 %

Differences are acceptable and the simulation is validated for the reference. The modification of the reference permits to estimate the VIP integration impact in one wall or more. The appliance used can integrate VIP in the door, the top and partially in the sidewalls because of the compressor niche. In order to maintain the mechanical rigidity of the cabinet, the VIP does not occupy all the wall thickness; the use of 60% of the thickness is a good compromise. Two walls cannot be easily modified: the back wall because of the condenser, and the bottom because of the appliance weight.

Different combinations of modifications are simulated and the results are presented in *Table IV*.

Table IV indicates that the energy saving potential of the VIP option is significant. This gain cannot be calculated using a steady state simulation of heat loss because of the strong dynamic behavior of the cycle coefficient of performance (COP). The mean COP of the cycle decreases with the compressor running time. The energy saving due to the heat loss reduction is amplified by the compressor running time reduction thus the mean cycle COP increases.

Table IV: Simulation results for different scenarios of integration.

Scenario	G (W/m^3.K)	COP	Energy saving (%)
Reference	4.82	1.08	-
Door only	4.5	1.1	8
Side walls only	4.25	1.1	13
Side walls and top	3.92	1.13	21
Side walls, top & door	3.68	1.15	26

6.2 Demonstrator Realization And Experimental Results

The predictive calculations presented promising energy savings. The realization of a demonstrator permitted to confirm the results. For this purpose, the VIP location is cut out in the walls of the reference upright freezer (*Figures 5a & 5b*).

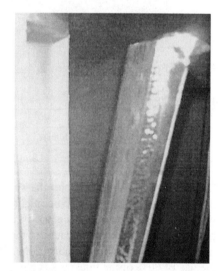

Figure 5a: VIP location   **Figure 5b:** Integration of VIP in a side wall

Three different integration scenarios are tested using the modified demonstrator. Test results compared to the calculations are presented in *Table V*.

Table V: Experimental results for three scenarios of integration.

Scenario	Measured energy saving (%)	Calculated energy saving (%)
VIP in the door	6.8 %	8 %
Door and top	8.3 %	10.3%
Door, top & one side wall	14%	-

The first and the second results are very close to calculations. The difference observed may be caused by the bad contact between the VIP and the residual RPU. This will not happen if the RPU is foamed with the VIP installed. The last scenario could not be simulated because the actual version of ENEREF® does not permit non-symmetric sidewalls.

Validation of the first two tests allows generalizing the simulation results. The energy saving calculated is technically possible.

6.3 Life Cycle Cost Analysis

An economical evaluation of the VIP is performed in this section. The main difficulty consists in the evaluation of the over cost. Two approaches are available.

For both approaches, assumptions are as follows.
- Appliance lifetime: 15 years,
- Discount rate: 5% and
- Electricity price: 0.55 FF/kWh (which is the price in France).

The first approach is proposed by AHAM [AHAM93]. It considers a material over cost of $15\$/m^2$ and an investment over cost of $10\$/unit$. The life cycle cost analysis based on this approach is performed and results are given in *Table VI.*

Table VI: Life cycle cost analysis (first approach).

Description	Purchase price	Purchase price diff.	Energy consumption	Energy saving	Life cycle cost	Payback period
	FF	FF	kWh/an	(%)	FF	(Year)
Reference	3000	0	502	0	5866	0
Door	3136	136	462	8	5774	7,6
Side walls	3162	162	439	13	5667	5,45
Door and top	3232	232	398	21	5504	4,65
Door, top &side walls	3259	259	371	26	5377	4,06

With this approach, the payback period is acceptable and the VIP option seems to be viable.

138

The second approach proposed by some manufacturers is more conservative. The proposition is to increase the manufacturing cost with an additional value of 15$/m² for the use of VIP. When applying the usual margins, this over cost becomes 45$/m² added to the purchase price. The calculations corresponding to this assumption are given in *table VII.*

Table VII: Life cycle cost analysis (second approach).

Description	Purchase price	Purchase price diff.	Energy consumption	Energy saving	Life cycle cost	Payback period
	FF	FF	kWh/an	(%)	FF	(Year)
Reference	3000	0	502	0	5866	0
Door	3216	216	462	8	5853	13.83
Side walls	3302	302	439	13	5808	11.73
Door and top	3518	518	398	21	5790	12.36
Door, top &side walls	3596	596	371	26	5714	10.94

By using this assumption, the payback periods have doubled. This assumption allows judging that the VIP option is not viable.

7   Conclusion

In this paper, the new method presented to characterize thermal resistance of VIP allows to have real data to be used in predictive calculations. These data are used to evaluate VIP technology in a European freezer. The calculated energy savings are significant (about 26%) for a coverage of 4 walls. With appropriate design a higher coverage may be possible and thus more energy savings.

Two opposite economical approaches are used to calculate the payback period. The two assumptions can lead to contradictory conclusions. The first approach considers the VIP option as viable but the second rejects it. The VIP technology allows to reduce energy consumption without changing the appliance geometry, which is a real advantage compared to options like increasing wall thickness.

Acknowledgements

The results presented in this paper are extracted from a study funded by ADEME, the French Agency for Environment and Energy Management. SAES Getters spa and DOW Chemicals have supplied the tested VIPs.

9   References

[EurDir96]   *European Council Directive 95/97 of September 3 1996 concerning requirements for energy efficiency of refrigerators, freezers and combined domestic electric appliances.* Official Journal n°L236, 1996, pp 36-43.

[Dow99]   Dow Chemicals, technical documents, 1999.

[Man97]   Paolo Manini, *The COMBOGETTER as a key component in the vacuum insulated panels technology*, Vuoto, 4-5/1997.

[Toyo99]   Toyo Aluminium, technical documents,1999

[Pon91]   Francesco De Ponte, Sorin Klarsfeld, *Thermal conductivity of insulating materials*, Techniques de l'Ingénieur. 1991.

[ASTM99]   *Standard test method for thermal performance of building assemblies by means of a guarded hot box.* ASTM C236-89 (1993)e1, 1999

[EN99]   ENEREF® a software for the simulation of energy consumption of refrigerators, freezers and combined refrigerators-freezers. Center for Energy Studies, ARMINES.

[AHAM93]   *Joint comments of the AHAM, NRDC, ACEEE, NYEO , CEC, PGE and SCE relating to energy conservation standards for refrigerator/freezers.* Docket No. EE-RM-93-801

Socio-Technical Networks and the Sad Case of the Condensing Boiler

Nick Banks

Environmental Change Institute, University of Oxford

Abstract. This paper uses ideas from actor network theory to characterise the structure of the UK domestic heating industry. This analysis is then applied to a discussion of the diffusion of condensing boiler technology. This innovative and proven design saves substantial amounts of energy yet its penetration into the homes of many European states remains pitiful. As such it remains a potentially important, but as yet unused, weapon in the struggle to heat our homes more sustainably. Quantitative survey analyses and interviews with the range of actors from builders and trades associations to heating installers and householders are used to characterise the relationships between actors. Some tractable policy recommendations are set out in the context of the UK government's attempts to increase the spread of condensing boiler technology.

1 Introduction

Conventional boilers lose around one third of the energy of the fuel in the flue gas. Condensing boilers (CBs) have a larger or secondary heat exchanger and are arranged so that not only the "sensible" heat of these flue gases is extracted but also the latent heat of vaporisation. This causes a condensate to form. In this way, boilers with an efficiency approaching 100% can be produced. The best that could be achieved with conventional boilers is under 80% (ETSU, 2000). The technology to do this is not new: a boiler with a heat exchanger capable of condensing flue gases was first designed in the 1930's. However the use of "town gas" at that time (made from driving off the volatile gases from coal) produced a highly corrosive condensate that quickly destroyed the iron heat exchanger surface. The development of stainless steels and aluminium alloys provided a corrosion resistant material whilst a widespread switch to natural gas in some EU states produced less acidic flue gases. Consequently a condensing boiler for domestic use became a viable technological proposition and the Dutch gas industry was able to produce a working prototype in 1979.

In most northerly EU states, space and water heating constitute the largest share of the domestic energy load. As such, widespread adoption of this technology offers potentially huge and cost effective energy savings in those countries with extensive gas infrastructures. For the EU, it is estimated that if condensing boilers were installed instead of the best of conventional boilers from 1990 onwards, the emission of 20 million tonnes of CO_2 would have been avoided by 2005 (IER, 1998). Clearly, given commitments under the Kyoto agreement, this is a politically significant figure. At a household level, the adoption of a CB also seems desirable. Although CB units are more expensive, paybacks can be relatively quick - between one and four years. Despite the evident advantages, penetration of this technology, particularly in some countries, has been dismal. The Germans and the Dutch remain notable exceptions where CBs have 18% and 60% of the market for new installations respectively. Both have taken different routes to encouraging uptake. In Germany this has been through firm regulation of boiler emissions whilst in the Netherlands there has been an extensive series of subsidy schemes, backed up by installer training programs and national advertising campaigns.

Ostensibly, the Dutch success story should be the model which the UK, with its similar gas supply network, should follow. However, the UK has had a number of subsidy and information campaigns running since in 1991 and yet the fraction of new condensing boiler sales in the housing market remains very small - at around 5%. What is more, 80% of these sales are to local authorities and housing associations rather than private individuals. Ten years after the introduction of the Dutch promotion programmes, CBs were already taking some 30% of the market of new sales. Are Dutch householders really so different from the British or can the different penetration rates be explained by structural factors and the orientations of other actors?

The conventional policy paradigm - the rational action model, would tend to assume that the disparity in adoption rates is due to differences in the information environment. Social scientists have powerfully criticised such a view – again and again pointing out that technology transfer is a thoroughly social processes (for a review see Lutzenhiser, 1993) involving issues of values, norms and identity. This basic insight is now increasingly elaborated into descriptions of the organisational and institutional networks in which the technologies are embedded. This entails an appreciation of technology as an actor in an evolving system of actors – all connected by social, institutional and regulatory relationships. Much like a body rejecting a foreign antibody technologies will not diffuse until the system as a whole is able to accept them - Schwartz Cowan's (1989) work on the diffusion of the domestic stove being a well worked case in point. More recently, a sociology of domestic technologies has begun to emerge which draws on insights from science studies – particularly the work of Bruno Latour. This work focuses on the properties of the technology itself in actively shaping the context into which the technology is introduced. The technology becomes an active player in the same sense that other actors are active in the network. The technology "speaks" and imposes its own agenda through its design. As such, in some views, it makes no sense to speak of the social on the one hand and the technical on the other. They are woven seamlessly together, changing and being changed by one another; evolving the possibilities and potentials for other technologies and ways of life yet to emerge. This virtual space is sometimes called a socio-technical landscape.

In what follows, an analysis of the diffusion of the condensing boiler is offered which tries to view the technology as such an actor in system or socio-technical relations. Each actor in the chain of specification of heating systems is isolated and their social, economic and institutional orientations to one another described. This was achieved by conducting interviews lasting at least one hour with a number of representatives of each actor group (between 5 and 10 interviews were conducted for each group) and the analysis of a questionnaire of householders who had recently changed their heating systems (n = 522). This results in a network diagram as shown in Figure 1.

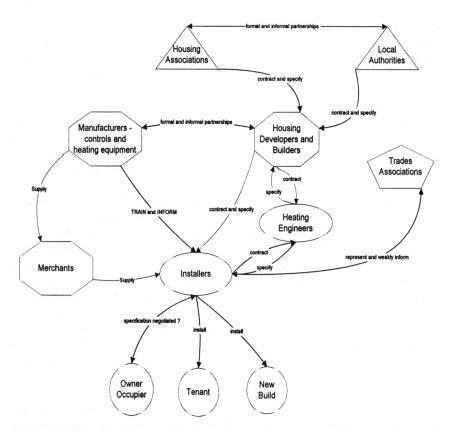

Figure 1: Network of actors implicated in the specification of domestic heating equipment

2   Householders

Positive attitudes to efficiency are virtually unanimous but generally qualified by the extent to which investments in efficiency will save money or add comfort rather than solely benefiting the environment.

> I mean we all try and do our bit for the environment to a certain extent. So I do try to be as energy efficient...actually its probably more money motivated actually.
>
> DH

In contrast to the widely shared idealism of heating the home in an environment friendly manner an interest in innovative heating techniques is confined to only a few as shown in
Table I.

Table I: Interest in innovative heating techniques

Agree Strongly	Agree Slightly	Neither Agree nor Disagree	Disagree Slightly	Disagree Strongly
9	10	11	34	37

This response is not entirely surprising. Adoption of an innovation is, almost by definition, confined to a minority. It requires some measure of personal risk and that the individual either has a strong normative motivation for the adoption and/or an appreciation of the relative advantages of the adopted system (Rogers, 1983).

Two of those interviewed had bought condensing boilers. Both were established professionals in a technical area - one a micro-electronics engineer, the other a physicist. Following installation both had problems with their respective boilers. MT entered into a lengthy correspondence with the manufacturers concerning, amongst other things, the technical information provided in the installation instructions whilst the other went to the point of designing a home-made electronic device which solved the boiler's control problems. It is likely that these kinds of skill and awareness give prospective buyers the confidence to invest in such technologies regardless of countervailing forces. They are the "early adopters", having a certain amount of technical confidence, financial resources to take a small risk and networks of friends and peers who would sanction such behaviour.

Such characteristics are not randomly distributed throughout the population. Early adoption is invariably associated with individuals from higher socio-economic groups (e.g. Fisher and Price, 1992; Hackett and Lutzenhiser, 1990). Condensing boilers seem no exception. The UK's Energy Savings Trust, the body responsible for overseeing the government's strategy on the promotion of the condensing boiler, compared those that had purchased a condensing boiler and obtained a rebate through the trust with those that had purchased a conventional boiler during the same period. A disproportionately high number of scheme participants were from socio-economic groups A and B (EST, 1994).

Buying a replacement boiler is generally considered a "distress" purchase. The questionnaire sought to gauge how distressing the purchase of a replacement boiler really is. Surprisingly, only 35% of respondents described the situation as an "emergency" whilst 70% felt that they had time to gather information prior to the work being done. This suggests that the need to replace a system immediately, without opportunity for prior consultation other than with the installer, only occurs in a third of cases. Hence, there is the potential for innovative or efficient technologies to be assessed and considered by the majority. Despite this opportunity 71 % had *only* used information and advice provided by the installer when choosing the particular type of equipment to be installed. Evidently, most householders are dependent on the advice of the installer in deciding the specification of their new heating system. Since only around half of respondents obtained more than one quote for the work and given a general reliance on installer advice (to the point where price may become a secondary issue) it is apparent that there is a great opportunity for an enthusiastic installer to encourage the householder to invest a little more in efficient equipment.

2.1 Conclusion

There seems to be a stubborn assumption in the policy making community that when presented with information all groups will use it in a logical way to come to the optimal decision. Although it is undoubtedly true that a few highly competent individuals will come to considered decisions and then ensure that they are executed it is unlikely that this is the case for many or most. This is particularly evident in the specification of heating systems. Much will depend on recommendation, hearsay, what is available, what is trusted etc. What is more, even assuming a fully knowledged and motivated householder, many of the most important decisions influencing what is installed will be taken by the installer. What then, are the orientations of this actor group?

3 Installers

There are a number of constraints on what the installer can choose. Most obviously, there is the issue of cost. The installer knows that the householder may get more than one quote for the job so it is important to put in a good price. However, self-employed plumbers seem to get much of their work through recommendation and therefore can trade on their reputations as well as the bottom line.

Installers feel that there is a number of boiler manufacturers in the marketplace all making good quality products and that all comparable boilers are similarly priced. How then do they choose between them? Each installer seemed to have a preference for one or two manufacturers depending on what kind of boiler was being installed. For instance, two mentioned "Worcester" as the best for combination boilers, whilst "Baxi" make great "back" boilers. Other than such pieces of "insider knowledge" and a respect for established brands, particular manufacturers tend to be selected based on individual experience of the products over the years. This is because reliability of the new installation is paramount for an installer. An installer does not want be called back to fix something. It is not good for one's professional reputation and the time spent is not usually chargeable.

> We tend to stick to what we know. The last thing we want is for them to say, "this boiler you've just installed isn't working" because we don't get paid for going back and the customer doesn't want it. To be honest we just want to put it in and forget it and carry on with the next one. You don't try things just for the cost - to try and save £30-40 on a boiler. RP

Each boiler is different and there is a limit to the detailed understanding that an installer can have of all manufacturers and all models. It makes sense to specialise in a few for which one knows that one has the right tools and spares should anything go wrong.

> [Interviewer] So once you have got training in a certain range of products you don't have to keep going back learning about other boilers?
> [Respondent] That's right and of course we keep a range of spares for that one make of boiler. Whereas if you are going to start using say, Potterton or Glowworm you are going to need to keep a range of their spares as well because they all like to make things a little differently - they are not interchangeable.
> JH

> We do an awful lot of work for Baxi. All the Baxi boilers I know inside out. My tools are not necessarily geared to Baxi, but I know what tool to pick up for a particular part of a boiler. I just naturally do it because I know that screwdriver will fit that and I know that adjustable will fit that whereas with another make I might be fumbling around until I get the right tool.
> DR

Evidently, these orientations will encourage a rather conservative orientation to new products and techniques. Indeed, despite generally positive attitudes to the notion of making heating systems more efficient attitudes to the condensing boiler were equivocal. All knew of the technology and most had fitted one. Even those who had not fitted one before did not perceive a difficulty in doing so: creating the drain for the condensate does not seem to be an issue.

There is a perception that there is no demand for the CB and, bearing in mind that householders may ask for a number of quotes, it makes no sense to suggest one thereby quoting a price that may be up to several hundred pounds more - unless given a clear lead to do so by the householder.

> If you stick another £300 on for a condensing boiler that can make the difference between getting you to do the job or not. So unless they actually stress a condensing boiler, you don't price for it.
>
> DR

Despite these perceptions of the importance of the quote, evidence presented above suggests that installers may have considerably more leeway than they imagine. In addition to the extra cost there also seems to be a perception that the condensing boiler is not really appropriate for ordinary domestic use where boilers don't have to work hard enough for long enough to get into "condensing mode" and hence generate the savings.

> And certainly a condensing boiler isn't much more efficient or cheaper to run, unless it is in full condensing mode all the time which really doesn't happen.
> [Interviewer] What conditions would allow that to happen?
> If it was very cold and the boiler was working hard all the time...I think the big sell on condensing boilers is their green properties rather than savings on running costs.
>
> JH

This is actually a myth (and a seemingly very widely held one). The extra energy saving through capturing the latent heat of vaporisation (thereby causing the flue gas to condense) is small compared to the extra sensible heat gained over conventional boilers due to the CB's larger area of heat exchangers. Consequently, CBs are substantially more efficient than conventional types whether they are causing flue gas to condense or not. Given a perception of a marginal energy saving when installed in domestic properties it is hardly surprising that the substantial extra costs do not seem justified.

> No-one ever asks you for one [a condensing boiler] unless they are really clued up and I've never come across one myself. But anyway,

anything average like a terrace or a semi, condensing boilers are of
no benefit because of the extra cost - big houses, yes.

JD

A perception of lack of demand and misconceptions over the possible savings do
not explain reticence to install the condensing boiler when the householder makes
a clear and unprompted request for the product. Both of the CB buyers amongst
those interviewed had encountered extreme resistance to the products.

..I found a perceived hostility amongst many heating installers to
condensing flue boilers when I made enquiries. Very striking that. I
received the impression that they were expensive, unreliable, dodgy,
untrusted technology.

MT

This resulted in one of these individuals installing the boiler himself,

[Interviewer] So you were thinking of actually installing it yourself?
Yes - I did do that. And there is a good reason for that. I phoned
about seven heating installers about installing condensing boilers of
which only two would actually consider installing them. Five of
them varied from being, "are you sure you know what you want" to
"Oh no - we don't install those". So they were questioning me about
my choice of boiler which I was quite surprised at.

DF

In sum, a vicious circle seems to be operating. Installers don't suggest CB's to
householders because they are, with good reason, conservative. In addition the
boilers cost a little more and installers do not perceive any overt demand for them
(because there is almost none). Therefore, householders who may have been
interested are not made aware of the CB as an option when discussing the
proposed work with the installer. Consequently, a conventional boiler is installed
where, had the condensing type gone in, it is likely that the householder would
have told others thus helping to stimulate further demand in the wider
marketplace. In addition, through installation of the condensing type and
observation of its reliable functioning, installer conservatism would begin to be
broken down. Instead, demand for this technology remains small, economies of
scale cannot be levered and the so the price of the CB remains relatively high.
Consequently, installers do not suggest them and so on. This seems to be an
instance where both demand and supply need to be simultaneously encouraged so
that the market takes off. At present, the two sides act negatively oh each other.

4 Manufacturers and Merchants

Most of the big manufacturers make condensing boilers and they want to see the market grow for a higher value product. In addition to this commercial imperative a stream of regulation is beginning to emerge from Europe which will drive up the average efficiency of all boilers made. As one boiler manufacturer put it, "the future is condensing!" The manufacturers of control systems are also particularly keen on promoting the efficiency credentials of their product and bemoan the lack of electrical skills amongst most installers who shy away from installing these systems. It is quite clear that manufacturers see their principle marketing target as the installer.

> So the installer is what we call "pulling product from the merchants shelf". So a lot of our marketing activity is actually aimed at the installer and it's all about trying to get them to buy our products. Because they are, from our point of view, the specifier of the product.
>
> KT

It is felt that once the installer is comfortable with a particular manufacturer that they will tend to stick with that manufacturer. Interviews with installers confirm this. Brand loyalty is often fostered at an early stage - those systems that are trained on as an apprentice tend to be what is specified as a matter of course in later life. Consequently, the manufacturers are very keen to be involved in the provision of training materials. Indeed, one respondent mentioned that his company had provided, free of charge, the boilers for 44 technical training colleges.

> These people are going to train on our boilers so when they go away and start specifying boilers for themselves they will pick the ones they know – that's the long term thinking in this.
>
> NS

A range of other techniques are employed to encourage the installer to choose one manufacturer over another including points schemes, sophisticated web-sites, help-lines offering advice and promotions in the trade press and at the builders and plumbers merchants. The latter technique is not seen as particularly effective given that the decision about boiler type is usually taken before arrival at the merchants.

> Merchants, although they probably won't admit it – they influence about 10% of installers that come across their counters. They could switch sell about 10% of people which is very low.
>
> BF

As such it seems that the merchants simply act as a counter and distribution system for the manufacturers – taking a credit risk on the products and being rewarded accordingly. Consequently, the merchants have no great brand loyalty: so long as installers are requesting a particular boiler in large enough numbers they will continue to stock it.

It seems that the manufacturers do not see a commercial conflict between their efforts to stay in business and the promotion of energy efficiency. They actually work together very well. The manufacturers would like to sell more boilers and therefore would like to see an increase in the turnover of the boiler stock. Even a conventional modern boiler will be substantially more efficient than the older boiler it replaces. They would also like to sell more of their higher value products such as the condensing boiler. The stumbling block identified by all is the installer who is considered to be conservative and not a sufficiently good salesman to convince the householder of the benefits of energy efficiency. Installers also suffer from something of an image problem in the UK – 60 % of the sample felt they could not, "trust an installer I didn't know to do a good job at a reasonable price". This lack of faith in the profession can act as a profound disincentive in taking the initial steps to improve ones heating system (Joseph Rowntree, 1998)

It is considered that this situation will only get worse in the medium term. The workforce of installers is ageing with very few new apprentices coming into the industry. Those that do come are "often not the brightest students". Hence, a conservative old guard remains whilst those few new boys that are in the industry may not have quite the initiative and appetite for further training that they otherwise might. Hence the manufacturers are very supportive of organisations such as the Institute of Plumbing who wish to see the average plumber becoming more professional and for the craft professions in general to be given a higher status. It is thought this will encourage higher calibre apprentices back into the trade who will be better at selling the benefits of greater efficiency.

5   Housing Developers

It seems that beyond complying with the regulations, energy efficiency or greenhouse gas reduction is not a priority when designing, building and equipping new homes. Designing in efficiency may impose extra costs which sales managers and managing directors do not feel will add value to the house in the eyes of a potential buyer. This may be partly because many new homebuyers are also first time buyers. Such purchasers are generally looking to move on after a few years having got to the first rung of the property ladder. Consequently, considerations of the cumulative running costs of the heating systems over the longer term may be less salient.

> If you are talking about energy efficiency, of course you could have it much higher than it is today and the cost of it could be measured in a few hundred pounds, but the market won't pay for it
>
> GFD

There remains the issue of the invisibility of many energy efficient building practises and products such that the *idea* of the efficiency must be sold rather than a tangible feature of the house such as UPVC window frames.

> [referring to the invisibility of efficient building design] You can't touch so how do you sell energy efficiency. You know - its not something you can say, "see that – that's energy efficient – that's going to save you £300 pounds a year."
>
> DM

Innovation in building practises is also suppressed by mortgage lenders.

> We have another problem which is lenders – if we build a product that they won't lend on we can't sell the product. So even though people want to buy our products we have to persuade the lenders to lend money on it
> [Interviewer] Why would a lender refuse?
> Because they feel that the house is built in such a way that they would not be able to resell it in ten years time should that mortgage go bad.
>
> DM

This reticence to lend on something unusual ultimately depends on the lenders' perception of what the house-buying market finds desirable. At present it is thought that most buyers are extremely conservative in their tastes. Certainly, it is established that the appearance of the home gives potent information about what the likely lifestyle and social standing of the occupants may be. This extends right down to the texture of the walls and the type of brick used (Sadalla and Sheels, 1993).

> You could build something that looked like something out of 2001 which is fine until you try and sell it because you find that people don't want to live in a house like that.
>
> DM

This leads to a rather paradoxical conclusion. Although, on the one hand, the marketing of efficiency is hampered by its invisibility on the other it is precisely because of its invisibility that it can be incorporated in design without putting off a conservative buying public.

The situation is analogous to that of electric utilities in the US supplying electricity generated with renewable sources. At present only 2% do so (some 85 utilities nationally). When the decision-making of utility managers who do supply renewably generated electricity is compared to the non-adopters it is found that there is little difference in the technical knowledge of the systems but large differences in what is termed "familiarity". This leads to the conclusion that simply bombarding utility managers with further information will fail. Instead, smaller scale projects should be encouraged which "familiarise" managers with the technology in question (Kaplan, 1999). There would seem to be a role here for housing associations and local authorities. If they could be encouraged to specify greenhouse gas reduction then the resulting familiarisation could create policy change when developers are building for themselves. The same effect would presumably work with subcontractors actually doing the work when they are in a specification role.

6   Policy Conclusions

For replacement of heating systems in existing housing it is clear that decisions about the appropriate system type and make or model are generally left to the installer. Installers have good reasons to be conservative but, this research suggests, they have less reason to be overly concerned about suggesting equipment that may be slightly more expensive. Once recommended it seems that the installer is usually able to suggest more or less what he sees fit. In addition there seems no reason why installers should not, as a matter of course, make two quotes – one for the more efficient system and one for the conventional package. Householders can then choose and the installer does not risk losing the business.

Hence, an enthusiastic and trained installer is potentially the key to more efficient heating installations. Hence a training course should be established for installers. This should cover both the technicalities of CB installation and develop an enthusiasm for selling the idea of greater efficiency to the householder. The development of an A to G label rating for new boilers would also raise awareness of efficiency as an issue with installers. Such a label would also assist other specifiers such as architects, local authority housing officers and indeed the small number of householders who do take a more active role in specification.

However it is not thought that the offer of additional training is not something that many or most installers will jump at. Even the existing training for certification as a "competent" person fit to perform gas installations is considered a chore. Perhaps ways can be found to further incentivise installers to attend the course – for instance manufacturers might feel able to link attendance with their points schemes. In the longer term, the trade associations' objective of registering all plumbers seems laudable indeed. This would increase the professionalism and standing of the trade. Consequently, a greater number of (higher calibre) apprentices may be encouraged back into the plumbing industry who in all likelihood would be trained on the newer, more efficient systems and who would consequently have a more positive orientation to the issue of efficiency in their working lives.

Difficulties with training installers switch the policy onus back to the householders who must be encouraged to ask for efficient systems. Indeed, the use of information strategies aimed at the householder rather than the installer is particularly important in creating demand for new boilers when the existing system has not yet broken down but is simply old and inefficient – an "early replacement" strategy. A modern boiler will always be more efficient than its older replacement regardless of whether it is condensing or not. Pursuing this strategy removes the onus on the installer to suggest the most efficient of the new systems available. Creating demand for new systems will not be easy. At present only around 5% of installations are motivated in this way whilst other evidence not presented here suggests that many householders will tolerate even barely functioning heating systems indefinitely (Fawcett *et al*, 2000). This suggests that the main route to early replacements will be through refurbishments, extensions and kitchen refits and targeted information.

What is more, there is much evidence to suggest that those that can least afford to have inefficient heating systems are also usually the least receptive to institutional initiatives in this respect. There is a nexus of issues concerning stake-holding, trust in institutions, equity, shared agenda's and social exclusion are at the heart of social marketing. In certain context's ones orientation to institutions goes a long way in explaining contrasting adoption rates of efficient technologies between different strata within a society (Banks, 1999) or in responding positively to energy labels on cold appliances between this country and say the Netherlands and Denmark (Cool Labels, 1998; IER, 1998).

Although environmental attitudes are apparently similar in these countries (SCPR, 1999), their societies are much less stratified and more democratic (at a local level) than our own (Harrison *et al*, 1996). Consequently, there tends to be greater acceptance of institutional agendas at the grass roots level and correspondingly greater uptake of institutionally sponsored initiatives (Cames, 1999).

Householders seem particularly receptive to the idea of improving the infrastructure when moving to a new home. This suggests that the forthcoming sellers' pack to be prepared by vendors for distribution to potential buyers should include some form of energy audit. Equally kitchen centres, DIY stores and builders and plumbers merchants would be appropriate places to leave leaflets and posters. The alternative to information strategies is regulation on emissions levels allied with a monitoring regime - as practised in Germany.

7 References

[1] Banks, N. (1999) Causal models of household decisions to choose the energy efficient alternative: the role of values, knowledge, attitudes and identity. *ECEEE proceedings 1999. Panel 3 Paper 24.*

[2] Cames, M. (1999) Differences between environmental consumption and perception between different social groups. *ECEEE proceedings 1999, Panel 3 paper 20.*

[3] Winward, J. and Schiellerup, P. (1998) Cool Labels: the first three years of the European Energy Label Energy and Environment Programme, Environmental Change Institute, University of Oxford. Save Contract: DG XVII 4.1031/E/97-001

[4] ETSU (2000) Domestic central heating and hot water: systems with gas and oil fired boilers. Good Practice Guide 284 UK Government Best Practice Program, HMSO

[5] Fawcett, T., Lane, K., Boardman, B. (2000) *Lower Carbon Futures for European Households* Environmental Change Institute, University of Oxford. Save Contract: 4.1031/Z/97-181

[6] Fisher, R.J. and Price, L.L. (1992) An investigation into the social context of early adoption behaviour. *Journal of Consumer Research*, Vol.19, 477-486

[7] Hackett, B. and Lutzenhiser, L. (1990) Social Stratification and Appliance Saturation. In *Proceedings of the 1990 ACEEE Summer Study on Energy Efficiency in Buildings*. Vol.2, 61-68. American Council for an Energy Efficient Economy.

[8] Harrison, C., Burgess, J., Filius, P. (1996) Rationalising Environmental Responsibilities: a comparison of the lay publics in the UK and the Netherlands. *Global Environmental Change*, Vol.6, No.3, 215-234

[9] IER (1998) Evaluation and Comparison of Utilities and Governmental DSM Programs for the Promotion of Condensing Boilers. University of Stuttgart. Save contract: XVII 4.103\2\96-136

[10] Kaplan, A.W (1999) Generating interest, generating power: commercialising photovoltaics in the utility sector. Energy Policy, Vol.27, No.6, 317-330

[11] Joseph Rowntree Foundation (1998) Make Do and Mend: explaining home-owners' approaches to repair and maintenance. Policy Press.

[12] Lutzenhiser, L. (1993) Social and Behavioural Aspects of Energy Use. *Annual Review of Energy and the Environment*, 18, 247-89

[13] Rogers, E. M. (1983) *Diffusion of Innovations*. 3rd Edition. Free Press, New York

[14] Sadalla, E.K. and Krull, J.L. (1995) Self-Presentational Barriers to Resource Conservation. *Environment and Behaviour*, Vol. 27, No.3, 328-353

[15] Sadalla, E.K. and Sheels, V.L (1993) Symbolism in Building Materials: Self-Presentational barriers and cognitive components. *Environment and Behaviour*, Vol.25, No.2, 155-180

[16] Schwartz Cowan, R. (1989) The Consumption Junction: A proposal for Research Strategies in the Sociology of Technology in *The Social Construction of Technological Systems*. Eds. W. Bijker, T. Hughes and T. Pinch. pp 261-281. The MIT press

[17] Social and Community Planning Research (1999) British and European Social Attitudes the 15th Report. Ashgate

[18] Southerton, D. and Shove, L. (1999) Paper for 'Consumption, Everyday Life and Sustainability', ESF TERM Programme, Lancaster University, United Kingdom, March 1998.

International Heat Pump Status and Policy Review

J. Bouma

IEA Heat Pump Centre

Abstract. Heat pumps stand out because they are energy efficient heating and cooling devices and able to reduce global CO_2 emissions to as much as 6%, which is one of the largest potentials for a single technology. This paper discusses the results of studies carried out by the International Energy Agency Heat Pump Centre (HPC) concerning the international status of residential heat pump technology, markets, as well as efforts for accelerating the use of heat pumps in buildings. The impact of heat pump policy instruments is analyzed.

The paper highlights how energy efficiency of heat pump and air conditioner equipment has evolved during the past decade. A recent HPC energy life cycle analysis of heat pumps shows that they contribute only marginally to global warming.

1 Introduction

Different definitions of heat pumps are being used world-wide. For instance, in Japan a residential heat pump is usually a reversible air conditioner, which is used for space cooling and heating. For historical reasons the term heat pump has specific meanings in North America (reversible air-to-air space heating and cooling unit, or heat recovery chiller designed for heating, cooling or both functions in a commercial/industrial application). However, in Europe, with exception of the Mediterranean countries, a heat pump is usually referred to as a heating-only device. Moreover, reversible air conditioners are often not referred to as heat pumps. These different definitions make it difficult to provide a reliable and accurate quantitative overview of the market. In this paper heat pumps are defined as appliances that provide heating, or cooling and heating. The focus is on electric heat pumps in residences that provide space and/or water heating.

2 International Heat Pump Status

Currently (2000) there are more than 100 million heat pumps installed world wide. The vast majority are appliances for space (and water) heating and heating/cooling. It is estimated that the world air conditioning and air-cooled heat pump population is approximately 239 million units (1997). Figure 1 presents a breakdown by region of the estimated world demand for air conditioner and heat pump products for residential and small commercial buildings. The European heat pump market takes a minor position (5%) as compared to North America and Asia (Japan and China), which are the largest markets for heat pumps today. With sales of 17.5 million units in Asia, the dominance of this market is clear (56% of the more than 31 million total units world-wide).

1997 A.C. & H.P. Shipments

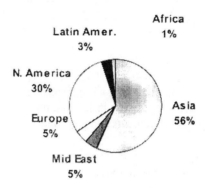

Figure 1: Estimated world demand for air conditioners and heat pumps (1997)

The current heat pump market in Europe can best be characterised by its wide diversity and different levels of maturity. It is generally in an early stage of development, with some exceptions (Sweden, Spain, Italy, Greece and Denmark). Determining factors are the climatic conditions, co-existence of a cooling and heating demand, and a favourable electricity-to-fossil fuel price ratio.

The European market is small compared to the USA, Japan and China, but in the south it is clearly tied with the air conditioning market. In central and northern Europe, heat pumps are often used for heating only purposes. Recent studies show that the market for air conditioning is on the move in several European countries, including those in central Europe.

Table I shows the installed heat pump stock (1999) in the residential sector. The more than 4.1 million residential heat pumps installed in the European region constitute only 4.3% of the world-wide residential heat pump stock. In some countries the main application of heat pumps is for water heating (Germany, Austria). The largest markets for residential heat pumps, in terms of installed stock, are in the south (Italy, Spain, and Greece), while Sweden is an important residential space heating market in northern Europe.

158

Table I: International residential heat pump stock (1999)

Country	Volume[1]	Country	Volume
Austria	146,000	Japan[3]	67,500,000
Canada[3]	450,000	Netherlands	21,500
China[3]	10,210,000	Norway	22,500
Denmark[3]	41,500	Spain[3]	1,060,000
Finland	15,000	Sweden	350,000
France[2,3]	147,000	Switzerland[4]	55,000
Germany[3]	480,000	UK[5]	15,500
Greece[3]	760,000	USA[3]	13,890,000
Italy[3]	1.060,000	Total	96,224,000

[1]incl. water heaters [2]excl. multi-family homes [3]estimate [4] excl. water heaters [5] mostly for swimming pools

In the past eight years the heat pump stock in Europe has grown at an average annual rate of 13% (world-wide the stock increased annually by more than 15%) [1]. This reflects the growing interest in heat pump technology, which is mainly driven by higher comfort requirements and environmental concern.

Table II gives an impression of the annual heat pump market (1999) in the residential sector in selected IEA countries.

Table II: Annual sales residential heat pumps (1999)

Country	Sales	Country	Sales
Austria	4,610	Japan	6,187,500
Canada	12,000-15,000	Netherlands	970
China	4,500,000	Norway	2,500
Denmark	600-800	Spain	85,000-95,000
Finland	700-900[1]	Sweden	18,670
France	7,000	Switzerland	7,000
Germany	4,150	UK	500-600
Greece	82,500	USA	1,293,400
Italy	137,000	Total	12,344,100-12,357,600

[1] estimate

The market development of residential and commercial heat pumps in selected European countries since 1993, relative to 1996, is depicted in Figure 2. The upward trend shown continues in 1999. In the countries shown, most heat pumps are residential units.

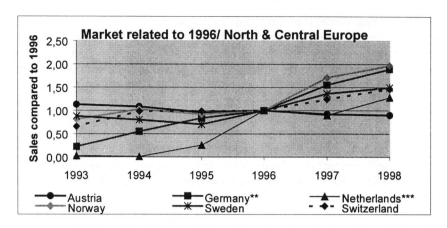

Figure 2: Residential/commercial heat pump market development

The heat pump market in the *USA* is strong and growing. Total annual shipments of unitary heat pumps and air conditioners have grown from just under 1 million in 1996 to almost 1.3 million in 1999. The majority of these systems are air-to-air. Their capacity ranges from 7-17 kW and they heat and cool an entire house. In 7 years time the total residential heat pump stock has risen from 7 million to almost 14 million. Single-room heat pumps currently only have a small share of the US market. However, a large potential market exists for minisplit-type systems as a retrofit measure in houses.

Japan is a very mature heat pump market, which has evolved from the world energy situation in the 1970s. In 1997 Japanese ductless heat pump and air conditioner shipments exceeded 7 million units, with heat pumps representing 94% of this total. This year shipments slightly declined (8 million in 1996) for the first time since 1993, mainly due to the recession and a mild summer and winter in recent years. More than 67 million heat pumps were installed in 1997.

The peak electricity demand in the summer has become a serious problem. This has a major impact on the load factor and reserve capacity. Consequently attention has shifted to cold/heat storage. The use of ice thermal storage type heat pump systems is now encouraged and supported by the federal government and major utilities. Dedicated products are being developed including air-to-air multi-split heat pump systems with ice storage.

Contrary to Japan, the market in *China* is not mature but most interesting because the market is developing rapidly and is enormous, perhaps the market with the largest growth potential in the world. Sales of room air conditioners alone grew from a rate of only a few hundred units in 1985, to nearly 5 million units in 1997. Ninety-two percent of all heat pumps are installed in houses. In 1999 the total residential heat pump stock installed was more than 10 million units. Several manufacturers are active in this market. In China alone, heat pump manufacturing capacity is outstanding and they are about the largest heat pump producing country in the world. However, production is not in balance with demand. The total air conditioner production capacity in 1999 was 17 million units, while the market demand was approximately 7.5 million units. However, the demand for room air conditioners is forecasted to grow, in China alone, to 16 million in 2007.

Virtually all installed residential heat pumps are electrically driven world-wide. A field-test with residential diffusion-absorption heat pumps commenced recently in the Netherlands and the results were encouraging. Unit optimisation and market introduction of these appliances will take a few more years.

On average, 77% of the residential heat pumps use outside air as the heat source. Ground-coupled (closed ground loops) heat pumps have gained considerable market share in 3 countries, notably Switzerland, Sweden and Austria (40%, 65% and 82% respectively). In the USA the Geothermal Heat Pump Consortium is actively promoting ground-source heat pumps. Ground water systems are applied much less in the residential sector. Ventilation air has found application as a heat source in Norway (41%) and Sweden (30%).

With the exception of France, Italy, Greece and Spain where air systems are used, the predominant heat sink in residences is water (radiators, ceiling, floor/slab coils). A trend of applying low temperature heat distribution systems (water, air) in new houses can be observed where such systems can be integrated at relatively low cost.

An indication of the significance of both the current heat pump market and the remaining potential is the share of the total space heat demand met by heat pumps. Considering the total space heat demand in the residential sector, the share met by heat pumps typically varies between 0-10% (in Europe between 0 and 5%). Japan stands out with more than 28% share.

3   Heat Pump Marketing

The energy efficiency of heat pump products has increased considerably over the years. This is the result of various developments and measures, not in the least from regulations (see next chapter). Equipment energy efficiency alone is not sufficient to make heat pumps successful in the market place. An important market impediment for residential heat pumps that needs to be removed is the lack of knowledge and skills of designers and contractors, and architects. Dedicated heat pump training and certification initiatives have been or are being taken in several European countries, and by the EU. In addition, heat pump associations and promotion organisations are developing schemes that are aimed at realising high quality heat pump installations and after-sales service. Training programmes usually include courses and practical training.

At EU-level quality heat pump installations are being stimulated through the SAVE II programme. One example is the project "Certification of Heat Pumping Technologies and Installers". Participants in the project were Eurovent Certification and a consortium of companies and training centres. The project's aim was to develop a system of certified heat pump products and installers, a pre-requisite for certified heat pump installations.

Recently the European Heat Pump Association (EHPA) was established. Its aim is to promote heat pump technology and stimulate developing the market in Europe. Members of the EHPA are mainly national and regional heat pump associations and promotion organisations. Typically in Europe the heat pump industry is fragmented with various small manufacturers serving local markets. The EHPA will be able to bundle and represent their interest at EU-level in areas such as energy efficiency and safety standards, training and certification and working fluids. Specific activities planned by the EHPA include setting meaningful targets and strategies for heat pump market penetration for electrically heated homes and developing a European product endorsement mark for residential heat pumps, based on the DACH scheme developed by Germany, Austria and Switzerland.

DACH is an association that is aimed at providing heat pump quality assurance. Heat pump test centres and promotion organisations in the three participating countries have joined forces and developed a heat pump quality label, which applies to the product. Manufacturers and test organisations define the requirements for the heat pump quality label. An important requirement is the minimum energy efficiency, or Coefficient of Performance (COP). As of 1 January 2001, new requirements will come into force; see Table III.

To qualify for the label, additional requirements need to be fulfilled:
• Clear installer and user manual;
• Two-year guarantee and 10 years parts supply;
• CE-mark conformity;
• On-site service within 24 hours.

Table III: COP-requirements for heat pumps (01.01.2001)

Heat pump type	Temp. sink/source (°C)	COP*
Brine-to-water	0/35	4.0
Water-to-water	10/35	4.5
Air-to-water	2/35	3.0

* standard EN 255

The label is a first step and has already become an important instrument for inspiring confidence in heat pumps. The next goal is to develop a quality label for the entire heat pump installation. Other countries where product certification has been practised for years are Sweden and Denmark.

4   Energy Efficiency Regulation and Labelling

An IEA-study [2] conducted in 1995 found that energy efficiency regulation of air conditioning and heat pump equipment was confined to only seven countries, including the two largest markets in the world for these products (USA, Japan). Product labelling requirements (energy efficiency and cost) existed in 1995 in the USA, Australia (some States), Canada, Israel, Mexico and South Korea.

US Federal energy efficiency regulations have been in force for many years with success. The shipment-weighted heat pump seasonal cooling efficiencies became higher than cooling-only units. Figure 3 illustrates this over a period of more than two decades (SEER: Seasonal Energy Efficiency Ratio in cooling mode; COP=SEER/3.413).

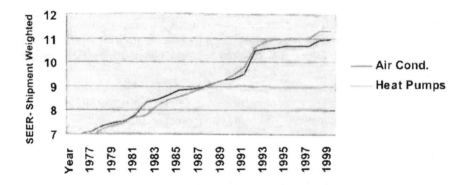

Figure 3: US unitary heat pump and air conditioner efficiency evolution, [4]

In 1992 new regulations for minimum energy efficiency of unitary air conditioners and heat pumps were mandated. Another tightening of the minimum energy efficiency regulations for residential (and commercial) appliances is underway. The rule is expected in late 2000. However, the levels and date of enforcement are not yet known.

The popularity of room heat pump and air conditioners and their potential for energy efficiency improvement are the reason that they have been singled out as two of the key end-uses requiring regulation. The new Top Runner Programme should ensure that the Japanese room unit market continues to be the global energy efficiency pace setter. The heat pump targets (63% COP improvement in 2004) are much stricter than the cooling-only targets (14% cooling COP improvement in 2007), because more units are sold (80% sales) and they have higher year-round energy consumption. The energy efficiency requirements for the different types of heat pumps on the Japanese market are depicted in Fig. 4. Apart from their challenging levels, the new targets are also notable in setting appreciably tougher efficiency thresholds for lower capacities than for higher capacities. These capacity-differentiated targets have been chosen, not because smaller-capacity units are more efficient, but because smaller units dominate the Japanese market and, in some cases, have already attained the minimum thresholds.

Electric household appliances such as air conditioners, heat pumps, fluorescent lamps, TVs, refrigerators and freezers will be labelled to show the energy-saving effects. Reference values for energy savings have been set for the years 2003-2006. The new labels show the extent to which the energy-saving effects approximate the reference value. Products that save more energy than the reference value appear with a different colour or design label for easy identification. Application of energy-saving labels is not mandatory for manufacturers.

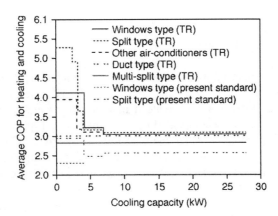

Figure 4: Top Runner energy efficiency targets for reversible room air conditioners (heat pumps) in 2002

5 Heat Pump Policies

Promoting heat pumps as an energy-efficient technology, and/or the rational use of electricity and gas, is of interest for governments as well as utilities. However, deregulation of the energy sector is influencing the attitude and strategies of utilities, which may have an adverse effect on their heat pump policy. A growing number of governments include heat pumps, in particular ambient heat, in their renewable energy policy, to save resources and thus benefit the environment.

Heat pumps typically face technical, economic and perception barriers. Policies are not only aimed at removing these barriers, which hamper diffusion of heat pumps, but also at stimulating the market. Different types of policy instruments are used in heat pump marketing concepts, all aiming to remove one or more barriers. Most instruments are applied temporarily. Common instruments for heat pump promotion include, [1]:

- Incentives
- R.D&D projects
- Energy efficiency regulations
- Labelling and certification
- Special tariffs
- Information
- Tax measures
- Heat (and cold) contracting
- Procurement competition

Governments and some utilities use *incentives* to stimulate heat pump diffusion. Incentives can take the form of subsidies, special energy prices, lease constructions etc. Subsidies can create a market rather quickly. However, that market is artificial, if not accompanied by other measures. Typically the market is unstable and may collapse when the incentive is removed. Governments can influence energy prices with *taxes*, which is a powerful instrument with considerable impact. Tax deduction is another vehicle to make the investment in a heat pump more attractive.

Technical barriers can be removed with *R.D&D* and governments often use the instrument. Examples of R.D&D are widening (new) heat pump applications and adapting a heat pump concept to the domestic market needs. Domestic industry and institutions may benefit from government-sponsored research. *Heat pump procurement* falls into this category too. It provides a powerful tool for the development of efficient and cheaper products and can stimulate industry to collaborate. Procurement has, for instance, been used in Sweden, Switzerland and the Netherlands, and has resulted in improved energy efficiency and cheaper products.

(Energy efficiency) regulation is an effective instrument, as has been demonstrated in the USA and Japan (see previous chapter). Besides for energy efficiency of appliances, regulations may also be implemented to stimulate ventilation heat recovery, low temperature heat distribution systems in new houses, or restricting use of direct electric heating. *Certification and labelling* of appliances are a useful way of informing the end user about product quality (and cost).

Lack of perceived value and confidence can be improved through *information and promotion*. However, information alone does not stimulate a market.

Utilities have some specific powerful instruments available to stimulate the use of heat pumps and retain customers: s*pecial tariffs, financial constructions* and *heat and cold contracting*.

Long-term strategic agreements (covenants) between industry, contractors, housing corporations, utilities and governments are another way of removing market and technical impediments and creating a basis for a healthy market. It is essential that each party commits itself to realising a pre-defined effort which can range from providing financial support, reducing the price of products or services, transferring knowledge and skills, improving the energy efficiency of a product, realising housing projects, etc.

The package of instruments to be used in a specific situation depends on many factors, such as climate, technical and economical barriers, maturity of the market etc. Successful programmes tend to be a mix of economic stimuli (e.g. incentives, special tariffs, lease arrangements) and information campaigns (e.g. promotion, education and demonstration projects). Unfortunately, the impact of policy instruments is not easy to monitor and market evolution is difficult to correlate.

An overview of heat pump programmes, schemes and initiatives is given in Table IV (1997).

Table IV: National heat pump initiatives (G: Government; U: Utility)

Country	Type of initiative	Prog.	Country	Type of initiative	Prog.
Austria	-	U	*Norway*	Renew. energy	G
Canada	Regulation, labelling	G/U	*S. Korea*	Regulation, labelling	G
Denmark	Renew. energy	G	*Spain*	Incentives	G/U
France	-	U	*Sweden*	R.D&D, labelling	G/U
Germany	Renew. energy	G/U	*Switzerland*	R.D&D, incentives	G/U
Italy	Energy recovery	G/U	*UK*	R.D&D	G
Japan	Regulation	G/U	*USA*	Regulation, R.D&D, labelling	G
Netherlands	Renew. energy	G/U			

6 Environmental Benefits

In 1999 the HPC released the results of an Energy Life Cycle Analysis on Heat Pumps [3]. In this study the environmental benefits were analysed using the newest knowledge base and insights. The energy chain considered included the energy consumed by the appliance, energy for mining, transportation, conversion, distribution and the energy for manufacturing and maintaining the appliance. The impact of refrigerant leakage and release at the end of service life was considered as well. For heat pumps the conclusions were that:

- Energy of operation dominates the CO_2 equivalent emissions. The other sources of energy use account for 15-20% of total CO_2 equivalent emissions;
- CO_2-emisions from power generation are the most important factor for the impact of electric heat pumps on global warming. With the growing supply of green power and dismantling of conventional fossil fuel power stations, the overall CO_2 emissions from using electric heat pumps will reduce drastically;
- Heat pumps offer CO_2 emission reductions of 30-50%, compared to fossil fuel heating boilers, using electricity generated from the same fuel;
- The specific CO_2 equivalent emissions caused by refrigerant losses from heat pumps range from 0-11%, depending on refrigerant type and equipment design.

7 Conclusions

In 1999 almost 100 million residential heat pump units were installed world-wide. The international heat pump market is developing steadily and is growing at a rate of more than 10% annually. Strong market growth occurred in Sweden, Switzerland, Austria and Greece. The residential heat pump markets in the USA and China are surging. Japan is a very mature heat pump market and is the largest in volume.

Considering the total space heat demand, up to 5% in Europe is met by heating heat pumps (10% world-wide). In individual countries the share can be up to 28% (Japan).

The European heat pump market is highly diverse as a result of widely varying factors that influence the market such as climate, energy prices, building standards, appliance cost, regulations etc.

Policy support is required in most heating-mainly markets, qualifying heat pumps as a sustainable space and water heating technology.

Energy efficiency regulations and energy labelling are in principle powerful instruments to improve appliance efficiencies, stimulate markets and improve the perceived value of heat pumps.

8 References

[1] *International heat pump status and policy review 1993-96.* HPC-AR7. July 1999. IEA Heat Pump Centre.

[2] *Heat pump energy efficiency regulations and standards.* HPC-AR4. June 1996. IEA Heat Pump Centre.

[3] *Environmental benefits of heat pumping technologies.* HPC-AR6. March 1999. IEA Heat Pump Centre.

[4] Reedy R., Groff G. *Heat pumps for space conditioning – A half century of progress.* May 2000. To be published.

Innovative Heating - Air Conditioning System with Electric Heat Pump in Residential and Services Sectors

Sergio Zanolin

De' Longhi

De' Longhi system is an innovative Direct Water-Loop System called "Evoluzione". It has been studied to improve the quality and stability of indoor conditions of comfort.

The system consists of an air-water heat pump with automatic heat output control.

New technology has been applied for this purpose to the cooling and water circuits, by inserting an electronic double-acting lamination valve and a variable speed pump both controlled by a microprocessor. The microprocessor used in these units has software which fully controls all parts of the chiller and detects even the slightest variation in temperature both indoors (user) and outdoors, in order to speed up adjustment of the cooling cycle parameters to the changes in temperature and humidity. The heat pump can quickly assess the change in heat in the environment, adapting the effective output to the new requirements in order to ensure a high level of comfort in the rooms.

As already stated, by controlling all the points of the cooling circuit and calibrating the effective outputs, this type of heat pump needs no inertial water storage to respond to the changes in heat loads. The software is also designed to prevent the formation of particles of ice, by injecting well-calibrated amounts of warm gas at certain moments, thereby almost totally eliminating the reversal of the cooling cycle of the actual unit in order to carry out defrosting. This results in better efficiency when operating at very low temperatures.

Figure 1 shows a diagram of the cooling circuit with the winter cycle configuration. Table I gives the rated technical characteristics of the De' Longhi model with incorporated heat pump.

Table I: Rated technical data of the De' Longhi air-water heat pump

Cooling capacity	5.8 kW
Compressor absorbed power	2.0 kW
Total absorbed power	2.2 kW
Heat output	6.8 kW
Water flow rate in cooling mode	0.8 m^3/h
Water flow rate in heating mode	1.0 m^3/h
Air flow rate	3500 m^3/h
Refrigerant	R22
Sound pressure	55 db (A)

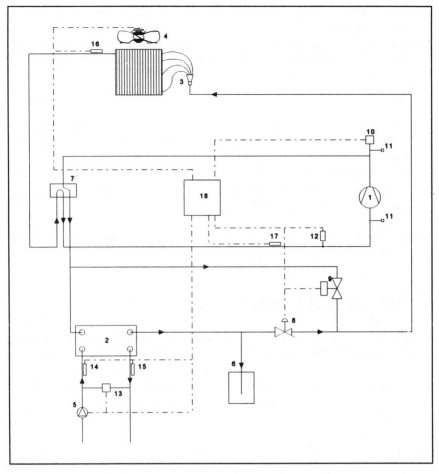

Figure 1: Diagram of the cooling circuit with winter configuration

LEGEND:
1 Compressor
2 System water heat exchanger
(plate-type exchanger)
3 External heat exchanger (finned
exchanger)
4 Fan
5 System pump
6 Liquid receiver
7 Cycle reverse valve
8 Electronic thermostat
9 By-pass valve
10 High pressure safety switch

11 Filling valves
12 Low pressure transducer
13 Water differential pressure switch
14 System water inlet temperature
transducer
15 System water outlet temperature
transducer
16 Condensation temperature transducer
17 Compressor intake gas temperature
transducer
18 Electric control panel

Tests were carried out in 1999 at the University of Cagliari and at Cesi (Enel - Electricity Board) in Milan.

Both units were left in operation for one summer and one winter season. They were located in two different geographical areas in order to embrace all possible conditions of temperature and humidity.

Two different types of tests were carried out on the two units and more precisely:

- At the University of Cagliari, efficiency tests and comparison with units currently on sale that have equal power and inertial storage units. The indoor comfort level was also assessed.
- At the Cesi laboratory through Enel in Milan, a unit was installed in a sample house and all the operating parameters were measured, calculating the seasonal efficiency of the actual unit.

Some graphs and tables showing the results of the tests were provided by the above-mentioned bodies and are given below.

1   University of Cagliari

The system is located in Cagliari in the Engineering Department of the Territory and provides air-conditioning for all the rooms (see Figure 2) of the Technical and Energy Physics Sector situated on the ground floor, the exposure and geometrical characteristics of which are given in Table II. Being installed on the east side, the heat pumps are exposed to the sun for a few hours in the morning.

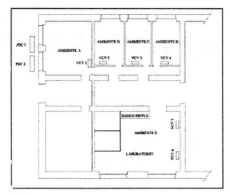

Figure 2

Room	A	B	C	D	E
Exposure	West	South	South	South	North
Length [m]	5,50	3,50	3,05	3,20	10,05
Width [m]	6,25	5,08	5,08	5,08	7,05
Height [m]	4,70	4,70	4,70	4,70	4,70
Surface [mq]	34,37	17,78	15,50	16,25	70,80
Volume [mc]	161,56	83,56	72,82	76,40	333

Table II

2 Cooling Mode

The operating graphs of the Eran/P unit are given below, compared to a standard chiller.
It can be seen that the Eran/p keeps the system supply water temperature constant with consequent environmental comfort.

172

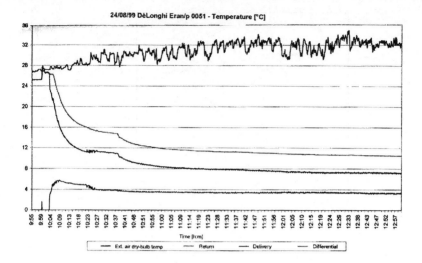

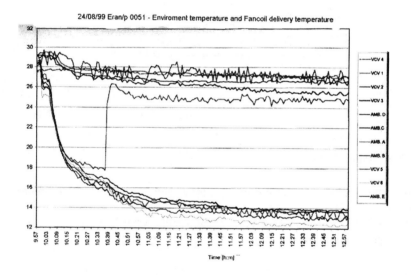

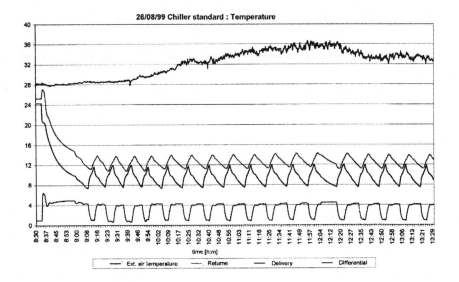

26/08/99 Chiller standard : Temperature

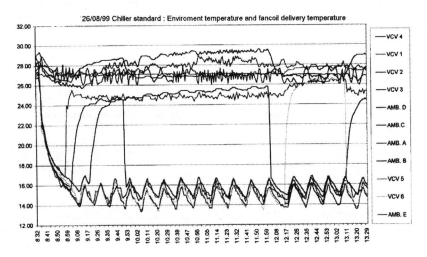

26/08/99 Chiller standard : Enviroment temperature and fancoil delivery temperature

174

3 Heating Mode

The operating graphs of the Eran/P unit are given below, compared to a standard chiller.
It can be seen that the Eran/p keeps the system supply water temperature constant with consequent environmental comfort.

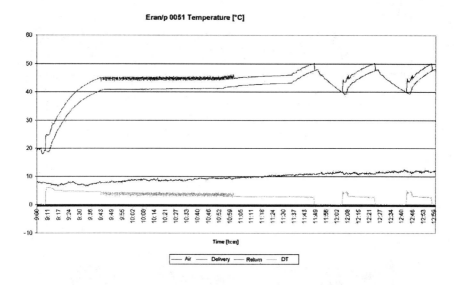

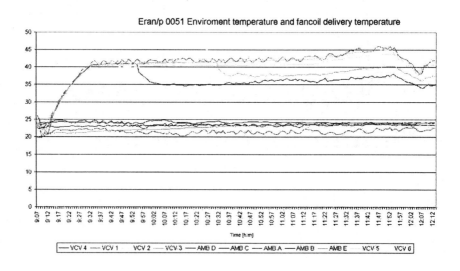

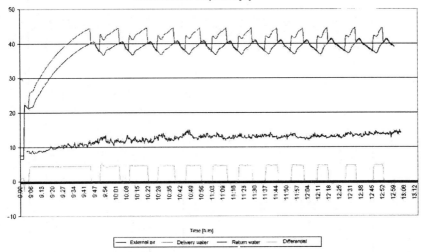

Chiller Standard - Temperature [°C]

Time [h.m]

External air — Delivery water — Return water — Differential

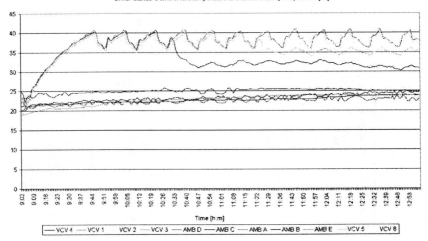

Chiller standard enviroment temperature and fancoil delivery temperature [°C]

Time [h:m]

— VCV 4 — VCV 1 — VCV 2 — VCV 3 — AMB D — AMB C — AMB A — AMB B — AMB E — VCV 5 — VCV 6

Day	External air temperature max °C	External air temperature min °C	Ext. Air temp. average value °C	ext. air UR %	Total COP Heater pump + fancoil	COP Heater pump	COP Compressor	Enviroment Temperature °C	Enviroment UR%
01/01/00	4.4	-4.5	-0.6	74.5	2.32	2.38	2.76	20.9	27.8
02/01/00	4.7	-4.8	-0.7	78	2.22	2.28	2.65	20.7	28
03/01/00	6.2	-5.2	-0.6	83.6	2.28	2.36	2.75	20.6	28.9
04/01/00	10.1	-4	1.36	79.2	2.2	2.24	2.61	21	30.6
05/01/00	8.9	-1.7	2.54	82.1	2.26	2.29	2.7	21	33.1
Weekly average value			0.4	79.48	2.256	2.31	2.694	20.84	29.68

Day	External air temperature max °C	External air temperature min °C	Ext. air temp. average value °C	External UR %	Total COP Heater pump + fancoil	COP Heater pump	COP compressor	Enviroment temperature °C	Enviroment UR %
03/02/00	11.6	2.7	6.6	89.9	2.69	2.81	3.5	20	44.8
04/02/00	9.8	2.9	4.7	97.3	2.47	2.57	3.13	20	42.9
17/02/00	14.6	1.8	7.6	24.8	3.03	3.21	3.93	19.8	24.4
18/02/00	11.3	-0.2	3.7	36.6	2.36	2.47	2.95	19.9	21.5
Weekly average value			5.65	62.15	2.89	2.765	3.3775	19.925	33.4

Operazione "Caldaia Sicura"
Maintenance and Inspection Campaign for Heating Systems

M. Macaluso and C. De Masi

ANEA – Agenzia Napoletana Energia Ambiente
The Naples Agency for Energy and the Environment

Preface. The program "Caldaia Sicura" is a complex aimed at improving safety and energy efficiency of heating systems and to reduce emissions. It involves the major players of local government in the sector of energy and environment, the trade associations, fuel suppliers and the local energy agency (ANEA). So as the program has been carrying out by dept of Municipality of Naples ("Servizio Progettazione e Valutazione Ambientale" - Mr. B. Sciannimanica; "Servizio Attività Amministrative" - Mrs. M. Aprea); by Craftsmen Associations (Confartigianato, CNA, CASArtigiani, CLAAI); by ANEA (Agenzia Napoletana Energia Ambiente) the Naples Agency for Energy and the Environment.

ANEA is an independent, non-profit organisation aiming to promote the rational use of energy and disseminate renewable energy sources. ANEA belongs to the RENAEL, the national network of Italian local energy Agencies and started in 1997 thanks to European Union co-funding, in the framework of the SAVE II Programme, and thanks to local financial support from the Municipality of Naples, Naples Public Transport Company (ANM), Suburban Mass Transit Company (CTP), the Electricity Board (ENEL) and the Employers' Association of the Province of Naples. In 1998 Naples Water Resources (ARIN) and Naples Natural Gas Company (Napoletanagas) joined the Agency. Main activities of ANEA are information, training, applied research and technical assistance for citizens, SMEs and Public Administration.

1 Introduction

Whilst much work has already been carried out to improve energy efficiency in many sectors over recent years, installed heating systems remain largely old fashioned and inefficient. The "Caldaia Sicura" campaign aims to improve safety and energy efficiency of heating systems and to reduce the emissions by stimulating the market in a comprehensive way. There are currently over 100,000 heating systems (90% of which are domestic boilers) in the territory of Municipality of Naples (population: 1,100,000). Most of such systems are used throughout the year for the production of hot water. There is, therefore, a significant opportunity to make large savings in both energy consumption and carbon dioxide emissions by encouraging the maintenance of boilers and the introduction of efficient and well-controlled systems.

There are currently two main barriers to innovations in energy supply and demand: first, installers often lack appropriate training geared towards energy efficiency; second, household users are not only unaware of potential savings and tax benefits tied to systems replacement, but also fail to take due account of the rudiments of safety regulations regarding heating installations and their maintenance.
The project has been also submitted in the SAVE 2000 Program (jointly with the Energy Agency of Belfast, Klagenfurt and Dublin) and is at the moment under evaluation.

2 Action

The programme takes its start from the application of Italian legislation (DPR 412/93 and recent update DPR 551/99) which gives to the local authorities the responsibility to verify (through technical inspections) the working conditions of heating systems. For a temporary period is allowed that local authority can substitute the inspections with a declaration signed by the owner/user or by the responsible of heating systems; in a second phase the Municipality will start a massive inspection campaign to verify the heating systems (a little sample of self declared systems and the 100% of the rest).

Under the application of Italian legislation (DPR 412/93 and recent update DPR 551/99), Naples City Council has initiated mandatory declaration of current heating systems.

According with the DPR 412/93 every heating system who falls into the municipal territory of Naples should be declared to the Municipality of Naples through the presentation of a standard form duly filled in each part. For the smaller boilers (less than 35 kW of power, that is a common boiler for a standard household) the presentation of declaration is free of charge; for the greater boilers (more than 35kW) there are different fees depending on the power. The declaration can be returned to the Municipality of Naples, to the Trade Associations or to ANEA directly by the owner of system or via installers.

By end-2000, when the self declaration phase will end, a database will be available and will provide valuable information about many characteristics of boilers as: location, age, power, efficiency, fuel and so on.

The financial resources coming from the self declaration phase will be used to cover the expenses of following phase (control phase) when a sample (about 5%) of self declared boilers will be controlled by technicians of Municipality. All heating systems for which the self declaration has been not presented a control campaign (100% of systems in two years) will be executed charging the owners.

At the same time "Caldaia Sicura" campaign addresses all aspects of the market including advice for citizens, energy advice and training for the installers of heating trade, training courses for the technicians which will control the heating systems. For the realisation of project the creation a list of firms is ongoing, certified and recognised by the City Council, which meet minimum technical requirements and guarantee a "political" price set for standard maintenance operations.

In details the campaign consists in the following activities.

1. Establishment of local steering groups; the local steering group oversees the local implementation of the programme and is composed by local authority, trade associations, consumer groups, fuel suppliers and by local energy agency (ANEA).

Trade Associations play a fundamental role in the *caldaia sicura* program above all in a large town as Napoli. The trade associations will inform directly the associated firms on the procedures to follow for the enforcement of national legislation. Moreover trade associations will sign an agreement with Municipality to set a standard price for the heating systems maintenance operations. But the most important aspects for a success of the campaign is the opportunity offered to installers to his the one who return the declaration to the Municipality (so called "third responsible"). In fact the declaration filled by the installers is - not only a "form duly filled" but - a kind of official certificate of good working conditions of heating systems. This new role gives to the installers more responsibility and will improve the awareness amongst installers and will increase the capability of firms.

2. Stimulating the application of existing tax incentives: Current tax benefits for home improvements leading to greater efficiency are rarely claimed by householders. At present, 36% of renovation costs (including the installation, replacement costs and related building costs) are tax deductible over five years following a straightforward administrative procedure involving both the building contractor and the householder. This opportunity is not well known among the citizens, and its diffusion will improve the demand of efficient heating systems and will generate interest and awareness on energy efficiency.

3. Designing and delivering of a promotional campaign directed at home owners: a promotional campaign is developed, in partnership with the local players to provide an appropriate level of energy efficiency advice to householders about heating to householders and simultaneous improvements in safety. This phase is intended to improve the heating system by carrying out technical surveys in actual households and to ensure the best use of system to gain maximum efficiency savings and to encourage further improvements in energy efficiency. At this scope all the players involved in the program activated their "communication channels". ANEA has designed, jointly with the Municipality of Naples, the campaign to the citizens through the realisations of poster and leaflets; ANEA, besides, activated its Web site and its information bureau.

4. Establishing a procedure for provision of energy advice to home owners: whilst a major investment is required to install or upgrade efficient heating systems, one aspect which is often neglected is the provision of energy efficiency advice to householders explaining the most efficient way to operation the system including the heating controls. Some very simple advice regarding efficient energy behaviour can be given to householders during maintenance and/or installing operations (i.e. insulation, rational use of domestic appliances, and so on).

5. Training for installers and technicians: Experience highlights that the heating market must be addressed in a comprehensive way if change is to take place. Whilst the previous phases are targeted primarily at the householder to stimulate demand for energy efficient heating, the firms operating in the heating sector must have the capacity and ability. Two types of courses will be developed and implemented for heating professionals:
- Course for installers: a short course for the staff of firm responsible of maintenance of heating systems to set the procedures of a correct maintenance procedures as set by the actual legislation (DPR 412/93 and DPR 551/99). This action involves all the players of the programme.
- Course for technicians who will make inspections on heating systems: a more intensive course will be held, according with the national legislation, with the overseeing of National Body for Energy and Environment (ENEA), intended to give the correct training to those technicians which will visit the heating systems to verify their correct working conditions.

6. Realisation of a heating system database
Using the declarations and the data coming from the fuel suppliers it will be possible to realise a database which will contain the characteristics of heating systems located in the municipal territory of Napoli. The main objectives of database is to address the inspecting visit and to establish a regular maintenance culture amongst householder and the heating trade. The analyses coming from the database will give information about the quality of heating systems in Napoli and of percentage of improvement in energy consumption and reduction pollutants.

3 Evaluation and Perspectives

Previous work carried out in Napoli showed that methane gas consumption, only for domestic use in 1998, amounted to about 90 million of m^3. The common situation in other Italian towns shows that about 27% of installed gas boilers have an energy efficiency below 86% (legal limit). Considering a realistic percentage of a 5% improvement in efficiency by carrying out proper servicing and, in some case, the substitution of older boilers, we have the following savings: 60.300 GJ of energy, 3.140.000 Euros, 192.000 tons of CO_2. So that this initiative can be comprise into the existing national strategies with regard to CO_2 reduction targets and improvement in energy efficiency.

Other important aspects is tied to the high number of heating systems (about 100.000) to control. Therefore is expected an improvement of work and of quality level for heating trade (and for manufacturers of boilers and accessories) and the creation of new job opportunity both for the installers and for the technicians who will make inspections.

Design for Energy Efficiency as a Basis for Innovations in Kitchen Appliances
"Ex Nihilo Nil Fit, Nil Fit Ad Nihilum" [1]

Bas Flipsen Msc.[1], Jasper Koot Msc.[2] and Geert Timmers Msc.[1]

[1] TNO Institute of Industrial Technologies
[2] Advanced Industrial Design Engineering

1   Introduction

Problem definition

The energy consumption of consumer products receives increasing attention in the product development of consumer and professional products. More and more products receive an energy label, often forced by EU legislation, showing their impact on the energy bill to the consumer. Therefore, designers and product managers want to know the energy performance in an early stage of the product development process. Energy efficient products are not just cheaper to use, they are smarter in more ways: less energy use requires less insulation or less cooling capacity (think of the air-conditioning in modern offices). Furthermore, electronic components tend to live longer in cooler surroundings, thus improving reliability and overall quality of the product.

At the moment four important trends can be seen, which stimulate the increase of the energy consumption in households [2]:

1. Increasing diversity in electrified products within a household;
2. Increase of functionality, like extra ICT related features, portability and remote control;
3. Sleep conditions as a standard feature;
4. Increase of add-on products (set-top boxes) for the use of services.

This increase of energy use has a negative effect on our environment, meaning that meeting the Kyoto agreements will be increasingly difficult. The increase of energy use can be tackled by means of [2]:

1. Decreasing and differentiating the demand for energy by changing the needs of the consumer;
2. Increase of sustainable energy sources;
3. Re-use of energy, e.g. heath or fluid flows;

4.   Use of less scarce energy sources like nuclear energy;
5.   Energy-efficient product development.

By looking at energy efficiency of products from the beginning, product designers and developers can seriously contribute to the reduction in energy use.

At the moment little is done about Design for Energy Efficiency (DfEE) within the companies, due to a number of motives, e.g. the energy-labeling schemes and green procurement actions, the energy-efficiency of products is becoming more of an issue.

Aim of Design for Energy-Efficiency

TNO Institute of Industrial Technologies (hereafter referred to as TNO) has developed – in co-operation with Atag Kitchen Group – a strategy for energy-efficient design of kitchen appliances [3]. Design for Energy-Efficiency can be used as a tool for designers to help them analyze existing products, and optimize new concepts in an early stage during the design process.

DfEE has been developed to improve products with a main impact in the usage phase of their life cycle. Furthermore the design strategy dealt with underneath is meant to calculate, analyze and visualize the energy use of a specific products and its sub-divisions.

2   Design for Energy-Efficiency (An Approach)

The approach towards Design for Energy-Efficient – DfEE – can be split up in three major phases:

1.   *Phase I Initiation*: in this phase the aim of the company, the R&D costs and the potential energy-improvement is analyzed. With this knowledge it can be decided if DfEE is an important issue for the company;
2.   *Phase II Analysis*: the basis for DfEE is a mathematical model of the product. The different energy flows, like electricity and heat, are calculated and mapped. With the calculations and the model, the main design parameters can be filtered and altered to reach the ultimate energy-efficient re-design of the product;
3.   *Phase III Design and testing*: this phase focuses on the major energy leaks. With the help of the calculations done, new innovative and simple alterations are proposed. A re-design is made which will be evaluated by building a prototype or tests.

To help the product developer with DfEE, he/she will be assisted with Energy Strategies and design directrix.

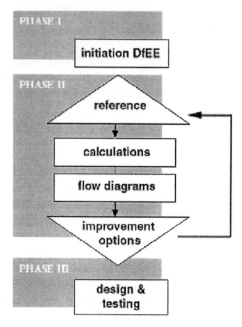

Figure 1: An approach towards DfEE.

Phase I: Initiation

During the initiation phase the importance of energy-efficient product development is mapped. The importance will be based upon: the costs of implementation, market value of DfEE, and legislation now and in the future [4].

Phase II: Analysis

TNO will make an analysis model of the reference product. This reference product is often an existing version of the product. If no reference product is available, the technical layout of the new product is used to predict its energy use. New energy-models of products can often be based on existing analysis models.

To analyze correctly the user profile is of extreme importance. These user profiles of the energy consumption are as close as possible to reality.

With the help of the analysis model and the (standard) user profile, the energy flows in, through and out of the product are calculated. These energy flows are for example electricity, heat or fossil-fuel, and will be mapped in flow diagrams and graphs. The theoretical minimum and the efficiency of the reference and new improved alternatives can be calculated.

Phase III Design and testing

First of all simple alterations will be recommended. Second the main energy-users within the product are tackled. New technologies and innovative alterations will be thought of to lower the energy use even more.

The potential energy-improvements are founded by means of flow diagrams and concrete numbers. Recommendations and improvements will be given on component level, materials, material characteristics, and technology and system infrastructure. The new improved product will be optimized and can be compared with the old design.

Finally the new design has to be tested by means of building a prototype.

Energy Strategies

With the help of *Energy Strategies,* designers are encouraged to evaluate their design from an energy-efficient point-of-view. They will be encountered with different perspectives on the product or problem. These energy strategies are based on design directrix e.g.: decrease the amount of conversions; reconsider stand-by functionality; et cetera. At the moment this field is still in research.

3   Case Study 'Electric Oven'

In a joint operation, Atag Kitchen Group, University of Technology Delft (the Netherlands), Gastec and TNO have worked on new kitchen concepts. Within this project the energy use of several kitchen appliances were calculated, evaluated and re-designed. Examples are the range hood, the electric water heater, and the electric oven [3,5,6,7].

Figure 2: The Atag electric oven (reference).

The current Atag electric oven is a modern oven with forced air circulation. In normal use conditions, about 20% of the energy input is directly used to heat the meal in the oven. The other 80% is lost due to heat storage in the oven material and losses to the surroundings. The electric oven of Atag can be 56% more energy efficient during use. For the electric oven two prototypes have been build to evaluate their energy use, one for short and one for long term planning.

Phase I: Initiation

Upcoming energy-labeling of the electric oven made energy consumption one of the major issues during the design of new electric ovens. Furthermore a benchmark with two competing ovens showed a rather poor performance. Atag Development decided to look more closely to the energy efficiency of their ovens.

The goal for the new Atag electric oven was class A for the short-term and class AA for the probable long-term future. The energy label for ovens is in development, and will be introduced at the end of 2001. In this long term project, Atag only paid attention to reduction of the energy losses, the core heating technology was unchanged, nor were the user habits studied.

Phase II: Analysis

A reference model of the oven has been made, including a reference of the usage profile. The reference model is based on the sub-parts in the oven and the electronic scheme.

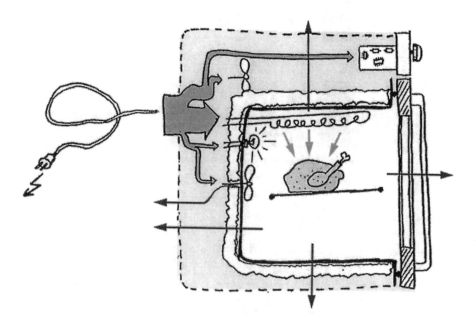

Figure 3: Energy flow diagram of the energy in, through, and output.

From the reference model, an analytical model is extracted which simulates the existing oven. The model is set up in such way that the most important design parameters can be altered easily. The designer can change parameters in the model in order to find the ultimate energy-efficient re-design of the oven. New technologies, like vacuum insulation, can be introduced by altering the analytical model. In the next image the calculated results are shown. The energy potential for vacuum insulation is large in comparison to the reference and the class A oven.

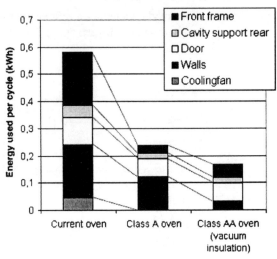

Figure 4: Energy used theoretically (pre-heating is excluded).

The usage profile is based on the incidental use of the oven, pre-heating and cooking for one hour. On the other hand the whole user phase is taken into account. This means the stand-by energy of the clock is also taken into account.

The energy use of the oven and the net-efficiency, depends heavily on the food that is prepared and the country in which it is used. From a survey conducted by *Save Working Group Domestic Ovens* [8] the oven usage in Finland is three to four times (185 times per year) higher than in the Netherlands (47 times per year). In the Netherlands the working period of the oven is approximately 38 minutes. The standby time in the Netherlands per oven is 8730 hours per year.

Phase III: Design and testing

Using DfEE, TNO has:

- visualized energy flows and losses in the oven;
- identified parts or constructions responsible for specific energy losses;
- designed a number of product improvements, and calculated their influence on the overall energy use.

The use of the energy analysis has led to a new list of requirements, which leads to a design of a 56% more energy-efficient oven. Important improvements are amongst others:

- An energy-efficient power supplies for the electronic clock and display. The standby losses are reduced (14%);
- Use of vacuum insulated panels instead of glass wool (10%). This improvement also reduces cooling for the electronics. The cooling fan and channel can be left out;
- Diminish of the oven thermal mass, using thinner plate for the inner case (10%);
- Thermal disconnection of the inner case with the front and outer case (9%).

The next step in the product development was building of two prototypes, for the class A and class AA respectively. In the next image the test results of a single heating is shown. Only the usage of the oven is measured, and the pre-heat phase is excluded. No standard tests has been applied yet. The image shows that the class A oven uses approximately 25% less energy than its reference, and the class AA uses even 50% less. Furthermore the pre-heat phase for the class AA oven now uses more energy than the usage phase.

The difference of the theoretical values (middle graph of the previous image) and the real values from the oven (next image) are due to simplification of the reality and idealization of the process.

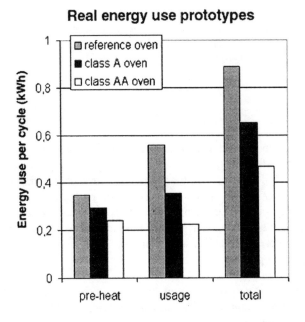

Figure 5: Energy used of the class A and AA prototypes (pre-heat, usage and total).

Figure 6: Class AA prototype.

Design Strategies

The design strategies are used to come to energy-efficient solutions for the oven. They give the designer a broad overview of possibilities ranging from easy to implement solutions to innovative design changes.

4 Conclusions

At the moment the Design for Energy-Efficiency method is still under development. We have been using the method to re-design existing kitchen appliances, like the kitchen boiler, oven and the range hood. It is shown that this approach helped the designer to get a quick overview of problems with the existing products. An energy analysis of the product can give the designer a good overview of bottlenecks in the existing design, and possibilities to improve the product. To get an even faster result TNO is working on Design Strategies for energy-efficient product development.

Within 1,5 man-year two prototypes of redesigns have been built of the electric oven. Both these ovens will probably meet the new A-label design requirements. One of the ovens uses a new innovative vacuum insulation and is designed to even exceed these requirements. DfEE fits well into the design process and helps the design engineer to tackle this problem successfully.

At the moment TNO is still working on the Design Strategies and different tools to help the designer to implement energy-efficient design more easily.

Endnotes

1   TNO Institute of Industrial Technologies
      Oostsingel 209
      PO box 5073
      2600 GB Delft
      the Netherlands
2   Advanced Industrial Design Engineering
      Jaffalaan 9
      2628 BX Delft
      the Netherlands

5   References

[1] quotation by Majer R, Italy, 1842.
[2] Wajer BPF, *An overview of energy related technical issues in home electronics*, Novem BV, Sittard (NL), 1999.
[3] Flipsen, SFJ & Timmers, G, *Energy-Efficient Product Development*, in Dutch Advanced Industrial Design Engineering, TNO Institute of Industrial Technologies, Delft (NL), 1999.
[4] Van der Horst, T & Zweers, M, *Environmental oriented product development: various approaches to success*, International Conference of Engineering Design, The Hague (NL), 1993.
[5] Flipsen, SFJ, *Design of a rest-heat boiler*, Atag Kitchen Group, TNO Institute of Industrial Technologies, University of Technology Delft, Gastec, Delft, 1999.
[6] Zijlstra, J, *Environmental benchmark of baking ovens*, Seppelfricke, University of Technology Delft, TNO Institute of Industrial Technologies, Delft, 1999.
[7] Koot, J, *Design of an energy-efficient oven*, under development, Atag Kitchen Group, Advanced Industrial Design Engineering, TNO Institute of Industrial Technologies, Delft, 2000.
[8] Zijlstra, J, *Energy-efficient baking oven*, under development, Atag Kitchen Group, University of Technology Delft, TNO Institute of Industrial Technologies, Delft, 2000.
[8] Groot, M.I., e.o., *Country specific real life energy consumption ovens*, Save Working group domestic ovens, 2000.

Energy Use of Central Heating Pumps, Appliance Efficiency at the Component Level

Senior Engineer Nils Thorup

Grundfos Management, Research department, DK-8850 Bjerringbro

Summary. Because the energy consumption of one pump can be many thousands kWh per year the energy saving potential in the pump industry is often focused on larger pumps.
 Smaller pumps such as circulator pumps used in central heating or cooling systems, only consume a few hundred kWh per year each but all together the energy consumption is 116TWh. In European perspective the energy saving potential is significant with up to 19.400 GWh within the existing systems.
Life Cycle Cost (LCC) calculations, labelling e.g. can be tools to promote energy efficiency in the heating systems

1   Background

The heat or cooling demand in a building varies significantly all through the year. Consequently the flow rate of the system also varies. For uncontrolled circulators control valves within the system achieve the variable flow requirement .The load profile in flow during the heating system for a typical heating installation is as shown in Figure 1.

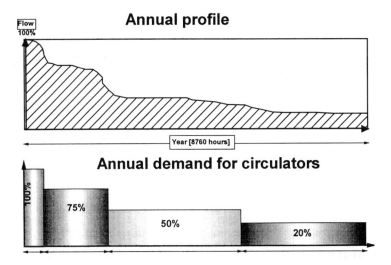

Figure 1: The demand for heating varies through the year, and, therefore, the demand of output of the pumps also varies.

The installed base of circulators in Europe are, approximately, 114 millions circulators. One hundred millions of small circulators (p<100W) and 14 millions of large Circulators (p>100W)

The circulators are installed either as a component in a system (for example wall hanged boiler) or the pump can be installed as a "stand alone" product. Until a few years ago, the state of the art of circulators was non-controlled one-speed circulators.

These circulators were replaced of 3 or 4 manually set speed circulators. The problems with pumps is, that the will often be installed in the highest speed, and the setting will not be changed who is necessary to reach the energy saving potential.

Specifically for a circulator as a stand-alone product with on/off regulations of the system, the unnecessary energy consumption can by far exceed the energy, which is needed to provide the output.

An external controller, which gives the capability of adapting the output from the circulator to the system demand, sometimes controlled larger circulators.

The new state of the art is "self adapting" pumps, with integrated speed control

Even though the circulator can be the most energy-consuming device in the household the owner is not aware of it.

The energy saving potentials are huge within self-adapting pumps, but to obtain savings, the industry has to convince the installers and the owners of the benefit and energy savings with variable speed driven circulators.

2   Self Adapting Circulators

The self-adapting circulators are based on two technologies, voltage regulations used in smaller pumps, and frequency controlled pumps used in larger pumps. Frequency controllers are normally used as external controllers.

The friction loss in the piping system depends on the flow. The required pump pressure decreases when the flow decreases. However, an uncontrolled (constant speed) circulator will increase the pressure when flow decreases. This results in unnecessary hydraulic losses at partial load. It is possible to reduce these unnecessary hydraulic losses by reducing the speed of the pump and thereby the pump pressure.

3   The European Perspective for Speed Controlled Circulators

Even that circulators in general are small pumps, with a low energy consumption per pump, the European perspective for saving energy with speed controlled circulators are huge.

The estimated installed base in Europe is given in Table I by dividing the circulators into two groups (small and large circulators).

In general voltage-controlled motors control the small circulators, and large circulators are controlled by a frequency-converter. The installation of circulators varies from one application to another and in the way a circulator is controlled or not controlled.. Therefore the energy saving potential is different from application to application. But an estimated average is 140 kWh for a small circulator and 1047 kWh for a large circulators. In some applications the energy saving potential is up to 60%

Table I: The estimated installed base of circulators in Europe, and the number, which can be speed controlled with energy saving as a result

.Circulators	Total numbers	Share which can be speed controlled
Small pumps p < 100 W	100 millions.	60%
Large pumps p > 100 W	14 millions.	75%

The price of electricity varies between the countries and from one user group to another. A large number of the circulators are installed in households who pay another price for electricity, than the industries.

An estimated consumer price has been set to 0,125 Euro per kWh.

Table II: The payments connected to change from "not controlled" circulators to "controlled" circulators, when the installed circulators have to be replaced.

Number of circulators which can be replaced with a speed controlled pump in millions		Additional price per unit [Euro]	Total investment [10^9 Euro]	Saved energy per year [GWh]	Unit price of 1 GWh [Euro]	Saved money per year [10^9 Euro]
Small	60	75	4,50	8.400	125.000	1,05
Large	10,5	500	5,25	11.000	125.000	1,38
Total			9,75			2,43

The traditional Pay Back calculation of the investment

The additional initial investment cost if replacing an old unregulated circulator with a new self-adapting circulator will for small circulator be, approx., 75 Euros and approx., 500 Euros for a large circulator.
This will give a total investment of 9.750 million Euros.
The annual saving potential will be set to 8.400 GWh for the smaller circulators is 11.000 GWh for the large circulators .

With an estimated price of at 125.000 Euros per GWh the annual saving potential is 2,34 billions Euros per Year.
If the traditional pay back calculation is used, the pay back time is, approximately 4 years. A pay back time, which often is decisive for not making the investment.

The Life Cycle Cost calculation
Using EUROPUMP LCC-guideline for calculating the Net Present value for the investment can be achieved.
The LCC can be made as a $\triangle$ -calculation, were only the payments who is different from
The sections, which not are filled out, represent the same cost for the two alternatives.

Input: **All cost in million Euro**

Initial more investment saving:	9.750
Installation and commissioning cost:	-
Energy price today (/kWh):	0,125
Routine maintenance/year:	-
Repair cost every 2nd year:	-
Operating cost/year:	-
Other yearly costs:	-
Down time cost/year:	-
Decommissioning cost:	-
Environmental cost: Disposal etc	-
Lifetime in years:	10
Average power saving of equipment in kW:	2,21
Operating hours/year:	8.760
Interest rate, %:	4,0%
Inflation rate %:	2,0%

Output

Present LCC-value:	- 12.032

There is a saving potential in Europe on 19.400 GWh/year by changing the circulators from uncontrolled pumps to self-adapting pumps in existing systems. The total Life Cycle Cost of this investment is –12.032 Euros or in another way- A Net Present Value on 12 billion Euros for the European consumers.

The marked situation:
The basic objective is to encourage the use of more efficient pumps within the EU. In a market driven economy that can mainly be done by stimulating end user to

buy more efficient pumps as well as stimulating the total process from pump manufacturer to end user to buy efficient pumps.

Today the end user do not take the decision of which type of pump, will be installed in the heading system. Those decisions are normally taken by the installers, consultants, contractors, e.g., who has no interest in the energy-saving during use.

An example for the buying process is given in the figure below.

Commercial Construction Customers & Partners

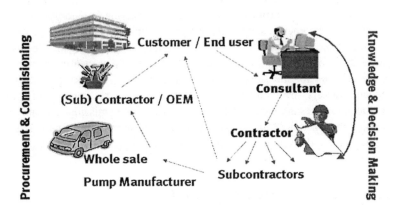

For the end user more necessary tools could give him information's on how to select a high efficient pump without spending much time. Among others can be mentioned:

- Minimum efficiency Standard for circulators
- Labelling Schemes
- General procurement advises and promoting the LCC- method

4   Conclusion

The energy saving potential in the Circulator marked in Europe is in the size of 19.400 GWh.

A circulator is a "unknown component" for many consumer so to reach the potential of 19.400 GWh/year a set of promotion tolls will be necessary.

Those tools can be labelling, Minimum efficiency Standard for circulators, Labelling Schemes, General procurement advises and promoting the LCC-method.

Natural Gas for Domestic Appliances in Austria. Future Perspectives and the Potential of Energy Efficient Technologies

Herbert Ritter and Georg Benke

E.V.A., the Austrian Energy Agency

Beginning with an overview of current trends on the Austrian heating market, the future impacts of energy efficient technologies – focussing on energy efficient gas appliances – on the energy supply are analysed. For space and water heating, gas condensing boilers are considered to be the future "state of the art" technology. Therefore gas condensing boilers are compared to other heating technologies which are relevant for the Austrian heating market. In the cooking sector natural gas-fuelled cookers are compared to electric ones. The comparisons of the different technologies are based on energy efficiency and emissions and are carried out by using the TEMIS Model[1], taking into account the different supply structures (mix of thermal, hydro and imports) in Austria depending on the season.

1 The Austrian Heating Market

Most Austrian households (925 200 households, which equals 28.7% of all Austrian households) use fuel oil as their primary energy source, directly followed by natural gas (see **Error! Reference source not found.**). In the last few years the Austrian natural gas supply system has been continuously expanded, mainly towards densely populated areas near large towns and major pipelines. In 1980, 352 000 households used natural gas as their primary energy source; this value increased to 882 300 (27.4% of all Austrian households) in 1999. This trend is supposed to go on in the next few years. In households natural gas is mainly used for heating purposes (space heating, water heating), followed by cooking purposes. In total, Austrian households used about 1.6 million m^3 of natural gas in the 1996/97 heating period, spending about 9.8 billion ATS. In addition about 500 million m^3 was used in the commercial sector. On 1 July 1996 an energy charge of 0.6 ATS/Nm^3 of natural gas was introduced in Austria. In addition, 20% value added tax is levied on the total amount.

In 1999 wood was used as a primary energy source by 521 200 Austrian households (16.2%). District heating is now used by 488 200 households (15.2%),

---

[1] Total Emission Model for Integrated Systems; the TEMIS was developed by the German Oeko-Institute

which reflects the strong increase in the use of this energy source over the last 20 years. For 286 000 Austrian households (8.9%) electricity is the primary energy source, followed by coal, which is used by 120 200 households (3.7%). The importance of coal for households has decreased over the last few years. This trend is still going on.

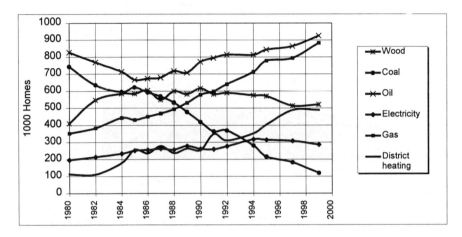

Figure 1: Number of Austrian households supplied with different energy sources (1980-1999), /Lit 2, Lit 3/

Figure 2 illustrates the energy consumption for heating purposes (space and water) for Austrian households in the years 1990 to 1998 and gives a forecast till 2010. The general trend is that the total energy consumption for heating purposes will decrease over the next few years. In 1998, 226 000 TJ was used for heating purposes, and it is supposed that this value will decrease by about 16% to 190 736 TJ in 2010. The reasons for this are technological improvements in heating systems and the improved insulation of old and new buildings.

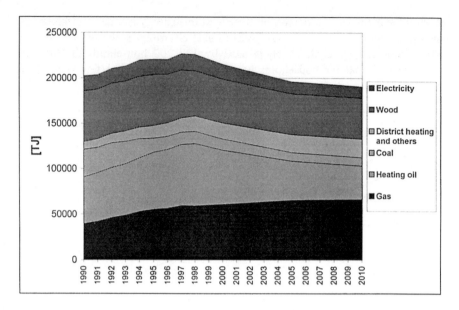

Figure 2: Scenario of energy consumption for heating purposes in Austrian households; /Lit 1/

Heating oil, which amounts to 68 450 TJ in 1998, will decrease to 36 800 TJ in 2010. This decrease will be substituted mainly by natural gas (59 200 TJ in 1998, which will increase to 64 600 TJ), and district heating (16 700TJ in 1998, which will increase to 20 700 TJ). So natural gas will be the most important energy source for heating purposes in Austria in the near future. Therefore it will be important to concentrate on natural gas appliances and to investigate their energy saving potential and their contribution to reducing greenhouse gases (GHG). Based on these results, supporting measures and instruments have to be developed in order to achieve environmental goals.

2   Heating with Natural Gas in Austria

In Austria, 882 300 homes are heated with natural gas. About 200 000 of them are heated with individual gas heaters, 392 300 with central heating systems serving one floor of a building, and 290 000 with central heating. Of the approx. 80 000 heating systems sold in Austria per year, about 48 000 are gas heating systems, 15 000 to 16 000 of which, in turn, are condensing boilers. About 75% of these condensing boilers are used in new buildings that have been optimised energy-wise for the use of such boilers. Experts estimate that there are currently approx. 130 000 condensing boilers on the market in Austria, which corresponds to a market share of approx. 15%. The break-through of condensing boilers in Austria only began around 1992.

Austria is thus rather at the end of the scale as far as the market penetration of condensing boilers is concerned. In Holland, for example, it is estimated that approx. 70% of gas boilers are condensing boilers, whereas in Switzerland the market share has already reached 50% as far as new devices are concerned. To understand this development, however, one has to consider the fact that approx. 2/3 of all gas boilers in Austria are located in Vienna alone, where the number of old buildings is extremely high. These old buildings, however, are to a large extent unsuitable for the use of condensing boilers. The existing building structure in Vienna is still largely dominated by brick chimneys, which require considerable effort and expense in order to be adapted for the use of condensing boilers. In addition, there is no subsidy available at all for the Viennese market, which is the largest of its kind in Austria. The market share in the field of new gas boilers is estimated at a mere 4% in this area.

For this reason, any reflection upon the Austrian situation has to take regional aspects – and thus the different subsidisation schemes – into account:

In Vorarlberg, condensing boilers already account for almost 90% of all new units, in Styria the respective figure is 80%, and in Upper Austria, although subsidies are rather low, a market share of a remarkable 70% has been reached. In Burgenland (no subsidisation) the figure is 40%, in Salzburg 30%, and in the Tyrol and Carinthia 40%. No data are available on the situation in Lower Austria.

3   Gas Condensing Boilers

The most frequent kind of gas condensing boiler in Austria is an 18 to 24 kW wall-mounted unit (followed by 7 to 12 kW units) with an integrated 90 litre hot-water tank; the main area of utilisation (75%) are newly-built detached houses.

During the burning process, heaters produce exhaust fumes which contain a considerable proportion of steam. In conventional burners these fumes are released into the environment through the chimney. In the case of gas condensing boilers, the remaining heat contained in the exhaust fumes is recovered via the condensation of the steam. This is effected by guiding the exhaust gases through heat exchangers. Theoretically, an additional energy content of 11% is made available in the case of natural gas[2]. The waste gas condenses, producing an "acidic" condensate (with a pH value of approx. 4.1 to 4.4). Since the condensate is acidic, additional standards have to be met by the chimney (it must be moisture-proof) and by the condensate collection device.

---

[2] There are also condensing boilers for heating oil. The additional energy content in these cases, however, is only 6%.

3.1 Efficiency

Because of their specific technological features, condensing boilers usually have an efficiency of up to 107%. Over the next few years, an increase to up to 109% can be expected owing to the use of matrix burners. Even if condensing boilers are in most cases advertised as providing an efficiency of 107%, this figure must be expected to be lower in practice. On the one hand, the 107% mentioned above refers to furnace efficiency, without taking stand-by losses etc. into consideration. Furthermore, a condensing boiler is often used in situations that do not guarantee optimum efficiency. This means that in most cases the heating system is not designed to provide the required low-temperature system. It has to be pointed out, however, that exhaust gases are cooled down and condensed in this case as well – though only partially. This means that, compared to conventional gas boilers, additional energy can be saved. According to expert opinion, given optimised utilisation, one can today assume an annual utilisation rate of approx. 94% for atmospheric burners and 106% for condensing boilers (laboratory values). The high annual utilisation rate of condensing boilers results from the fact that these boilers show even higher efficiency when operated at part load. In practice, however, the values are lower. For this reason, one has to expect efficiency rates of approx. 90% for atmospheric burners and of 98-102% for condensing boilers (in optimised new buildings).

3.2 Additional Costs

The additional costs for condensing boilers have been estimated at about ATS 10 000, or an additional 30% for small systems. Aside from these costs, additional costs for chimney renovation have to be taken into account if modernisation is planned (installation of a special tube). Additional costs are accrued above all by operating an exhaust ventilator. It is recommended that special emphasis be placed on dimensioning and on regulating operating hours, as these often constitute a weak spot in the system. In practice, it can be assumed that the average ventilator used requires approx. 40 Watt and that it is in operation for approx. 1,500 hours per year. This means that approx. 60 kWh is required per year. Operating costs for circulating pumps are also accrued for normal gas heating systems and therefore need not be mentioned separately here. Condensing boilers installed in new buildings are expected to pay for themselves in less than 5 years, due to current price reductions. In low-energy houses, however, it may take up to ten years owing to reduced energy demand.

3.3 Obstacles to Product Introduction

Besides additional investment costs for gas condensing boilers, experts have stated that further market penetration is sometimes prevented by professionals who often do not have the necessary know-how for the installation of condensing boilers in old buildings. A crucial detail in this context is the renovation of the existing chimney. In practice, even interested final customers are advised by some professionals not to install gas condensing boilers.

3.4 Subsidies

In the following, we will give an overview of the existing specific subsidies for gas condensing boilers. This overview takes into account only those means of subsidisation that exclusively offer an additional incentive for the utilisation of gas condensing boilers. Subsidies for condensing boilers are either granted by public institutions (*Länder*) or by the local gas supplier. So far, the legislator has not provided any rules or regulations that might be considered as a means of subsidising or promoting the increased utilisation of condensing boilers.

Table I: Overview of subsidies for gas condensing boilers in Austria

Land	Subsidies
Vienna	no subsidies for gas condensing boilers
Lower Austria	In Lower Austria, the *Land* grants a 15% subsidy for replacing boilers. The utilisation of a gas condensing boiler increases the maximum subsidy by ATS 5 000.
Upper Austria	An additional investment subsidy of ATS 2 040 is paid to new gas customers by *Ferngas Oberösterreich*, the regional gas supplier, for the use of a gas condensing boiler.
Burgenland	no subsidies for gas condensing boilers
Styria	In Styria, the utilisation of gas condensing boilers is strongly promoted by the gas supplier. If a condensing boiler is used, costs per m^3 gas are reduced by a total of 0.6 ATS. This equals a cost reduction of approx. 10% . In addition, *Ferngas Steiermark* grants an investment subsidy of ATS 10 000.
Salzburg	The *Land* of Salzburg provides a powerful subsidising instrument for gas condensing boilers. Since 1996, the only gas boilers subsidised by the *Land* in the course of renovation activities have been condensing boilers. As of March 2000, this regulation will also be applicable to new buildings. From January 2000, the renovation subsidy will be increased from ATS 6 000 to ATS 9 000 for every boiler replaced. In addition, a subsidy of ATS 5 000 is granted by *Stadtwerke Salzburg*, the local utility from the city Salzburg.

Carinthia	In the service area of *KELAG*, there is an investment subsidy that is dependent on the boiler rating. This subsidy is ATS 3 600 for devices of up to 50 kW and increases to ATS 9 600 for devices of more than 255 kW.
Tyrol	The subsidisation system of the Land provides extra plus points for the utilisation of gas condensing boilers in new buildings. These points may increase the amount of the subsidy granted. A specific subsidy of ATS 5 000 is available from *TIGAS*, the regional gas supplier. In the service area of *Innsbrucker Stadtwerke*, a considerable proportion of the basic annual charge (approx. ATS 1 600 per year) is waived if condensing boilers are used.
Vorarlberg	In Vorarlberg, a priority campaign for boiler replacement was carried out in 1998 and 1999. The only boilers subsidised during this campaign were condensing boilers. In these cases, subsidies of up to ATS 15 000 were granted by the Land, while an additional ATS 7 000 was provided by the gas supplier. A subsidy of yet another ATS 1,500 was given if chimneys were renovated at the same time.

4   Emissions and Energy Requirement

All calculations of emissions and of the so-called cumulated energy requirement (CER) are done by using the TEMIS (Total Emission Model of Integrated Systems) 4.0 alpha model in combination with a specific Austrian data set. The TEMIS model maintains a comprehensive database on environmental characteristics of energy technologies and determines environmental impacts on a life cycle basis. It takes into account not only the direct impacts of a certain process or technology but also those of transport, fuel conversion, etc., as well as the material input for production and for setting up the required infrastructure. TEMIS analyses and compares airborne gases and greenhouse gas emissions as well as internal costs associated with investment in and operation of all kinds of energy technologies including their life cycles.

In the present analyses of Austrian heating systems and cooking appliances the focus is on environmental issues such as total CO_2 emissions and on the so-called cumulated energy requirement of the whole process chain. Cost aspects are not considered. In order to get an idea about the energy intensity of a process, the CER is determined. The CER is defined as the sum of all primary energy inputs (including the energy that is needed to produce the materials) that are needed to deliver a certain product or service. The CER takes into consideration the total supply chain and allows a comprehensive overall comparison between different options. The CER is automatically calculated by TEMIS.

5 What Can Be Achieved by Using Best Technology Heating Systems?

In the following chapter the impacts of the usage of best technology heating systems are investigated with regard to their overall GHG reduction potential and their CER. The reference year is 1998 (see chapter 1: The Austrian Heat Market), where a total of 226 000 TJ was used for heating purposes.

For every energy source (oil, gas, coal, wood, electric heating, district heating) a specific reference technology is defined which represents an average status quo of the heating systems installed nowadays (given in TEMIS). Based on this reference case it was assumed that each "status quo" heating system will be substituted by the best heating technology available on the Austrian market (e.g. atmospheric gas heating systems are substituted by modern gas condensing boilers, etc.). The results calculated by using the TEMIS model are given in Figure 3 and Figure 4.

In Figure 3 the overall CO_2 equivalents for the whole energy supply chain are illustrated. The CO_2 equivalents are marked according to location so that it is possible to distinguish between Austrian and non-Austrian CO_2 equivalents. A technological change to best technology - especially for gas and wood heating systems - would lead to significant GHG reductions (-15.8% for gas and -24.2% for wood). A reduction of 5.2% for oil heating systems and of 10.3% for coal heating systems could be gained by using only the best available technology. Generally it can be said that almost all reductions are obtained on site, so they are directly related to Austria.

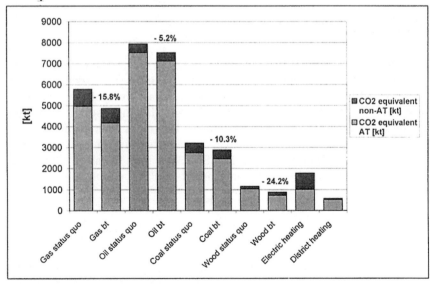

Figure 3: CO_2 equivalent for heating purposes

The biggest contribution to GHG reduction could be gained by using gas condensing boilers instead of atmospheric gas heating systems (minus 916.2 kt in absolute terms). The fact that gas will be the main energy source in future, as well as the high GHG reduction potential reflect the importance of well-aimed measures fostering the use of gas condensing boilers.

The different values for the CER, which is related to the various heating technologies and to the corresponding best available technologies, are illustrated in Figure 4.

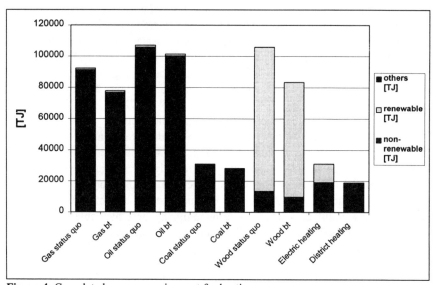

Figure 4: Cumulated energy requirement for heating purposes

For all "fossil fuel systems" (gas, oil, coal), by far the greatest part of the CER is of course caused by non-renewable energy sources. But it is remarkable that the CER for wood heating systems also shows a significant share of non-renewable sources. This is caused by the production of the wood heating system itself and by the supply and preparation of the fuel source. This means that the fuel source "wood" is only to a certain extent a GHG neutral energy source. For district heating systems a high proportion of the CER comes from so-called "other" fuel sources, where landfill gas, energy from incineration plants, etc. are subsumed.

6 What is More Environmentally Friendly for Cooking - Gas or Electricity?

The TEMIS model allows a comparison between different cooking technologies. In particular, the question whether electric or gas cookers are more energy efficient and/or environmentally friendly is of interest. For this reason different cooking technologies (electric, gas, wood) are compared, taking into account the different electricity supply structures (mix of thermal, hydro and imports in summer, in winter and as an annual average) representing the Austrian situation. The basis of this comparison is 1 kWh "final cooking energy".

The Austrian electricity supply structure is characterised by an extraordinarily high proportion of hydroelectric energy (about 60% on average over the year, about 44% during the winter season and about 87% during the summer season). The imports vary from about 9% in the summer season to about 24% in the winter season and come for the most part from Polish coal power plants.

Figure 5 summarises the results of the TEMIS calculations. 1 kWh from gas cookers causes about 453g CO_2 equivalent, which remains more or less constant over the year.

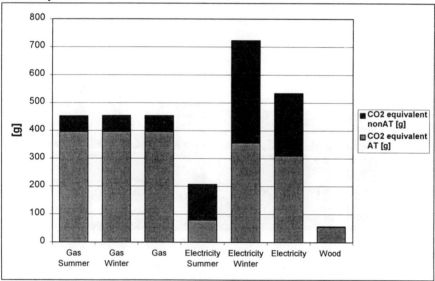

Figure 5: CO_2 equivalent for cooking

Very slight differences between the seasons are caused by the electricity that is needed to operate the gas supply system (compressors, pumps, etc.). 1 kWh from electric cookers causes about 206 g CO_2 equivalent during the summer season and 723 g during the winter season, which leads to about 535 g on average over the year. It is remarkable that around 50% of the CO_2 equivalent is caused by the imports, which only contribute a small amount to the Austrian electricity supply system. So from a general point of view, cooking with gas is better with regard to GHG than electric cooking. However, cooking with wood, which is still rather common in Austria, is the most environmentally friendly option (56 g CO_2 equivalent for 1 kWh).

A rather similar situation is reflected by calculating the CER. For 1 kWh cooking energy from gas cookers the CER is 1.9 kWh, fairly independent of the season. For 1 kWh electric cooking energy the CER amounts to 1.9 kWh in the summer season and reaches a maximum of about 3 kWh in the winter season. The CER for electric cooking (1 kWh) is 2.5 kWh when taking into account the average over the year of the Austrian electricity supply structure (see Figure 6). The CER for electric cooking is characterised by a high proportion of renewable sources (hydro, 62% annual average), whereas the CER for gas cooking is caused almost totally by fossil energy sources. The CER for wood cookers is about 2.5 kWh but comes for the most part from renewable sources. So if you consider the total CER, cooking with gas is the best option. But if you concentrate on the non-renewable share of the CER, wood cooking (0.1 kWh) is the best option, followed by electric cooking (1.6 kWh).

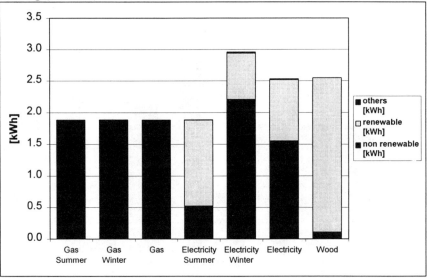

Figure 6: Cumulated energy requirement for cooking

7   Trends and Results

The Austrian natural gas market is currently experiencing a period of change which is particularly valid for the heating sector (space, water). The major trends of the Austrian heating market for the domestic sector are described below:

- The total energy consumption for heating purposes will decrease by about 16% by 2010. The reasons for this are technological improvements in heating systems and the improved insulation of old and new buildings.
- Over the next few years (by 2003), solid fossil fuels (e.g. coal) will almost disappear from the Austrian heating market in the domestic sector. The released market potential will be shared by other energy sources, with natural gas playing an important role.
- Natural gas will be the most important energy source for heating purposes in Austria in the near future.
- Due to the ongoing liberalisation, gas suppliers are experiencing increased cost pressure and will only penetrate this market if they meet economic efficiency criteria.
- The specific heat output of buildings is decreasing (to 5 kW) in new buildings, which will have an impact on heating technologies. Heating capacity will no longer depend on heating requirement but on short-term hot water need.
- Gas condensing boilers are already a "customer-accepted" technology and will become the standard technology in the Austrian heating market in the next few years.
- With regard to GHG, cooking with gas is better than electric cooking.
- For cooking purposes, electrical energy will remain the dominant energy source. Complicated installation and "lack of mobility" of natural gas-fired cookers are the main obstacles.
- Market instruments are necessary to accelerate the market introduction of new energy efficient technologies.

8 Literature

/Lit 1/ Austrian Institute for Economic Research; *Energieprognose bis zum Jahr 2010 mit einem disaggregiertem Strukturmodell*, Vienna; February 1996

/Lit 2/ E.V.A.; *The Market Situation for Condensing Boilers in Austria*; Vienna 1999

/Lit 3/ E.V.A.; *The Austrian heating supply structure*; internal working paper; Vienna; May 2000

/Lit 4/ E.V.A.; *Umsetzung der EU-Gasbinnenmarktrichtlinie in Österreich*; Final Report; Vienna; February 2000

/Lit 5/ Statistics Austria; *Energieverbrauch der Haushalte 1996/97 – Ergebnisse des Mikrozensus Juni 1996*; Vienna 1998

/Lit 6/ TTS-Institute; *Efficient Domestic Ovens*; Final Report of the SAVE II Project 4.1031/D/97-047, Helsinki 2000

Cogeneration for Energy Saving in Household Applications

Massimo Dentice d'Accadia, Maurizio Sasso and Sergio Sibilio

DETEC - Università degli Studi di Napoli Federico II

Abstract. During the last decade, small-scale combined heat and power systems are becoming a viable alternative to conventional power supply and boiler-based heating system in many types of applications. With regard to domestic sector, the use of combined Heat and Power generation on micro scale (< 15 kW electric output), *micro-CHP* for short, is currently relatively uncommon, but the market availability of gas-fuelled generating equipment, together with a significant number of current R&D projects, confirms the large potential for micro-CHP development, that was up to now limited to niche applications. In this paper attention is paid to the problems derived by the transfer of this technology, well known in the industrial field, to small-scale applications. One of the greatest obstacle is the match between thermal and electric energy outputs of the micro-CHP and the load profiles of end users. Consequently, some energy residential appliances such as dishwasher, clothes washer, water heater, will be considered in the paper, both in their traditional configuration (electricity driven) and in alternative more efficient configurations (thermal and electricity driven). Some energetic and economic analyses are presented. Finally an experimental apparatus set up to study the integration of a micro-CHP system with usual household appliances and heating equipment is shown.

1 Introduction

Many circumstances are expected to contribute greatly to the diffusion of micro-CHP systems. As well as reducing energy costs, gas fired CHP also reduces pollutant emissions compared to centrally generated electricity; this is a result of both the higher efficiency of CHP systems and the use, in urban regions, of a cleaner primary source such as natural gas, characterised by a low carbon/hydrogen ratio and negligible sulphur content.

However, there are a number of barriers to the introduction of micro-CHP in domestic and light commercial applications. Space heating is required for only a short period of the year, and even during the heating season heating systems often operate for short periods of the day, and the electrical demand profile is not always well matched to that of heat. For example, a residence may require maximum thermal output during a cold winter night while electrical demand may be minimal or maximum electrical output may be required for residential cooling during hot summer day when there is little or no need for thermal energy. Despite these obstacles, micro-CHP offers some particular advantages which emphasise its role as a potentially viable future energy supply option; the natural gas is readily available, most users already have their own hot water distribution system, in some countries it would appear attractive to consider selling surplus electricity and finally, due to the millions potential users, mass production techniques may be used to reduce system manufacturing costs.

Micro-CHP systems basically consist of an engine, a generator, heat exchangers, and electronic controls; heat from the water-cooled engine is combined with heat recovered from an exhaust heat exchanger to supply hot water for space heating and domestic hot water supply.

One basic element needed to create a marketable residential cogeneration system is mechanical feasibility, that must encompass all physical elements of the system, i.e. user interface, reliability, service intervals, acoustics, etc.. Ideally, the micro-CHP package should replace the boiler providing space and domestic hot water heating, so it should be as quiet and compact as existing heating boilers. Moreover, the micro-generator should require no more service visits than the boiler it replaces. It should also have high reliability and no need for expensive spare parts. However, the greatest obstacle is the match between thermal and electric energy outputs of the micro-CHP and the load profiles of end users. Consequently, some energy residential appliances such as dishwasher, clothes washer, water heater, will be considered in the paper, both in their traditional configuration (electricity driven) and in alternative more efficient configurations (thermal and electricity driven).

Another problem that affects the design of an effective micro-CHP/user configuration is that the type of application, residential and light commercial, doesn't allow the use of MCHP components and/or appliances with an high first cost. As a consequence, the design must be addressed to mass-produced components, also if that implies a decrease of the energetic efficiency.

There are many different technologies which could be used for micro-CHP applications, but the internal combustion (I.C.) engines packages represent the more widely diffuse units in the small-scale applications. Thus, they presently appear as the best candidates for these applications. The prime mover in a cogeneration system is a major consideration. Gas engines are widely available but it's difficult to achieve long service intervals; they are also noisy and prone to vibrations. Despite these shortcomings, today's available engines are relatively inexpensive and highly developed and continues to undergo significant levels of research and development, aimed at reducing maintenance, increasing efficiency and reducing emissions. The real problem for the lower end of cogeneration range consists in the limited choice of an appropriate liquid-cooled engine. In fact, small gas engines are usually cooled by air: this involves a lower heat recovery capability in CHP applications.

The opportunity of introducing an efficient power supply system, such as a cogenerator, in residential and light commercial market, derives by the energetic weight of these sectors in the overall energy balance of developed countries. In

Italy, commercial and residential sectors are responsible for about 30% of the global energy consumption. As for electric energy, in particular, figure 1 shows the distribution of consumption among agriculture, transport, industrial, commercial and residential sectors: the last two have been consuming from 1996 to 1998 about 40% of the total electric energy [ISTAT 1999]. From an economic point of view, in the monthly economic balance of an italian household, gas and electricity expenditures contribute for only 4% of the total. Therefore if the introduction of an efficient energy conversion system seems to be strategic for the national energy balance, it clashes with the low influence of energy dependent expenditures of the end users.

In spite of the above mentioned problems, many research institutes, manufacturers and gas companies, all over the world, are involved in R&D activities on micro-CHP and great attention is paid to the new technologies, such as: Stirling engines, thermoelectric and thermophotovoltaic systems, fuel cells [Bartholomeus P.H.J 1998, Bontempi 2000, Cirillo N.C. 1998, Dann R.G 1998, Dentice d'Accadia 1996, Dentice d'Accadia 1998].

2   Energetic Considerations

In order to evaluate the potential energy saving of a micro-cogenerator it is important to compare the primary energy used by the Micro-CHP with that used in conventional systems that supply the electric and thermal demands of the user. In figure 2 the energy saving, referred to the primary energy used by a boiler (PER = 0.800) and a power plant (PER = 0.374), is shown as a function of theoretical thermal and electric efficiencies of the Micro-CHP. For an electric efficiency of 20% a micro-CHP allows an energy saving only when its thermal efficiency is greater than 38%. It is important to underline that the efficiencies must be defined with reference to the energy actually used by the end user and not with reference to a control volume that includes the system only, as available by manufacturers data. If waste heat is fully recovered, using for micro-CHP performance suitable values for small scale electric output (η_{el}=0.20 and η_{th}=0.60), the equipment allows to save about 22% of the primary energy, compared to the conventional system.

In order to optimise the match between the micro-CHP and the thermal and electric users, an analysis of residential appliances has been performed. Attention has been pointed on domestic Electric Storage Water Heater, ESWH, domestic Washing Machines, WM, and finally on household DishWasher, DW basically for two reasons:

a)   they consume a significant part of household electricity. The energy efficiency of ESWH, WM and DW is analysed by manufacturers, research groups, national and international energy authorities that are finding the technologies to reach the optimum balance between energy consumption and overall working performances, [SAVE 1998, CECED 1997, CECED 1999, Mc Mahon J 1995];

b)   they allow to shift the energy requirements from electricity to thermal energy: in fact the energy supplied to ESWH, WM and DW is mainly used to produce hot water, usually by means of electric resistances. There are commercially-available equipment that are thermal and electrical driven and therefore can be linked to alternative energy suppliers such as boilers and/or micro-CHP.

Referring to topic (a), the total European Union electricity consumption by ESWH in 1997 was 87 TWh, and about 15% is due to household consumption. About 30% of the 142 millions EU's households use this equipment. The energy consumption of the estimated 120 millions WMs installed in EU amount to about 38 TWh, this is approximately 2% of the total EU electricity consumption. In Italy, ESWH, WM and DW are responsible for about 45% of the average annual household energy consumption and, considering the penetration of the these appliances, about 70% of the total energy consumption due to appliances. In the USA, WM and DW that meet the standards set in the National Energy Conservation Act consume about 30% of total annual energy consumption of typical domestic appliances.

Referring to topic (b), it can be noted that, for a ESWH, the net energy supplied, taking into account stand-by losses, is used to heat the water. Starting from cold water at 10 °C, in order to supply 100 litres of hot water at 60 °C, the average European citizen consumes 36 litres of hot water each day, that is about 6 kWh of energy. However, during a whole 24-hour period, average stand-by losses range from 1 to 2.5 kWh, depending on insulation thickness, thermal conductivity of insulation material, geometry of the ESWH. In a WM typical wash cycle (at 60 °C) about 85% of the total energy requirement is used to heat water and only 15% to other electric devices. For a bio cycle of a DW, 55/65 °C, only 10% of the energy supplied is not used to hot water production.

In order to evaluate the potential energy saving of thermal and electric activated appliances, market-available WMs and DWs are tested in both traditional and hybrid operation mode, that is, powered by electric network, and with electric input in addition to hot water feed at 45 °C. For the DW, no restraint are introduced on the temperature level of input hot water and every appliances can be linked to the hot water pipe instead of cold pipe to avoid water heating by electric resistances. For the WM, a control of the temperature must be introduced and therefore appliances are available with two pipes for water inputs. In Table I the results of some tests are summarised. DW and WM are considered for a typical wash cycle with three possible energy inputs:

c)   TS: electrically driven with a power plant efficiency of 37.4%;

d)   AS#1: driven by the electric network and by a boiler with 80.0% efficiency;

e)   AS#2: coupled with a micro-CHP with $\eta_{el}=0.20$ and $\eta_{th}=0.60$. The micro-CHP is designed to supply the electric requirements of the appliance, consequently a surplus of thermal energy is dissipated at every cycle.

With the DW, a great quantity, 47%, of the total energy input is supplied by the external thermal source, boiler or micro-CHP system. Consequently, the alternative systems, AS#1 and AS#2, perform better than the traditional one, TS.

The AS#1 has a PER of 0.500 and allows an energy saving of about +25%. The AS#2 is only slightly more efficient than the TS with equivalent PER for the two configurations.

For the above mentioned restraints on the temperature level of washing water also double-pipe WMs use intensively the electric resistance to warm water. Therefore, only a 17 % of the energy requirement is in the form of thermal input. Also in this comparison the AS#1 performs better, with an energy saving of 14%, whereas the micro-CHP dissipates a great amount of the available thermal energy (about 93% of its thermal output).

From the results of Table I it is evident that a micro-CHP coupled exclusively with a DW or with a WM performs significantly worse than its potential efficiency (PER of about 0.800). In order to optimise the coupling CHP-user, the heat output should bet be fully used. To this aim the surplus of heat output can be used to warm the ambient and/or to supply domestic hot water: for every cycle of DW and WM coupled to a micro-CHP, the energy needed to warm 80 litres of water from 20 °C to 60 °C is available. Table II resumes the results of the comparison among different energy systems used to supply the energy requirements related to a cycle of the DW, to a cycle of the WM and to the production of 80 litres of hot water at 60 °C. It is important to observe that the traditional electric appliances are very inefficient and a potential energy saving of about 54% is possible with an integrated micro-CHP based system.

The technical feasibility of matching the most widespread appliances to a micro-CHP is linked to the availability of suitable water reservoirs: nowadays, storage water heaters thermally and electrically activated are available on the market. In order to evaluate the energetic and economic feasibility of cogeneration in residential applications, an analysis of energy demand profiles of an house (extension 120 m^2) has been performed:

f)   from the annual electric-load duration profile it can be derived that the duration of peak loads is very short, hence the energy used at these level is actually quite small. In Italy, about a 70% of the domestic electric-energy consumption is found to occur below 1.4 kW, in agreement with [Dann R.G. 1998] (referred to the UK);

g)   heat demands depend on site location: three sites in the North, Middle and South of Italy have been considered and consequently three annual thermal load duration curves have been defined.

A computer simulation of a micro-CHP system was developed which used these energy demand profiles. The typical cogeneration operation modes have been analysed (thermal-match, electric-match). The results show that it is preferable to select the thermal-match operation mode: the cogenerator is sized to meet the thermal needs of the house and the electric energy in excess could be exported to other users through the public electric grid.

In figure 3 the results of the simulation are shown. The micro-CHP (PER=0.800) performs everywhere better than the traditional electric and thermal sources (PER=0.622) with an average annual energy saving of about 22%. Due to a longer annual operation time (North = 2611h/y, Middle = 2125 h/y, South = 1762 h/y) in the north the equipment allows to save about 21.103 MJ/y.

3 Economic Considerations

Many variables influence the economic performance of a cogeneration system, among these: energetic efficiency, time of operation, unit cost of electric energy (electricity rates), fuel cost (gas rate), maintenance costs, capital costs. The simple pay-back period, SPB, comparing the installed costs with the net annual savings for the cogeneration systems, has been used for a preliminary evaluation of the micro-CHP in residential application. In figure 4, the results of the economic analysis are reported: for two sites, North and Middle, the simple pay back has been evaluated as a function of the incremental capital cost of the cogeneration system with respect to the traditional one (costs are in Euro, 1 Euro = 1 USA Dollar) per electric output power and as a function of the Gas rate. A micro-CHP in the north site allows a payback period ranging among 0.9 to 4.8 years while in the middle site the SPB range is 1.6 – 11 years. If a SPB less or equal to 3 years is assumed, from data of Figure 4 it is possible to derive that for low heating loads (middle site), even for a small increment of the capital cost the economical feasibility of a micro-CHP must be supported by a low gas price, less than 3.25 Euro/m^3. With the same restraint on the SPB, a micro-CHP application can be advantageous also with an increase of the capital cost of about 1000 Euro per kW$_e$ with a longer heating season.

4 Conclusions

The large diffusion of gas-fuelled cogenerators in residential and light commercial applications strictly depends on the following key-points. From the point of view of users, efficiency, reliability and capital cost of the package as a whole are definitely needed to gain customer acceptance. In addition, the interaction of the micro-CHP with the building, heating and electrical systems, the economic operating strategy and electrical grid connection issues are other problems that will need to be addressed. From the point of view of gas utilities, this technology should be accompanied by service offered by the gas utility itself; the company could bear the cost of financing, installing and operating the generator, and try to have that the occupant does not have to worry about cogenerator operation and management. Finally, as soon as technological problems will be solved, the efforts to diffuse micro-CHP will have to join with a national policy considering the principle of incorporating environmental costs into energy-pricing and promoting the installation of high-efficiency and low-emission equipment.

At this moment, 71 million european houses are supplied with natural gas, and the European Commission recognises the advantages of cogeneration and have made the increased cogeneration capacity a key part of his CO_2 reduction strategy. In [Dentice d'Accadia 1998] a potential energy saving of about 200.000 tep/y, about 16% of the total energy requirements, has been evaluated if 500.000 micro-CHP units will replace the usual energy-supply equipment in Italy. In [Hu S.D. 1995] it has been estimated that the lower and the upper boundaries for cogeneration, usually using natural gas as fuel, in the residential and commercial sector would be 1% (7.95 GW) and 2% (15.9 GW) of the total USA generation capacity in 2000. The cogeneration potential in these sectors can be significant because of their large share of USA total energy use (37%) and the low temperature-level of their heat requirements.

Finally a R&D project is in progress for simulation and experimental validation of micro-CHPs for residential and light commercial applications. A research institute (DETEC Università degli Studi di Napoli Federico II), a standby unit industry (BRUNO Srl) and a gas utility (NAPOLETANAGAS Spa) are involved in the program. An experimental apparatus, has been designed and built to evaluate the performance of MicroCHP modules in different operation modes, to this aim the test facility is divided in two sections: the first contains the MCHP module (<5 kW electric output) and permits its performance evaluation, the second contains specific household systems in order to evaluate both their energetic efficiency and their matching with cogeneration units, figure 5. To find a better energy saving configuration, domestic appliances (DW, WM) and water heater are both electricity or thermal and electricity driven; furthermore pure thermal and electrical loads can be introduced too.

Acknowledgements

This work was supported by a grant from Regione Campania by means of Legge 41/1994. Many thanks to ing. A. Volpicelli and Sig. C. Riccardo of Merloni Elettrodomestici, Teverola (CE)-Italy, that supplied a thermo-electrical WM and their technical support.

5 References

[1] Bartholomeus P.H.J. et al, "Microcogeneration is coming into pratice", *Proceedings International Gas Research Conference*, pp. 1-10, San Diego, USA, 1998

[2] Bontempi A., "Nuove Tecnologie", *L'Installatore Italiano*, n. 60/00, p. 5, 2000

[3] CECED, "Voluntary commitment on reducing energy consumption of domestic washing machines", *Final report*, 1997

[4] CECED, "Voluntary commitment on reducing energy consumption of household dishwasher", *Final report*, 1999

[5]   Cirillo N.C., Cozzolino R., Nacca A., "Un prototipo di microcogeneratore da 3 kWe: realizzazione e risultati", *Proceedings 53° congresso ATI, pp.1235-1246*, 1998

[6]   Dann R.G., Parsons J.A., Richardson A.R "Microcogen – Cogeneration for the home", *Proceedings International Gas Reserch Conference*, pp. 12-22, San Diego, USA, 1998

[7]   Dentice d'Accadia M., Sasso M., Sibilio S., Vanoli R., "The future perspective of micro-CHP, *Proceedings III° International Conference Energy and Environment Towards the Year 2000*, Vol. I, pp. 407-414, Capri, 6-8 giugno 1996

[8]   Dentice d'Accadia M., Sasso M., Sibilio S., Vanoli R., "La microcogenerazione diffusa", *Proceedings Workshop ENEA Le innovazioni Tecnologiche e gli effetti sugli scenari futuri nell'industria del gas*, Roma, 1998

[9]   Hu S.D., *"Cogeneration"*, Reston Publishing Com, Inc, Reston Virginia, USA, 1995

[10] ISTAT, *Annuario Statistico Italiano*, 1999

[11] Mc Mahon J, Pickle S., "Appliance efficiency standards", CBS Newsletter, Spring 1995

[12] SAVE, "Analysis of energy efficiency of domestic storage water heaters", EU SAVE Project Final Report, 1998

Nomenclature

DW	Dishwasher
ESWH	Electric Storage Water Heater,
P	Power, kW
PER	Primary Energy Ratio, -
SPB	Simple Pay Back, y
WM	Washing Machines

Greek Symbol

η	efficiency, -

Subscript

el	referred to electric energy
in	referred to energy/power input
T	total
th	referred to thermal energy

Table I: Energy balance for electric and heat activated DishWasher and Washing Machine for different energy sources. The power plant efficiency is 0.374, the boiler efficiency is 0.800, the micro-CHP has η_{el}=0.20 and η_{th}=0.60.

Appliance	DW			WM		
Energy Source	ELECTRIC	ELECTRIC + BOILER	M-CHP	ELECTRIC	ELECTRIC + BOILER	M-CHP
RequiredElectric Energy [kJ]	5447	2859	2859	3460	2701	2701
Required Thermal Energy [kJ]	-	2571	2571	-	560	560
Total Required Energy [kJ]	5447	5430	5430	3460	3261	3261
Waste Thermal Energy [kJ]	-	-	6006	-	-	7543
Primary Energy [kJ]	14564	10857	14295	9251	7922	13505
Energy saving [%]	-	+25	+1.8	-	+14	-46
PER [-]	0.374	0.500	0.380	0.374	0.412	0.241
_PER [%]	-	+34	+1.6	-	+10	-36

Table II: Energy balance considering a cycle of the DW, a cycle of the WM and the production of 80 litres of hot water at 60 °C. The energy is supplied by a power plant (efficiency = 0.374), or by a power plant (efficiency = 0.374) and a boiler (efficiency = 0.800), or by a micro-CHP (η_{el}=0.20 and η_{th}=0.60).

Energy Source	ELECTRIC	ELECTRIC + BOILER	M-CHP
Total Required Energy [kJ]	22456	22240	22240
Primary Energy [kJ]	60043	35715	27800
Energy saving [%]	-	+40	+54
PER [-]	0.374	0.623	0.800
PER [%]	-	+67	+114

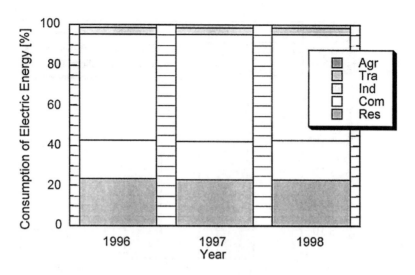

Figure 1: Distribution of electric energy consumption in Italy, 1996-1998, among Agricultural, Transportation, Industrial, Commercial and Residential users

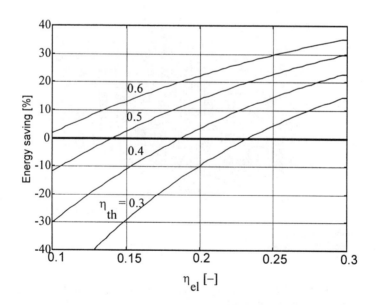

Figure 2: Primary energy saving using a micro cogenerator instead of a traditional equipment to supply thermal and electric energy.

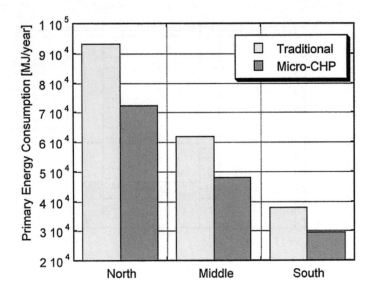

Figure 3: Annual primary energy consumption using a traditional equipment and a micro cogenerator by geographical region.

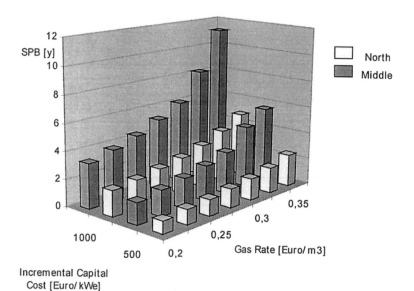

Figure 4: Simple Pay Back period vs gas rate and incremental micro-CHP first in two different geographical sites.

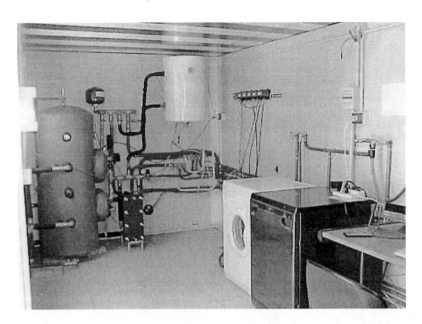

Figure 5: Part of the test facility with domestic appliances (DW, WM) and two water heaters (80 litres and 200 litres). All the equipments are both electricity or thermal and electricity driven.

An Efficient Induction Motor Vector Controller for Washing Machine Applications

Stefano Frattesi[1], Roberto Petrella[2] and Marco Tursini[2]

[1] SPES s.c.r.l
[2] Dept. of Electrical Engineering, University of L'Aquila (Italy)

Abstract. This paper presents an efficient controller for washing machine applications employing an on-line energy optimizer. The control scheme is based on a standard rotor flux field-orientation for three phase induction motors. The torque current command is given by the speed regulator whilst the flux current command is given by the energy optimiser. The proper level of flux is selected in order to minimise the motor input power for any speed/torque operating condition. The proposed system does not require any additional measurements with respect to a standard vector controller. Experimental investigations are presented based on a TMS320F240 DSP controller, an IGBT inverter and a four poles, 500 W induction motor for washing machine applications. The system performance and the efficiency improvements achieved by the proposed controller are presented and discussed.

1 Introduction

In the last few years the reduction in the cost of power electronics, the improvements in reliability of power devices and the availability of low-cost high performance micro-controllers have made it possible to consider drive options which could not have been economically applied to domestic appliances in the past. Among them the inverter-fed three-phase induction motor has been proved to be suited for washing machine applications due to possibility of variable speed operation, high maximum speed and good matching of the speed/torque characteristics. Moreover, as compared to the commutator motors, generally employed in front-loaded washing machines, three phase induction motors do not have maintenance and noise problems caused by the commutators brushes.

As far as the energy requirements, it is well known that the induction motor is a high efficiency machine when working close to its rated operating point (typical value of 75% for a standard 1.1kW line operated induction motor is achieved). Unfortunately variable speed/variable torque operations can lead to efficiency drop, which cannot be avoided by the improvement in the machine design. Nevertheless, energy efficiency improvements can be obtained by a proper control strategy of the power converter.

Several controllers have been proposed in literature, based on different approaches, such as Power Factor Control, Loss Model Based Control or Input Power Minimising (e.g. [Famouri 1991]). They all rely on the consideration that efficiency improvements can be obtained by controlling the balance between the copper and iron losses, i.e. the electromagnetic losses. This balance can be controlled by selecting the flux level in relation to the torque and speed of the machine.

Thus for any specific torque and speed there is a specific flux level which minimises the total losses and maximises the efficiency. Unfortunately there are no general analytical results that provide information on the best flux level for specific torque and speed conditions. The copper and iron losses depend on the saturation characteristics of the machine and hence the problem is strongly non-linear. The selection of the proper flux level in each individual case can be achieved by means of machine models including electromagnetic losses, but direct measurement of motor performance is probably the best guide as it assures independence with respect to model uncertainties and parameters variations.

In this paper an efficient induction motor vector controller is presented employing an on-line energy optimiser. After a brief recall of the efficiency calculation in induction machines, the control scheme based on a standard rotor flux field-orientation and the efficiency optimiser is presented. The torque current command is given by the speed regulator whilst the flux current command is given by the energy optimiser. The proper level of flux is selected in order to minimise the motor input power for any speed/torque operating condition.

Experimental investigations are presented based on the TMS320F240 DSP controller, an IGBT inverter and a four pole, 500 W induction motor for washing machine applications. Test results showing the system performance and the efficiency improvements achieved by the proposed controller are presented and discussed.

2 Field Oriented Control-Based Efficiency Controller

In this section, the theoretical background on efficiency optimisation in Field Oriented Control (FOC) induction motor drives is recalled.

A. Efficiency Determination

The motor efficiency can be determined by the general definition:

$$\eta = \frac{P_{out}}{P_{in}} \tag{1}$$

where:

$$P_{out} = \omega_r T_e \tag{2}$$

is the "output power", given by the product between the motor torque T_e and the rotor (mechanical) speed ω_r, and

$$P_{in} = P_{out} + P_{loss} \tag{3}$$

is the "input power", given by the sum of the output power and the internal losses P_{loss} which in turn includes the copper and iron losses.

Assuming the T-form steady-state equivalent circuit of the induction motor in Figure 1, the following impedances can be defined [Slemon 1989]:

$$\dot{Z}_0 = R_c \, // \, j\omega M \tag{4}$$

$$\dot{Z}_2 = R_r \frac{\omega}{\omega_2} + j\omega L_{\sigma r} \tag{5}$$

$$\dot{Z}_{eq} = R_s + j\omega L_{\sigma s} + Z_0 \, // \, Z_2 = R_{eq} + jX_{eq} \tag{6}$$

where the synchronous speed (ω), the slip speed (ω_2) and the rotor speed (ω_r) are related by the well known relation:

$$\omega = \omega_r + \omega_2. \tag{7}$$

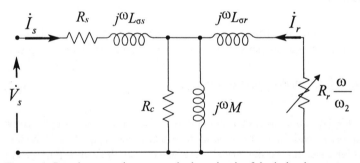

Figure 1: Per phase steady-state equivalent circuit of the induction motor.

The motor losses can be expressed as:

$$P_{loss} = 3R_{eq}I_s^2 \tag{8}$$

and the motor torque as:

$$T_e = \frac{3p}{\omega_2} R_r I_r^2 = \frac{3p}{\omega_2} R_r I_s^2 H_{12}^2 \tag{9}$$

being:

$$\dot{H}_{12} = \frac{\dot{Z}_0}{\dot{Z}_0 + \dot{Z}_2}. \tag{10}$$

The motor efficiency (1) is obtained by the previous relations as:

$$\eta = \frac{\omega_r \left(\dfrac{3p}{\omega_2} R_r H_{12}^2 \right)}{\omega_r \left(\dfrac{3p}{\omega_2} R_r H_{12}^2 \right) + 3 R_{eq}} \tag{11}$$

i.e.:

$$\eta = \eta \left(\omega_2, \omega_r, motor\ parameters \right). \tag{12}$$

One can notice that the motor efficiency is independent on the stator current amplitude.

Assuming constant speed (ω_r) operation, equation (12) can be derived with respect to the slip speed (ω_2) in order to achieve the maximum efficiency:

$$\left. \frac{\partial \eta}{\partial \omega_2} \right|_{\omega_r = const} = 0 \quad \rightarrow \quad \omega_{2\eta}. \tag{13}$$

Solution of (13) in a closed form is quite difficult. Nevertheless, the analysis of (12) allows to affirm that the slip speed that assures maximum efficiency ($\omega_{2\eta}$) will be a function:

$$\omega_{2\eta} = \omega_{2\eta} \left(\omega_r, motor\ parameters \right). \tag{14}$$

Figure 2 shows the typical efficiency plots versus the slip speed at different rotor speeds (from eq. (12)) and the corresponding maximum locus (eq. (14)). The last one, in a particular case, can be numerically evaluated once the motor parameters are known.

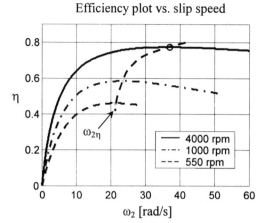

Figure 2: Efficiency plots at different speed of the induction motor.

B. Influence of Iron Loss

In the previous exposition, the equivalent resistance accounting for iron loss (R_c) has been considered as a constant. Really, this parameter is a quite complex function of synchronous and slip speed. Assuming the iron losses expressed as:

$$P_{loss,iron} = c_e \Phi_m^2 (\omega + \omega_2) + c_h \Phi_m^2 (\omega^2 + \omega_2^2) \qquad (15)$$

where c_e and c_h are the loss coefficients due to eddy currents and hysteresis respectively, and Φ_m the magnetising flux, the equivalent resistance can be derived as:

$$R_c = \frac{E^2}{\frac{1}{3}P_{loss,iron}} = \frac{3\omega^2}{c_e(\omega + \omega_2) + c_h(\omega^2 + \omega_2^2)} \qquad (16)$$

being $E = \omega\Phi_m$ the modulus of the voltage across the parallel of the magnetising inductance and the iron loss equivalent resistance (see Figure 1).

It is easy to notice that, once the loss coefficients are identified and the speed relation (7) is used to eliminate the synchronous speed, the motor efficiency and the slip speed at maximum efficiency can be expressed by similar relations than (12) and (14) respectively.

C. Rotor Flux Field Oriented Control

In the case of field orientation on the rotor flux, the impressed stator current splits into the flux producing component I_d and the torque producing component I_q, Figure 3.

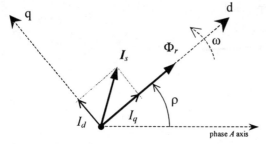

Figure 3: Rotor flux orientation.

At steady state operation, assuming to neglect the iron loss, the current components are related to the slip speed by the following relation [Novotny 1997]:

$$\omega_2 = \frac{R_r}{L_r}\frac{I_q}{I_d} \tag{17}$$

whilst the torque expression is given by:

$$T_e = 3p\frac{L_m^2}{L_r}I_d I_q . \tag{18}$$

Assuming to operate at maximum efficiency and field orientation, equations (14) and (17) are both true, from which one achieves:

$$\frac{I_q}{I_d} = \omega_{2\eta}(\omega_r, motor\ parameters)\frac{L_r}{R_r} . \tag{19}$$

This relation shows that, for any given operating speed, it exists a proper ratio between flux and torque current components which maximises the motor efficiency. This ratio depends on the motor parameters that affect (19).
Such kind of result was presented by Garcia in [Garcia 1994], who assumed some simplifying hypothesis (such as neglecting the leakage inductances) and obtained the following ratio:

$$\frac{I_q}{I_d} = K(\omega_r) = \sqrt{\frac{R_s(R_c + R_r) + (M\omega_r)^2}{R_s(R_c + R_r) + R_c R_r}} . \tag{20}$$

In the result section, it will be shown that this formula is in good agreement with the numerical calculation of eq. (19).
It should be noted that the torque-to-flux current ratio which optimises the efficiency does not correspond to the one which maximises the torque per stator ampere. In fact, the analysis of (18) demonstrates that this is achieved by setting:

$$\frac{I_q}{I_d} = 1.$$ (21)

3 Drive Scheme

The closed loop controller, which automatically assures maximum efficiency operation is an improved version of the one presented in [Kirschen 1987]. The scheme is based on a rotor field oriented controller where the flux (i_d^*) and torque (i_q^*) command currents are provided respectively by an *efficiency optimiser* and the speed regulator, Figure 4.

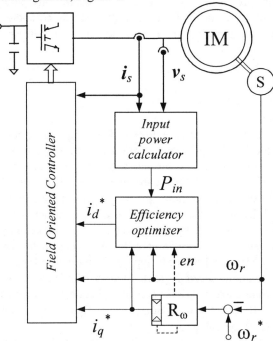

Figure 4: Block diagram of the field oriented controller with efficiency optimiser.

The system employs the current and voltage signals for on-line measurement of the motor input power. The efficiency controller calculates, as a first attempt, i_d^* as a function of i_q^* by using the relation (20) from Garcia. Then, the controller starts to make small step changes in the flux command current (i_d^*) by setting the value which results in minimum input power. Since the output power is held constant by the speed regulator (for a fixed load torque), the system will always seek the maximum efficiency.

The changes of the flux current command are enabled at constant speed operation only by the enable signal (*en*). Changes of (i_d^*) are made slowly with respect to the response time of the drive in order to avoid perturbation of the steady state operation. During transients, i.e. when a large speed error is detected, the enable signal is set to "false" and the efficiency optimisation is disabled. In this case, the flux command current is increased to a level that produces high torque per stator ampere to assist in responding to the speed error.

With respect to the solution presented in [Kirschen 1987], the use of the pre-calculated relation (20) from Garcia allows to speed-up the search of the maximum efficiency operating condition. With respect to the approach of the same Garcia in [Garcia 1994], the inclusion of an auto-tuning engine allows to take into account parameter variations and model inaccuracy.

One can notice that the proposed system does not require any additional measurements with respect to a standard vector controller. In case the stator voltage is not measured in a particular implementation, the corresponding command can be used.

4   Results

In this section experimental investigations are presented, aiming to give an idea of the performance obtainable by the proposed controller, especially in terms of efficiency gain. The rotor flux oriented controller with efficiency optimiser has been implemented on a TMS320F240 DSP controller. It has been tested on a three-phase four pole, 500 W induction motor for washing machine applications, rating 196 V-170 Hz, fed by an IGBT inverter. Tests at constant speed and constant torque have been carried out by using a commercial brake. Figure 5, Figure 6 and Figure 7 show the test results at different speeds in the constant torque region, respectively 550 and 1000 rpm (typical washing speeds) and 4000 rpm (which falls in the spinning speed region). Figures labelled (*a*) show the values of the flux and torque currents (I_q vs. I_d locus) obtained by means of the efficiency optimiser at different torque values, while figures (*b*) show the corresponding efficiency vs. torque plots.

Particularly, in figures (*a*) the basic line locus from Garcia and the maximum efficiency locus numerically achieved from (19) are draft for comparison.

One can notice that the locus achieved by the efficiency optimiser, which accounts for the actual input power, is a little different from the basic locus from Garcia, which is obtained on a purely theoretical base. Nevertheless, the global behaviour is confirmed, and the locus from Garcia, which in turns confirms to be a good approximation of the theoretical maximum efficiency locus, seems to be a good choice for the first attempt to efficiency optimisation. For the sake of comparison, in figures (*b*) the efficiency vs. torque plots achieved by the efficiency optimiser are draft together with the corresponding ones (experimental) obtained by the usual strategy which sets the flux current command to be a constant.

This comparison is also reported in Figure 8 where, for each of the tested speed, the efficiency gain is shown as a function of the load torque.

Finally Figure 9 resumes the results in terms of efficiency plots (*a*) and percentage efficiency gain (*b*). The maximum efficiency gain is obtained at low torque-low speed operation, where it reaches the 40% (550 rpm, 0.25 Nm in Figure 9 (*b*)).

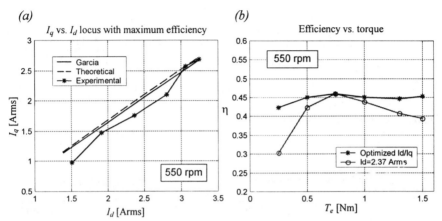

Figure 5: Test results at 550 rpm.

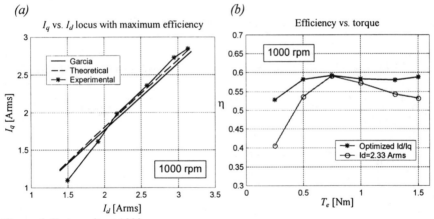

Figure 6: Test results at 1000 rpm.

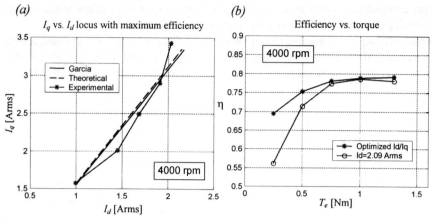

Figure 7: Test results at 4000 rpm.

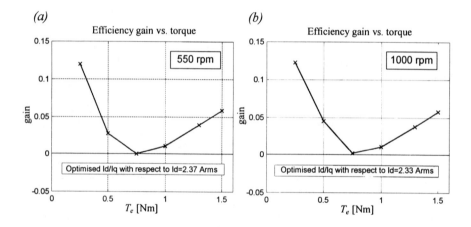

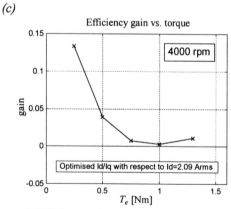

Figure 8: Efficiency gain.

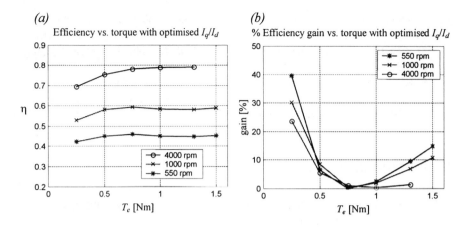

Figure 9: Efficiency and efficiency gain with optimisation.

5 Conclusions

In this paper an efficient induction motor vector controller is presented, using an on-line energy optimiser. The torque command current of the field oriented controller is given by the speed regulator, whilst the flux command current is given by the energy optimiser. The proper level of flux is selected in order to minimise the motor input power for any speed/torque operating condition. The proposed system does not require any additional measurements with respect to a standard vector controller. The maximum efficiency gain is obtained at low torque-low speed operation, where it reaches the 40%. In washing machine applications, where the induction motor works at very different speeds (such as washing and spinning speeds) and different load conditions, the optimisation of the motor efficiency by a proper command strategy represents a concrete possibility for energy saving.

6 References

[1] **P. Famouri**, J. J. Cathey, "Loss Minimization Control of an Induction Motor Drive," *IEEE Transactions on Industry Applications*, Vol. 27, No. 1, January/February 1991, pp. 32-37.

[2] **G. R. Slemon**, "Modelling of Induction Machines for Electric Drives," *IEEE Transactions on Industry Applications*, Vol. 25, No. 6, November/December 1989, pp. 1126-1131.

[3] **V. Novotny** and T. A. Lipo, "*Vector Control and Dynamics of AC Drives*," Oxfor Science Publications, 1997.

[4] **G. O. Garcia**, J. C. Mendes Luis, R. M. Stephan and E. H. Watanabe, "An Efficient Controller for an Adjustable Speed Induction Motor Drive," *IEEE Transactions on Industrial Electronics*, Vol. 41, No. 5, October 1994, pp. 533-539.

[5] **S. Kirschen**, D. V. Novotny and T. A. Lipo, "Optimal Efficiency Control of an Induction Motor Drive," *IEEE Transactions on Energy Conversion*, Vol. EC-2, No. 1, March 1987, pp. 70-76.

7 Nomenclature

x , X , $\dot{X}$	Space vector, stedy-state space vector, phasor for variable $x(t)$
$\overset{*}{}$	Commanded values
v_s , i_s	Space vectors of stator voltage and current
$\dot{V}_s$, $\dot{I}_s$, $\dot{I}_r$	Phasors of stator voltage, stator and rotor currents
Φ_r , Φ_m	Rotor, magnetising flux
ω_r , ω_s , ω_2	Rotor, synchronous and slip speed
T_e	Electromagnetic torque
d-q	Rotor flux fixed reference frame
I_d , I_q	d-q components of the stator current
R_s , R_r , R_c	Stator and rotor resistance, equivalent resistor for iron losses
$L_{\sigma s}$, $L_{\sigma r}$, M	Stator, rotor and magnetising inductance
E	Voltage across the magnetising inductance
$L_r = L_{\sigma r} + M$	Rotor inductance
η	Motor efficiency

An Improved Permanent Magnet Synchronous Motor Drive for Household Refrigerators

Francesco Parasiliti, Roberto Petrella and Marco Tursini

Department of Electrical Engineering – University of L'Aquila (Italy)

Abstract. Permanent Magnet Synchronous Motor (PMSM) is a practicable alternative to mains-supplied single-phase Induction Motor (IM) for the new-generation energy-saving household refrigerators. The requirements of a compressor system do not allow the use of mechanical sensors on the rotor shaft. Hence, speed and rotor position should be evaluated by means of suitable sensorless techniques.

In this paper, the performance of a sensorless field-oriented control of a PMSM is presented, with particular emphasis on efficiency. The sensorless scheme includes a Sliding Mode Observer and a Kalman Filter, the former used to detect the instantaneous value of the motor back-EMF, the latter to identify the back-EMF fundamental (from which the position information is derived) and the rotor speed.

The scheme is implemented on a TMS320F240 DSP controller. Test results carried on at different operating conditions confirm the proposed scheme is well suited for high-efficiency domestic refrigerators.

1   Introduction

In despite of its low cost, reliability and roughness, the mains-supplied asynchronous motor starts to become obsolete for the new-generation domestic refrigerators due to the energy-saving requirements. On the other hand, the massive production enables the compressor re-design: the present trend is to increase the cylinder displaced-volume and reduce the shaft speed. Then, the choice of motor is not constrained by the internal available room of existing compressor.

Due to the inherent high power factor and the absence of field losses, PMSMs are now taken into account for a wide range of energy saving applications including domestic appliance and automotive. All these aspects lead to consider the PMSM as practicable alternative to mains-supplied single-phase IM.

The requirements of a compressor system do not allow the use of mechanical sensors on the rotor shaft, then both speed and rotor position must be evaluated by means of suitable sensorless techniques. The choice of the estimation strategy affects the drive performance, the hardware requirements and cost. For these reasons the research efforts in this field are addressed to the development of solutions being accurate, robust but simple and possibly involving a minimum hardware.

Among the proposals in this field, the approaches using state observers seem to give the best performance. They require the use of a relatively accurate motor model, the measurement of the motor currents (system output) and the knowledge of the feeding voltages (system input), from which the rotor position and speed (state variables) are estimated.

In this paper, the authors present a sensorless field-oriented control of a PMSM. The scheme includes a Sliding Mode Observer and a Kalman Filter, the former used to detect the instantaneous value of the motor back-EMF, the latter to identify the back-EMF fundamental (from which the position information is derived) and the rotor speed. This solution allows improving the accuracy and robustness of position estimation thus achieving higher motor efficiency in a wide speed range. An open loop starting procedure has also been implemented given the estimated variables are not available near zero speed.

The proposed drive has been tested on a 141 W PMSM for compressor applications. The obtained results show the compliance of the proposed sensorless controller with household refrigerators efficiency requirements, its robustness with respect to load and mains voltage variations, as well as start-up capabilities.

2　New Generation of Motor-Drive Systems for Refrigerators

The recent standards and recommendations in the field of refrigerator applications have pointed out some basic demands for the new generation motor–drive systems:

➢ replacement of R14-type fluids with "green" types (CFC-free) forced to somewhat recover the efficiency loss of the thermodynamic cycle by increasing the energy-saving features of other system components (e.g., motor-drive, compressor, etc.);

➢ the market and European Commission requirements and recommendations about energy-saving have been asking for smarter management of thermodynamic cycles;

➢ recommendations about a "more clean" electrical environment entail new efforts to contain the mains-side transients and harmonics.

In this situation, despite of its low cost, reliability and roughness, the classical solution employing mains-supplied IMs starts to become obsolete. In fact, from the motor–drive point of view, these demands imply to focus on electronically controlled high–efficiency motors, such as PMSMs.

Nevertheless, depending on the application, two solutions seem to promise the best overall performance: Variable Speed Drives (VSDs) with IMs and Rated Speed Drives (RSDs) with PMSMs.

2.1 Variable Speed Drives

Variable Speed Drives are usually preferred in <u>commercial refrigerator applications</u>. In this field, the motor-drives power generally is above 0.5 kW and high COPs are met by switching the thermal circuit branches. VSDs allow adapting the operating speed to the variations of the thermal circuit so as to achieve the best system efficiency.

As regards to the motor choice, the following performance and constraints must be considered:

➢ due to the high torque required by the compressor during variable speed operation ($\approx$1:2 typical speed range), very-high torque-vs.-volume (Nm/vol.) and torque-vs.-current (Nm/A) figures are wished;

➢ the motor must usually be fitted into the internal available room of existing compressor. As for commercial refrigerators, the low production quantities generally do not justify investments to redesign the compressors;

➢ in case of PMSMs use, rare-earth permanent-magnets have to be forgotten as too expensive, and standard-ferrite is the unavoidable choice (up to today, copper/iron saving enabled by rare-earth magnets would not justify the costs increasing);

➢ sensorless operation of the whole motor-drive system is required.

Based on the foregoing statements (i.e., max. Nm/vol. and Nm/A figures, min. copper and iron amount, standard-ferrite magnets), the PMSM cannot easily be justified on commercial refrigerators. The general conclusion is that the controlled induction motor represents the best choice for such kind of application.

2.2 Rated Speed Drives

Rated Speed Drives are usually preferred in <u>domestic refrigerator applications</u> (fridge). In this area, for the sake of simplicity and cheapness, the thermal circuit is fixed. Contrary to the commercial refrigerators, the motor–drive system and the compressor play a more fundamental role for the cycle COP.

Thus, the whole system is optimised to give the best efficiency at the operating speed, while the motor–drive is on–off switched according to the thermostatic demand.

As for the motor choice, one can consider that:

➢ the massive production enables the compressor re-design (the present trend is to increase the cylinder displaced-volume and reduce the shaft speed). Then, the choice of the motor is not constrained by the internal available room of existing compressor.

238

> ➤ in domestic fridge, the motor-drives power is under 0.5 kW. PM synchronous motors, equipped with standard-ferrite magnets, can assure the required Nm/vol. and Nm/A figures, with efficiency greater than Induction Motors.

All these factors lead to consider the PMSM as practicable alternative to mains-supplied single-phase IM.

3   Sensorless Drive Scheme

The sensorless drive scheme for compressor application is shown in Figure 1. It is based on field-orientation in the synchronous dq frame aligned with the rotor flux, while the rotor position observer is arranged in the stationary $\alpha\beta$ frame.

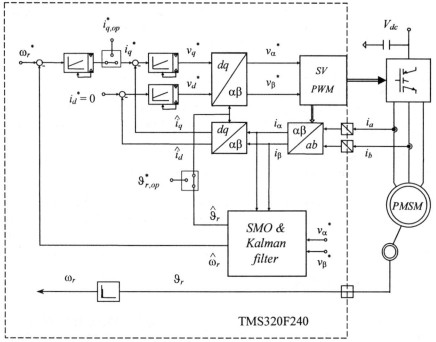

Figure 1: Sensorless vector control scheme for PMSM

The rotor position and speed are computed by means of an estimation of the induced back-EMF components in the Sliding Mode Observer & Kalman Filter block, using the instantaneous values of the motor phase currents and reference voltages. The sensorless control strategy reveals an excellent robustness to parameters deviations. Nevertheless, the accuracy of the observed variables at very low speed is poor. Then, an open-loop start-up procedure is implemented in order to move the motor up to a speed value where the observer gives good estimations.

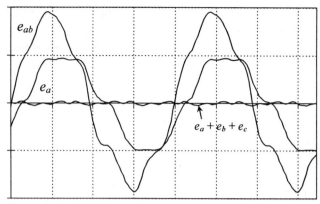

Figure 2: Back-EMFs waveforms

During the start-up only the current control loop is enabled. The motor starts following an "open-loop" (*op*) reference position $\vartheta^*_{r,op}$ whilst a current amplitude of fixed value $i^*_{q,op}$ is imposed to the motor.

3.1 Estimation Principle

The sensorless strategy assumes that the motor back-EMF could be non-sinusoidal. An example is given in Figure 2, which reports the phase (e_a), line-to-line (e_{ab}) and zero sequence component for the motor considered in the paper.
Considering the motor behaviour in the $\alpha\beta$ stator reference, one can evidence the 1st harmonic ($e_{\alpha1}$, $e_{\beta1}$) of the instantaneous motor back-EMF (e_α, e_β) as follows:

$$e_\alpha(\vartheta_r) = e_{\alpha1}(\vartheta_r) + \sum_{n=2}^{\infty} e_{\alpha n}(n\vartheta_r)$$

$$e_\beta(\vartheta_r) = e_{\beta1}(\vartheta_r) + \sum_{n=2}^{\infty} e_{\beta n}(n\vartheta_r)$$

(1)

where:

$$e_{\alpha1}(\vartheta_r) = -k_e\omega_r \, \sin\vartheta_r$$

$$e_{\beta1}(\vartheta_r) = k_e\omega_r \, \cos\vartheta_r$$

(2)

and k_e is the 1st harmonic back-EMF constant.

The rotor speed (ω_r) and position (ϑ_r) detection is achieved by three stages in cascade, Figure 3:
➤ the Sliding Mode Observer estimates the instantaneous $\alpha\beta$ components of the motor back-EMF, using the motor terminals currents and voltages measurement;

➢ the Kalman filter detects, from these signals, the instantaneous value of the 1st harmonics and the rotor speed information;

➢ finally, the rotor position is calculated from (2) by means of an inverse trigonometric function, e.g.:

$$\vartheta_r = arccos \frac{e_{\beta1}}{\sqrt{e_{\alpha1}^2 + e_{\beta1}^2}} \tag{3}$$

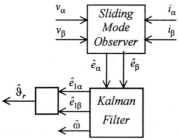

Figure 3: Rotor speed and position detection scheme

3.2 Sliding Mode Back-EMF Observer

The Sliding Mode Observer for the estimation of the back-EMF in a PMSM is shown in Figure 4, [Parasiliti 1997]. It refers to an extended (fourth order) αβ state model based on the motor voltage equations, where the back-EMF components (considered as disturbances) are modelled by the additional equations:

$$\dot{e}_{\alpha\beta} = 0 \tag{4}$$

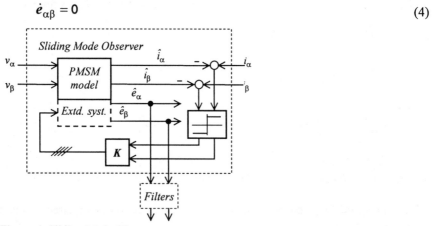

Figure 4: Sliding Mode Observer

From this assumption, the SMO assumes the following structure:

$$\hat{\dot{x}} = A\,\hat{x} + B\,u + K \cdot sgn\left(y - C\,\hat{x}\right) \tag{5}$$

where:

- $\hat{x} = [\hat{i}_{\alpha\beta}, \hat{e}_{\alpha\beta}]^{\mathrm{T}}$ vector of extended state variables (currents and back-EMFs);
- $y = i_{\alpha\beta}$ output vector (motor currents);
- $u = v_{\alpha\beta}$ input vector (motor voltages);
- A, B, C extended system matrices;
- K gain matrix.

The system matrices are constant, i.e. the extended system is linear. The main feature of the SMO (e.g. with respect to the classical Luenberger Observer for linear systems) is the use of the sign of the output error instead of the actual value. This gives the SMO an inherent "high gain" feature even by small feedback gains, which increases the robustness to parameter deviations.

As a drawback, the estimated variables are affected by a high frequency ripple, which prevents their use in the control. In [Parasiliti 1997] such a ripple is removed by means of simple smoothing filters (shown in Figure 4). Apart from the problems related to the filters' design (choice of the cut-off frequency, off-sets in digital implementation), this solution introduces a phase shift between the estimated and the actual back-EMFs. This leads to a not negligible angular error when the filtered back-EMF signals are used to calculate the rotor position.

Such a problem is overcome by the use of the Kalman filter in [Parasiliti 1999]. In fact, the Kalman filter removes both the ripple introduced by the SMO and the high order harmonics of the back-EMF estimates, without the introduction of relevant phase shift with respect to the fundamentals of the actual signals.

3.3 Kalman Filter

The Kalman filter is an optimal recursive algorithm that provides the minimum variance state estimation for a time-varying linear system. It is able to tolerate system modelling and measurement errors, which are considered as noise processes in the state estimation. Its extension to non-linear systems, the Extended Kalman Filter (EKF) does not assure the minimum variance estimate. Moreover, no convergence proof can be given. Nevertheless, the approach behaves well in most situations, as demonstrated by numerous applications.

In particular, the Kalman filter can be used as filter itself providing that a proper state model is arranged for the case problem [Labbate 2000]. This kind of approach is followed, in the present application, to extract the back-EMFs fundamental from the SMO estimates. In the hypothesis that the rotor speed $\square_r$ can be considered constant during the sampling period ΔT, a discrete time-equivalent state model is derived as follows:

$$\tilde{x}_{k+1} = f(\tilde{x}_k) + w_k \tag{6}$$

$$\tilde{y}_k = \tilde{C}\,\tilde{x}_k + r_k \tag{7}$$

where:

- $\tilde{x}_k = [\hat{e}_{\beta 1}, -\hat{e}_{\alpha 1}, \hat{\omega}_r]_k^T$ vector of state variables;

- $\tilde{y}_k = [\hat{e}_{\beta}, -\hat{e}_{\alpha}]_k^T$ vector of measured quantities;

- w_k, r_k modelling and measurement errors vectors;

$$f(\tilde{x}_k) = \begin{bmatrix} \tilde{x}_1 cos(\tilde{x}_3\,\Delta T) - \tilde{x}_2 sin(\tilde{x}_3\,\Delta T) \\ \tilde{x}_1 sin(\tilde{x}_3\,\Delta T) + \tilde{x}_2 cos(\tilde{x}_3\,\Delta T) \\ \tilde{x}_3 \end{bmatrix}_k \tag{8}$$

$$\tilde{C} = \begin{bmatrix} 1 & 0 & 0 \\ 0 & 1 & 0 \end{bmatrix} \tag{9}$$

By the comparison of (1) and (2) one can notice that the harmonics contents of the back-EMF is included in the measurement error in this schematisation. The system model for filtering is non-linear as regard to the state. Thus an EKF algorithm is arranged.

4   Results

The hardware used to implement the PMSM sensorless scheme is based on a Texas Instruments TMS320F240 Digital Signal Processor (DSP) controller and an integrated IGBT based Intelligent Power Module, Figure 5.

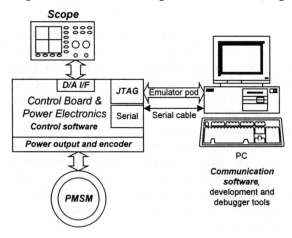

Figure 5: Experimental drive system

The main features of the control system include:
- inverter switching-period of 75□s (about 13.3 kHz);
- current control cycle: 150 □s;
- computation time of the whole control algorithm: less than 140 □s;
- computation time of the sliding mode observer: about 15 □s;
- computation time of the extended Kalman filter: about 90 □s.

During the development of the control program, an incremental encoder has also been used to measure the actual rotor position and speed.
The parameters of the tested motor are resumed in Table I.

Table I: Motor rated parameters

Power	141W
Frequency	135 Hz
Voltage	76.4 V rms
Current	1 A rms
Torque	0.5 Nm
Pole pairs	3
Stator resistance	7.3Ω
Synchronous inductance	0.045 H
Back-EMF constant	29.7 mV$_{rms}$ / rpm

Figure 6 shows the motor start-up and the speed transient from 0 to 2700 rpm at no-load. One can notice the fast convergence of the position estimation (Figure 6a) even during the first electrical period. The commutation to the closed loop control is visible roughly at the half of the second period, after which the reference position is not updated anymore. The corresponding speed transient is shown in Figure 6b. It takes about 5s to reach the rated speed, a time largely inside the specifications of the application.

The position and speed estimation errors , experienced by numerous tests, are less than 10 degrees (position) and 1% (speed) in all the operating conditions. Moreover the control system seems to be robust to mains voltage as well as motor parameters variations.

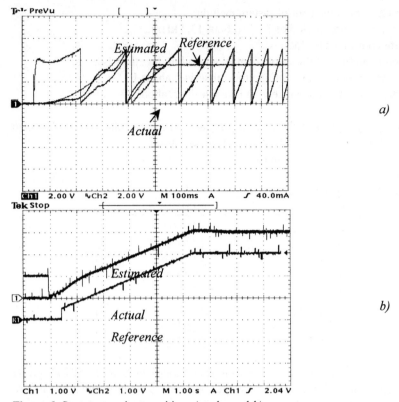

Figure 6: Start-up transient: position *a)* and speed *b)*

The system efficiency has been investigated by using a commercial brake which allows to set the operating load torque, while the speed is imposed by the drive system. The efficiency plots vs. speed and torque are shown in Figure 7 and Figure 8 respectively. As the motor copper losses are strictly related to the stator

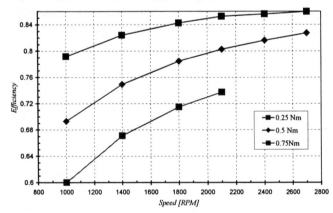

Figure 7: Efficiency vs. speed plots

current, i.e. the load torque, better efficiency is achieved at low load. For the same reason at constant load torque, the efficiency increases with the speed as the output power increases. Figure 8 can be useful to chose the operating condition of the considered motor which achieves the compliance to the efficiency specifications for compressor systems.

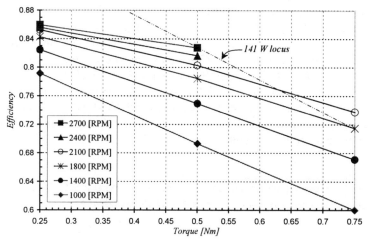

Figure 8: Efficiency vs. torque plots

5   Conclusions

A sensorless PMSM drive suitable for domestic (fridge) application has been presented and tested. The overall performances include:
- rated/max. shaft power: 141/240 W (steady-state);
- rated-vs.-peak instantaneous torque: 1:2;
- speed preciseness: rated speed ±1 %;
- motor-drive efficiency: up to 86 %;
- mains voltage range: 187÷264 Vrms;
- sensorless operation (i.e., 3 wires between motor and controller);
- compliance with European Standards.

From a rough estimate, a similar system (with optimised compressor and thermal circuit) could allow the following end-user performances:

- energy consumption: 0.6÷1.8 kWh per day;
- energy saving: 0.2÷0.5 kWh per day (compared to a standard compressor);
- cost amortisation-time: 2÷5 years (in Europe).

246

6 References

[1]   **F. Parasiliti**, R. Petrella, M. Tursini "Sensorless Speed Control of a PM Synchronous Motor by Sliding Mode Observer", *IEEE International Symposium on Industrial Electronics (ISIE'97)*, Vol.3, p.1106-1111, Guimaraes, Portugal, July 7-11, **1997**.
[2]   **F. Parasiliti**, R. Petrella, M. Tursini "Rotor Speed and Position Detection for PM Synchronous Motors Based on Sliding Mode Observer and Kalman Filter", *Proc. of the 8th European Conference on "Power Electronics and Applications" (EPE'99)*, Lausanne (Suisse) September 7-9, **1999**.
[3]   **M. Labbate**, R. Petrella, M. Tursini "Fixed point implementation of Kalman filtering for AC drives: a case study using TMS320F240xDSP ", *Proc. of the 3rd The 3rd European DSP Education & Research Conference*, Paris (France) September 20-21, **2000**.

This work has been carried out in the ambit of the project M.U.R.S.T. CO.FIN. '98 *"Permanent magnet synchronous motor drives for household refrigerators"*.

Voltage Regulator for Single-Phase Asynchronous Motor

Andrea Bianchi and Davide Martini

MagneTek S.p.A., Italy

1 Introduction

For single-phase asynchronous motors used in cold appliances (freezers, refrigerators, air conditioning, etc.) it is possible to reduce over 15% the energy consumption, by feeding the motor with a voltage level that minimizes its electrical losses, especially when it works with a torque significantly below its maximum value [1].

In fact, maximum efficiency voltage for single-phase asynchronous motors usually differs from the actual operating line voltage ($220V_{AC}$), for the following reasons:
- to guarantee motor start-up at the minimum operating voltage ($187V_{AC}$), the design of the motor is usually optimized close to this minimum value (start ability design criteria)
- the working torque of the motor is not constant during the whole on-cycle of the compressor, but varies from 100% at the start-up to about 75% at steady state.

As a result, motors used in compressors, work significantly below their design capacity and draw more power than is required for operation, and therefore a great deal of energy is wasted.

The most common low-cost, simple solution available up today to regulate an AC voltage was the scr/triac line-frequency phase control, whose major disadvantages are :
- step down regulation only
- high total harmonic distortion of the output voltage,
- low efficiency operation,
- high dv/dt that generates EMI noise and discharges inside the motor.

2 Energy Recovery Regulator

A new cost-effective Energy Recovery Regulator (E.R.R.) has been developed by MagneTek, to feed the motor always with the optimum voltage level that reduces electrical losses in the motor itself, and whose main features and advantages, with respect to the above-mentioned solution, are:

- A built-in microprocessor control, to sense the energy needs of the motor and to adjust automatically the voltage level to the motor for any variation of the working torque that may occur during the on-cycle of the compressor.
- Both step-down and step-up regulation
- Quasi-sinusoidal output, with low total harmonic distortion.
- No additional EMI noise generated, as the whole system works at the same frequency as the line voltage.
- Over 99% efficiency and Power Factor improvement.
- Possibility to add a Run Capacitor to further improve efficiency.
- A circuit to disconnect the PTC, after the compressor start-up is provided, in order to further reduce the power consumption and to guarantee the start ability after a brief power interruption.

3   Circuit Description

The Energy Recovery Regulator circuit application is shown in Figure 1

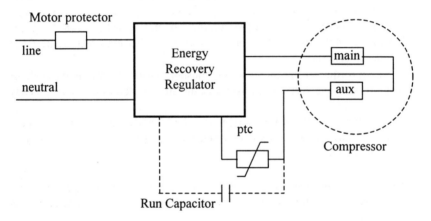

Figure 1: E.R.R. circuit application

The optimum ac voltage level to feed the motor is achieved by adding or subtracting a square wave voltage to the line voltage. The energy necessary to make the system work, is obtained by storing in a low voltage (max 100V) high value bulk capacitor (100μF to 1000μF) part of the reactive energy of the motor. The square wave voltage, generated by four low voltage mosfets, has the same frequency as the line voltage, but phase shifted, in order to recuperate the inductive energy of the motor.

Since the square wave voltage can be either added or subtracted to the line voltage, the ERR output can be regulated not only below the line voltage level, but also above it. This behavior could be particularly useful to improve the "start ability" of the motor when operating on those underdeveloped geographical areas where the line voltage could be much lower than 15% of the nominal value.

The ERR is based on the principle schematics shown in Figure 2:

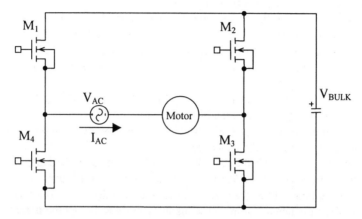

Figure 2: E.R.R. principle schematics

Mosfets M_1 and M_3 are driven simultaneously by the same signal while Mosfets M_2 and M_4 are driven simultaneously by a second signal, 180° phase displaced with respect to the first. Both signals are at the same frequency as the line voltage but properly phase displaced with respect to it.

Line Voltage, Motor current and Square wave, and V_{BULK} are shown in Figure 3

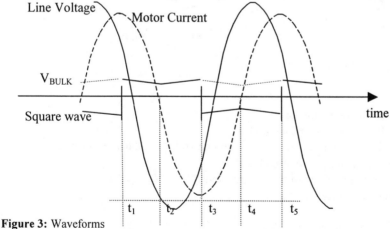

Figure 3: Waveforms

The motor current is lag with respect to the line voltage due to the inductive nature of the motor.

Within the time period $[t_1, t_2]$ $M_1=M_3=$ON and $M_2=M_4=$OFF and the motor current I_{MOT} discharge the energy stored in the bulk capacitor.

Within the time period $[t_2, t_3]$ $M_1=M_3=$ON and $M_2=M_4=$OFF and the current I_{MOT} change sign and charges the bulk capacitor.

If $(t_2 - t_1) < (t_3 - t_2)$, the voltage on the capacitor (V_{BULK}) increases because the charge stored within the time period $[t_2, t_3]$ is greater than what discharged during the time period $[t_1, t_2]$; vice versa, if $(t_2 - t_1) > (t_3 - t_2)$, the bulk capacitor discharges. It is understood that V_{BULK} cannot become negative because the diode internal to the mosfet.

Within the time period $[t_3, t_4]$ $M_1=M_3=$OFF and $M_2=M_4=$ON and the current I_{MOT} discharge the energy stored in the bulk capacitor.

Within the time period $[t_2, t_3]$ $M_1=M_3=$ON and $M_2=M_4=$OFF and the current I_{MOT} change sign and charges the capacitor

Under the hypotheses for the current I_{MOT} to be sinusoidal, the balance is reached when the driving signals are 90° phase displaced with respect to I_{MOT}

Since the rotating field of the motor is generated by the first harmonic of the voltage, the ERR mathematical model is made by considering all the parameters to be sinusoidal, and also the value of the bulk capacitor to be sufficiently high to assume its voltage across to be constant, for every cycle of the line voltage. Therefore, based on these hypotheses, the following vector diagram results :

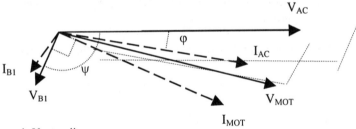

Figure 4: Vector diagram

Where :
V_{AC} is the line input voltage
V_{MOT} is the voltage to feed the motor
ψ is the phase displacement between the line voltage and the driving signal
φ is the impedance phase of motor.
V_{B1} is the first harmonic of the square wave ($V_{B1} = V_{BULK} \cdot 4/\pi$)
I_{B1} is the motor current due to V_{B1}

At the start up, $V_{BULK} = 0$, $V_{MOT} = V_{AC}$ and $I_{MOT} = I_{AC}$

The phase displacement between the driving signal and I_{MOT} is greater than 90° and the bulk capacitor starts charging; the vector V_{B1} start increasing its amplitude displacing the phase of V_{MOT} with respect to V_{AC} and consequently also the phase of I_{MOT} with respect to I_{AC}.

The balance is reached when the amplitude of the bulk voltage V_{BULK} (and therefore the vector V_{B1}) is such to make a 90° phase displacement between I_{MOT} and the driving signals.

Based on this concepts, it is possible to obtain the following relationships between the bulk voltage V_{BULK} and the motor voltage V_{MOT} as a function of ψ :

For $\quad \varphi+\pi/2 < \psi < \varphi+3\pi/2$

$$V_{BULK} = V_{AC} \cdot \frac{\pi}{4} \cdot \frac{\sin\left(\psi - \varphi - \frac{\pi}{2}\right)}{\cos\varphi}$$

$$V_{MOT} = \sqrt{V_{B1}^2 + V_{AC}^2 + 2\cdot V_{B1}\cdot V_{AC}\cdot\cos\psi}$$

For $\quad 0 < \psi < \varphi+\pi/2 \quad$ and $\quad \varphi+3\pi/2 < \psi < 2\pi$

$$V_{BULK} = 0$$
$$V_{MOT} = V_{AC}$$

V_{BULK} as a function of phase displacement ψ, is shown in Figure5.

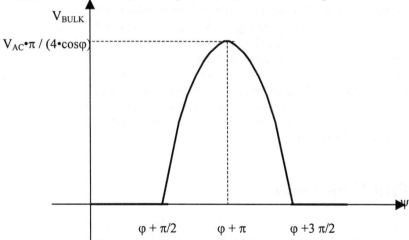

Figure 5: V_{BULK} as a function of ψ

$V_{MOT.}$ as a function of phase displacement ψ, is shown in Figure 6,

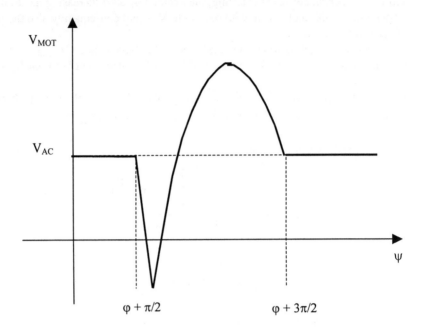

Figure 6: V_{MOT} as a function of ψ

It is understood from graphs of Figure 5 and 6, that V_{BULK} may range from zero to high voltage values even higher than V_{AC}. On the other hand, by working with a phase displacement ψ a little higher than $\varphi+\pi/2$, is possible to regulate V_{MOT} at a level lower than the V_{AC}, by keeping the V_{BULK} within acceptable values. Similarly, by working with phase displacement ψ a little lower than $3\varphi+\pi/2$, it is possible to regulate V_{MOT} at a level higher than the V_{AC}.

By assuming, for example $\cos \varphi=0.6$, we obtain for V_{MOT} about 0.8 time V_{AC}, with a V_{BULK} lower than 100V. This result is particularly important because it would allow to design and build the Energy Recovery Regulator by using low voltage semiconductors and passive components, which are known to be low cost, top performances and most reliable.

4   ERR Logic Control

As explained in the previous paragraph, the bulk voltage $V_{BULK,}$ and consequently the motor voltage V_{MOT}, are controlled by the phase displacement ψ between the line voltage V_{AC} and driving signal of the mosfet.

As the purpose of the E.R.R. is to regulate the motor voltage V_{MOT} to the optimum value in order to minimize the losses into the motor it is connected to, a double close loop control is implemented. The first to calculate that particular voltage level, and the second to maintain it.

Figure 7 shows the block schematics for the ERR logic control.

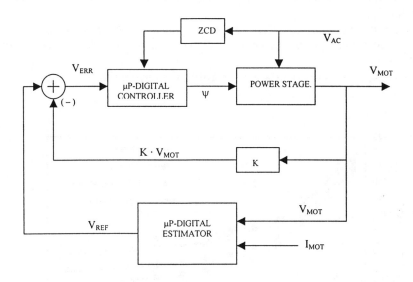

Figure 7: Block schematics for ERR logic control

Within the first close loop, the digital estimator, through its built-in microprocessor control and a powerful algorithm, senses the energy need of the motor, by monitoring instantaneously the motor voltage (V_{MOT}) and current (I_{MOT}), and calculates the maximum efficiency working point of the motor and provides the proper closed loop voltage reference (V_{REF}).

Within the second close loop, the motor voltage V_{MOT}, properly attenuated, is compared with a voltage reference V_{REF} to obtain the error voltage V_{ERR}. The digital controller, for a given V_{ERR}, figures out the phase displacement ψ with respect to V_{AC}, besides the right amplitude for the square wave to be added or subtracted to line voltage in the power stage.

The Zero Crossing Detector (ZCD) synchronizes the digital controller with the line voltage V_{AC} to assure the square wave to have the same frequency as V_{AC}

254

5 Experimental Results

The Energy Recovery Regulator as described in this paper has been tested on a number of different refrigerators and freezes using compressors in the range from 80 to 200W to verify the validity of the theory presented.

Two different control algorithms to regulate V_{MOT} were considered :

- Standard : the motor voltage V_{MOT} is kept fixed during the on-cycle of the compressor, to the value that provides the maximum efficiency of the motor under full torque.
- Smart: the motor voltage V_{MOT} changes continuously its value during the on-cycle of the compressor to automatically compensate the torque variations of the motor.

Power consumption curves, measured on a 80W freezer, are shown in Figure 8.

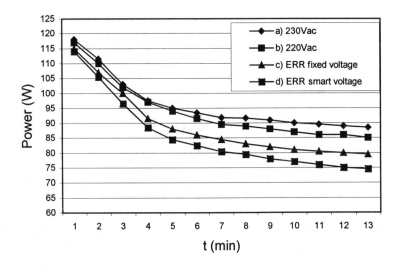

Figure 8: Power consumption curves

Power consumption curves a) and b) refer to measurements carried out without the ERR device, with an input line voltage V_{AC} of 230V and 220V, respectively. The ptc is connected between the main and auxiliary windings of the motor to improve the start ability, and remain connected during the whole on-cycle.

Power consumption curve c) refers to measurements carried out with the "standard" version of the ERR (fixed voltage, 195V, during the on-cycle) and shows an improvement of about 7% with respect to those obtained at 230V. The ptc is automatically disconnected after the start-up.

Power consumption curve d) refers instead to measurements carried out with the "smart" version of the ERR (voltage to the motor changed automatically to minimize the losses) and shows an improvement of 11% with respect to those obtained at 230V. The ptc is automatically disconnected after the start-up.

Similar power consumption measurements were also carried out at 240Vac (Great Britain line voltage) and improvement of 10% and 14,6% were achieved, for the standard and the smart versions respectively.

5 Conclusion

An innovative high efficiency circuit topology has been proposed for an AC Voltage Regulator to reduce the energy consumption of asynchronous motor used in refrigeration compressors.

Two major causes were known to be responsible for the degradation of the motor efficiency : the design start ability criteria and the variation of the working torque during the compressor on-cycle. Through a powerful built-in microprocessor control, two algorithms (standard and smart) have been implemented to reduce or minimize their effects, and therefore to measure their individual contributions.

Test results carried out on 80W freezer has proven that the best benefits of the Energy Recovery Regulator were achieved with smart version, due to variable nature of the working torque of compressors. In fact, the higher is the working torque variation during the on-cycle (combi unit, with a single compressor and valve) the higher is the energy reduction.

New energy efficiency standard (labeling) programs, will request cold appliance manufactures to further reduce the energy consumption of their equipment, and therefore always   more sophisticated cost-effective µP-based electronic controls should be introduced on the market, to achieve efficiency improvements of over 20% (at 220 V_{AC}) for compressor using single-phase asynchronous motors. With voltage regulation to the motor only, as described in this paper, we may not reach this goal and therefore the possibility of speed control of the motor should be also take into account. MagneTek is committed on this new program, whose experimental results would be shortly available.

9 Reference

[1] Frank J. Nola : *Save Power in AC Induction Motor* – NASA Technical Briefs, Summer , 1977 – pp 179, 180

Energy Use of U.S. Consumer Electronics at the End of the 20th Century

Karen Rosen and Alan Meier

Lawrence Berkeley National Laboratory

Abstract. The major consumer electronics in U.S. homes accounted for over 10% of U.S. residential electricity consumption in 1999, which is comparable to the electricity consumed by refrigerators or lighting. We attribute 3.6% to video products, 3.3% to home office equipment, and 1.8% to audio products. Televisions use more energy than any other single product category, but computer energy use now ranks second and is likely to continue growing. In all, consumer electronics consumed 110 TWh in the U.S. in 1999, over 60% of which was consumed while the products were not in use.

1 Introduction

Until recently, energy analysts have dismissed the electricity use of consumer electronics as an insignificant portion of U.S. residential energy use (Meier et al. 1992). This view has changed as the stock of these products increased, and as research revealed higher than expected energy consumption levels for many of these devices. Recent U.S. research has focused on the electricity use of specific consumer electronics products (Rosen & Meier 1999a; Rosen & Meier 1999b; Rosen et al. 2000). Studies in Europe (Siderius 1995; Meyer & Schaltegger AG 1999; Sidler 2000), Australia (Harrington 2000), New Zealand (EECA 1999) and Japan (Nakagami et al. 1997) have produced similar results. To date, however, there have been no comprehensive estimates for the energy use of the entire category of consumer home electronics – that is, video, audio, telephone, and computer-related products.

We report here a preliminary estimate of total electricity consumption for the U.S. consumer electronics end use in 1999. Products included in this study are listed in Table I.

Table I: Consumer Electronic Products Included in this Study

Video	Audio	Set-top Boxes	Telephony	Office
Analog TV	Component	Analog cable	Answering Device	Computer
Digital TV	stereo	Digital cable	Cordless phone	Printer
VCR	Compact stereo	Digital satellite	Cordless/Ans. Dev.	Fax/Copier
DVD player	Portable stereo	Game console	Cell phone charger	Peripherals
	Clock radio			

NOTES: TV = television; VCR = videocassette recorder; DVD = digital versatile disc. 'Analog TV' includes TV/VCR combos. 'Computer' includes monitors.

2 Data Gathering Approach

The ideal approach to estimating energy use would employ a long-term appliance-monitoring program involving a statistically representative group of homes. Findings of this program could then be extrapolated to the whole country. Studies in France (Sidler 2000) and New Zealand (EECA 1999) have used this approach; however, this approach is both expensive and time consuming. Instead, we resorted to a bottom-up approach combining direct measurements of individual devices and survey data. The approach consists of four major steps for each device:

1. Determine the principal operating modes
2. Estimate the time spent in each mode
3. Measure the power use of actual products in each mode
4. Estimate the number of units of that product in U.S. homes

These steps are briefly described below. Details of the methodology can be found elsewhere (Rosen & Meier 1999a; Rosen & Meier 1999b; Rosen et al. 2000).

2.1 Power Modes

Many consumer electronics operate in different modes (or states) during normal usage. Typical TVs have just two modes, but an audio or computer system may have a dozen. An accurate estimate of energy consumption requires an estimate of the energy use for each mode. This entails identifying common modes, then estimating average power levels and usage for each.

This study covers a wide variety of product types, so we adopted a much cruder classification of modes than would be appropriate for a single appliance. This simple classification is consistent with the quality of the other data used in this study. The operating modes assumed for this report are listed in Table II.

Table II: Common Consumer Electronic Modes

Mode	Description
Active	The unit is performing a requested service, e.g. record, play, talk, etc.
Charge	The battery charger provides current to the battery
Idle	The unit is on but not Active. This includes computer Sleep modes.
Standby	The unit is plugged in and appears off to the user
Disconnected	The unit is unplugged

Note that Idle may occur (1) after the user switches the unit on but before requesting a service, or (2) after the service ends (e.g. end of tape or automatic "Sleep" modes) but before the user switches the unit off. Some consumers never switch off units with idle modes.

2.2 Time Spent in each Mode

Of the three types of data needed for a bottom-up estimate of energy consumption, the usage profile is usually the most difficult to acquire. Conventional surveys are not helpful in estimating usage patterns for the Standby and Idle modes, which are critical for accurate estimates of unit energy use. Office equipment is particularly complicated because there are so many Standby and Idle modes and such a great range in the power consumption of the different modes.

Estimating usage patterns for this investigation involved a variety of data sources and usage assumptions that varied from product to product. In general, Active usage values were based on consumer surveys, while the remaining values were estimated from anecdotal data or best guesses.

2.3 Power Measurements

We compiled measurements of appliance power consumption from many sources, including stores, homes, and repair shops. Most measurements were performed by LBNL staff using the following procedure:
1. Identify the operational modes to measure
2. Plug the power cord of the unit into the power meter
3. Record the Standby power draw before switching the unit on
4. Record the power draw, in watts, of the unit in all other modes studied
5. Switch the unit off, and confirm the Standby mode power consumption

We used a true RMS wattmeter that was custom-built for measuring low power levels with high resolution (0.1W) and accuracy ($\pm0.5\%+0.1$W). This meter displays average power every second. Where readings fluctuated, we visually estimated the average power draw based on the readings. Readings typically fluctuate only one or two tenths of a watt.

2.4 Number of Units in U.S. Homes

We employed two methods in estimating the number of units in U.S. stock; one used ownership surveys and the other recent sales data. In most cases, the two methods resulted in similar stock estimates. Where results differed significantly, we chose the estimate that seemed more probable.

2.5 Energy Consumption

To calculate the average unit energy consumption (UEC) for each product type, we weighted the arithmetic mean of the measured power levels to reflect the average usage profile, and then multiplied by the number of hours per year. National energy consumption values were calculated as the product of UEC and the number of units nationwide.

3   Results: National Energy Use of Consumer Electronics

Using the methods described in the previous section, we collected over one thousand product measurements. Table III shows the average measured power levels in watts (W), estimated usage patterns, and average annual UEC values.

Table III: Usage, Power, and Unit Energy Consumption of Consumer Electronics

Product	Units Measured	Standby		Idle		Charge		Active		UEC
		W	Time	W	Time	W	Time	W	Time	kWh/yr
Analog TV	372	4.6	84%					75	16%	140
Digital TV	14	8.8	84%					177	16%	320
VCR	126	5.9	77%	13	24%			17	4%	74
DVD	18	4.1	72%	15	24%			17	4%	64
Component Stereo	119[b]	3.0	65%	43	16%			44	19%	150
Compact Stereo	19	9.8	72%	20	18%			22	10%	110
Portable Stereo	22	1.8	51%	4.9	13%			6.1	6%	17
Clock Radio	33	1.7	99%					2.0	2%	15
Analog Cable Box	42	11	78%					12	22%	95
Digital Cable Box	5	23	78%					23	22%	200
Satellite Receiver	31	16	78%					17	22%	150
Game Console	12	1.1	78%					7.9	22%	23
Answering Device	27	3.2	99%					3.6	1%	28
Cordless Phone	30	2.6	35%			3.6	60%	3.1	5%	28
Cordless/Ans. Dev.	23	1.1	35%			3.7	60%	3.1	5%	24
Cell Phone Charger	7	1.0	75%			5.0	5%			9
Computers	13	2.2	35%	80	50%			125	5%	410
Printers	4	3.6	39%	12.1	50%			361	1%	100
Fax/Copiers	5	4.0	0%	30	99%			100	1%	270
Peripherals	20	--	--	--	--			--	--	--

NOTE: Peripherals include a variety of products, so average values are not appropriate.

Estimates sensitive to our assumptions are discussed below. Estimates of stock and national energy consumption, by mode and total, are given in Table IV.

Table IV: Number of Units and National Energy Consumption of Consumer Electronics

Product	Units in 1999	Standby Energy	Idle Energy	Charge Energy	Active Energy	Total Product Energy Use	Share of U.S. Residential Electricity Use in 1999
	M	TWh/yr	TWh/yr	TWh/yr	TWh/yr	TWh/yr	%
Analog TV	220	7.3			23.5	30.8	2.7%
Digital TV	1	0.1			0.3	0.3	0.0%
VCR	130	5.2	3.7		0.7	9.6	0.9%
DVD	3	0.1	0.1		0.0	0.2	0.0%
Component Stereo	75	1.3	4.6		5.5	11.4	1.0%
Compact Stereo	50	3.1	1.6		0.9	5.6	0.5%
Portable Stereo	70	0.6	0.4		0.2	1.2	0.1%
Clock Radio	130	1.9			0.0	1.9	0.2%
Analog Cable Box	40	2.9			0.9	3.8	0.3%
Digital Cable Box	3	0.5			0.1	0.6	0.1%
Satellite Box	13	1.5			0.4	2.0	0.2%
Game Console	54	0.4	0.0		0.8	1.2	0.1%
Answering Device	77	2.1			0.0	2.1	0.2%
Cordless Phone	87	0.7		1.6	0.1	2.5	0.2%
Cordless/Ans.Dev.	35	0.1		0.7	0.0	0.8	0.1%
Cell Phone	70	0.5		0.2	0.0	0.6	0.1%
Computers	61	0.4	21.2		3.3	24.9	2.2%
Printers	74	0.9	3.9		2.3	7.2	0.6%
Fax/Copiers	10	0.0	2.7		0.1	2.8	0.2%
Peripherals	186	1.9	0.7		0.5	3.2	0.3%
Total		32	39	2.6	40	113	10%

4 Discussion

We briefly discuss the results below for each product and for the category as a whole. We also include discussions of confidence in our estimates.

4.1 Video Products

Televisions consume more energy than any other product in the consumer home electronics category. We are confident in the accuracy of this estimate, because it is based on over 300 measurements and extensive survey data. The digital TVs in our database draw twice the Active power of the analog sets because the average digital TV is about twice the size of the average analog TV. The apparent average Standby power of digital TVs may also be misleadingly high. Excluding three outliers, the average Standby power of the measured digital TVs was nearly two watts *lower* than that of analog sets measured.

VCRs and DVD players are nearly identical with respect to energy use of individual units, but because there are forty times more VCRs than DVDs in U.S. homes, overall energy consumption of VCRs is much higher. VCRs and DVDs spend 96% of time in Standby and Idle modes, where we estimate that 90% of energy use occurs. The distribution of usage within this 96% is the greatest source of uncertainty here, because surveys cannot easily collect this information.

4.2 Audio Products

Component stereos have the highest per-unit energy consumption in this product group, with compact stereos taking a close second. One source of uncertainty for the component stereo estimate is the extent to which they are connected to video systems. When connected, the components may be Active whenever the video system is Active. In homes where this occurs, Active energy use increases dramatically. Note also the large energy consumption that occurs in Standby/Idle modes for compact audio systems. Portable stereos and clock radios are among the lowest energy consumers in our study, with UEC values under 20 kWh/yr.

4.3 Set-Top Boxes

The power needs of cable and satellite boxes in Standby and Active modes are considerable – and nearly identical. As a result, the set-top boxes often consume more energy per year than the TVs they supplement. Digital cable boxes have particularly high UEC values. At 200 kWh/yr, these units use more than twice the energy of an analog box and 40% more energy than the average TV set.

We observed a wide range of cable and satellite set-top power requirements. Traditionally, service providers in the U.S. rent set-top boxes to customers. We measured set-top boxes from each of several service providers. Unfortunately, we could not determine which boxes were most frequently supplied to customers, so we can not be confident about the distribution of power needs in the set-top box stock. As a result, our estimates have a high degree of uncertainty. In any event, the market is changing so rapidly that greater certainty will be difficult to achieve.

Some of the newer set-top boxes not included in this study are remarkably power-hungry. For example, one new personal video recorder has no power switch and consumes 60 W at all times. Another new product, a combination high definition TV/satellite receiver box, used over 30 W in all modes. These devices contribute little to current U.S. residential electricity use because there are so few, but could have a noticeable effect over the next decade as they proliferate.

4.4 Telephony Products

Telephones, answering machines, and closely related devices use very little power – typically less than 5 watts regardless of the tasks they are performing. But because there are over 250 million in the U.S., total energy consumption adds up to 6 TWh/yr.

4.5 Office Equipment in the Home

Results for office equipment products are based on incomplete information. We are relatively confident in the Active values, but less confident about estimates for other modes because usage values were unavailable. As a result, we were forced to estimate usage for the Idle, Standby and Disconnected modes based solely on anecdotal evidence. We expect to improve our estimates in the near future, following further power measurement campaigns and usage surveys.

Computers (including monitors) were responsible for about 25 TWh in 1999, or a little over 2% of residential electricity use. Thus, the energy use of computers is now second only to TVs within the consumer home electronics category.

Printer energy use ranked second in the home office group at 7.2 TWh. We combined ink-jet and laser printers to form this group, but energy use for these two printer technologies differ markedly, with ink-jet printers using significantly less energy than laser printers.

Combination facsimile/copier machines accounted for less than 3 TWh in 1999. These products are becoming harder to distinguish because multi-function devices provide printing, copying, faxing, and scanning features. Future studies will probably combine all the paper-processing devices into one product group.

The average computer has three or four separate peripherals. Together, power speakers, scanners, network hubs, external disk drives, external modems and power strips consumed over 3 TWh of electricity in 1999. This estimate is probably low because we did not include all peripherals. For example, digital subscriber line (DSL) modems, which hardly existed when we began this study, have already achieved significant market penetration. We found one DSL modem setup, supplied by the local phone company, with a UEC of over 70 kWh/yr.

264

4.6 Total Electricity Use

The national electricity use of each of these products is relatively small: no single product accounts for more than 3% of residential electricity consumption. Combined, however, electricity use of the twenty different product groups investigated here exceeds 10%. This equals the electricity used by lighting or refrigerators (Office of Building Technology 1999).

Televisions still consume more electricity (31 TWh/yr) than any other consumer electronics device, but computer energy use (25 TWh/yr) is close behind and increasing faster.

There is considerable uncertainty associated with the energy use of each product. The largest source of uncertainty is the time operated in the Standby and Idle modes. Nevertheless, actual energy consumption is likely to be higher than our estimate because several consumer electronic products were not included.

The data presented in this report comprise a preliminary estimate of the electricity use for each of twenty distinct consumer electronics products. In the future, however, it will be more difficult to categorize these devices because the trend is towards combining the functions of several devices into a single product. For example, new set-top boxes incorporate features found in VCRs and personal computers. Components are often shared between video and audio applications. Telephone and fax services often rely on personal computers for all or part of their functionality. We believe future analyses will need to treat consumer home electronics in a more unified manner because it will be difficult, if not misleading, to estimate the energy use for individual products in this category.

5 Conclusions

We presented here preliminary estimates of electricity use by consumer electronics in the U.S. residential sector. Consumer electronics now account for at least 10% of residential electricity use. This is similar to the electricity consumed by all residential refrigerators or lighting. There is strong evidence (though not presented here) that energy consumption in nearly all home electronic product groups is still growing rapidly, especially in the home office product group.

This is the first time that energy consumption of home office equipment has been estimated in a procedure consistent with other consumer electronics devices. Its inclusion revealed that the national energy use of home computers is now second only to TVs. Energy use of computers could easily overtake televisions in the next decade, although the distinction may be completely blurred by then.

This study also reveals the complexities of estimating the energy use of devices with multiple operating modes. TVs represent an old-style appliance, where the majority of energy use occurs during the Active mode. Computer systems represent the new multi-mode device, which may have as many as a dozen significantly different power modes. While energy use estimates often include only Active modes, our study shows the importance of including the non-Active modes, which comprise nearly two-thirds of consumer electronics energy use.

In the future, many consumer electronics devices will be combined with others into a single product. We believe future analyses will need to treat consumer home electronics in a more unified manner because it will be difficult, if not misleading, to estimate the energy use for individual products.

Acknowledgement

This work was supported by the Assistant Secretary for Energy Efficiency and Renewable Energy, Building Technologies, of the U.S. Department of Energy under Contract No. DE-AC03-76SF00098.

6 References

[1] EECA. 1999. Energy Use in New Zealand Households, Report on the Year Three Analysis for the Household Energy End Use Project (HEEP). Wellington (New Zealand): Energy Efficiency and Conservation Authority.

[2] Harrington, Lloyd. 2000. Study of greenhouse gas emissions from the Australian residential building sector to 2010. Canberra: Prepared by Energy Efficiency Strategies, Inc.

[3] Meier, Alan, Leo Rainer and Steve Greenberg 1992. "The Miscellaneous Electrical Energy Use in Homes." *Energy - The International Journal* 17(5): 509-518.

[4] Meyer & Schaltegger AG. 1999. Bestimmung des Energieverbrauchs von Unterhaltungselectronikgeraeten, Buerogeraeten und Automaten in der Schweiz. St. Gallen (Switzerland): Meyer & Schaltegger AG.

[5] Nakagami, Hidetoshi, A. Tanaka, Chiharu Murakoshi, et al. 1997. "Standby Electricity Consumption in Japanese Houses" In the *Proceedings of* First International Conference on Energy Efficiency in Household Appliances. Florence (Italy): Association of Italian Energy Economics.

[6] Office of Building Technology. 1999. *BTS Core Data Book*. Washington, D.C.: U.S. Department of Energy.

[7]   Rosen, Karen and Alan Meier. 1999a. *Energy Use of Televisions and Videocassette Recorders in U.S. Homes*. LBNL-42393. Berkeley, California: Lawrence Berkeley National Laboratory.

[8]   Rosen, Karen and Alan Meier. 1999b. *Energy Use of Home Audio Products in the U.S. (Draft)*. LBNL-43468. Berkeley: Lawrence Berkeley National Laboratory.

[9]   Rosen, Karen and Alan Meier. 2000. *Energy Use of Set-top Boxes and Telephony Products in the U.S.* LBNL-45305. Berkeley: Lawrence Berkeley National Laboratory.

[10] Siderius, Hans-Paul. 1995. *Household Consumption of Electricity in the Netherlands*. Delft (Netherlands): Van Holsteijn en Kemna.

[11] Sidler, Olivier. 2000. *Campagne de mesures sur le fonctionnement en veille des appareils domestiques*. Report No. 99.07.092. Sophia-Antipolis (France): ADEME.

An Energy Efficiency Index for TVs

Hans Paul Siderius[1] and Robert C. Harrison[2]

[1] Van Holsteijn en Kemna BV
[2] Consumers' Association Research and Testing

Abstract: Since standby power consumption of TVs has been decreasing rapidly over the last years, the total energy consumption of a new TV is dominated by the consumption in the on-mode. It is well known that power consumption in the on-mode depends on various features of a TV, e.g. screen size, scan rate, luminance and sound level. Thus, contrary to standby power consumption, the on-mode power consumption of TVs can not be compared directly to determine which is the most efficient.

Earlier SAVE research showed that it is possible to model power consumption of TVs in the on-mode by a set of equations taking into account energy relevant features. This model was used to develop an energy efficiency index for TVs. This index enables comparison of TVs on efficiency regardless of the features.

The parameters of the model have been tuned by using a large set (102 samples) of measurement data of modern TVs. It is shown how the energy efficiency index can be used for an energy label for TVs and other labelling schemes, and what the impact is on energy consumption when setting certain minimum levels.

Furthermore, an outlook is provided to other applications of the energy efficieny index, e.g. for computer monitors.

Key words: consumer electronics, TVs, energy efficiency, energy labelling

1 Introduction

Televisions are amongst the most popular consumer electronic products in the world. In Europe almost every household has one TV and the number of second and third TVs is growing. Life cycle analysis shows that energy consumption of TVs in the use phase accounts for about 90 % of total energy consumption during the life time of a TV (Behrendt, et.al., 1998). Therefore it is worthwhile to investigate the energy consumption of TVs and possibilities to decrease energy consumption.

The energy consumption of a TV is the sum of the energy consumption in the various modes: off, standby and on. Standby power consumption of TVs has been decreasing rapidly over the last years: from 7.5 W in 1995 to 3.7 W in 1999 (average new TV, figures based on report on EACEM Voluntary Agreement). Thus, the energy consumption of a new TV is dominated by the consumption in the on-mode.

268

Energy consumption of TVs in the on-mode is the result of power consumption times the time the TV is in the on-mode. Since it is not feasible to try to reduce the time people are watching TV, efforts have to concentrate on reducing power consumption in the on-mode. However, it is well known that power consumption in the on-mode depends on various features of a TV, e.g. screen size, scan rate, luminance and sound level. Thus contrary to standby power consumption, the on-mode power consumption of TVs can not be compared directly to determine which is the most efficient. Therefore in this paper an energy efficiency index will be introduced. This index enables comparison of TVs regardless of the features.

This paper is structured as follows. First some attention is paid to the basis for the energy efficiency index: a model for the power consumption of a TV in the on-mode (section 2). Next in section 3, the concept of an energy efficiency index is presented. In section 4, some applications of an energy efficiency index and impacts on energy consumption in certain situations will be given. Eventually, section 5 contains conclusions and recommendations.

2 A Model of TV On-Mode Power Consumption

2.1 Introduction

In the SAVE project 'Analysis of Energy Consumption and Efficiency Potential for TVs in the on-mode' (EC-DGXVII 4.1031/D/97-023) a model for TV on-mode power consumption was developed. Since this model is the basis for the energy efficiency index, this section will provide a short overview. For further reference on the background, please see section 3.5 and Appendix V of Huenges Wajer and Siderius (1998).

The power consumption model is based on detailed power measurements on 5 models, complemented by an analysis of circuit diagrams of 14 other TVs. Measurements and analysis were carried out by Consumers' Association Research and Test Centre in the U.K.

Two types of the model were developed: a standard base case and a practical base case. In the standard base case, the luminance and sound level have been fixed to standard values (according to EN50301). In the practical base case the luminance level and sound level reflect levels used in practice. Since the use of the energy efficiency index is to compare TVs under standardized conditions, in this paper the standard base case model will be used.

2.2 A Power Consumption Model

The following equations describe the power consumption in the on-mode of the standard base case $P_{on, sb}$:

$$P_{on,sb} = \frac{P_{proc} + P_{sb,audio}}{\eta_{power\ supply}} + \frac{P_{tube} + P_{screendrive}}{\eta_{SMPS}} \qquad [2.1]$$

With

$$\eta_{power\ supply} = \eta_{SMPS} \times \eta_{power\ distribution} \qquad [2.2]$$

$\eta_{power\ supply}$	the overall efficiency of the power supply of the TV
η_{SMPS}	the efficiency of the (main) switched mode power supply
$\eta_{power\ distribution}$	the efficiency of the power distribution (power rails, voltage regulators)

$$P_{proc} = P_{basis} + P_{digital} \qquad [2.3]$$

P_{proc}	power consumption of small signal processing [W]
P_{basis}	power consumption of analogue small signal processing [W]
$P_{digital}$	power consumption of digital processing [W]
$P_{sb,\ audio}$	power consumption of large signal audio under standard conditions [W]

$$P_{tube} = \alpha_{tube} \times screenarea \qquad [2.4]$$

P_{tube}	power for illuminating the tube [W]
α_{tube}	tube coefficient [W/dm^2]
screen area	area of the screen [dm^2]

$$P_{screendrive} = \alpha_{screen} \times [0.80;\ 0.87]_{ws0;1} \times screen\ size + \Delta_{scanrate} \qquad [2.5]$$

with

$P_{screendrive}$	power consumption of the screen drive circuitry [W]
α_{screen}	coefficient of screenwidth [W/cm]
$[0.80;\ 0.87]_{ws0;1}$	0.80 in case of screen with 4:3 aspect ratio
	0.87 in case of widescreen (16:9 aspect ratio)
screen size	screen diagonal [cm]
$\Delta_{scanrate}$	impact of scanrate of 100 Hz [W]

The equations for the standard base case power consumption will be used to calculate a reference power consumption for a TV with certain features. For the calculation the following features are relevant:

- screen size (in cm)
- screen format: 4:3 or 16:9
- scan rate: 50 Hz or 100 Hz
- digital processing: yes or no

Since these features are also (important) items for consumers, data on these features is readily available in each product catalogue or brochure.

For the parameters in the equations, the following reference values will be used (see Table I). These reference values are based on the results of the detailed measurements. The values were confirmed by applying the model to a large set (102 samples) of measurement data of modern TVs. These TVs represented in 1999 in the U.K. a market share of more than 70%. All measurements were carried out under the same conditions (according to EN50301) by Consumers' Association Research and Test Centre (CART). The data set can be found on http://www.mtprog.com.

Table I: Reference values parameters

Parameter	Description	Reference value
$\eta_{\text{power supply}}$	overall efficiency of the power supply	0.75
η_{SMPS}	efficiency of the (main) switched mode power supply	0.825
P_{basis}	power consumption of analogue small signal processing	6 W
P_{digital}	power consumption of digital processing	9 W
$P_{\text{sb, audio}}$	power consumption of large signal audio	6 W
α_{tube}	tube coefficient	0.38 W/dm^2
α_{screen}	coefficient of screenwidth	0.75 W/cm
Δ_{scanrate}	impact of scanrate of 100 Hz	23 W

3   The Concept of an Energy Efficiency Index

Energy consumption of TVs varies with characteristics (features) of the TV. The concept of the energy efficiency index relates the energy consumption of the appliance for which the index is calculated to a reference energy consumption. This energy efficiency index E_i can be defined as:

$$E_i = \frac{E}{E_R}$$ [3.1]

where
E   is the energy consumption of the TV for which the energy efficiency
          index is calculated [kWh], based on a standard measurement method
E_R is the reference energy consumption [kWh]
E_i is the energy efficiency index

The reference energy consumption E_R is the energy consumption of an average
TV with the same (energy relevant) features as the TV for which the index is
calculated.
Thus, a TV with the same energy consumption as the reference TV has energy
effiency index equal to 1.00, a TV that is more efficient than the reference TV has
an energy efficiency index < 1.00, and a TV that is less efficient has an index >
1.00.

The energy consumption is calculated taking into account the various modes of the
TV: on, standby, off. Thus a duty cycle of e.g. 24 hours (1 day) is used in which
each mode has a representative share of the total time. Using a duty cycle, in stead
of only the on-mode consumption, does not only give better insight into the total
energy consumption of a TV, but also gives manufacturers more flexibility on
what items they want to improve the efficiency in order to reach certain targets.

The energy efficiency index is calculated in 3 steps.

Firstly the reference value for the TV is calculated for a duty cycle of 24 hours
which takes into account the various modes (standby, on, off) of the TV.

$$E_{24hrs,R} = \sum_{i=1}^{n} P_{i,R} \times t_{i,R}$$ [3.2]

And

$$\sum_{i=1}^{n} t_{i,R} = 24\,hours$$ [3.3]

The reference power consumption for the various modes is determined as follows.
The reference power consumption in the off-mode ($P_{off,R}$) is 0 W. The reference
power consumption in the standy mode ($P_{sb,R}$) is based upon the Figures of the
EACEM Voluntary Agreement results : 4 W. The reference power consumption in
the on-mode ($P_{on,R}$) is calculated with equations [2.1] to [2.5] using the reference
values of Table I, and taking into account the relevant features.

The period that the TV is in a certain mode ($t_{i,R}$) is derived from EU averages, based on Huenges Wajer and Siderius (1998): 4 hours per day in the on-mode and 20 hours per day in the standby-mode. These values will be used for calculating both the E and the E_R because the energy efficiency index should be independend from (specific) consumer behaviour. So,

$$E_{24hrs,R} = P_{on,R} \times 4 + 80 \tag{3.4}$$

Secondly, the actual power consumption for the various modes P_i of the TV is measured (according to standard measurement conditions, e.g. EN50301) and the energy consumption for the duty cycle E_{24hrs} is calculated:

$$E_{24hrs} = \sum_{i=1}^{n} P_i \times t_{i,R} \tag{3.5}$$

In the current situation, power consumption in standby-mode and in on-mode have to be measured. These values are already reported in current product catalogues of manufacturers.

Thirdly, the energy efficiency index is calculated:

$$E_i = \frac{E_{24hrs}}{E_{24hrs,R}} \tag{3.6}$$

Figure 1 shows the distribution of the values for the energy efficiency index of the 102 sets measured by CART. The average value is 0.92, the minimum value is 0.48, the maximum value is 1.68, and the standard deviation is 0.22.

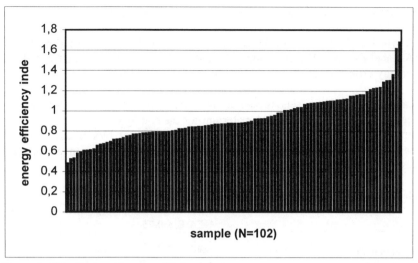

Figure 1 : Distribution of energy efficiency index

4 Applications of an Energy Efficiency Index

The following applications of an energy efficiency index are discussed in this section:
- comparative information on energy consumption
- energy quality mark
- A to G energy label
- minimum efficiency standards

4.1 Comparative Information on Energy Consumption

By publishing the energy efficiency index of a TV, the consumer gains insight into the relative efficiency of the appliance. At the moment product documentation only gives power consumption for the on-mode and standby-mode. From the energy efficiency index, the consumer can immediately conclude whether the TV has an average efficiency ($E_i \approx 1$), is more efficient than average ($E_i < 1$), or is less efficient than average ($E_i > 1$).

In databases on the Internet with product information there should be a possibility to rank TVs on their energy efficiency index with the most efficient on top.

4.2 Energy Quality Mark

Energy quality marks, e.g. the GEA-Label or Energy Star, at the moment only include the standby power consumption of TVs. Applying the energy efficiency index, an energy quality mark can easily include power consumption in other modes, i.e. the on-mode.

GEA will from 1-1-2002 use the energy efficiency index as the criterium for the GEA-Label for TVs. The criterium will be 0.75, i.e. a TV can get a GEA-Label if the energy efficiency index is 0.75 or less.

Already 20 % of the sets measured by CART (reflecting the 1999 market situation) complies with the GEA criterium for the energy efficiency index. Table II shows percentage of compliance for several categories of TVs. In each category at least 10 % of the models can comply, up to almost 30 % of the conventional (no widescreen, not digital) TVs.

Table II: Evaluation of GEA-Label energy efficiency index criterium

Category	percentage complying with GEA $E_i \leq 0.75$	base (n=)
all TVs	20%	102
conventional TVs	29%	48
widescreen TVs	11%	54
digital TVs	19%	26
small TVs (screen size: < 50 cm)	23%	26
medium TVs (screen size: 50 to 63 cm)	n=2	7
large TVs (screen size: > 63 cm)	17%	69

4.3 EU Energy Label (A to G)

One of the disadvantages of a voluntary energy quality mark is that it is gives no information on appliances that do not carry the label. For these appliances the consumer does not know whether the appliance does not meet the label criteria or whether the manufacturer has not applied for the label. With a (mandatory) A to G energy label the relative efficiency is clearly indicated. The energy efficiency index can be very useful to establish an energy label with an A to G scale. In fact it uses the same methodology as is used for the energy label for dishwashers or washing machines.

Based on the distribution of the energy efficiency index (Figure 1) the following category boundaries can be suggested (Table III). Table III also shows the percentage of appliances in each label category.

The impact of all TVs shifting a category upwards in efficiency (except for TVs in category A) can be estimated to improve the average efficiency index by 12 % : from 0.92 to 0.81.

Table III : Suggested EU energy label category boundaries and percentage of TVs in each category

Label category	Category boundaries	% TV in category (n=102)
A	$E_i \leq 0.64$	8.8
B	$0.64 < E_i \leq 0.76$	10.8
C	$0.76 < E_i \leq 0.88$	32.4
D	$0.88 < E_i \leq 1.00$	12.7
E	$1.00 < E_i \leq 1.12$	18.6
F	$1.12 < E_i \leq 1.24$	10.8
G	$E_i > 1.24$	5.9

4.4 Minimum Efficiency Standards

Minimum efficiency standards imply that all appliances on the market must comply with a certain minimum efficiency level. This minimum efficiency level can be implemented by a certain value of the energy efficiency index. Looking at Table III, $E_{i, MES} = 1.12$ (no TVs with F and G label on the market) or $E_{i, MES} = 1.00$ (no TVs with E, F and G label on the market) could be suggested as minimum efficiency levels.

If appliances with an energy efficiency index higher than the minimum level would be improved just to the minimum level, the average efficiency index would decrease from 0.92 to 0.90 (with $E_{i, MES} = 1.12$), or from 0.92 to 0.86 (with $E_{i, MES} = 1.00$). If appliances with an energy efficiency index higher than the minimum level would be improved to the average efficiency index of the remaining products, the average efficiency index would decrease from 0.92 to 0.85 (with $E_{i, MES} = 1.12$), or from 0.92 to 0.79 (with $E_{i, MES} = 1.00$).

5 Conclusions and Recommendations

An energy efficiency index is a useful tool to compare energy efficiency of TVs, independent from the features of the TV. Basis for the energy efficiency index is a set of equations that model the power consumption of TVs in the on-mode. The complexity of calculations for an energy efficiency index for TVs in not higher than the energy efficiency index used for e.g. the EU energy label for refrigerators.

An energy effciency index can be used in several policy options: comparative information on energy consumption, energy quality mark, EU energy label (A to G) and minimum efficiency standards.

An energy efficiency index gives consumers an easy way to conclude whether a TV has an average efficiency ($E_i \approx 1$), is more efficient than average ($E_i < 1$), or is less efficient than average ($E_i > 1$).

An energy quality mark can be used to highlight those appliances that are very efficient, e.g. have an energy efficiency index of 0.75 or less.

A (mandatory) EU energy label for TVs with an A to G scale would give consumers comprehensive information on the energy efficiency and energy consumption of a TV, without limiting consumer choice on the market. Consumers can still choose a TV with a G-label if they like this model better. Figure 2 gives an impression of an EU energy label for TVs.

Energy	Television	
Manufacturer	XYZ	
Model	ABC	
Efficiënt 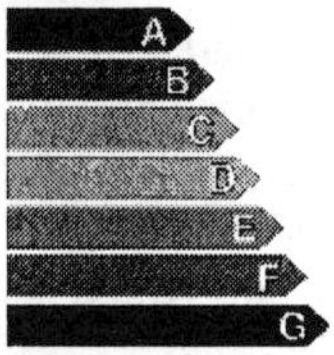 A B C D E F G Inefficiënt	B	
Energy consumption kWh per day (24 hours) based on EN50301	0,45	
Screen size (cm)	82	
Widescreen	yes	
Scan rate	100 Hz	

Figure 2: Impression of EU energy label for TVs

Given the research already carried out, and the availability of a standard measurement method (EN50301), it is expected that an EU energy label can be implemented relatively fast.

Experiences from the white goods industry show that an energy label is not only a good tool to improve the efficiency of the products on the market, but also guides product development. Thus, an energy label for TVs could also enhance the importance of energy efficiency in product developments of TVs. The following examples illustrate that this is an important aspect. Since several years research and development work for large screens (screen size > 1 m) is carried out. The first types, 42" ($\approx$ 1.07 m) plasma display panels, that are on the market now, do not only have a high price (about 10.000 to 15.000 EURO) but also have a high power consumption in the on-mode (about 300 W). This would result in an energy efficiency index of 1.75, which is higher than the maximum value resulting from the measurements. On the other hand, screens that apply LCD technology have a much lower power consumption, but still struggle with larger screen sizes. Experimental 15" ($\approx$ 38 cm) LCD screen TVs have a power consumption of about 35 W. This would result in an energy efficiency index of 0.51. Thus the energy efficiency index clearly indicates the impacts of different technologies and sets a target for future developments.

The energy efficiency index is not a static index; a regularly revision of the index is necessary. If the market accepts the challenge of an energy label by producing and buying ever more efficient TVs, in a few years most TVs will be in categories A, B and C. In that case, the values for the parameters need a careful review to reflect improvements in technology.
Furthermore, new standby modes, e.g. standby-active, are emerging. They can be easily incorporated in the duty cycle by measuring the power consumption and defining a representative period of time for these modes.

Finally, the concept of an energy efficiency index can be extended to other products in the consumer electronics and office equipment sector. Preliminary research shows that the concept can also be applied to CRT and LCD monitors. In these cases the power consumption in the on-mode can be described by a linear equation with screen size as an independent variable.

6   References

[1]   Behrendt, S., Kreibich, R., Lundie, S., Pfitzner, R., Scharp, M. (1998), Ökobilanzierung komplexer Elektronikprodukte, Springer-Verlag Berlin Heidelberg, p. 145
[2]   Huenges Wajer, B.P.F, P.J.S. Siderius, Analysis of Energy Consumption and Efficiency Potential for TVs in the on-mode, contract EC-DGXVII 4.1031/D/97-023, Novem, November 1998

Whole-House Measurements of Standby Power Consumption

J.P. Ross[1] and Alan Meier[2]

[1]University of California, Berkeley, USA
[2]Lawrence Berkeley National Laboratory, USA

Abstract. We investigated the variation in standby power consumption in ten California homes. Total standby power in the homes ranged from 14–169 W, with an average of 67 W. This corresponded to 5%–26% of the homes' annual electricity use. The appliances with the largest standby losses were televisions, set-top boxes and printers. The large variation in the standby power of appliances providing the same service demonstrates that manufacturers are able to reduce standby losses without degrading performance. Replacing existing units with appliances with 1 W or less of standby power would reduce standby losses by 68%.

1   Introduction

Standby power use now occurs in many modern appliances. These appliances can not be turned "off" without being unplugged or continue to draw power while not performing their primary purpose. Standby power has become a growing concern in the international community. Studies in Germany (Rath et al. 1997), Japan (Nakagami et al. 1997), the Netherlands (Siderius 1998), and the United States (Meier et al. 1999, Huber et al. 1997) have found that standby power accounts for as much as 10% of national residential electricity use. These studies (and others) have resulted in standby measurements of thousands of appliances, but few measurements of total standby power consumption in individual homes. To our knowledge the Jyukankyo Research Institute in Japan (Murakoshi 2000) and ADEME in France (Sidler 2000) have conducted the only studies of whole-house standby power consumption, but the results have not been widely circulated. Whole-house measurements of standby provide important perspectives on the variation of standby electricity consumption in individual homes and the likely impact of policies aimed at reduction.

We report here the results of standby measurements in ten Northern California homes. The objectives of the study were to measure standby power in a large number of homes, estimate potential savings from reductions in standby power use, and test the accuracy of alternative measuring techniques. The results from this study will help formulate policies and programs to reduce standby power in California and the United States.

2 Methods

Ten single and multi-family units were recruited from a range of income levels with occupancy varying from one to five persons. This group is too small to accurately represent the range of California homes, but it encompasses a great diversity of situations. In each home, we surveyed the appliances, measured each appliance's standby power consumption, and briefly measured the house's total electricity consumption, with all appliances switched off. In addition, we surveyed occupants about their future appliance purchases.

While several definitions for standby power have been proposed, there is no agreed-upon procedure for measuring it. With few exceptions, we measured the minimum power while the appliance was plugged into the mains. This is consistent with the procedure used for measuring most Energy Star consumer home electronics. Appliance standby power was measured at the lowest level with a Power Line Watt Meter, model PLM-1-PK ($\pm$0.2W). Televisions, VCRs, stereos, and cable boxes were plugged into the watt-meter and switched off at the power button before being measured. This insured that the standby power was measured at the lowest point, as the standby draw of this mode is often lower than when the appliance is turned off with a remote control. Computers were measured in the mode that the owner specified as the most frequently used setting, generally the off mode for towers and the sleep mode for monitors. Cordless telephones were measured with the handsets removed from the chargers. Every appliance was photographed for further analysis.

The total standby power for each home was calculated by summing the individual appliance measurements. This measurement was verified by comparison with the electricity use indicated by the home's utility kilowatt-hour meter for approximately eight minutes while all appliances were either off or in standby mode. All appliances that could automatically cycle on, such as a refrigerator or water pump, were unplugged for the whole-house measurement. The utility meter measurement captured loads that could not be unplugged, such as security systems and outdoor motion sensors. In two cases it was impossible to unplug the refrigerator, in which case it was turned off.

If there was a large discrepancy between the two measurements, we searched for missing loads. These two methods allowed us to calculate both an upper and lower limit of total household standby power consumption. The sum of standby power from all household appliances comprised the lower limit while the home utility meter measurement comprised the upper limit. The annual electrical consumption was tallied from individual monthly utility bills, or obtained from the local utility.

We were also curious about future changes in standby. We asked the occupants, "What do you foresee as the next three appliances you will purchase?" This information was used to estimate the percentage of new appliances which will have standby.

3 Results

Ten homes were investigated and 190 appliances with standby were identified and measured. The results are summarized in Table I. The majority of these appliances fit into one of three categories: entertainment, communications and computer hardware. Table I shows the household appliances, occurring in at least three houses or drawing more than 5 W. These results can be compared to other compilations of measurements (Meier et al. 1999, Huber 1997). The ranges in observed standby are also similar to those reported by Meier et al.. In most cases, we found that the service provided by the appliance was identical, even though the maximum value was often four times greater than the minimum value.

Table I: Measured appliance standby loads

Appliance	Average Load (W)	Minimum (W)	Maximum (W)	Number Measured
Entertainment				
TV	6.4	2.5	12	16
Set top box	10.2	1.5	23	3
VCR	5.3	1.3	11.3	13
Music box	5.2	1.3	10	8
CD player	2.2	0	6.8	6
Receiver	2.8	0	8.8	7
Tape player	1.0	0	2.3	5
Communications				
Phone	2.1	0.6	3.5	19
Answer machine	2.2	1.8	2.9	3
Fax	5.0	3.1	6.6	5
Computer				
Tower	1.2	0	2.3	8

Monitor	2.0	0	5.9	8
Printer	4.2	1.7	11.5	6
Subwoofer	6.9	4	10.8	3
Laptop charger	4.5	1.10	19.6	7
Copier	5.1	0.3	9.8	2
Miscellaneous				
Microwave	2.8	1.6	3.9	7
Clock	1.0	0.6	2.2	13
Furnace	5.0	5.0	5.0	1
Telephone system	24.5	24.5	24.5	1

Our approach also permits a presentation of the data by home (Table II). We found that the average standby load in the ten homes was 67 W, but ranged from 14 to 169 W. The households had an average of 19 appliances with standby power, ranging from 0.3 to 24 W. Standby accounted for 5% to 26% of total annual electricity consumption. The appliances with the largest standby losses were televisions, set-top boxes and printers.

Each home's annual electricity consumption, based on utility bills, is also displayed. Average annual electricity use, 6769 kWh/yr, was commensurate with the Northern California average of 6287 kWh/yr (PG&E 1995). All ten homes used natural gas for space and water heating.

Table II: Residential standby loads

Home	Annual Electricity Use (kW-hr/yr)	Standby Power (W)	Standby Power as % of Annual Electricity Use	Number of Appliances Measured	Percent Reduction in Standby with 1 Watt Target
1	4531	47	9%	12	-75%
2	4977	39	7%	15	-61%
3	1260	14	10%	5	-64%
4	20060	144	6%	46	-68%
5	6665	75	10%	21	-72%
6	5470	48	8%	15	-69%
7	5658	169	26%	32	-81%
8	6126	61	9%	16	-74%
9	5819	36	5%	12	-67%
10	7122	41	5%	19	-54%
Average	6769	67	9%	19	-68%

Nine homeowners responded to the survey of their next appliance purchases. Seven listed their next three purchases, while two listed only one appliance. Of the appliances listed, 70% will have standby power, while another 9% may have standby power (furnace and clothes dryer).

4 Discussion

There was close agreement in the quantity of standby power measured by the two independent procedures used in this study (Figure 1). The only discrepancies were found in the homes in which it was impossible to unplug the refrigerators. Thus we regard the data presented as an accurate first estimation of standby power consumption in California homes.

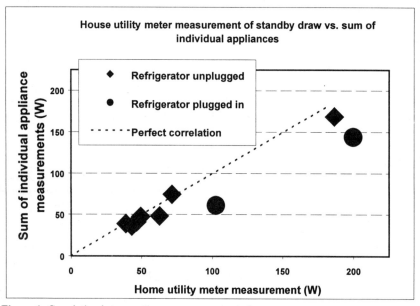

Figure 1: Correlation between two measurement techniques

Does standby power consumption correlate with other features of the homes or characteristics of the occupants? We were unable to collect reliable data on some of the more likely factors, such as floor area and income; however, we were able to compare annual electricity use to measured standby. This relationship is shown in Figure 2. Standby power appears to increase with total electricity consumption, although there is wide variation at any given level of electricity consumption. The home at 20,000 kWh/year has a swimming pool and hot tub (whose pumps are major consumers of electricity). It also has more than twice the average number of appliances with standby, leading to a fraction consistent with the other homes. A simple regression indicates that each 1000 kWh/year of non-heating electricity consumption correlates to about 8 W of standby.

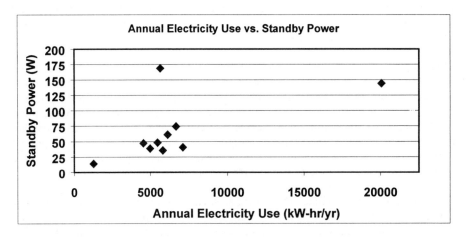

Figure 2: Annual electricity consumption versus standby power

The 67 W of average standby observed in these homes is higher than the national average of the United States estimated at 50 W by Meier et al. 1999. The observed fraction of total electricity consumed by standby (9%) is nearly twice the 5% estimated by Meier for the United States. This inconsistency may arise because Northern California homes consume relatively little electricity for heating. Electricity consumption is higher in other areas of the United States where resistance heating is used for space and water heating, making standby power a lower fraction of total electricity use. Larger and more detailed studies are required to determine if the homes measured here are unusually high or if the earlier estimate was low.

It has been proposed that standby power be limited to 1 W (Meier et al. 1999). This study allows us to estimate the potential savings from this policy goal (Table II). If all the existing appliances were replaced with units drawing only 1 W, standby power consumption would be reduced by an average of 68%. Note that, for many of the appliances, units drawing less than 1 W already exist and were observed in the ten homes. At least 70% of new appliance purchases in these homes will have standby. A policy encouraging manufacturers to reduce standby to 1 W or less in new appliances would significantly reduce the annual electricity consumption of homes used for standby power.

5 Conclusions

A study of ten homes cannot provide definitive evidence of the magnitude of standby power consumption. However, it can provide new insights to the scope of the problem and the opportunities for reducing it.

This study of ten homes suggests that earlier estimates of standby understate the actual value (or at least for California). The homes surveyed comprise a diverse sample representing the range of typical California residential electricity use. By employing two separate measurement strategies that bracket the true amount of standby, these results have an unusually high degree of confidence. Whole-house standby measurements based on short-term measurements of the house's utility meter proved to be reasonably accurate. Future studies of standby power may be able to rely on the utility meters instead of time-consuming measurements of individual appliances.

This study also demonstrates the potential savings by reducing standby. Similar appliances in these homes had wide ranges in standby power consumption, even though they all provide essentially the same service. Implementation of a 1-W ceiling on standby could reduce standby energy consumption as much as 68%.

Acknowledgement

This work was supported by the Assistant Secretary for Energy Efficiency and Renewable Energy, Office of Building Technologies, of the U.S. Department of Energy under Contract No.DE-AC03-76SF00098.
We would also like to thank the Energy Foundation for its support of J.P. Ross.

6 References

[1] Huber, W. 1997. *Standby Power Consumption in U.S. Residences.* LBNL-41107. Lawrence Berkeley National Laboratories, Berkeley, Ca. USA
[2] Meier, A., Rosen, K. 1999. *Leaking Electricity in Domestic Appliances.* Proceedings of 50th International Appliance Technical Conference, West Lafayette, Indiana, Steering Committee of the IATC.

[3] Meier, A., Huber, W., Rosen, K. 1998. *Reducing Electricity to 1 Watt*. ACEEE Summer Study of Energy Efficiency in Buildings. August. Pacific Grove, California.

[4] Meier, A., Huber, W. 1997. *Results from the investigations on leaking electricity in the USA*. First International Conference on Energy Efficiency in Household Appliances, Nov. 10-12, Florence, Italy.

[5] Meier, A., Ranier, L., Greenberg, S. 1990. *Miscellaneous Electrical Energy Use in Homes*. Energy, V.17 No. 5: 509-518.

[6] Molinder, O. 1997. *Study on Miscellaneous Standby Consumption of Household Equipment*. Prepared for the EU under contact 4.1031/E96-008. (EU-DG XVII, June 25) Brussels, Belgium.

[7] Murakoshi, C. Personal communication. May 2000

[8] Nakagami, H., Tanaka, A., Murakoshi, C. 1997. *Standby Electricity Consumption in Japanese Houses*. Jyukanko Research Institute, Japan

[9] Rath, U., Hellmann, R. 1997. *Klimaschutz Durch Minderung von Leerlaufverlusten bei Elektrogeraten*. Umweltbundesamt, Berlin, Forschungsbericht 20408541, UBA-FB 97-071.

[10] Siderius, H. 1998. *Standby Consumption in Households*. Van Holsteijn en Kemma, Delft, The Netherlands.

[11] Siderius, H. 1995. *Household Consumption of Electricity in the Netherlands*. Van Holsteijn en Kemma, Delft, The Netherlands.

[12] Sidler, O. 2000. *Campagne de mesures sur le fonctionnement en veille des appareils domestiques*. Sophia-Antipolis (France), ADEME.

Drivers of Market Transformation in Domestic Lighting[1]

Diana Ürge-Vorsatz and Jochen Hauff

Central European University, Budapest, Hungary

Abstract. Hungary has experienced one of the most remarkable market successes over the past five years in a key energy-efficiency technology: compact fluorescent lighting. While market shares of compact fluorescent lamps (CFLs) were negligible half a decade ago, today residential CFL market penetration exceeds that in many industrialised economies, ranking Hungary among the eight countries in Europe with the highest penetration rates. Since substantial efforts are being invested internationally to promote the proliferation of CFLs, the understanding of the Hungarian success can bring us closer to an effective planning of programs and policies designed to transform the markets of energy efficient technologies around the world. Therefore, the paper's goal is to provide an insight into the driving forces which contributed to this outstanding market success, and to investigate how the findings can apply in designing market transformation programs aimed at increasing the penetration of cost-effective energy efficient technologies internationally.

The paper presents the results of nationally representative residential surveys, and a large number of in-depth interviews with households, industry and other market participants. The market success is analysed in detail and differences in CFL penetration among the market segments provide an important clue for understanding which market barriers are the key in hampering market transformation, and which factors contributed to the overcoming of these barriers. Based on the findings on the drivers of the market success the authors draw lessons for the design of effective market transformation programs.

1 Introduction

For the design of effective market transformation (MT) programs, beyond focusing on the end-user, it is crucial to understand the dynamics of the market the program aims to target. For market transformation programs addressed to influence penetrations of particular efficient end-use technologies, it is valuable to examine the driving forces of the transformation in particular success stories, when end-use markets have been transformed in a significant way, with or without the help of strategic MT or DSM programs. The identification of the key drivers behind major autonomous market transformations can show us which market forces or characteristics are the key to important changes in markets of energy-efficient technologies, and can help our understanding of how such transformations could be induced by programs. In addition, the authors believe that it is easiest to introduce new MT agenda items to markets which are already undergoing dynamic change. Thus, dynamic changes in end-use technology markets should be paid close attention to, and should be utilised as vehicles for reaching environmentally or economically motivated market-related public goals. When no such dynamic development is taking place in the market in an autonomous way, MT programs can consider focusing on market characteristics and market forces which played a driving role in dynamic market transformations elsewhere.[2]

This paper seeks to analyse the driving forces behind a unique market success of a popular MT/DSM end-use technology target: the compact fluorescent lamp (CFL) in Hungary. Despite the fact that lighting usually represents only a fraction of national electricity consumption, the CFL is an ideal target for energy efficiency programs for several reasons (Vorsatz 1996). Since all over the world the majority of lightpoints in homes are still served by incandescent lamps, there is a massive potential for lamp replacements (Vorsatz, J.G.Koomey et al. 1997; Palmer and Boardman 1998). In summary, while there are several concerns with the CFL as an identical replacement of the incandescent lamp, it has rightfully for long been a very popular target of energy efficiency programs around the world.

The effectiveness of these programs has been the subject of much discussion (Plexus Research 1995; Bergstrom 1997; Martinot and Borg 1999). Despite the large number of CFL oriented programs, the lighting marketplace has been transforming slowly, if at all, from the GSL to the CFL for years, sometimes even decades. For instance in the US, the home of one of the largest number of DSM programs targeting the CFL, less than 10% of households have owned a CFL in 1994 (Vorsatz 1996). The exceptions to low CFL market penetration rates in the mid-90s were a few Western European countries including Germany, the Netherlands, and Denmark, where CFL market penetration in 1995 reached values of 50%, 56% and 46% of homes, respectively, owning at least one CFL (Kofod, Naser et al. 1996; Palmer and Boardman 1998).

Hungary, similarly to the rest of Central and Eastern European (CEE) countries, was no exception to the general trend: CFLs were barely known in the early 1990s, and only few households owned a CFL even in the mid-90s. However, by 1997, every 5th household used a CFL, which ranked Hungary in the top 8 countries in Europe (and thus likely in the top dozen worldwide) in terms of CFL penetrations (Palmer and Boardman 1998).

This paper describes this rapid market transformation, examines the possibly key driving forces behind the swift success, and derives the implications for the design of market transformation programs worldwide.

2 Methods

The findings in this paper rely on several years of research related to the lighting market in Hungary by the first author and a team of students of the Central European University.   The first key component of the research was a representative market survey of 2400 Hungarian households evaluating CFL ownership, awareness, purchasing behaviour and barriers to CFL market success in 1997, repeated in 1999. The second component of the research consisted of in-depth consumer interviews related to consumption attitudes, behavioural patterns and awareness related to the CFL, as well as to motivations for purchasing or not purchasing CFLs.   Third, an industry survey was conducted to understand the dynamics of the CFL marketplace in Hungary.

3 Market Transformation in Hungarian Domestic Lighting

As mentioned above, the average Hungarian, similarly to the average CEE citizen, did not know about CFLs at the beginning of the 90s (Kazakevicius, Gadgil et al. 1999). However, CFLs, or "energy saving lamps", as they are commonly referred to in Hungary, started to gain popularity during the mid-90s. In 1997, 8 out of 10 Hungarians knew what a CFL was, and one out of five households (19%) owned a CFL.

This relatively high market penetration was reached in only a few years. Figure 1 shows that there was a dramatic increase in CFL purchases in 1995. 83% of those who owned a CFL in 1997 bought their first CFL in or after 1995, only 5% of CFL owners, or less than 1% of all households, bought their first CFL before 1992.

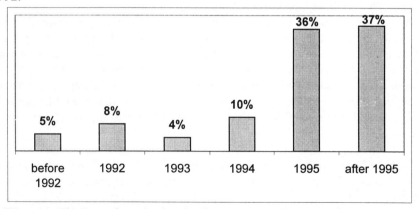

Figure 1: The year of purchasing the first CFL

This led to a dynamic transformation of the CFL market in Hungary: market penetration jumped from less than 5% in the beginning of 1995 to 19% in 1997. Awareness of the CFL has risen from less than half of the respondents in 1996 to 80% in 1997. While such rapid market transformations have been experienced in other cases, this was mostly due to large-scale programs. For instance, in the case of Poland, market penetrations increased from 11% to 33% in 4 years mainly as result of the Poland Efficient Lighting Project (PELP) (Navigant Consulting Inc 1999). Another example for a swift increase in market penetration of CFLs is the UK, where supplier subsidies and government sponsored give-aways of CFLs had a significant impact on the CFL penetration: the share of households which owned at least one CFL increased from 10% in 1993 to 23% in 1997. Awareness concerning CFLs in the UK increased from 50% of households to 75% during the same period (Martinot and Borg 1999). These comparisons let the Hungarian case stand out as a remarkably fast change, so that it is worth looking into the reasons for and details of this rapid success.

4 The Current Hungarian CFL Landscape: Success Stories and Unresponding Market Segments

This section reviews the CFL market picture in more detail, to reveal the details of the transformation in various market segments. This analysis sheds light on the drivers of the dramatic changes.

The awareness concerning CFLs was found to be very high in almost all population groups: at least 75% of people from all population segments (by settlement type, geographic location, gender, and income level) were aware of CFLs. Two population groups differed significantly: only 62% of the elderly (those above 60), and less than half, 47% of the least educated (those who have not completed primary school) could say what a CFL was.

However, the picture is not so uniform from the perspective of ownership. As displayed by Figure 2, CFL ownership was significantly higher in some market segments than the average of 19% in 1997. The highest disparity was according to the level of education. While only 6% of those with no complete primary school education had a CFL, close to half, or 44% of all households with a college or university degree opted for the energy efficient alternative of the GSL. Already a high school degree implied that the CFL penetration was double of that among households with only a primary school degree.

While some may conclude that this is probably due to a higher income level among the better educated, the correlation between education level and income is not necessarily direct in Hungary. This is shown by the ownership distribution according to income group. Figure 2 shows that a higher income level did not mean higher CFL ownership, although there is some correlation in the extremes.

The strong correlation between the level of education and CFL ownership is confirmed by both the 1999 repeat survey and another representative market survey conducted by Szonda Ipsos for Philips Hungary Ltd (Szonda Ipsos 1999).. In bth studies, the level of difference between the most and the least educated in terms of CFL ownership is the same: five-fold. According to the Philips survey, more than half (51%) of respondents with a university degree owned a CFL in 1999[3].

The trend between income level and CFL ownership is also supported by the Philips survey: there is not a direct correlation between the level of wealth and the decision to use an efficient lighting technology. While the lowest percentage of CFL usage is recorded in the poorest households and the highest among the richest, the correlation, although slight, is in fact reverse in the medium income groups: the more affluent the households are, the fewer of them installed a CFL[4].

Additional socio-demographic factors influencing CFL ownership include the size of household. Logically, the larger the household the longer hours lighting is needed, thus replacing incandescent lamps by CFLs is more economic. This is clearly supported by the survey results: almost three times as many households with 4 persons use CFLs as single households. However the trend of increasing CFL penetration with increasing household size is broken for households with more than 4 persons: about 40% less of them use this efficient lighting technology. This is probably due to the fact that large families are the most likely to have liquidity constraints, and therefore may not be able to afford the purchase of these lamps.

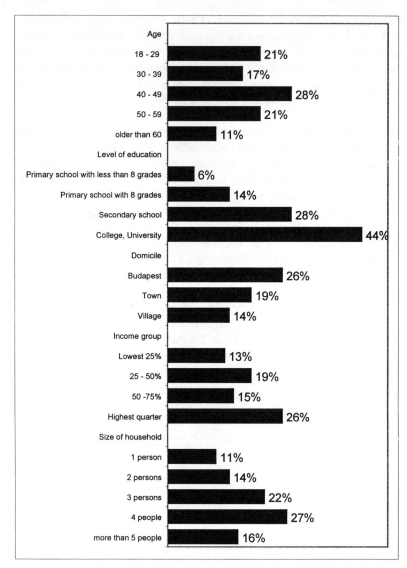

Figure 2: Ownership of CFLs according to social-demographic characteristics

CFLs are similarly accepted and used by all age groups, besides the group who is most in need of this technology due to its long lifetime and economic savings potentials - the elderly. While close to one-third (28%) of 40 – 49 year olds declared that they own a CFL, only 11% of those above 60 use them. In addition to potential liquidity constraints this can also be attributed to the fact that they are the least informed about the existence of this energy-efficient alternative to the incandescent lamp.

In summary, the Hungarian domestic lighting market has experienced a major transformation towards the CFL between 1995 and 1997. However, the success of the CFL was constrained to certain market segments. For instance, more than 7 times as many highly educated people own CFLs as those with the lowest level of education. A household in Budapest is twice as likely to use a CFL than a rural household. The elderly, who are probably most in need of expenditure saving and long-lifetime technology, has one of the lowest awareness and ownership levels in the society. Market penetration rates are half in the Southern and Eastern regions than those in economically most developed regions of Budapest and the West.

5 Analysis: Drivers of Market Transformation and Implications for Market Barrier Taxonomy

5.1 Programs Affecting the CFL Market in the Discussed Period

The first logical question is: were there any major market transformation or other programs which have affected the CFL market in such a profound way? It is important to note that there were no *major,* nationwide CFL programs taking place in Hungary during the discussed period, to which this success could be solely attributed. But there were a large number of small-scale energy-efficiency programs initiated by various organisations, the industry and the government that included the promotion of the CFL to various extents. The vast majority of these programs concentrated on awareness raising. Industry initiatives ranged from information and advertising campaigns, professional and community education programs, articles and advertisements in newspapers and professional publications, trainings, to promotional sales at low prices. From the non-governmental (NGO) community, there were several domestic energy efficiency educational campaigns, some in cooperation with the government, many of which incorporated lighting and the CFL.

While the aggregated effect of these smaller-scale awareness raising activities are estimated to be considerable in raising awareness levels to 80%, the authors are not aware of any *significant* market transformation efforts which were targeted at overcoming other market barriers, such as the high first cost.

5.2 Analysis of Potential Drivers of the Market Success

It is clear from the discussion above that the educational efforts of a wide range of market players has resulted in a major increase of the awareness of the CFL in the mid 90s in Hungary. What was the reason behind this active promotion of the CFL by suppliers in Hungary? If we compare Hungary to other CEE countries, we think that few of them experienced the same activity in advertising campaigns.

As a result, CFL awareness in Hungary appears to be higher than in other countries of CEE, such as Lithuania (Kazakevicius, Gadgil & Urge-Vorsatz 1999). An important driving factor was the presence of GE Lighting Tungsram, traditionally a key player in lighting innovation and product manufacturing in Hungary and around the world. Tungsram dominated the Hungarian residential lighting market in the early 1990s. However, with the liberalisation of markets, other lighting manufacturers have entered the marketplace. As it may have been more difficult to attract consumers in the well established incandescent market, the competitors may have considered it less difficult to obtain higher market shares in a newly emerging market: that of the CFL. Therefore, a strong competition started between three CFL suppliers in Hungary in the mid 1990s: Philips, Osram and Tungsram. The competition has resulted in aggressive advertising campaigns, trying to preserve Tungsram's leading market role in Hungarian domestic lighting. In the authors' opinion, the high level of awareness of the CFL, therefore, can largely be attributed to the fierce competition of the three manufacturers.

However, a high level of awareness does not necessarily guarantee a market success. For instance, in the US 69% of the population in a survey recognised the CFL in 1991 (Macro Consulting 1992), but less than 10% owned them. In addition to our general understanding of market mechanisms, this is also supported by the fact that while the educational/promotional activities were more or less continuous since 1992, the major success in market penetrations was experienced in the period of 1995 to 1997. What happened during this period?

Aside from advertisement campaign, the competition among the three main CFL suppliers in the Hungarian market also triggered decreases in CFL prices. The average market price[5] for a CFL decreased from around HUF 1350 in 1997 to HUF 1125 in 1999, accounting for a nominal price decrease of roughly 17%. This decrease in prices can safely be assumed to have started already in 1995, although reliable data is not available.

At the same time, however, there was a significant decrease in incandescent lamp prices, so that the ratio of CFL vs. incandescent lamp price remained relatively high: Still in 1997 CFLs where about a factor of 20 more expensive than a 60 W incandescent lamp. This exceeds the factor of 10 - 16 found in West European countries with high CFL penetration (Palmer and Boardman 1998, p.21), but is significantly lower than the factor of 50 cited by the same source for East European countries. This shows that despite the absence of large scale market transformation programs, the Hungarian price ratio of CFLs to incandescent lamps was relatively close to a level found favorable to the market success of CFLs in other countries. Nevertheless, the still high factor of 20 makes it clear that the decreasing price of CFLs can not be the sole explanation for their success in Hungary.

As cost savings through reduced use of electricity is one of the frequent reasons for the decision to buy a CFL mentioned in consumer interviews, a look at the electricity price development in Hungary over the past years might help to understand the market success of CFLs. As is evident from Figure 3 below, nominal electricity prices increased more than tenfold during 1990 and 1998, the price increase in real terms amounted to roughly 140%. Nominal and real price development can be interpreted as conducive to increased household interest in CFLs. Nominal price jumps are likely to trigger psychological effects, where people strive to protect themselves from rapidly rising energy bills. Combined with a stagnant average wage, expenditure for energy including electricity placed an increasingly tangible burden on the average Hungarian household since 1995. Hence they had good reasons to consider the purchase of electricity saving technology such as CFLs.

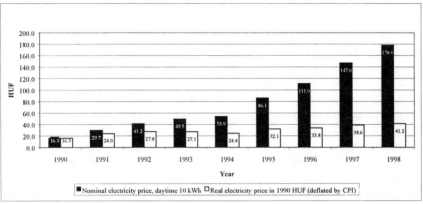

Figure 3: Nominal and real term electricity prices to households in Hungary. Source: KSH 1995, 1997 and 1999

Since most people did (and still do) not possess sufficient knowledge on how to conserve energy, with the exception of the CFL, the "energy savings lamp" has therefore become the symbol of energy conservation, and the only widely known and available technology for those eager to cut their energy bills.

There are other factors that are worth given consideration to when analysing driving forces, including technological "sexiness". Although only 6% of users admitted in the survey that they purchased the CFL because it was "modern", and 6% because it was "appealing, fashionable", our in-depth interviews reveal that the real motivation is often not the financial savings. 36% of respondents named aesthetics as one of the factors for using the CFL. There were several households that replaced GSLs in location where this substitution was clearly not economically justified. Many respondents mentioned arguments in in-depth interviews of focus groups such as "That is the modern lighting technology. No-one buys a black-and-white TV either...", or "My neighbours had them" and "They have a good quality of light". The non-economic motivations probably dominate in the highest income group, where CFL penetrations are the highest, while economic incentives are the weakest. The technological appeal as a driving factor maybe supported by the gender distribution of CFL owners: almost 50% more men claimed to own a CFL than women.

296

6 Conclusion: Implications for Market Transformation Program Design

In conclusion, the authors of this paper think that the Hungarian case underlines the significance of information as the main barrier to CFL success - and agree with Veitch (1994), who stressed the high significance of advertisement and Kjoerulf, (1997) making the case for improved customer information in the Danish case. The authors emphasise that awareness does not necessarily imply an understanding of the real benefits, i.e. the longer lifetime, and the understanding of the real magnitude of energy saving. The difficulty of understanding of the extent of economic benefits probably applies for other energy-efficient technologies with a high first cost premium or different lifetimes as well.

The implications of the prime importance of the information barrier to the design of market transformation programs are several. Our research suggests that advertising and information campaigns, educational programs and other awareness raising activities are the most important tool for the promotion of the CFL (and perhaps it is also worth investigating to what extent this applies for other energy-efficient technologies as well). However, since a high level of awareness is only a necessary but not a sufficient condition, it can only be effective if there is a real interest among the population already in place to save energy.

In addition, awareness raising has a different success rate in different population groups, mainly depending on the level of education. While those with high levels of education can understand and be convinced easily of the benefits, others need more sophisticated education of the economic benefits when the benefits are more complex (such as different lifetimes, very high first cost difference, etc.). Due to the complex nature of energy-efficiency cost/benefit calculations, it can be expected that not all population segments will be able to understand the real benefits. Hence, for these market segments different educational methods can be effective: those which concentrate on single, easily understandable benefits, such as the longer lifetime.

Even the most aggressive and wide-spread information campaigns will be ineffective in certain population groups, for instance those adults out of school who do not read and watch limited TV. For these population groups different, innovative marketing methods need to be applied, which are specially targeted to this market segment.

Acknowledgements

The authors would like to thank the following institutions and companies for funding this research: Central European University, GE Lighting Tungsram, and the Global Environment Facility. The authors would specially like to thank

Monika Juhasz of Philips Hungary for her invaluable help on CFL market data, and Laszlo Beck of Medián for his readiness to answer endless questions.

7 References

[1]   Bergstrom, W. (1997). The DSM Programmes and Lighting Activities of the Danish Electric Utilities. *Right Light 4*, Copenhagen, DEF.

[2]   Blumstein, C., S. Goldstone, et al. (1998). A Theory-based Approach to Market Transformation. *1998 ACEEE Summer Study*, Asylomar, CA, American Council for an Energy Efficient Economy.

[3]   Golove, W. and Eto, J. 1996. Market barriers to energy efficiency: a critical reappraisal of the rationale for public policies to promote energy efficiency. Energy & Environment Division, Lawrence Berkeley National Laboratory. LBL 38059. Berkeley.

[4]   Geller, H. and Leonelli, P. 1997. Energy-Efficient Lighting in Brazil: Market Evolution, Electricity Savings and Public Policies. In Right Light 4 Vol. 2, 205-210. *Proceedings of the 4th European Conference on Energy-Efficient Lighting*. Association of Danish Electric Utilities and the International Association for Energy Efficient Lighting. Stockholm.

[5]   Hollander, J.M. and Schneider, T. R. 1996. Energy-efficiency: issues for the decade. *Energy* 21 (4): 273-287.

[6]   Kazakevicius, E., A. Gadgil, et al. (1999). "Residential Lighting in Lithuania." *Energy Policy* 27(10): 603-611.

[7]   Kjoerulf, F. 1997. Transforming the CFL Market By Consumer Campaigns. In Right Light 4 Vol. 2, 145-147. *Proceedings of the 4th European Conference on Energy-Efficient Lighting*. Association of Danish Electric Utilities and the International Association for Energy Efficient Lighting. Stockholm.

[8]   Kofod, C., L. Naser, et al. (1996). *Market Research on the Use of Energy Saving Lamps in the Domestic Sector*. Lyngby, DEFU.

[9]   Központi Statisztikai Hivatal (KSH) Various issues. *Statistical Yearbook of Hungary*. Various years. Budapest.

[10] MACRO Consulting (1992). Perceptions of Compact Fluorescent Lamps in the Residential Marketplace. Palo Alto, EPRI.

[11] Martinot, E. and N. Borg (1999). "Energy-efficient lighting programs. Experience and lessons from eight countries." *Energy Policy* 26(14): 1071-1081.

[12] Navigant Consulting Inc (1999). Evaluation of the IFC/GEF Poland Efficient Lighting Project CFL Subsidy Program. Final Report.

[13] Palmer, J. and B. Boardman (1998). DELight. Domestic Energy Efficient Lighting. Oxford, Environmental Change Unit.

[14] Pesic, Radmilo, and Diana Ürge-Vorsatz (2000). "Lessons from the Restructuring of the Hungarian Electricity Industry." *ACEEE Summer Study on Energy Efficiency in Buildings.* Forthcoming.

[15] Plexus Reseach (1995). 1994 Survey of Utility Demand-Side Programs and Services. Palo Alto, CA, Electric Power Research Institute (EPRI).

[16] Szonda Ipsos (1999). Piacfeltáró elemzés a fényforrások használatáról, vásárlásáról, Philips Hungary Ltd, Confidential.

[17] Veitch, J. 1994. The Psychology of Choices. *IAEEL newsletter* 2/94

[18] Vorsatz, D. (1996). *Exploring US Residential and Commercial Electricity Conservation Potentials: Analysis of the Lighting Sector.* Environmental Science and Engineering; and Energy and Resources Group. Los Angeles and Berkeley, University of California.

[19] Vorsatz, D., J.G.Koomey, et al. (1997). *Lighting Market Sourcebook,* Lawrence Berkeley National Laboratory.

[20] Weber, L. 1997. Some reflections on barriers to the efficient use of energy, *Energy Policy* 25 (10): 833-835

Endnotes

[1] This paper is a revised and updated version of the paper presented at the ACEEE Summer Study conference in August 2000.

[2] The authors of this article consider efforts from the side of the CFL market participants towards market transformation, such as advertising campaigns, as part of the autonomous transformation of the market

[3] In fact this penetration maybe even higher in reality, if we consider the highest educational degree in the household versus that of the respondent: since the households were represented by the respondents, it is possible that some additional households with a college degree (through the spouse) own CFLs, which are currently classified in categories with lower educational levels.

[4] The Philips survey used 5 income groups, so mid-income groups here are represented by three-fifth of the society.

[5] The data reflect nominal average prices to consumers (including 25% VAT) across all CFL producers and lamp types and result from market surveys (Juhász, Personal communication).

Compact Fluorescent Torchieres: A Case Study in Market Transformation

Chris Calwell

Ecos Consulting

1 Introduction

The halogen floor lamp or torchiere was invented originally in Italy in the mid-1980s. Recognizing a longstanding consumer preference for indirect light, its creators married a powerful new technology, the high lumen linear halogen lamp, with classic Italian fixture design. The resulting product was sleek, simple, portable, and dimmable, making it an extremely versatile means of illuminating a home or office. The combination proved immediately popular with customers in Europe and the United States, though initial prices of $150 or more kept the products beyond the reach of many buyers.[1]

In the early 1990s, Chinese and Taiwanese manufacturers began to replicate the basic halogen torchiere design and export it to the United States at far lower prices. Sales increased in direct proportion to declining prices. In fact, in 1995 and 1996, about 20 million halogen torchieres sold each year in the U.S. – a remarkable 11% of all light fixtures sold in the country for all purposes.[2] Sales were also significant in other parts of North and South America and in Europe.[3] By 1998, halogen torchieres were already consuming more electricity in the U.S. and the Netherlands each year than compact fluorescents were saving.[4]

These halogen products were successful primarily because they provided a great deal of light for a very low purchase price. Customers liked the color of the light and the convenience of a very bright, portable fixture. But by the end of 1996, four trends began to emerge that would dramatically alter the course of halogen torchiere sales: declining product quality, the rise of safety problems, the creation of an ENERGY STAR® labeling program, and the expanded role of electric utilities in residential lighting.

2 Declining Product Quality

By the mid-1990s, many torchiere manufacturers switched to very low cost components in the hopes of achieving more competitive retail price points. Though this allowed home improvement centers like Home Depot and HomeBase to sell the products for as little as $12, it also increased the likelihood of early lamp failure and other problems with product quality and performance.

The halogen lamps provided with the torchieres by many Asian exporters were found to operate far below the benchmark specifications for 300 and 500 halogen lamps. Instead of providing 20 lumens per watt or more, the lamps actually operate below the efficiency level of typical incandescent lamps, often providing only 10 to 12 lumens per watt at full power and substantially less when dimmed.[5]

The relatively high cost and frequent failure of these halogen bulbs, along with the special requirements for their safe handling, caused some consumer dissatisfaction with halogen torchieres. More importantly, it created a market opportunity for a longer-lived, more convenient alternative.

3   The Rise of Safety Problems

At the same time, fire departments, insurance investigators, and government regulators began to link the products to an unusually high number of fires. Television news programs and newspapers also covered this subject extensively, particularly after the torchiere fire that gutted the apartment and destroyed virtually all of the possessions of famous jazz musician Lionel Hampton in January 1997.

The halogen bulbs within torchieres operate at temperatures of 350 to 570 degrees C – hot enough to ignite paper, cloth, or wood and many plastics within minutes of contact. Not surprisingly, lamp surface temperature correlates rather closely with lamp wattage, which means that 300 and 500 watt halogen bulbs exceed a fire risk threshold that compact fluorescent and most incandescent products remain below (Figure 1).

The quartz pressurized lamps have also failed catastrophically during operation, causing injuries to people nearby or fires on adjacent carpets and bedding. The highest wattage products seem to be responsible for the majority of catastrophic lamp failures, causing the U.S. Consumer Product Safety Commission (CPSC) to recall an assortment of 500 watt models and Underwriters Laboratories (UL) to withhold listing of products exceeding 300 watts.

To date, more than 435 fires, 114 injuries, and 34 deaths have been attributed to halogen lamps in the U.S. Nearly half of these fires are the subject of ongoing or pending litigation, creating powerful incentives for retailers like Wal-Mart and Home Depot to cease the sale of the halogen products.[6]

U.S. safety regulations now compel the use of metal "safety cages" attached to the bowls of halogen torchieres to keep combustibles away. The products also now utilize numerous warning labels, thermal cutoff switches, tipover shutoff switches, and tempered glass guards to reduce UV emissions and contain exploding lamp fragments (Figure 3). These changes have increased the cost of the products, reduced their aesthetic appeal, and made them less convenient to use.

Although Canada has also imposed similar safety requirements on halogen torchieres, European governments do not appear to have imposed strong regulations on the products. The lower risk of litigation outside of the U.S. may also help explain the smaller degree of government involvement.

4   The Launch of ENERGY STAR® Labeling

In early 1997, the U.S. Environmental Protection Agency (EPA) announced the creation of a new consumer labeling program – ENERGY STAR® -- to help consumers locate and purchase high quality, energy efficient residential light fixtures. This created for the first time a national marketing platform from which to sell better products. It shifted the focus away from product *price* and toward product *value*, emphasizing the environmental, financial, convenience, comfort, and safety advantages of compact fluorescent fixtures.[7]

The ENERGY STAR® program has continued to grow, encompassing more than 50 fixture manufacturers and hundreds of qualifying products. About ten of these companies offer qualifying torchieres (see www.energystar.gov for more information), and more than 1 million ENERGY STAR® torchieres have been sold to date.[8]   In addition, the U.S. Department of Energy (DOE) has launched a parallel ENERGY STAR® effort to label screw-based compact fluorescent lamps (CFLs), further increasing the label's visibility in the residential lighting market.

ENERGY STAR® torchieres typically employ one or more compact fluorescent lamps of the following types: square (55 watt 2D), circular (30 watt Circline or 58 watt 2C), or quad (two or three 26 watt lamps). While early designs often provided less light than 300 watt halogen torchieres, most current models provide at least 3,500 nominal lumens – an amount roughly comparable to typical halogens. Typical retail prices range from $30 to $80. ENERGY STAR® torchieres do not appear to have made major inroads in European markets yet, though Dijkstra offers a European model utilizing the General Electric 2D lamp and a Tridonic dimmable ballast.[9]

5 Electric Utilities' New Focus on Fixtures

In addition, electric utilities began to expand their efforts to encourage the use of more energy efficient residential lighting. Rather than focus only on CFLs, the utilities began to implement programs specifically for ENERGY STAR® torchieres and hard-wired fixtures. These dedicated fixtures with permanent ballasts and pin-based lamps tend to offer greater and more long-lived energy savings than CFLs, while often improving product performance and appearance.

The utilities also joined forces to address entire regional markets simultaneously. With larger combined program budgets, they have been able to:

- offer incentives directly to manufacturers to achieve greater retail price leverage
- work with major national retailers to secure regional cooperative advertising
- hire field staff to train salespeople and create eye-catching merchandise displays, and
- build sustained marketing and educational campaigns co-branded with ENERGY STAR®.

6 Out with the Old, In with the New...

The convergence of these four trends has led to a dramatic decline in U.S. halogen torchiere sales (Figure 2). However, a tremendous number of the products – perhaps 50 million -- remain in use from past sales. Most of these products are not equipped with the array of safety features required by UL since late 1998.

In the quest to capture greater energy savings, electric utilities are finding that it is not enough to encourage their customers to purchase ENERGY STAR® models. As long as the 300 to 500 watt halogen lamps remain in use in customers' homes, they continue to increase utility bills and pose a safety hazard to their users. So instead, utilities are creating promotional events in which customers return halogen fixtures for free recycling and a discount toward the purchase of ENERGY STAR® models. More than 25 of these "Great Torchiere Turn-Ins" have now been conducted around the United States in places as diverse as California, Oregon, Washington, Idaho, Montana, Wisconsin, Illinois, New York, Massachusetts, and Connecticut. In addition, a number of universities and military bases have conducted similar turn-ins at their own locations.

The model for most of the events is fairly similar. A few days prior to the event, the utility teams with a local fire department to conduct a press conference demonstrating the safety dangers associated with older halogen torchieres. This can be done by draping a towel or sheet over the top of the fixture, which will normally burst into flames within 15 to 60 seconds. It has also been done by frying an egg on a pan atop the halogen torchiere, which requires about two to three minutes with a 500 watt lamp and perhaps five minutes with a 300 watt lamp. After the demonstration, utility and community representatives encourage residents to bring their halogen torchieres to a central location or retail store on the following weekend. Utilities normally provide incentives of $10 to $20 apiece for the ENERGY STAR® torchieres, and may provide an additional $5 credit for each halogen torchiere returned.

7 Turn-In Results

At the first such event, held in Milwaukee, Wisconsin in October of 1998, more than 700 ENERGY STAR® torchieres were sold in 90 minutes. The local Home Depot retail store subsequently ordered an additional 10,000 units, and sold them through three local retail stores in three months.[10] Subsequent turn-ins have generated even greater responses. Retail store-based turn-ins in the city of Portland, Oregon (Figure 4) yielded 8,500 halogen units recycled and over 11,000 ENERGY STAR® models purchased.[11]

By far the most important aspect of halogen torchiere turn-ins is the awareness they create to stimulate subsequent sales. While single-day events may lead to the recycling of 500 to 2000 halogen torchieres and the sale of an equivalent number of ENERGY STAR® models per location, they can generate retail sales in the months thereafter of tens of thousands of ENERGY STAR® models through word of mouth.

Turn-ins also play an enormously useful role in providing hands-on consumer education. Most customers who come to these events believe that halogen torchieres can be upgraded to fluorescent models with a simple change of a lamp. It appears that they have been trained by years of utility programs to assume that any existing lamp can simply be unscrewed and replaced with a screwbased CFL. Once they understand that ENERGY STAR® fixtures employ a fundamentally different approach, including a dedicated ballast and pin-based lamp, they are often receptive to purchasing other energy efficient lighting options for their home. In fact, an 11 day turn-in event recently conducted at a shopping mall in Milford, Connecticut yielded sales of 15,000 CFLs in addition to the nearly 5,000 ENERGY STAR® torchieres sold.[12]

Customers also come to the events understandably confused about the relative efficiency of halogen lighting. After years of hearing the message that both compact fluorescent and, to a lesser extent, halogen lighting are superior efficiency choices to incandescent lighting, they are naturally confused to learn that halogen torchieres are frequently even less efficient than typical incandescent fixtures. It helps in many cases to explain that both incandescent and halogen lamps convert 90 to 95% of their electricity immediately into heat, leaving only a small fraction to produce visible light. This point is profoundly obvious to those who watch the egg fry or the towel burn atop a halogen lamp, especially if they can then compare that experience to the relatively cool temperatures they feel above a compact fluorescent lamp.

8   Lessons Learned

The most powerful lesson that emerges from the torchiere experience in the U.S. is that the non-energy benefits of efficient lighting are often vastly more important to consumers than the energy benefits. While it is true that a $40 investment in an ENERGY STAR® torchiere may yield lifetime savings in replacement lamps and energy bills of more than $400, the safety advantages of the products remain more immediate and palpable to most purchasers.[13]

Utility and government managers of efficiency programs need to remember that the public rarely shares their singular enthusiasm for energy efficiency. Utility bills are only one of many pressing concerns a typical family faces each day, and few families can observe a direct link between usage of a particular light fixture and the size of the check they must write each month to their utility.

However, to the extent that energy efficient products meet genuine needs and solve real problems that consumers face, they will be successful. Thus, for people with small children or pets, ENERGY STAR® torchieres may solve a safety problem. For people in hot climates with inadequate air conditioning, ENERGY STAR® torchieres may solve a comfort problem. For the elderly no longer able to drive or walk to the store to buy replacement light bulbs, ENERGY STAR® torchieres may solve a convenience and reliability problem.

Universities, housing authorities, military barracks, hotels, and other "institutional residential" buildings have in many cases been drawn to ENERGY STAR® torchieres for a similar reason. They welcome the financial savings from lower energy bills, but place a higher premium on occupant safety and low labor and maintenance costs. The safety risk is especially acute in these multi-family situations, where a single fire could lead to massive property loss or injury.[14]

Finally, it is also important to recognize that market transformation is the product of numerous forces at work simultaneously in the marketplace for particular products. While investments by utilities have been important in introducing customers to energy efficient alternatives to halogen torchieres, those programs would have enjoyed little success without the marketing platform and quality benchmark offered by the federal ENERGY STAR® program.

At the same time, the role of the legal and regulatory communities has been enormous and often overlooked. The insurance companies for America's largest discount retailer, Wal-Mart, and one of the largest manufacturers of halogen torchieres, Cheyenne Lighting, settled a single fire and injury claim from a halogen torchiere in April 1999 for $11 million.[15] That single incident was more influential in Wal-Mart's decision to stop selling halogen torchieres and begin selling ENERGY STAR® alternatives than all the money the utilities had offered for rebate and promotion programs.

By the same token, the U.S. Consumer Product Safety Commission possesses the legal authority and weight of evidence to bar the sale of halogen torchieres entirely, or to recall products that cannot meet current safety standards. Yet the agency remains unwilling to exercise it.

Until they choose to act, consumers will continue by the thousands to participate enthusiastically in turn-in events, grateful for the opportunity to replace a known fire hazard with a product that is safer, more convenient, more comfortable, and less expensive to operate. It is this natural enthusiasm for a better product – for a true solution – that will continue to propel the transformation of the torchiere market forward.

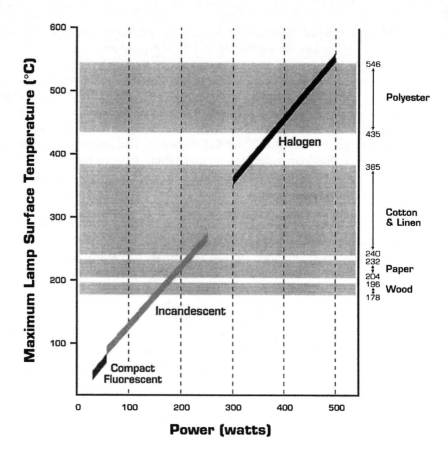

Figure 1: Comparing Lamp Operating Temperatures to the Ignition Temperatures of Common Household Materials

307

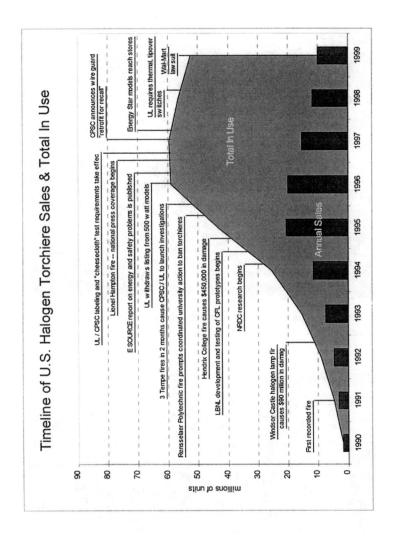

Figure 2

308

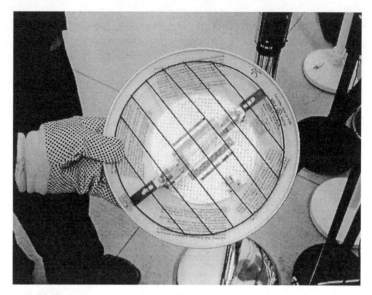

Figure 3: Warning labels and safety equipment found on a recently manufactured halogen torchiere

Figure 4: Fire-damaged torchiere from retail turn-in event

Endnotes

[1] For more information about the development and early history of halogen torchieres, see Chris Calwell and Evan Mills, "Halogen Torchieres: A Look at Market Transformation in Progress," *Right Light 4*, International Association for Energy Efficient Lighting, Volume 1, 1997, pp. 41-47.

[2] Chris Calwell and Chris Granda, *Halogen Torchiere Market Transformation: A Look at Progress to Date and Future Strategies*, prepared for the Natural Resources Defense Council by Ecos Consulting, September 10, 1999, p. 2.

[3] Marco Kavelaars, "Halogen Torchieres in the Netherlands – A Threat to Energy Savings?" *Right Light 4*, International Association for Energy Efficient Lighting, Volume 1, 1997, pp. 55-58.

[4] Marco Kavelaars, pp. 55-58.

[5] Erik Page and Michael Siminovitch, "Photometric Assessment of Energy Efficient Torchieres," *Right Light 4*, International Association for Energy Efficient Lighting, Volume 1, 1997, pp. 49-53.

[6] Chris Calwell and Chris Granda, pp. 8-9.

[7] For more information about value-based marketing of residential lighting, see Chris Calwell, Chris Granda, Lois Gordon, and My Ton, *Lighting the Way to Energy Savings: How Can We Transform Residential Lighting Markets?*, prepared for the Natural Resources Defense Council by Ecos Consulting, December 1999, pp. I:19-21, II:14-15, II:30-32.

[8] Chris Calwell and Chris Granda, p. 8.

[9] Personal communication, Viktoria Schmidt, General Electric Lighting, Budapest, Hungary, May 29, 2000.

[10] Chris Calwell, "Customers Turn Out For Torchiere Trade-In," *Home Energy*, March/April 1999, pp. 32-35.

[11] Personal communication, Maggie Nilsson, Ecos Consulting, May 1, 2000.

[12] Personal communication, Janet Leishman, Applied Proactive Technologies, Mary 24, 2000.

[13] For more financial background on torchieres, see www.lightsite.net.

[14] Chris Calwell, "Are Hot Halogens History? Energy Star alternatives are gaining ground in commercial buildings," *Energy User News*, February 2000, pp. 10-11.

[15] Chris Calwell, *Energy User News*, p. 11.

The European Design Competition for Dedicated CFL Fixtures: A Successful Example of Market Transformation

Paolo Bertoldi

European Commission Directorate General Energy & Transport

Abstract. Market research has indicated to achieve durable market transformation and increase the use of CFLs in the residential sector, it is essential to develop and market attractive and well designed CFL dedicated lighting fixtures. To this end a major European design competition for designers, students and fixture manufacturers has been launched at the beginning of 1999. The main challenge for competition participants is to produce innovative and attractive design solutions aimed at the residential market for lighting fixtures dedicated to pin-based CFLs. The key technical design feature is that fixtures embody the ballast for the CFLs, thus making the retrofit of an incandescent lamp impossible. One of the main expected results of the competition is to boost the market for pin-based CFLs, nowadays almost not present at points of sale for residential customer, because there are almost no domestic fixtures able to take them. The competition has attracted a very large number of participants, representing 19 European countries, and about 200 designs have been proposed, including very well know designers and the largest European lighting fixtures manufacturers. The winning models have been be shown at the largest European fair on domestic lighting, Euroluce in April 2000. The competition will be followed by a European-wide marketing and promotion campaigns for winning models, planned for the fall 2000.

1 Introduction

In the context of the Kyoto Agreement, the European Community and individual Member States are looking for cost-effective measures to reduce CO_2 emission and combat climate change. To this end the European Commission under the SAVE and PACE programmes has pursued several actions to improve energy efficiency of equipment in the domestic, commercial and industrial sectors. These actions include labelling and classification schemes, minimum efficiency standards and negotiated agreements, and technology procurement.

In particular, the Commission has investigated the possibility to introduce energy efficiency actions in domestic lighting. Total domestic lighting consumption in the 15 Member State of the European Union (EU) is about 90 TWh, i.e. about 15% of all residential electricity use. Moreover this consumption it is predicted to raise to 105 TWh by 2020, largely because of the growth in household numbers.

A major investigation on the lighting consumption in the EU, the DELight study (ECU 1998) reported that: "The average number of light bulbs is 24 per house across the EU. The majority (at least 70%) are incandescent, with the remainder being fluorescent (strip or CFLs) and halogen lamps. In Germany, Sweden and Italy, there are more halogens than CFLs in the installed stock." The study confirmed that there is a growing trend towards an increased use of halogens lamps both in ceiling fixtures (e.g. low voltage dicroic lamps in light spot) and in upright floor standing luminaires ("torchieres").

Electric lighting is used in practically all households throughout Europe and represents a key component of peak electricity demand in many countries. There is already a well developed energy-efficient technology available on the market, in the form of compact fluorescent light bulbs (CFLs), that could deliver substantial savings. Such savings could be accessed quickly due to the rapid turnover of lamps in the stock - the challenge is to get the more efficient technology installed and guarantee the savings.

Compact fluorescent light bulbs (CFLs) use at least 60% less electricity than the traditional incandescent lamps while lasting ten to twelve times as long and can therefore deliver substantial savings in terms of both electricity and money. Integral ballast CFLs, with a screw or bayonet base, currently represent the best opportunity to achieve significant electricity savings in residential lighting since they are the most energy-efficient technology suitable for use in fixtures already in the home. Pin-based CFLs are also available. These have a separate ballast either in a screw or bayonet based adapter (modular system) or incorporated into the fixture (dedicated system).

The EU lamp manufacturing industry is dominated by three large multi-national companies, common to both the residential and commercial sectors, whereas luminaires are manufactured by over 1000 companies in the EU, often specific to the residential sector (called also the "decorative" sector), moreover this sector is characterised mainly by national manufacturers. Successful collaboration between these two industries has been already demonstrated by the rapid development of the market for low voltage halogen lamps, which require specific fixtures.

2 Barriers to Overcome

As indicated in the DELight study "One of the main reasons for not owning CFLs is the lack of well-designed fixtures suitable for CFL use in the residential sector. So the purchase of a new fixture is likely to add to the stock of inappropriate fixtures. Moreover consumer after having tried a CFL often switch back to incandescent bulbs".

Although one of the main reasons for not owning CFLs was that they were too expensive, now CFL price has been dramatically reduced. Today the main reason is the fact that most lighting fixtures for the domestic sector are designed for incandescent or halogen lamps. In most of the current fixtures the CFL would not fit nor give an appropriate light output.. Even current owners of CFLs need assistance in recognising the opportunities for installing CFLs in their fixtures.

Moreover consumers lack confidence in the durability and continuity of CFL technology. The range of CFLs on the market is confusing and they do not know how to choose the appropriate one for their fixtures. There is little collaboration between the bulb and fixture manufacturers in developing a range of well-designed fixtures suitable for CFL use in the residential sector, so the annual purchase of a new fixture is likely to add to the stock of inappropriate fixtures.

Moreover in several cases the beneficiary of promotion rebate schemes would most often switch back to the incandescent lamp when the CFL would fail, this because either have a spare incandescent lamp at home or they can find it easily at the local supermarket.

3 The Way Forward

After a considerable number of promotions and rebate schemes, promoted mainly by utilities and European manufacturers, about 135 million CFLs are used in European homes. However it shall also be noted that only 30% of household in the EU have at least a CFL, with those households that own them having an average of three or four. Increasing ownership further will need a continuing level of policy support. However, if the full savings available in this sector are to be realised, a coherent strategy is required to transform the lighting market. Market transformation is a well-established strategic approach, utilising a combination of policies, such as education, labels, rebates, procurement and standards, to speed up the introduction of energy efficient technologies into the home. This approach is currently less well developed with domestic lighting than with appliances. To this end the Commission has put in place a number of new policy instruments, including mandatory energy labelling of lamps and a new major promotion campaign for integral CFLs.

One of the most important developments to ensure a sustainable growth and use of CFL is to develop the market for dedicated CFL fixtures, which is now basically not existent in the residential sector. To this end collaboration between the lamp and fixture manufacturers has to be promoted to ensure the availability of a sufficient range of suitable fixtures within the next five years. In parallel with this, promotion of integral ballast CFLs needs to be continued in the short term because of the current lack of dedicated fixtures. The underlying aim of any approach must be to build a positive image of CFLs to lay the foundation for the successful transfer to dedicated fixtures.

A two component strategy has been chosen by the European Commission to promote efficient lighting in the domestic sector. The first is to promote the integral CFL though a new European wide campaign, sponsored by the European Commission and the European association of the electricity industry (Eurelectric). However to achieve the long term goal of transforming the domestic lighting market and having a large penetration of CFL in each household, the best approach is to develop and to help the initial market penetration of CFL dedicated luminaires.

4   How Best to Transform the Market?

Many of the problems associated with the use of CFL in the existing fixtures could be avoided through the use of fixtures designed for pin-based CFLs. Dedicated fixtures optimise the light distribution and performance of CFLs and improve the cost-effectiveness of installation (pin-based CFLs are cheaper than the integral ballast versions), as well as guaranteeing the energy savings. . As experienced with other equipment there are a number of policy actions to transform the lighting fixture market and create a sustainable use of CFL. These include procurement, labelling, efficiency requirements. After discussion with experts and careful analysis it was agreed that in order to stimulate the introduction of energy efficient luminaires in the residential market, which is very fragmented, the best action was to promote the design competition. With this actions new model were to be designed and produced, and marketed. This was also to match interesting design (something high on the private purchasers' list with energy efficiency which does have a very low profile with lighting customer in the residential sector). Moreover if successful the competition would break a vicious loop. Since there are on the market only a very limited number of luminaires for the domestic sector using pin-based CFL, this type of lamps are not usually available in retail outlets. By creating a demand for pin-based CFLs, it is hoped that this type of lamp would become common available in the EU shops and supermarket chains. Moreover there is a lack of suitably designed fixtures for the residential sector, representing an energy-saving opportunity that has not yet been fully exploited. On the contrary dedicated fixtures are rather common in the professional/service sector (e.g. offices, hotels, etc.), where a price premium is demanded for this type of luminaires.

It is also important to notice that the current price of pin-based CFL is about half of the integral CFL; moreover this type of lamp will last at least the double of the integral CFL, and therefore it will generate much less waste and therefore less environmental impact.

5 The Competition

The aim of this European design Competition is to foster and to promote the design, production and marketing of attractive, well-designed dedicated fixtures, i.e. fixtures that can take only pin-based CFLs.

The competition entry design should give the most innovative and attractive solutions to the presentation of this technology, in particular the selected designs must look good decoratively, give an aesthetic lighting impact, and exploit new design, materials and technology.

Competition entries can range from modern to classical luminaires. The products must be suitable for the retail decorative market. Emphasis should be on well designed and mass produced products rather than 'one-off' architectural schemes.

All competition entries must: 1) use a ballast which must be part of the luminaire; must not use retrofit lamps;[1] 2) show market research undertaken and demonstrate the application of this research to the design; 3) identify materials used and why they were chosen; 4) show an understanding of product pricing including materials, assembly, packaging and distribution; 5) show creative use of lighting technology. Products already in production, which are converted to take only pin-based bulbs, are eligible for the competition.[2] The technical specifications have been kept as simple as possible and include the following three points: i)   only class A, B1 and B2 ballasts are allowed (no C or D class ballasts are permitted as per the CELMA, the Committee of the European Luminaire Manufacturer Association, classification scheme(CELMA 1997)); ii)only class A and B pin-based lamps are allowed, according to the new legislation for energy labelling of lamps; and iii) the luminaire must be fit for use in the residential sector. The competition is open to all manufacturers, designers and students. It is rather important that large and well know manufacturers participate in the competition as they are the only one able to guarantee the production and marketing of the competition entry models. However it was experienced in previous design competition for lighting fixtures that students and professional designers are likely to come up with new and innovative ideas. However it is not always possible or

---

[1]   Screw based or bayonet type CFL.
[2]   The products with non pin-based bulbs have to be withdrawn from production within one year.

guaranteed that original design would at the end of the competition be put in production.

The EU competition covers five product categories for the domestic sector using as indicated only pin-based fluorescent bulbs. The chosen categories are: ceiling luminaires, walls luminaires, floors luminaires, tables luminaires and outside luminaires. In addition each product category is further divided in two (retail-)price intervals in such a way to cater for both ends of the market. In particular the floor category is intended to stimulate design, which would replace the halogen torchieres.

The competition consists of two phases. In the first phase the design flatwork together with an estimate of production costs and the potential market is submitted to the jury for selection. The selected finalists participate in the second phase, where they are requested to submit prototypes. These prototypes will undergo a second selection, which shall include more detailed estimates of production costs, market analysis and preliminary production plan.

The competition awards will be given to those prototypes ("winning products"), which have complied with the requirements of the competition, are safe and are deemed worthy of introduction into the market (according to price, production and aesthetic criteria) .

It is felt that the right balance between the need to have attractive designs and marketable products was reached with a jury composed by well known professional designers and retailers, the latter having a feel on which product can be sold and would be accepted by customers (especially taking into account the price). To judge he submitted design an independent jury, composed of 10 representatives of professional designers and retailers, has been established.

6 The Awards

The competition awards were discussed with the European manufacturer association and other experts. The main award for the winning models is the publicity deriving from having won a European design competition and having being exhibited at the Euroluce fair. This is particularly true for manufacturers, which looks for extra publicity for their models, and for professional designers who could then get their model purchased/produced by manufacturer and they would establish themselves in public recognition. This is because in Europe (or certainly in some part of Europe product by famous design/companies are best-sellers, as example of famous design one can quote Philip Stack, and one of a famous producer Artemide). Moreover, the award of the competition includes the use on the winning products of the European Design Excellency Logo for Energy Efficient Fixtures and on consumer awareness campaigns at the European and national levels. This is particularly appropriate for the winning designs that become products. The winning products were exhibited at the Euroluce Fair, held in Milan (Italy) in April 2000. Euroluce is the largest and more important European fair for decorative (i.e. residential) lighting, being visited by about 200000 people (mainly involved in the lighting business). The Commission is also investigating the possible to show the winning models at other exhibitions and at other national Lighting Fairs.

For students there was also a cash price (of 1000 € for the first and 500 € for the second classified together with a paid trip to Euroluce). However the major incentive for students and professional designers is to find a suitable manufacturer to produce their models after the Euroluce show.

7 The Promotion and Marketing Campaign

During the lighting commercial season 2000/2001 which will start in September 2000 both European and national promotion activities, in at least five EU Member States (United Kingdom, Italy, Sweden, Denmark, The Netherlands) will take place. The national energy agencies in collaboration with the national lighting association will be responsible for these co-ordinated campaigns. Discussion are underway to have similar campaigns also in France and Germany.

Promotion and marketing activities will consist of: at European level in further exhibition of the winning models at the EU sponsored 2nd international conference on energy efficiency in household appliances in Naples in September 2000 and at World Expo 2000 in Hanover in October 2000 and to other national lighting exhibitions.

At national level the following activities will be carried out: distribution of the catalogue of the winning models; articles and press coverage in lighting and design magazine and in DIY/home furnishing television programmes; promotion at national lighting and home furniture exhibitions; market launch events by lamp and luminaire manufacturers; intensive discussion with retailers to persuade them to stock the winning models and to run special promotions on this models, including dedicated space in the retail shop to energy efficiency in lighting; and finally consumer awareness promotional campaigns. The detailed national lighting campaigns are now under detail design, at European level an advisory board for this action has been established consisting of retailers, manufacturers and other experts in the sector. The total cost of the promotion campaign will be around 250000 € and the European Commission will pay half of it, while the other half will come form national governments.

8 The Competition' Results

While the full impact of the competition can be evaluated only after the promotion campaigns to start in the autumn 2000, so far the competition has been rather successful in terms of participants and the quality of products. The competition was officially launched at the Hanover fair (the largest European lighting fair in 1999) in April 1999. More than 10000 brochures were distributed to manufacturers, students and professional designers, several announcement were made in the specialised press. About 650 participants registered for the competition by July 31st 1999, and 140 participants have submitted the required flatwork by the deadline of December 31st 1999 for about 200 different models of luminaires. The actual number of models submitted was below the initial expectation, but it was understood that to prepare the flat work required a lot of efforts without any guarantee of success, and not many people especially designer were ready to undertake the risk. The quality has been very high and after the first selection 62 models were exhibited at the Euroluce fair. The final models belong to 52 different participants so divided : 18 manufacturers, 20 professional designers and 12 students. The participants represent 12 European countries. The 18 manufacturers include some of the most well known and largest ones at European level. 27 models (9 by professional designers, 11 by manufacturers and 7 by students) were declared winners during the public award ceremony on April 12th. The competition stand was visited by a large number of visitors and about 5000 competition folders were distributed at Euroluce during the fair. The overall competition budget up to include the Euroluce fair was of around 100000 €. The next step is to ensure that the winning products will be manufactured and marketed. A big efforts is now underway to try to marry the design by students and professional designers to manufacturers, in such a way that they could be produced and marketed.

9 Conclusions

This is the first serious tentative at European level to transform the residential lighting market and drastically reduce incandescent lighting in a sustainable manner. The Commission and the other organisations (CELMA, Assoluce, The Lighting Association, iSaloni, Eurelectric, DEA, ENEA, NOVEM, STEM, ETSU), who have supported this competition, are aware that this is a difficult task, but the competition and the following marketing campaign will create awareness in the lighting business and in residential consumers. It will show a new way to design and produce luminaires. The critical point will be to convey the message to the large number of visitors at Euroluce (or any other Lighting Fair) and then to reach the largest number of players in the lighting field, and residential customers in future promotional campaigns. Given the successful (9 at least so far) of the design competition, a new competition will be organised in two year time; the future competition will perhaps open to other new promising energy savings lighting technologies (e.g. L.E.D.). Of course the full evaluation of this action will be possible only after the marketing campaign has been completed and changed in the market place will be observed. Perhaps in terms of models sold the competition will not make such an important impact, but in term of raising awareness in the lighting industry it will have a long lasting impact. The full documentation and pictures of the winning models are now available on the web at the following web address: www.etsu.com/eulightdesign.

10 References

[1] Bertoldi, P. 1996. "European Union Efforts to Promote More Efficient Use of Electricity: the PACE Programme" *in Proceeding of the ACEEE 1996 Summer Study on Energy Efficiency in Buildings*, pp. 9-11.Washington, D.C.: American Council for an Energy-Efficient Economy

[2]

[3] Bertoldi, P. 1997. "European Union Efforts to Promote More Efficient Appliances" *in Proceeding of the ECEEE 1997 Summer Study on Energy Efficiency*, pp. 1-2-id24

[4] DEFU 1996, *"Market research on the use of energy savings lamps in the domestic sector"*, DEFU, Lyngby, Denmark , ISBN: 1-874370 20-6, for the Commission of the European Community SAVE Program.

[5] ECU 1998, *"Delight"*, Environmental Change Unit - Oxford University, May 1998, ISBN: 1-874370 20-6, for the Commission of the European Community SAVE Program.

[6] C.E.L.M.A. 1997 "Classification of Ballast-Lamp Circuits for Energy Efficiency lighting" Milan, Italy, document available on request from the CELMA secretariat.

[7] Commission Directive 98/11/EC of 27 January 1998 implementing Council Directive 92/75/EEC with regard to energy labelling of household lamps.

Know Thy Customers: The Use and Value of Customer Segmentation in Marketing Energy-Efficient Lighting

Shel Feldman[1] and Bruce Mast[2]

[1] Shel Feldman Management Consulting
[2] Pacific Consulting Services

This paper begins with a discussion of an important marketing problem that faces those attempting to increase the penetration of energy-efficient appliances and lighting, just as it faces marketers of other products and services. We then introduce the use of needs-based customer segmentation and summarize the development of that approach for application in the utility industry through research sponsored by the Electric Power Research Institute. The report then describes the specific application used in this study and samples some of the important results obtained. The final sections offer several conclusions and recommendations for further consideration.

1 Problem

Utilities and consortia attempting to increase energy-efficiency have recently invested considerable hope and resources in strategies collectively labeled "market transformation." These programs involve market interventions intended to increase both customer demand for energy-efficient technologies such as lighting, refrigerators, and clothes washers, on the one hand, and the supply, distribution, and promotion of those products and services, on the other.

Several of these programs have begun to demonstrate moderate success, particularly in market penetration with certain limited portions of the general public. Nonetheless, program designers and implementers have tended to ignore the possibilities of either the potential cost-savings of limiting their efforts to specific, predisposed target markets or strategies that seek to expand coverage and penetration in markets that are currently under-addressed.

Over the course of the twentieth century, marketers of most consumer goods (and some business marketers, as well) learned that marketing to specific market segments can be more effective and more cost-efficient than mass marketing or simply producing a variety of products for different tastes. As noted in a representative marketing text (Kotler & Turner 1989, chapter 10.), "Target marketing helps sellers identify marketing opportunities better. The sellers can develop the right offer for each target market. They can adjust their prices, distribution channels, and advertising to reach the target market efficiently.

Instead of scattering their marketing effort ..., they can focus it on the buyers whom they have the greatest chance of satisfying..." (p. 273).

The first step in target marketing is to segment the customer base—to identify the relevant segmentation variables and consider segment characteristics in the planning process. But few energy-efficiency programs have taken advantage of target marketing or explored the value of customer segmentation.

2 Approach

In the late 1980s, the Electric Power Research Institute (EPRI) sponsored extensive research on the potential value of segmentation approaches to marketing energy efficiency. The work, conducted by National Analysts (1989; 1990; 1994), is outlined cogently in *Segmentation Marketing* (Berrigan & Finkbeiner 1992).

The researchers first considered several possible bases for segmentation (e.g., demographics, energy use, product attitudes). To fully meet customer needs, segmentation must address four basic questions: what, who, where, and why. "What" products and services to offer; "who" should the target audiences for those products and services; "where" should the products and services be advertised and marketed; and "why" customers behave as they do. Segmentation techniques that rely solely on geographic or demographic variables address the "what", "who", and "where", but often fail to meet the "why". For this reason, the researchers selected the needs-benefits approach. In this method, customers are divided according to their energy-related needs and the benefits they seek from the selection, purchase, and use of energy-related products, such as appliances and lighting. It allows the analyst and the program planner to understand the different needs of the several customer segments and to target related products and services to those segments through appropriate use of the marketing mix.

The initial research, conducted in 1986, started with a series of focus groups and in-depth interviews with customers and families to explore their needs, purchase decision-making process, and attraction to energy efficiency programs. Focus group results were used to develop a list of 141 statements of energy needs and preferences. These statements were then administered to 161 residential decision-makers in four diverse U.S. locations. The survey results were used to reduce the list to 48 statements. This reduced list was administered to a U.S. sample of 861 residential decision-makers. Results from the expanded survey were used to reduce the list down to 24 key questions. In 1988, EPRI validated the earlier results by administering the identical survey to a sample of 1,700 respondents. In 1994, EPRI revised and updated the original results, producing the CLASSIFY software used to generate the results discussed below.

The later analysis (National Analysts 1994) identified eleven distinct needs associated with general electricity usage:[1]

Table I: Energy-Related Needs of Residential Customers

Resource conservation	Safe appliances
Hassle free purchases	Increased comfort
High technology appliances	Enhanced security
Personal control	Low energy bills
Time saving appliances	Attractive appliances
Surge protection	

Based on respondents' needs profiles, the research found that residential customers fall into eight distinct segments:

- *Energy Reliants*, characterized by above-average needs for time saving appliances, enhanced security, personal control, attractive appliances, hassle free purchases, and high-technology appliances.
- *Enthusiasts*, characterized by above-average needs for low energy bills, increased comfort, surge protection, time saving appliances, resource conservation, enhanced security, safe appliances, personal control, and hassle free purchases.
- *Hassle Avoiders*, characterized by above-average needs for hassle free purchases.
- *Lifestyle Simplifiers*, characterized by above-average needs for resource conservation.
- *Middle Roaders*, characterized by above-average needs for personal control.
- *Resource Conservers*, characterized by above-average needs for low energy bills, increased comfort, surge protection, resource conservation, and safe appliances.
- *Technology Focused*, characterized by above-average needs for attractive appliances and high-technology appliances
- *Value Seekers*, characterized by above-average needs for high technology appliances

Using the results of CLASSIFY, marketers have provided value to program managers seeking to improve their use of market channels, product offerings, and promotional materials. (See, for example, National Analysts 1989.) We hypothesized that similar value could be obtained through the application of a segmentation approach to a market transformation program.

It may be noted that the CLASSIFY segments were defined in the context of resource acquisition programs. However, while the strategies of many energy-efficiency programs have shifted, from resource acquisition to market transformation, the underlying energy-related needs of customers have not. Accordingly, we decided to use the most recent (short) form of the CLASSIFY battery, at least in the initial test of our hypothesis, in evaluations of recent energy-efficient lighting programs.

3 Methods

Data collection for this project consisted of a 12-page survey mailed to 5,000 households in California and 2,500 households in the Pacific Northwest (Idaho, Montana, Oregon, and Washington). The survey collected information relating to consumers' levels of awareness and attitudes regarding energy-efficient lighting, their recent experiences shopping for lighting fixtures, fixture selection criteria, future purchase intentions, and basic demographic information. The survey included the reduced form of EPRI's Residential CLASSIFY™ Questionnaire, comprising 12 questions. A total of 900 valid responses were returned, including 591 for the California survey and 309 for the Northwest survey.[2]

Responses to the segmentation battery were analyzed with EPRI's CLASSIFY software, which identified the most likely segment membership for each respondent and appended that designation to the data file. Differences between the California and Northwest respondents were not systematic and the two datasets were combined for the remaining analyses. We report here the results of selected cross-tabulations and multiple t-tests analyzed through SAS.

4 Findings

This section first describes the results of the segmentation analysis and the relationship between segment membership and standard demographic categories. Next, it demonstrates the relevance of the segmentation to customer awareness and purchase of energy-efficient lighting products. Finally, the section reviews differences among the segments that can be used when developing the marketing mix. These include differences in information sources, shopping behavior, decision criteria, and reasons for purchasing.

4.1 Segment Distribution and Demographics

Application of the EPRI segmentation weights produced segment assignments for 809 of the 900 respondents (90%). The remaining 89 respondents could not be assigned to a single segment with confidence and were left unclassified and excluded from the remainder of the analysis. The distribution by segment for the remaining respondents is shown in Figure 1.

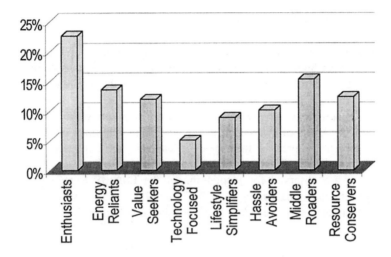

Figure 1: Distribution of Survey Respondents, by Segment

The segments do differ from one another with respect to standard demographic categories, but in ways that are difficult to capture through the use of individual categories. For example, *Value Seekers* tend to be younger (mean age, 38 years) than members of other segments and are twice as likely to be males as females. Moreover, they are least likely among the different segments to have been in their homes for more than five years and most likely to have young children. *Enthusiasts*, in contrast, tend to be older (mean age, 56 years) than members of other segments, least likely to have received a college degree, and have the lowest average income (median, $39,800) of all segments (overall median, $47,600). They are among those most likely to have been in their homes for more than five years and among those least likely to have young children. About two-thirds of *Hassle Avoiders* have also been in their homes for more than five years and are the least likely to have young children at home. However, they are found toward the middle with respect to the other demographic categories studied.

Thus, the relationships between segment membership and demographics are systematic. Moreover, the constellations of traits observed appear to be interpretable as factors that predispose customers toward certain needs and an interest in certain benefits. For example, the age, education, and income status of *Enthusiasts* seems consistent with their needs for low energy bills, increased comfort, enhanced security, and hassle-free purchases. But marketing based strictly on the demographics and without a focus on customer needs and wants would seem likely to miss the underlying promotional levers.

4.2 Awareness and Purchasing

Customer segments differ to a statistically significant degree, and in similar fashion, with respect both to compact fluorescent fixture (CFF) technologies and the national ENERGY STAR® program. First, Figure 2 shows that awareness of CFF technologies varies significantly by segment ($\chi^2 = 23.1$; $df = 7$; $p < .05$). *Energy Reliants*, *Value Seekers*, *Technology Focused*, and *Lifestyle Simplifiers* demonstrate higher levels of awareness, where awareness is defined either having compact fluorescent fixtures in the home or having seen pin-based compact fluorescent bulbs.

Similar patterns are observed for recognition of the ENERGY STAR logo from advertisements or product packaging ($\chi^2 = 31.7$; $df = 7$; $p < .05$). Awareness levels for both CFF technologies and ENERGY STAR are shown in Figure 2.

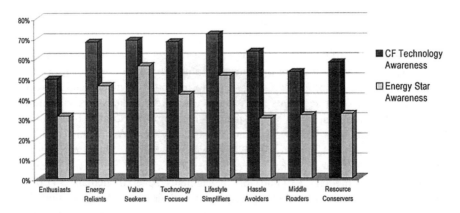

Figure 2: Awareness of Compact Fluorescent Technology and ENERGY STAR® Products, by Segment

4.3 Attitudes, Perceptions, and Beliefs

The segments differed significantly in their attitudes toward CFF products and in related beliefs and perceptions. The results are largely specific to the attitude or belief statements, however. That is, the segments do not differ in their perceptions or their overall favorability toward CFFs. (With one exception: *Resource Conservers* are significantly more favorable than are *Technology Focused*.)

For example, *Enthusiasts* are least likely of all segments to disagree with the statement that, "Compact fluorescent fixtures break down frequently." However, they are also among the two segments most likely to agree that, "Most compact fluorescent fixtures and bulbs reach full brightness almost instantly." Similarly, *Technology Focused* are least likely of all segments to agree that, "Compact fluorescent fixtures and bulbs are safer because they produce less heat." But they are among the segments most likely to disagree with the statement that, "The light from compact fluorescent bulbs looks unnatural."

These results suggest that detailed consideration will be required to identify the relationship between the needs of individual segments and their concerns about the energy-efficient technology that is being promoted. In turn, it will be necessary to develop promotional strategies that address those particular concerns.

4.4 Information Sources and Decision Factors

The segments differ in their reported use of various sources of information about lighting products. Figure 3 shows the differing frequency with which members of each segment identify books and magazines (χ^2 = 17.8; df = 7; p < .05), in-store displays of operating fixtures (χ^2 = 17.8; df = 7; p < .05), and fixture packaging (χ^2 = 16.9; df = 7; p < .05) as being among the top four most trustworthy information sources. *Value Seekers* are particularly likely to trust the information from books and magazines and from fixture packaging. *Energy Reliants* are more likely than members of other segments to trust the information they derive from operating fixture displays at the point of purchase. *Enthusiasts* are among the least likely to find any of these information sources trustworthy..

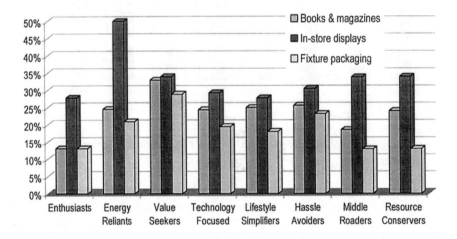

Figure 3: Trust in Information Sources, by Segment

Analysis of the factors that customers cite as most important to their purchase decisions is particularly useful for the development of advertising and promotions of energy-efficient lighting. Respondents were asked to select among five specific factors those that are the most important determinants of their lighting product decisions. Across the choices of all respondents, the following order emerged: Performance; Fire safety; Style; Purchase Price; and Operating Cost. Although the segments did not differ among themselves with regard to the relative importance of Performance, Fire Safety, and Purchase Price, they did differ significantly with respect to the importance of Style and Operating Cost (χ^2 = 59.8; df = 7; p < .05 and χ^2 = 21.0; df = 7; p < .05, respectively). Figure 4 shows the percentage of each segment that ranked each of those factors as being among their top two decision factors.

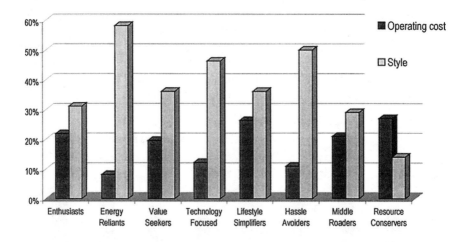

Figure 4: Importance of Operating Cost and Style as Major Fixture Selection Criteria, by Segment

Generally, the segments that are more likely to select Style as being among their top two decision factors are less likely to report that Operating Cost reaches that level of importance. Thus, for example, *Energy Reliants* are more likely than members of other segments to attend to Style as one of their critical decision factors and less likely than members of other segments to attend to Operating Cost. This negative relationship is so marked among *Resource Conservers* that they are more likely to include Operating Cost as one of their top two decision factors than they are to attend to Style.

4.5 Shopping Behavior

Members of the different segments also differ with respect to the number of different stores they believe reasonable to visit and the types of stores in which they buy lighting fixtures. They do not differ significantly, however, with respect to the reasons they give for their shopping.

Customers appear to have little expectation of being able to find the lighting fixture they desire without considerable shopping. Overall, about one-half (52%) of the respondents believe it is reasonable to visit three or more stores while shopping for light fixtures. This figure varies significantly among the segments (χ^2 = 14.4; df = 7; p < .05), ranging from a high of 70% among *Technology Focused* customers to a low of 42% among *Hassle Avoiders*. (The mean number visited ranges from 2.63 stores, for *Technology Focused*, to 2.02 stores, for *Hassle Avoiders*.) Their reported shopping behavior follows the same pattern, although it is at a lower level and the differences among segments do not quite reach statistical significance (χ^2 = 13.8; df = 7; p = .054).

In their shopping, customers make use of a variety of store types, keyed to a large extent by the type of fixture sought, whether an indoor fixture, an outdoor fixture, or a torchiere. Generally, respondents from all segments make similar usage of Lighting stores and showrooms; Hardware stores and lumberyards; Home-improvement centers; Furniture and home décor stores; and Builders or electrical contractors. However, the segments differ significantly from one another (χ^2 = 19.8; df = 7; p < .05) as to use of Discount and department stores. Specifically, almost one-half of the *Middle Roaders* (48%) use such channels, while only about one in six (15%) of the *Hassle Avoiders* do so.

5 Discussion

The data presented in this paper demonstrate that, with respect to energy-efficient lighting products, customer segments with different patterns of needs do differ from one another with respect to their awareness and purchase of those products. In addition, the segments differ as to their attitudes and perceptions, the sources of information they trust and the factors that guide their purchase choices, as well as the number and types of stores they frequent. This information can be used by program planners to improve their designs and implementation efforts so as to increase the speed and breadth of penetration for energy-efficient products.

One general tactic for cost-efficient marketing is to identify the segments that are both large and aware of CFF products and to develop plans for penetrating those markets even more effectively. In this study, however, it can be seen that the two largest segments, *Enthusiasts* and *Middle Roaders*, are the least likely to be aware of CFFs. Accordingly, consideration should be given to identifying the barriers to reaching those segments and addressing them through changes in the marketing mix. For example, one possibility for reaching *Middle Roaders* more effectively

328

might be to direct more program activities to discount and department stores. Lighting programs have tended to avoid these channels, but the results show that *Middle Roaders* tend to favor those outlets.

In applying the lessons of these data, it is important to consider all aspects of the marketing mix. For example, the importance of Style to segments such as *Energy Reliants* and *Hassle Avoiders* indicates the need for program designers to encourage the creation and promotion of a range of attractive CFF styles by manufacturers, rather than passively accepting the designs initially provided. Program designers might consider methods of targeting price promotions to groups such as *Enthusiasts*, who are particularly limited in their financial ability.

6 Recommendations

In brief, the success of market transformation programs is likely to be improved by target marketing, which would entail segmentation analysis and the use of segmentation information in program design and implementation. Market assessment studies should include the use of the CLASSIFY™ battery or an equivalent instrument that will enable analysts to provide program planners and implementers with appropriate guidance about potential target markets.[3]

Furthermore, program planning should include a forecast of the segments that are likely to show high penetration of the energy-efficient technology or service under consideration and those that are likely to show relatively low penetration. This should be supplemented by an analysis of the degree to which this forecast fits with a growth strategy and the development of a sustainable market. Based on this analysis, planning should develop the appropriate tactics for targeting the relevant market segments. Finally, program evaluators should review the degree to which the program has identified and taken advantage of the opportunities for targeting.

Endnotes

[1] The EPRI research also identified six specific needs related to lighting, including:
- *Improved aesthetics*
- *Low operating costs*
- *Improved household security*
- *Well lit space*
- *Low first costs*
- *Adaptable systems*

Time and survey space constraints prevented their use in this study.

[2] More complete reports of the surveys, along with depth interviews of key market actors and other related research may be found in Pacific Consulting Services & Shel Feldman Management Consulting 1998a; 1998b; 1999.

[3] Based on this study, CLASSIFY remains useful for the analysis of attitudes and behaviors relating to consumer responses to energy-efficient products. Although the nature of many programs has changed from a resource acquisition emphasis to a market transformation framework, the pertinent needs of customers are unlikely to have changed. Nonetheless, it may be useful to consider other segmentation approaches and to replicate or expand the current battery.

7 References

[1] Berrigan, J., & Finkbeiner, C. 1992. *Segmentation Marketing: New Methods for Capturing Business Markets.* HarperCollins: New York, New York.
[2] Kotler, P., & Turner, R.E. 1989. *Marketing Management.* Prentice-Hall Canada: Scarborough, Ontario.
[3] National Analysts. 1989. *Residential Customer Preference and Behavior: Market Segmentation Using CLASSIFY™.* Electric Power Research Institute: Palo Alto, California.
[4] National Analysts. 1990. *Residential Customer Preference and Behavior: CLASSIFY™ and PULSE® Technical Guide.* Electric Power Research Institute: Palo Alto, California.
[5] National Analysts. 1994. *CLASSIFY-PLUS V2.0 User's Manual.* Electric Power Research Institute: Palo Alto, California.
[6] Pacific Consulting Services & Shel Feldman Management Consulting. 1998a. *ENERGY STAR® Residential Lighting Fixtures Program Market Progress Evaluation Report #1.* Northwest Energy Efficiency Alliance Report (E98-016): Portland, Oregon.
[7] Pacific Consulting Services & Shel Feldman Management Consulting. 1998b. *ENERGY STAR® Residential Lighting Fixtures Program Baseline Assessment.* Pacific Gas & Electric Co. and Southern California Edison Co.: San Francisco, California.
[8] Pacific Consulting Services & Shel Feldman Management Consulting. 1999. *ENERGY STAR® Residential Lighting Fixtures Program Market Progress Evaluation Report #2.* Northwest Energy Efficiency Alliance (Report E99-035): Portland, Oregon.

Acknowledgements

The authors are grateful to Marian Brown of Southern California Edison for funding the analytic work described here and to Rich Gillman of Primen for use of the EPRI analytic engine. Thanks are also due to Heidi Hermenet of the Northwest Energy Efficiency Alliance, David Altscher of Pacific Gas and Electric Company and Bill Grimm of Southern California Edison Company for support of the initial data-gathering efforts. The interpretations and recommendations offered are solely the responsibility of the authors, however.

CFL Assembly at some Budgetary Consumers
A DSM Action Financed by CONEL/S.C. Electrica S.A

Camelia Burlacu

S.C. Electrica S.A., Bucharest, Romania

Abstract. In developed countries, it was demonstrated that, from the electricity supplier point of view, in many cases it more profitable to invest money at the consumer than in new electricity production, transport and distribution capacities.

The development strategies of Electricity Companies are more and more based on methods such Integrated Resources Planning (IRP), in which Demand Side Management (DSM) programs are considered a energy resource.

In 1998, the Board of Administration of RENEL (CONEL at this moment, the Romanian National Electricity Company) decided to use some funds for DSM actions. One of these actions was the acquisition and assembly (by lending) of 100,000 CFL (compact fluorescent lamps) instead of classical incandescent (tungsten) làmps (GLS) at some budgetary consumers (hospitals, orphanages, homes for old people, people with handicap, schools and others).

The idea was to help these budgetary consumers by reducing the invoice for consumed electricity.

Taking into account that the life of CFL is greater than the life of GLS, there are also reduced some other expenses ("upkeep") for the above mentioned budgetary consumers.

The paper presents:
- a comparison between CFL and GLS
- some statistics regarding this DSM action financed by CONEL/ELECTRICA (categories of end-users, places of CFL assembly, results).

1 Why DSM Action?

Electricity Companies from developed countries apply investment programs to consumers, taking into account that these investments are more profitable, from the economical analysis point of view, than those for new electricity generating, transport and distribution capacities.

These kinds of programs aim at Demand Side Management (DSM).

2 CFL – A Typical Example of DSM Action in Lighting

In interior lighting, 80 % cut-off of electricity consumption can be rapidly realized, thanks to their construction, by replacement of classical incandescent (tungsten) lamps (GLS) with compact fluorescent lamps (CFL) within existing luminaires.

In order to do not affect the users' visual comfort, CFL must have:
- at least the same luminous flux (1200 lm)
- the same colour temperature (2700 K)
- the same colour rendering group (1B),
compared to GLS.

Another CFL advantage is their life-time: 8, 10 or even 12 times longer than GLS.

Payback period of investment in CFL must be as short as possible. So, it is recommended to use CFL in places where artificial lighting must have daily using-time as long as possible.

Among CFL producers, one can mention the followings (in alphabetical order):
GENERAL ELECTRIC (GE) USA, GE-TUNGSRAM Hungary, MAZDA Japan-France, PHILIPS Holland, SYLVANIA USA, TESLA Czech Republic and others.

3 General Data

3.1 Initiator

At the beginning of 1998, the Board of Administration of *Romanian National Electricity Company* (CONEL) decided to earmark some funds for DSM action, one of these being that regarding 21W CFL instead of 100W GLS.

All the actions were carried out by ELECTRICA (*Romanian company for electricity distribution*), a part of CONEL.

3.2 Goal

The goal was the acquisition and assembly (by lending) of 100,000 CFL (21W) at some budgetary consumers (hospitals, orphanages, homes for old people, people with handicap, special schools and others).

The following aspects were in mind:
- for budgetary consumers:
  ▸ maintaining (if not bettering) the visual comfort
  ▸ reducing lighting costs: ➤electricity bill
  ➤investments (in GLS)
- for CONEL/ELECTRICA:
  ➤ reducing currency need by reducing fuel oil imports for electricity generating
  ➤ reducing power curve peak
  ➤ improving CONEL/ELECTRICA image.

3.3 CFL Acquisition

It was a public international bidding, organized according to the Romanian laws.

100,000 CFL were bought, having the following characteristics:
- nominal wattage: 21W
- nominal voltage: 230V
- frecquency:50 Hz
- nominal luminous flux: 1200lm
- colour temperature: 2700 K
- socket type: E27
- ballast type: electronic, incorporated
- life-time: 8000 h
- functioning position: any
- environmental temperature for CFL correct functioning: -30°C÷50°C.

4 Action Progress

4.1 End-Users

All types of consumers where CFL were assembled are presented in Table I.

CFL were delivered (by lending-on reports basis-) from April to June 1998 to almost 700 budgetary consumers, most of them being:

from which:

- medical units (47% of the consumers total amount), especially hospitals (35%)
- units for people with handicap (abandoned children, orphans, old people) (≈30%), especially orphanages (≈21%).

Note: Each ELECTRICA branch (from the counties) handed over CFL to more than one consumer, excepting Olt county where all the earmarked CFL were distributed at Slatina Municipal Hospital.

4.2 Concrete Way of Action Accomplishment

88,500 CFL were in fact assembled, the remaining 11,500 CFL being kept by the same budgetary consumers as spare (for replacement in 'out of order' case)

In Table II, it can be seen that most CFL were assembled in:
- medical units (≈68%), the majority being hospitals (≈61%)
- units for people with handicap (≈18%), the majority being orphanages (≈12%).

5 Statistics

5.1 Room Types

Most of CFL were assembled in the following room types (according to Table III)
- consulting rooms, treatment rooms emergency rooms (≈20%)
- halls, corridors (≈18%)
- wards (≈15%).

5.2 Daily Using-Time

In Table IV, CFL number depending on their daily using-time is presented. Most of CFL are used:
- 6 h/day (≈23%)
- 8 h/day (≈22%)
- 4 h/day (≈17%)
- 10 h/day (≈17%).

Taking into account the respective weights, the average value of one CFL daily using-time is: 6 hours/day.

Note: Daily using-time varies at the same room type (e.g.: consulting rooms – 2, 4, 6, 8 or 10 h/day) depending on window area of each room).

5.3 Annual Using-Time

In Table V, CFL number depending on their annual using-time is presented. Most of CFL are used:
- 365 days/year (≈55%)
- 280 days/year (≈25%).

Taking into account the respective weights, the average value of one CFL annual using-time is 320 days/year.
Note: Annual using-time varies at the same room type (e.g.: consulting rooms – 230, 260, 280 or 365 days/year) depending on window area of each room).

6 Results

At each budgetary consumer aimed at by this DSM action, consumed electricity is measured for the whole unit; a separate electricity meter for lighting does not exist.

This is the reason why the electricity savings were calculated taking into account the following data:
- CFL number
- wattage reduction (100-21=79W/lamp)
- using time (h/day and days/year).

So, the electricity savings are 1,122MWh/month (≈13,500MWh/year).

In Table VI, it can be observed that the greatest electricity savings were at medical units (≈75%), especially hospitals (≈66%).

7 Conclusions

CONEL/ELECTRICA purchased (by bidding) 100,000 CFL which were distributed at some budgetary consumers with great social impact (hospitals, orphanages, homes for old people, people with handicap, special schools and others). At these consumers, artificial lighting has the greatest weight within total electricity consumption.

From the total amount of 100,000 CFL, 88,500 CFL (21W) were in fact assembled instead GLS (100W), the remaining 11,500 CFL being kept as spare (at the sane consumers).

The using of CFL in fact assembled during 1,920 h/year (average value) ment:
- for budgetary consumers:
  ▸ maintaining (if not bettering) the visual comfort
  ▸ reducing lighting costs: ➢ electricity bill
                                      ➢ investments (for GLS)

  Electricity savings: 13,500 MWh/year
- for CONEL/ELECTRICA:
  ➢ reducing fuel oil imports for electricity generating
  ➢ (with 1.49 million USD/year)
  ➢ reducing power curve peak (with 7 MW)
  ➢ improving CONEL/ELECTRICA image
  by promoting an efficient electricity end-use technology with great using-time during the power peak period.

Table I: Consumers types at which CFL were assembled

Consumer type	Consumers of the same type	
	Number [pieces]	Percentage of the consumer total number [%]
Medical units:	**328**	**47.26**
hospitals	243	35.01
hospital-homes	32	4.61
health units	22	3.18
prophylactic sanatoria for lung diseases	11	1.59
polyclinics	9	1.30
health centres	6	0.86
watering sanatoria	3	0.43
Red Cross premises	1	0.14
inspectorates for public health	1	0.14
Units for people with handicap (abandoned children, orphans, old people):	**207**	**29.83**
orphanages, nurseries	143	20.62
homes for old people	46	6.63
societies for children, young people, old people	13	1.87
county offices for children protection	3	0.43
territorial state inspectorate for persons with handicap	1	0.14
Mutual Aid Fund for pensioners	1	0.14
Education units for normal children:	**109**	**15.71**
secondary schools	46	6.63
high schools	38	5.48
technical schools	9	1.30
kindergartens	9	1.30
crèches	7	1.00
Education units for children with handicap:	**42**	**6.05**
special schools	32	4.61
special home-schools	6	0.86
special kindergartens	4	0.58
Miscellaneous:	**8**	**1.15**
Archiepiscopate of Suceava and Rădăuţi	1	0.14
cultural clubs	2	0.29
others	5	0.72
TOTAL	**694**	**100.00**

Table II: Number of assembled CFL depending on consumer type

Consumer type	Number of assembled CFL [%]
Medical units:	**67.60**
hospitals	60.55
hospital-homes	3.13
health units	1.35
prophylactic sanatoria for lung diseases	0.92
polyclinics	0.74
health centres	0.44
watering sanatoria	0.14
Red Cross premises	0.05
inspectorates for public health	0.01
Units for people with handicap (abandoned children, orphans, old people):	**17.64**
orphanages, nurseries	11.92
homes for old people	4.71
societies for children, young people, old people	0.58
county offices for children protection	0.32
territorial state inspectorate for persons with handicap	0.107
Mutual Aid Fund for pensioners	0.003
Education units for normal children:	**10.53**
secondary schools	4.18
high schools	4.03
technical schools	0.78
kindergartens	1.11
crèches	0.43
Education units for children with handicap:	**3.51**
special schools	2.65
special home-schools	0.72
special kindergartens	0.14
Miscellaneous:	**0.72**
Archiepiscopate of Suceava and Rădăuți	0.40
cultural clubs	0.21
Others	0.11
TOTAL	**100.00**

Table III: Number of assembled CFL, depending on room type

Room type	Number of assembled CFL [%]
consulting rooms, treatment rooms, emergency rooms	19.64
halls, corridors	17.62
wards	14.75
rooms	9.08
bathrooms, toilets	9.07
classrooms, surgery, laboratories	8.10
studies	5.59
cellars, basements, balconies, gates, pharmacies, morgues	4.49
kitchens	3.28
warehouses, pantries, archives	2.38
canteens	2.32
wash-houses, dry-houses	1.67
laboratories for medical analysis	1.09
repair shops (within medical or educational units)	0.84
exterior (yards, entrances)	0.06
locker rooms	0.02
TOTAL	**100.00**

Table IV: CFL daily using-time, depending on room type

CFL daily using-time [h/day]	Room type	Number of assembled CFL [%]
6	bathrooms, canteens, cellars, consulting rooms, corridors, halls, kitchens, locker rooms, rooms, studies, toilets, treatment rooms, ware-houses, wards, yards	22.74
8	bathrooms, classrooms, consulting rooms, corridors, emergency rooms, halls, laboratories for medical analysis, kitchens, repair shops, studies, wards, warehouses, wash-room	21.49
4	bathrooms, canteens, classrooms, consulting rooms, corridors, laboratories for medical analysis, locker rooms, rooms, studies, wards, warehouses	17.27
10	bathrooms, classrooms, consulting rooms, halls, toilets, treatment rooms, wards	16.52
12	bathrooms, corridors, toilets, wards	9.81
2	archives, consulting rooms, corridors, rooms, studies	6.96
14	corridors, hall	1.99
16	cellars, corridors, emergency rooms, treatment rooms	1.86
24	corridors, emergency rooms	0.62
18	corridors, emergency rooms, toilets	0.62
20	basements	0.12
	TOTAL	**100.00**

Table V: CFL annual using-time, depending on room type

CFL annual using-time [days/year]	Room type	Number of assembled CFL [%]
365	balconies, basements, bathrooms, canteens, cellars, consulting rooms, corridors, emergency rooms, entrances, gates, halls, kitchens, morgues, pharmacies, rooms, toilets, wards, yards	54.39
280	consulting rooms, corridors, halls, studies	24.92
260	corridors, studies, rooms, toilets, ware-houses	9.56
210	classrooms, laboratories for medical analysis	5.64
230	corridors, repair shops, rooms toilets, studies	2.98
300	dry-rooms, laboratories for medical analysis, locker rooms, treatment rooms, wash-rooms	2.51
	TOTAL	100.00

Table VI: Monthly electricity savings, due to CFL assembly

Consumer type	Average monthly electricity savings [kWh/month]	[%]
Medical units:	**834,131**	**74.38**
hospitals	743,414	66.29
hospital-homes	24,056	6.61
health units	5,520	0.49
prophylactic sanatoria for lung diseases	3,972	0.35
polyclinics	2,847	0.25
health centres	2,056	0.18
watering sanatoria	1,967	0.18
Red Cross premises	160	0.02
inspectorates for public health	139	0.01
Units for people with handicap (abandoned children, orphans, old people):	**165,773**	**14.78**
orphanages, nurseries	115,801	10.33
homes for old people	44,327	3.95
societies for children, young people, old people	2,339	0.21
county offices for children protection	1,802	0.16
territorial state inspectorate for persons with handicap	1,470	0.127
Mutual Aid Fund for pensioners	34	0.003
Education units for normal children:	**84,192**	**7.51**
secondary schools	38,363	3.42
high schools	22,694	2.02
technical schools	17,712	1.58
kindergartens	2,008	0.18
crèches	3,415	0.31
Education units for children with handicap:	**35,269**	**3.14**
special schools	30,088	2.68
special home-schools	4,090	0.36
special kindergartens	1,091	0.10
Miscellaneous:	**2,106**	**0.19**
Archiepiscopate of Suceava and Rădăuți	1,659	0.15
cultural clubs	300	0.03
others	147	0.01
TOTAL	**1,121,471**	**100.0**

Promotion of Consulting Center Establishment for Energy Efficiency Lighting Systems Implementation in Kharkov (Ukraine)

Luis Andrada and Luciano Ranalli

CESEN

Abstract. The project was financed by the European Commission in the framework of the TACIS/BISTRO Programme aimed at funding short-term projects facing actual requirements in the recipient Countries.

The project was supported, at local level, by Kharkov Regional Administration.

1   Present Situation

1.1 Streets Lighting

During the last years, the streets lighting significantly got worse. The data of injuries increasing on the streets and the growth of offenses and traffic accidents testify indirectly these facts.

The share of energy consumption of exterior lighting was 48 mil kWh per year. In 1996, only 21 mil kWh were used up. The causes of a.m. are:

- reducing of operation time of exterior lighting
- unfinished reconstruction of streets lighting in some of the regions.

1.2 House Hold Lighting

The house sector of the city with 2.5 mil of population consumed more than half of all generated energy of the region and the house lighting played not the last role at that consumption.

The population is not aware of the questions of energy saving and this fact is considered to be one of the causes of effective expenditure of energy, including the lighting.

2 Results

The general objective of the project was to work out the recommendations for the establishment of a Consulting Group for energy efficient lighting systems implementation in Kharkov.

Main results included:

- The development of a realistic action plan, methodology and tools for Regional Consulting Group operating on the base of the experiences and current practice of relevant EU members;
- The provision and installation, through the above mentioned Group, of the energy efficiency equipment and measures to promote modern lighting systems implementation;
- The development of training actions directed at all levels of management and technical staff, who are likely to be involved to implementing energy efficiency lighting systems and for further dissemination in the regions;
- To formulate a strategy where Western European enterprises can investigate opportunities to establish joint ventures with Ukrainian companies in local manufacturing, transferring of technology and/or marketing of modern lighting equipment.

3 Identification of Main Barriers

The Consulting Group established in the framework of the project will have to face certain barriers and limitations in its activities.

Part of them are external obstacles to the activities of the Consulting Group. These obstacles and difficulties do not depend on the group staff, forms and methods of work.

The first and most difficult barrier is general economic crisis in Ukraine. Most enterprises lack working capital and carry heavy burden of taxes and debts. This relates to both state-owned and non-state companies. In these conditions it is not likely that enterprises will themselves finance their business development programmes including implementation of energy saving in lighting.

Specific barrier in the established mentality of local management based on poor awareness of energy saving advantages. However, this obstacle can be overcome by through information campaigns. The Consulting Group naturally connected with educational activities and participates in all training actions planned by the scientific and technical council on energy saving within Kharkov State Administration.

Chinese Lighting Energy Consumption and the Potential Impact of the Proposed Ballast Energy Efficiency Standard

Jiang Lin [1] and Yuejin Zhao [2]

[1] Lawrence Berkeley National Laboratory
[2] China National Institute of Standardization

1 Introduction

As China's economy grows and the standard of living continues to increase, China's energy consumption is growing rapidly. Lighting electricity consumption has reached 125 TWh in 1998 and will continue to grow rapidly in the near future. However, lighting efficiency levels remain low in China, thus lighting energy conservation has significant potential. To this end, China launched its Green Lights Program in 1996, which aims at saving energy as well as raising lighting quality through the promotion of high quality efficient lighting products.

In support of China's recently passed Energy Conservation Law and China's Green Lights Program, China State Bureau of Quality and Technical Supervision (SQBTS) proposed in 1998 to formulate the Ballast Energy Efficiency Standard, in order to raise fluorescent ballast efficiency, to guide industry's technical progress, and to enhance the competitiveness of Chinese ballast products. In contrast to previous standard development activities, SQBTS has decided to introduce internationally recognized analytical methodologies in the development of the ballast efficiency standard, and sought collaboration with the Lawrence Berkeley National Laboratory in U.S., which has conducted extensive economic and engineering analyses in support of US DOE's appliance efficiency standards effort.

The proposed Ballast Energy Efficiency Standard will set two key parameters: ballast minimum energy efficiency level and the required energy efficiency value for energy conservation product certification[1]. These parameters have different regulatory power: the former is mandatory, while the latter is voluntary. Products whose efficiency rating falls below the stipulated minimum efficiency will be eliminated, while those that pass the energy conservation product certification would receive special energy conservation label, indicating that such products have met the efficiency and quality requirement stated in the national energy efficiency standards. The energy conservation label essentially provides performance assurance for end-users.

In the sections below, a detailed engineering analysis is first presented for fluorescent lamp ballast products in China. Next, the economic and environmental impacts of the proposed Ballast Energy Efficiency Standard will be assessed.

2 Engineering Analysis

In this section, the current status of fluorescent lamp ballasts in China will be discussed, followed by a three step engineering analysis that includes product classification, analysis of existing product energy efficiency range, and the determination of minimum energy efficiency.

2.1 Classification of Fluorescent Lamp Ballasts in China

Product classification helps to determine the range of products that a standard covers, and should be set for the most widely used products. According to a report by China's Illumination Engineering Society (CIES), most fluorescent lamp ballasts in China are pre-heat ballasts in the range of 18 to 40 watts. Magnetic ballasts have the highest market share, roughly 65 million units in 1997, followed by electronic ballasts, about 30 million units. Ballasts with little sales in China such as hybrid ballasts are omitted from the consideration. Figure 1 summarizes the ballast production trends in China. Data up to 1997 are historical, and those beyond 1997 are forecast.

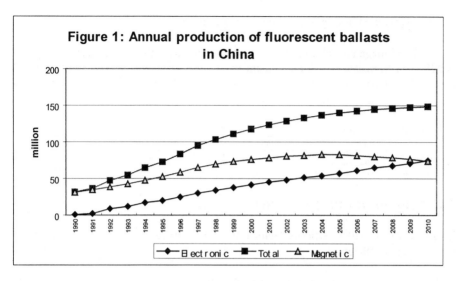

Figure 1: Annual production of fluorescent ballasts in China

Magnetic ballasts generally have lower efficiency than electronic ballasts, but cost less and have longer lifetime than Chinese made electronic ballasts. These two factors make it unlikely that electronic ballasts would overtake magnetic ballasts in China in the near future. Treating magnetic and electronic ballasts as separate categories will be beneficial to the development of energy efficient magnetic ballasts. A consensus was reached among regulatory and industrial stakeholders that magnetic and electronic ballasts should be classified as separate categories in this standard analysis. Table I below lists product classification used in this research.

Table I: Product classification

Magnetic ballasts	Electronic ballasts
40W	40W
36W	36W
32W	32W
30W	30W
22W	22W
20W	20W
18W	18W

Ballast performance data were principally collected through the following three channels:
- Sending questionnaires to major ballast manufacturers
- Purchasing ballasts from market and having them tested in national testing centers

- Obtaining testing data from national testing centers

A total of 95 magnetic ballast samples were gathered from the market near Beijing and Ningbo and from several key manufacturers. Energy efficiency data on additional 195 electronic ballasts were collected from the National Light Source Testing Center in Beijing. Table II lists sample sizes of different product classes.

Table II: Ballasts sample size

	18W	20W	22W	30W	32W	36W	40W
Magnetic	2	29	7	16	9	2	27
Electronic	4	14	3	19	3	16	119

2.2 Analysis of Energy Efficiency Levels

In this analysis, ballast efficacy factor (BEF), as defined below, is used to evaluate ballast energy efficiency,

$$BEF = 100 \times \mu / P \qquad (1)$$

where μ is ballasts factor, P line input wattage.

Ballasts factor is the ratio of lumen output of a standard lamp driven by the tested ballast to that of the same standard lamp driven by a standard ballast, under the same operating conditions. It measures the capacity of a ballast to produce lighting outputs with a standard lamp.

Ballast efficacy factor (BEF), defined as the ratio of ballast factor over ballast input wattage, measures the efficiency with which the ballast delivers lighting outputs. Since it not only measures lighting outputs but also power inputs to a ballast, the ballast efficacy factor is the more accurate measure of the energy efficiency of a ballast.

Due to sample size limitations, distributions of ballast efficacy factors are analyzed only for 40W, 30W, and 20W ballasts (presented in Figures 2,3, and 4, respectively). In general, the distributions of BEFs for electronic ballasts lay to the right of those for magnetic ballasts, indicating that electronic ballasts in China are substantially more efficient than magnetic ballasts.

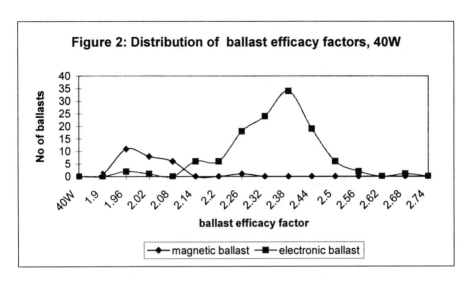

Figure 2: Distribution of ballast efficacy factors, 40W

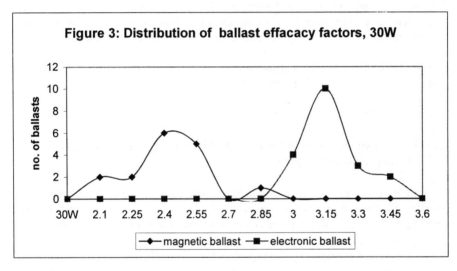

Figure 3: Distribution of ballast effacacy factors, 30W

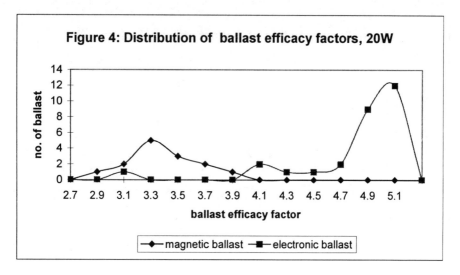

Figure 4: Distribution of ballast efficacy factors, 20W

For magnetic ballasts, the distributions of BEFs tend to be more flat. This indicates that there is a wide range of variations in ballast efficiencies, and substantial improvements can be achieved through technical innovations and standards.

For electronic ballasts, the distributions of BEFs are more concentrated. However, there are significant long left-tails, particularly for 40W and 20W electronic ballasts. The existence of such tails suggests that a significant minority of electronic ballasts in China has far lower (than normal) energy efficiency.

2.3 Determining Minimum Energy Efficiency and Efficiency Rating Classifications

Minimum energy efficiency is a mandatory requirement, and is the threshold often used to eliminate high energy consuming products. Efficiency ratings are indexes used to evaluate the efficiency of energy utilization, and are the basis for certifying energy conservation products. In setting such parameters, the following principles have been agreed upon by regulatory and industrial stakeholders:

1. Products, whose energy efficiency ratings are below the minimum energy efficiency, are considered high energy consuming products. Such ballast products do exist, as indicated by the left tails in efficiency distributions. These products typically have significant quality problems or use outdated technologies, and should be eliminated through energy efficiency standards. Such products account for approximately 12.5% of total ballast products.

2. Adopting internationally compatible notations, A, B, C, D, and E are used to represent efficiency levels, with A representing the highest efficiency
3. In setting efficiency classifications, most products should be set at level E, advanced products set at level B, and level A reserved as development target.
4. In order to encourage the development of high efficiency magnetic ballasts, magnetic and electronic ballasts are considered separately in their efficiency classifications. Because magnetic ballast are generally less efficient than electronic ballasts, the proportion of magnetic ballasts reaching higher efficiency ratings should be lower.
5. The number of products reaching higher efficiency ratings should decline as the rating goes up. The ratings are set at 12.5%, 65%, 85%, 95%, 100% for magnetic ballasts, and 12%, 50%, 75%, 90%, 100% for electronic ballasts.

The BEF values corresponding to above percentiles points are presented in Table III and Table IV for magnetic and electronic ballasts, respectively.

Table III: Magnetic ballast BEF classifications.

Proportions	12%	65%	85%	95%	100%
18W	3.396	3.460	3.483	3.495	3.501
20W	2.997	3.309	3.514	3.632	3.725
22W	2.878	3.092	3.306	3.481	3.569
30W	2.252	2.559	2.645	2.704	2.867
32W	2.332	2.367	2.426	2.543	2.615
36W	2.225	2.231	2.233	2.234	2.235
40W	1.993	2.067	2.108	2.120	2.289

Table IV: Electronic ballast BEF classifications.

Proportions	12%	50%	75%	90%	100%
18W	5.150	5.399	5.626	5.752	5.837
20W	4.284	4.837	5.000	5.051	5.099
22W	3.754	3.805	4.403	4.761	5.000
30W	3.110	3.208	3.356	3.459	3.537
32W	2.673	2.980	3.069	3.122	3.158
36W	2.402	2.540	2.626	2.666	2.695
40W	2.256	2.386	2.426	2.500	2.701

Running regression analyses on such percentile contour lines across wattage groups, a set of equations are obtained. Due to space constraints, the detailed results are presented elsewhere (Lin, 2000). These equations are then used to calculate minimum energy efficiency values for ballasts, which are set at level E. The resulting minimum energy efficiencies for fluorescent lamp ballasts are presented in Table V.

Table V: Minimum energy efficiency requirements for ballasts (ballast efficacy factors)

Type	18W	20W	22W	30W	32W	36W	40W
Magnetic	3.154	2.95 to	2.770	2.232	2.146	2.030	2.992
Electronic	4.778	4.370	3.998	2.870	2.678	2.402	2.270

3 National Energy and Environmental Impacts

Implementation of the proposed fluorescent lamp ballast energy efficiency standard would eliminate the most inefficient fluorescent ballasts in China. And the adoption of energy conservation products certification would further increase the average efficacy factors of fluorescent lamp ballasts in China. The increased average ballast efficacy factor would reduce China's electricity consumption and related environmental impact, and lower electricity tariff payment by consumers. This section evaluates such impacts of the proposed ballast energy efficiency standard on national energy consumption and pollution emissions.

3.1 Forecast of Ballast Stock and Energy Efficiency Levels

Fluorescent ballasts energy consumption is determined by the average ballast efficacy factor and ballast stock. Future ballast stock is calculated based on future ballast production, ballast retirement rates, and imports and exports. Given that the proposed standard is expected to come into effect in 2000, forecast is made from 2000 to 2009.

According to a report by China Illumination Engineering Society, most Chinese manufacturers determine their production based on sales. In addition, ballast imports and exports are roughly in balance, and therefore domestic production equals to domestic sales.

3.1.1 Stock of Electronic Ballasts

Average lifetime of electronic ballasts in China is about 13,000 hours. Assuming average daily lighting hours is 9 hours and average lighting days per year is 325 days; the average lifetime of electronic ballasts is 4.5 years. Further, the retirement rate for electronic ballasts is 0% for the first year, 0.1% for the second and third year, and 0.3% for the fourth and fifth year.

Production of electronic ballasts is projected to be about 41.5 million in 2000 previously, and assuming the growth rate of 9% until 2009, the stock of ballasts that will be impacted by the proposed standard is projected as in Table VI.

Table VI: Forecast of electronic ballast stock between 2000 and 2009

Year	Stock (million)	Year	Stock (million)
2000	41.5	2005	218.1
2001	86.9	2006	231.4
2002	130.8	2007	245.7
2003	173.5	2008	259.8
2004	205.5	2009	273.8

3.1.2 Stock of Magnetic Ballasts

Average lifetime of magnetic ballasts in China is about 40,000 hours. Assuming daily lighting hours is 9 hours and average lighting days per year is 325 days; the average lifetime of magnetic ballasts is about 13 years. Further, the retirement rate for magnetic ballast is assumed to be 0% between the first and seventh year, 5% for the eighth year, 10.5% for the ninth year, 16.5% for the 10th year, 22.5% for the 11th year, 30.9% for the involves year, and 100% for the thirteenth year.

Production of magnetic ballast is projected to be 76.1 milling in 2000 previously, assuming 2.5% growth rate until 2009, the stock of magnetic ballasts that will be impacted by the proposed standard is projected as in Table VII.

Table VII: Forecast of magnetic ballast stock between 2000 and 2009.

Year	Stock (million)	Year	Stock (million)
2000	76.1	2005	482.6
2001	154.3	2006	564.0
2002	235.1	2007	643.9
2003	316.6	2008	718.7
2004	399.7	2009	780.6

3.2 Forecast of Energy Efficiency Level

Implementation of the proposed standard would raise the average efficiency of ballasts in two ways: 1) by eliminating the least efficient ballasts, the standard would raise the average ballast efficiency; 2) the proposed standard could spur further efficiency gains among ballasts that already meet the proposed minimum efficiency. The first scenario is referred as the Minimum Impact Case, and the second as the Efficiency Impact Case in this study. The average BEF values for 40W ballasts under these two scenarios are summarized in Table VIII.

Table VIII: Average BEF values

Average BEF	Magnetic	Electronic
Current	2.05	2.37
Minimum impact case	2.06	2.40
Efficiency impact case	2.08	2.42

3.3 National Energy and Environmental Impact Analysis

In order to estimate energy saving and environmental impacts of the proposed ballast minimum efficiency standard, the average efficiencies before and after standard implementation are calculated. Together with estimated ballast stock and electricity tariff, the annual electricity and expenditure savings can be calculated. Based on electricity savings, reductions in CO_2 and SO_2 emissions are then determined. Emission factors are estimated from data presented in *China Energy Statistical Yearbook, 1991-1996*. Average electricity tariff is assumed to be 0.52 yuan/kWh with 1% growth rate, which is also very conservative. Annual lighting hours is assumed to be 3000 hours per year.

Under the minimum impact scenario, cumulative electricity savings would reach 5 billion kWh by 2009. That is equivalent to 2.7 million tons of coal, reducing China's CO_2 emissions by 1.35 million tons of carbon and SO_2 emissions by 54,000 tons. Electricity bill savings over the same period would amount to 2.8 billion yuans.

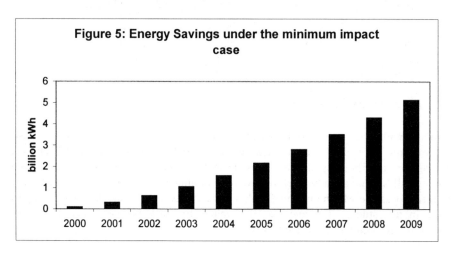

Figure 5: Energy Savings under the minimum impact case

Under the efficiency impact scenario, annual electricity savings would reach 12 billion kWh by 2009. That is equal to 6.3 million tons of coal, reducing China's CO_2 emissions by 3.2 million tons of carbon and SO_2 omissions by 126,000 tons. Electricity bill savings over the same period would reach to 6.6 billion yuans.

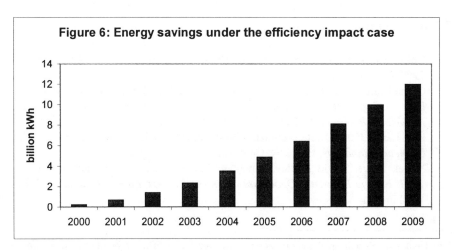

Figure 6: Energy savings under the efficiency impact case

A comparison of the results under the two scenario discussed here indicates that the energy saving and environmental impact would be doubled if average ballast efficiency can be raised by 1%. If all ballasts reach the top efficiency level available in the present market, the electricity savings would be much greater, so would GHG emission reductions.

4 Discussions

Raising ballasts energy efficiency is only one of the many ways that can save electricity in fluorescent lighting. To raise the overall efficiency of fluorescent lighting, the efficiency of fluorescent lamps and fixtures must be considered as well. In addition to raising lighting efficiency of each component of the lighting system, the system efficiency can be further enhanced through adjustment in total lumen outputs. This is particularly relevant when replacing magnetic ballast with electronic ballast, because electronic ballasts produce more light output even with the same lamp. By choosing electronic ballasts with lower ballast factors, such replacement could further reduced lighting energy consumption while maintain the same light output. The estimated electricity savings and GHG emissions reductions of the proposed ballast energy efficiency are substantial, despite very conservative assumptions in the average efficiency gains. This further highlights the fact that there are tremendous potentials in lighting energy conservation and resultant pollution reductions in China. Light up, but don't choke up.

5 Reference

[1] China Statistical Publishing House, 1998, *China Energy Statistical Yearbook 1991-1996*, Beijing
[2] China Illumination Engineering Society, 1998, *China Ballast Market Survey.*
[3] Lin, Jiang, 1999, *China Green Lights Program: A Review and Recommendations*, LBL Report, Berkeley, California
[4] Lin, Jiang, 2000, Draft, Technical support document for China's ballasts energy *efficiency standards*, Berkeley, California
[5] Nadel, Steve, David Fridley, Jonathan Sinton, Yang Zhirong, and Liu Hong, 1997, *Energy Efficiency Opportunities in the Chinese Building Sector*, ACEEE, Washington.

---

[1] In 1998, China established a certification system for energy efficient products, which aims at promoting such products through energy efficient labeling. This labeling scheme is similar to the Energy Star program in the US in that the certification is only granted to the most efficient products.

Energy and Carbon Impact of New U.S. Fluorescent Lamp Ballast Energy Efficiency Standards

Isaac Turiel, Barbara Atkinson, Peter Chan, Andrea Denver, Kristina Hamachi, Chris Marnay and Julie Osborn

Lawrence Berkeley National Laboratory

Abstract. Climate change policy requires generation of carefully considered estimates of possible energy and carbon savings from various policies. There is always uncertainty in such estimates; we describe how these savings estimates were arrived at for the case of energy efficiency standards for fluorescent lamp ballasts. Several standards scenarios are described in detail along with all the assumptions that had to be made. We worked closely with the ballast industry to develop all of the engineering data needed to estimate energy savings when magnetic ballasts are replaced with electronic ballasts. Current market data was collected from distributors to establish ballast prices.

The commercial and industrial lighting sector will be subject to new energy efficient lighting regulations beginning in 2005. These regulations affect ballasts (the market is mostly magnetic now) that operate T12 fluorescent lamps. The use of electronic ballasts will result in cumulative energy savings of about 1.3 to 5.2 exajoules (1.2 to 4.9 quads) of primary energy over the period 2005-2030. Businesses will reduce electricity costs by about 2.0 to 7.2 billion dollars (discounted to 1997 at 7% real) over the same period. Carbon emissions will be reduced by about 11 to 32 million metric tons over the period 2005-2020.

1 Introduction

In recent years, about 80 million fluorescent lamp ballasts were sold annually in the U.S. for use in commercial and industrial buildings. They come in four basic types: energy efficient magnetic and electronic ballasts operating T12 (38 mm or 1.5 inches diameter) or T8 (25 mm or 1 inch diameter) lamps. In the present U.S. market, almost all of the T8 lamps are operated by electronic ballasts whereas almost all of the four-foot T12 lamps are operated by magnetic ballasts. Since electronic ballasts are more efficient and their lamp/ballast systems are more efficacious, a large potential energy savings, can be tapped by transforming the market for ballasts operating T12 lamps.

In March of 2000, the U.S. Department of Energy (DOE) published a notice of proposed rulemaking for fluorescent lamp ballasts operating T12 lamps (Federal Register, 2000). The ballast standards were established through a multi-year process involving discussion and technical input from industry and efficiency advocacy groups (stakeholders). The inputs to the analysis were refined several times in response to stakeholder comments. Stakeholders reached consensus through negotiations .The ballast efficiency standard takes effect for ballasts incorporated in fixtures in 2005 and applies to all ballasts, including those for the replacement market, by 2010. This paper describes the methodology used to estimate national energy savings and carbon emission reductions from various standards scenarios. Further details can be found in a Technical Support Document (TSD) on the DOE web site:
(www.eren.doe.gov/ buildings/codes_standards/applbrf/ballast/html).

The U.S. standards use the metric of ballast efficacy factor (BEF), which is a measure of ballast system performance or efficacy, and is defined as the ballast factor in percent divided by the ballast input power in Watts (BEF = 100 x Ballast Factor/Input Power). The ballast factor (BF) for a subject ballast is the ratio of the light output of a lamp tested on the subject ballast to the light output of the same lamp tested on a "reference" ballast under identical environmental conditions. The use of BEF allows comparison of the efficacy of lamp/ballast systems using the same lamp with different ballasts; the inclusion of ballast factor accounts for the different light output provided by the systems. Standards in some other countries that consider ballast losses only appear to have not accounted for these potential differences in light output.

The proposed standards amend existing standards for ballasts under NAECA for F40T12, F96T12, and F96T12HO (high output) fluorescent lamp ballasts. The existing standards eliminated the standard magnetic ballast and allowed the "energy-efficient" or low-loss magnetic ballast. The minimum BEFs for the existing standards are listed in Table I. The "Existing Std" BEFs apply to all ballasts manufactured after January 1, 1990 and sold by manufacturers after April 1, 1991. The standards apply to ballasts operating at input voltages of both 120 and 277 with an input current frequency of 60 Hz. Although the regulated ballasts can operate reduced wattage or energy-saver lamps, all BEF tests are made with full wattage lamps. The standards do not apply to dimming ballasts, ballasts designed for use in ambient temperatures of 0° F or less, or ballasts that have a power factor less than 0.90 and are designed for use in residential applications. The "Proposed Std" BEFs are also shown in Table I. In effect, the proposed standards eliminate magnetic ballasts for the F40T12 and F96T12 categories, and still permit magnetic ballasts for F96T12HO lamps.

357

Table I: Minimum Allowable Ballast Efficacy Factors

Lamp (s)	Voltage	Nominal Lamp Watts	BEF Existing Std	BEF Proposed Std
1 F40T12	120/277	40	1.805	2.29
2 F40T12	120	80	1.060	1.17
2 F40T12	277	80	1.050	1.17
2 F96T12	120/277	150	0.570	0.63
2 F96T12HO	120/277	220	0.390	0.39

2 Methodology

The standards analysis was performed for DOE by the Lawrence Berkeley National Laboratory (LBNL). Another paper in the proceedings discusses the life cycle cost analysis. Analyzing the national impacts of proposed standards for fluorescent ballasts required comparing projected U.S. commercial and industrial lighting energy consumption with and without these standards. In this analysis, we projected interior lighting energy savings as well as net energy impacts of lighting systems on heating, ventilating and air conditioning energy use. Projections of energy consumption in the absence of standards are referred to as *base case projections*. These base case projections were compared to projections of energy consumption if each trial standard scenario were implemented. For each scenario, the difference between the energy consumption of the base case and the standards scenario projection was defined as the energy savings impact of the standard.

The time series of annual energy savings was used along with projected annual electricity prices to calculate fuel cost savings. Lighting equipment expenditures were also projected for the base case as well as for standards. Expenditures and fuel cost savings were used to calculate the national net present value (NPV) of each scenario.

The national energy savings and net present value of the scenarios were calculated using the National Energy Savings (NES) spreadsheet model developed by LBNL . Inputs to the model were: national shipments forecasts, electricity price projections, ballast system characteristics, percentages of different ballast types in fixtures, percentages of ballasts converted to different ballast types under standards, years of standards implementation, and social discount rate.

The NES model calculates the energy savings, fuel cost savings, equipment costs, and net present value derived from the replacement of magnetic ballast systems. The spreadsheet model (NES v_4) is available to interested parties from the U.S. Department of Energy (DOE) website (www.eren.doe.gov/ buildings/codes_standards/ Applbrf/ballast.html). The model is documented in the TSD, Appendices B and E. Users may adjust inputs and observe the resulting national energy savings and NPV. The model calculates the impacts from replacing magnetic ballasts with either electronic or cathode cutout (heater cutout) ballasts. In this paper, we report results using the input assumptions for the proposed electronic ballast standard as described below.

3 Data

Data and input assumptions including ballast shipments, ballast market segments, lighting/HVAC interactions, standards implementation dates, and percentages of base case ballasts converted to different ballast types in the scenarios, are discussed below. Data on ballast system characteristics (wattages, lifetimes, operating hours, light output, and equipment prices) and electricity price projections for the commercial and industrial sectors are shown in Chapters 3 and 4 and Appendices A and B of the TSD. Factors for conversion of site energy to source energy and social discount rate are discussed in Chapter 5.

3.1 Shipments

The NES model tracks magnetic ballast shipments from 1997 through 2030. (Savings are calculated beginning with the year the standard is projected to take effect.) It is important to note that the shipments in the NES model are only the magnetic ballasts projected to be used in the base case(s), and those ballasts (electronic or cathode cutout) that replace them in the standards scenarios. The model does *not* track the total ballast market, i.e. the electronic ballasts that would be installed in the base case(s).

The NES model shipments scenarios used as their starting point historical ballast shipments data provided by NEMA, the National Electrical Manufacturers Association (NEMA, 1995and NEMA Comment, 1999) for products covered by this analysis. Note that these shipments are lower than the number of total ballasts sold in the commercial and industrial sectors, since they do not include ballasts exempted from the standards. Figure 1 shows the annual shipments of magnetic and electronic ballasts from these data by year. The data show a modest overall increase in annual total shipments (with a small decrease in 1996), as magnetic shipments decrease (with the exception of a small increase in 1997). Magnetic ballasts also comprise a decreasing percentage of the total over time (with the exception of a small increase in 1996). Note that magnetic ballast shipments remained essentially constant over the last three years for which we have data.

Magnetic ballasts shipments were assumed to decline over time under two of the three base case scenarios. These shipments forecasts are described below.

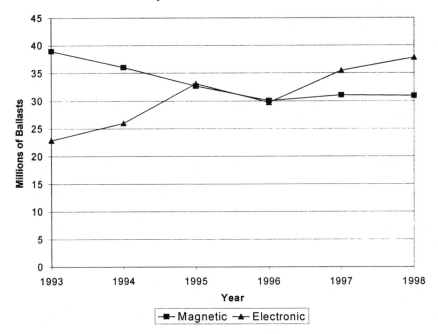

Figure 1: Shipments of magnetic and electronic ballasts (millions)

During the first half of the 1990s, the trend of decreasing magnetic ballast market share was caused in large part by incentive programs, such as utility demand-side management rebates and voluntary federal agency programs, which promoted electronic ballasts. Utility programs have diminished since 1995, but other non-regulatory programs as well as building codes were expected to continue to provide future incentives for installation of electronic ballasts.

The NES model tracks magnetic ballast shipments data for four major lamp-ballast combinations using T12 lamps: one-lamp F40T12, two-lamp F40T12, two-lamp F96T12, and two-lamp F96T12HO. The model accounts for one and two-lamp T12 ballasts used in 1, 2, 3, and 4-lamp fixtures. For the three-lamp and four-lamp T12 lamp-ballast combinations, there are very few magnetic ballast models. In the US, T8 magnetic ballasts have a tiny market share, while they have a larger share in some other countries. In the U.S., the T8 lamp does not perform well with the T12 magnetic ballast. Therefore, switching to T8 lamps requires a ballast change, whether magnetic or electronic ballasts are used. In this analysis, a simplifying assumption was made that the F40T12 and F96T12 lamp-ballast combinations operated in the commercial sector and that the F96T12HO lamp-ballast combination operated in the industrial sector.

In the standards scenarios, the T12 magnetic ballasts that would persist in the base case were assumed to be converted to electronic ballasts. "Converted" meant that the end user, instead of purchasing a new magnetic ballast, purchased and installed another type of ballast because magnetic ballasts were no longer available as the result of efficiency standards. Some scenarios included ballast/lamp system conversions using T8 electronic systems in place of T12 systems. The proposed standards reported in this paper assume nearly 100 percent T8 electronic systems under standards.

Energy savings resulted from conversion of magnetic ballasts to electronic ballasts over time. Total energy savings accrued for ballasts operating from the year in which standards were assumed to take effect, through the year 2030. For NPV, the electricity cost savings and equipment costs accrued past 2030. The equipment costs of ballasts purchased in the later years of our forecast would be associated with electricity savings beyond 2030. Therefore, electricity cost savings associated with any ballasts purchased before 2030 and occurring in years after 2030 were included in the total for the NPV calculations. For the same reason, the costs of replacement lamps purchased after 2030 to be used in ballasts purchased before 2030 were also counted in the equipment costs. Electricity prices, which were obtained from an Energy Information Administration (EIA) forecast (Annual Energy Outlook (AEO) 1999), were assumed to be constant at the 2020 price after the year 2020 (EIA, 1999). The DOE selected a 7 percent social discount rate for the national NPV analysis.

3.2 Lighting/HVAC Interactions

When lighting energy consumption is reduced through installation of more efficient equipment, there is a secondary impact on cooling and heating energy consumed by the building heating, ventilating, and air conditioning (HVAC) systems. Because less heat is given off by the more efficient lighting systems, less cooling is needed in some buildings or spaces while more heating is needed in other buildings or spaces. The annual HVAC impact varies by climate zone ånd building type. As explained in Appendix B, this analysis used a national average for HVAC impacts that included a cooling decrease for some climates and building types and a heating increase for others. This *net* national average impact was estimated to be an additional 6.25 percent of the total national lighting energy savings from each scenario. HVAC savings were not included in the total energy cost benefit and therefore were not in the NPV calculation.

3.3 Market Segments

The ballast market has three segments: (1) the new/major renovation market, (2) the replacement market, and (3) the early retrofit market. The new/renovation market consists of luminaires installed in new buildings or buildings where entire lighting systems are replaced; this is also called the original equipment manufacturer (OEM) market. The replacement market consists of ballasts replaced at the end of life (typically spot-replaced at failure). The standards scenarios reported here had different standards implementation dates for the new/renovation and the replacement markets.

The early retrofit market, which consists of ballasts replaced with more efficient models before they fail, was considered to be covered in the base case scenarios with their decreasing magnetic ballast shipments. This market is already installing efficient ballasts and would be little affected by the trial standards.

3.4 The Standards

The proposed standards apply to ballasts for the new and renovation market (sold as part of a fixture in the OEM market) manufactured as of April 1, 2005, sold by manufacturers as of July 1, 2005, or incorporated into luminaires by luminaire manufacturers as of April 1, 2006. Replacement ballasts manufactured as of June 30, 2010, must meet the standard.

The proposed standard will keep the BEF at the existing standard level for F96T12HO ballasts. However, a class of HO ballasts that was previously exempted from standards because of their operation at low temperatures will now be covered. At the time (1988) that the previous standards were established, energy-efficient magnetic HO ballasts were not designed to operate below 50 degrees F and therefore "cold temperature" HO ballasts (rated to start at 0 degrees F or lower) were exempted.

Currently, EEM cold temperature HO ballasts are manufactured, so the exemption will be removed for most of them (some ballasts used for outdoor signs and operating at -20 degrees F or lower will still be exempted). Under the proposed standards, most cold temperature HO ballasts will now have to meet the existing standard BEF; that essentially means they must be energy-efficient magnetic ballasts.

4 National Energy Savings and NPV

4.1 Shipments Scenarios

The proposed standards were analyzed using three base case shipments scenarios: decreasing shipments to 2015, decreasing shipments to 2027, and constant shipments. For Decreasing Shipments to 2015, the magnetic T12 ballast shipments decrease to 10 percent of their 1997 value by 2015, and remain constant at that level through 2030. This rate represented a 5 percent annual decrease from 1997 shipments levels. To account for the fact that without regulatory action some magnetic ballasts would continue to be used, shipments decreased to a base level (rather than zero) in 2015 and remained at this level through 2030. The base level was calculated as 10 percent of the magnetic ballast shipments in 1997 for each ballast lamp-ballast combination. For Decreasing Shipments to 2027, the magnetic T12 ballast shipments decrease to 10 percent of their 1997 value by 2027, and remain constant at that level through 2030. This rate represented a 3 percent annual decrease from 1997 shipments levels. An exception for the two declining shipments cases is that shipments for HO ballasts remain constant at 1 million per year for the whole analysis period (see Scenario Assumptions below). The third scenario, Constant Shipments, assumes that magnetic T12 ballast shipments remain essentially constant at 1997 levels throughout the period.

4.2 Scenario Assumptions

The proposed standards scenario assumed that electronic ballast standards took effect in 2005. There was a delay period of 5 years for the replacement ballast market. This delay did not apply to HO ballasts; the standard for all HO ballasts was assumed to be effective April 1, 2005. The new/renovation market was assumed to comprise 70 percent of magnetic ballast shipments and to become 100 percent T8 electronic ballasts under the standard. The replacement market was assumed to comprise 30 percent of magnetic ballast shipments and to become 95 percent T8 electronic and 5 percent T12 electronic.

4.3 National Energy Savings and NPV Results

This section presents the summary of results (see Table II) for the proposed standards scenarios with the three shipments base cases. We estimated that the base case cumulative fluorescent lighting energy consumption for the period from 2005 to 2030 was approximately 85 Quads or 90 exajoules (source energy) for the Decreasing Shipments to 2027 base case. The savings from the proposed standards for the 2027 base case was about 2.7 percent of this total estimated consumption.The total cumulative energy savings (with HVAC) ranges from 1.34 to 5.50 EJ.

Table II: Energy Savings and Net Present Value to Society of Standards for Fluorescent Ballasts Purchased from 2005-2030 (1997 Billion Dollars, Discounted to 1997 at 7 percent Real)

	Electronic Standards For Units Sold from 2005 to 2030 Discounted at 7% to 1997 (in billion 1997 $)		
SCENARIO	*Decr Shipments 2015*	*Decr Shipments 2027*	*Constant Shipments*
Total Exajoules Saved	1.27	2.45	5.17
Total Exajoules Saved w/ HVAC*	1.34	2.60	5.50
Total Quads Saved	1.20	2.32	4.90
Total Quads Saved w/HVAC*	1.27	2.46	5.21
Total Benefit	1.95	3.51	7.24
Total Equipment Cost	0.53	0.91	1.83
Net Present Value	1.42	2.60	5.41

*For energy savings only; Total Benefit and Net Present Value do not include HVAC savings.

5 Carbon and NO$_x$ Emission Reductions

The effects of proposed fluorescent ballast energy-efficiency standards on the electric utility industry were analyzed using a variant of the Energy Information Administration (EIA) National Energy Modeling System (NEMS) called NEMS-BRS, together with some exogenous calculations (NEMS, 1998). NEMS was used by the EIA to produce the *1999 Annual Energy Outlook* (AEO99), and NEMS-BRS is used to provide some key equivalent inputs to LBNL's standards analysis. Because electric utility restructuring is well under way, we can no longer assume the historical cost recovery regulation of utilities, which was the basis of previous utility impact analyses. Therefore, the electric utility analysis consists of a comparison between model results for a case comparable to the AEO99 Reference Case, as reported in the AEO99, and cases incorporating each of the ballast standards scenarios. Because the standards effects are small compared to the total size of the power sector, NEMS-BRS was not used directly. Instead, exploratory runs were conducted to estimate marginal effects, which were then used to calculate the small effects on utilities resulting from each standards scenario. The reduced electricity demand from ballast efficiency standards lowers generation from both coal and natural gas, but, because natural gas is more frequently the marginal fuel, it is usually affected to a greater degree.

Although energy savings from the proposed appliance standards continue through 2030, the effects of these savings are reported through 2020 because this is the time horizon of NEMS-BRS. Total carbon and NO_x emissions for each of the 3 shipments scenarios are reported in the table below. The annual carbon emission reductions range from 1 to 4.0 Mt (million metric tons) in 2020 and the NO_x emissions reductions range from 1.5 to 8.8 kt (thousand metric tons) in the same year.

Table III: Power Sector Emissions: Electronic Ballast Standard Scenario

NEMS-BRS Results						Difference from AEO99 Reference				
	2000	**2005**	**2010**	**2015**	**2020**	**2000**	**2005**	**2010**	**2015**	**2020**
AEO99 Reference										
Carbon (Mt/a)[1,3]	588.9	612.9	653.2	704.6	744.6					
NOx (kt/a)[2,3]	4,191.2	3,547.1	3,665.0	3,819.2	3,882.8					
Decreasing Shipments 2015										
Carbon (Mt/a)	588.9	612.8	652.7	703.8	743.6	0.0	-0.1	-0.5	-0.8	-1.0
NOx (kt/a)	4,191.2	3,546.2	3,663.0	3,817.0	3,881.3	0.0	-0.9	-2.0	-2.3	-1.5
Decreasing Shipments 2027										
Carbon (Mt/a)	588.9	612.7	652.4	703.1	742.6	0.0	-0.2	-0.8	-1.5	-2.0
NOx (kt/a)	4,191.2	3,546.1	3,662.1	3,814.8	3,878.8	0.0	-1.0	-2.9	-4.5	-4.0
Constant Shipments										
Carbon (Mt/a)	588.9	612.7	652.2	702.1	740.6	0.0	-0.2	-1.0	-2.5	-4.0
NOx (kt/a)	4,191.2	3,545.3	3,661.8	3,810.9	3,873.9	0.0	-1.8	-3.2	-8.4	-8.8

[1] Comparable to Table A17 of AEO99: Electric Generators
[2] Comparable to Table A8 of AEO99: Emissions
[3] All results in metric tons (t), equivalent to 1.1 short tons

Cumulative emissions savings over the 15-year period modeled are listed below for the three shipment scenarios.

Table IV: Cumulative Emissions Reductions (2005-2020)

Emission	Decreasing Shipments 2015	Decreasing Shipments 2027	Constant Shipments
Carbon (Mt)	10.9	19.0	32.1
NOx (kt)	34.0	59.6	103.4

366

Acknowledgment

This work was supported by the Assistant Secretary for Energy Efficiency and Renewable Energy, Office of Building Research and Standards, of the U.S. Department of Energy under Contract No. DE-AC03-76SF00098.

6 References

[1] Federal Register, 2000. Fluorescent Lamp Ballast Energy Conservation Standards,, Vol. 65, No. 51, March 15, 2000.
[2] Data submitted by NEMA, 1995. to Lawrence Berkeley National Laboratory, on January 6, 1995, January 29, 1997 and February 4, 1998.
[3] NEMA Comment #50, 1999. on the *Final Inputs to the Draft Analyses Related to Rulemaking to Consider New Efficiency Levels for Fluorescent Lamp Ballasts*, Attachment B, May 18, 1999.
[4] Energy Information Administration, *Annual Energy Outlook, 1999,: with Projections through 2020*, DOE/EIA-0383(99), December, 1998.
[5] NEMS, 1998. For further information about NEMS, see Energy Information Administration, *National Energy Model System: An Overview 1998*, DOE/EIA-0581(98), February, 1998.

Application of Uncertainty in Life Cycle Cost Analysis of New U.S. Fluorescent Lamp Ballast Energy Efficiency Standards

Isaac Turiel, Barbara Atkinson, Andrea Denver, Diane Fisher, Sajid Hakim, X. Liu, and Jim McMahon

Lawrence Berkeley National Laboratory

Abstract. Life cycle cost analysis is often used in evaluating potential energy efficiency standards for appliances and lighting. Usually, point estimates are used for each input variable. There is always uncertainty in such estimates; we describe how we developed the distributions representing the variability in electricity price, equipment price, operating hours and ballast lifetime for the analysis of energy efficiency standards for fluorescent lamp ballasts. We worked closely with the ballast industry to develop all of the engineering data needed to estimate energy savings when magnetic ballasts are replaced with electronic ballasts. A survey was performed to establish ballast prices. The life cycle cost (LCC) outputs are also in the form of a distribution, we discuss how to interpret these distributions.

These LCC results were one of the tools that were used by manufacturers and efficiency advocates to arrive at consensus energy efficiency standards. The commercial and industrial lighting sector in the U.S. will be subject to new energy efficient lighting regulations beginning in 2005. These regulations affect ballasts (the market is mostly magnetic now) that operate T12 (38mm or 1.5 inch) fluorescent lamps. The use of electronic ballasts will result in cumulative energy savings of 1.2 to 4.9 quads of primary energy over the period from 2005 to 2030. Businesses will reduce electricity costs by 2.0 to 7.2 billion dollars (discounted to 1997 at 7% real) and carbon emissions will be reduced by 11 to 32 million metric tons over the period 2005-2020.

1   Introduction

Life cycle cost (LCC) analyses are commonly used in evaluation of alternative efficiency standards (or technologies) for appliances and lighting equipment. Such analyses allow policy makers to determine the impact on consumers of purchasing a more efficient product. Generally, the more efficient product (or technology) is more expensive to purchase but less expensive to operate. LCC is the sum of purchase price and operating expense discounted over the lifetime of the appliance. Payback period is also often calculated. In most standards analyses, average point values of the inputs are utilized and some sensitivity analyses are performed for some of the most important inputs. This approach provides a single LCC output for each set of point input values.

There is always some uncertainty or variability in the values chosen for each of the input variables. For example, the incremental price for the more efficient technology has uncertainty associated with new technologies not yet incorporated into existing appliances and variability associated with technologies already found in existing appliances. The variability can be temporal or regional. Fuel prices can vary regionally and there is also uncertainty associated with forecasts of future fuel prices. In this paper, we discuss an improved method (which considers uncertainty and variability) of calculating LCC for evaluation of alternative efficiency standards or technologies.

2 Methodology

2.1 Introduction

LCC is the sum of purchase price (PC) and operating expense (OC) discounted at rate r over the lifetime (N) of the appliance. Installation costs, such as equipment and labor, are included in PC, and energy consumption and lamp replacements (and associated labor) during the life of a ballast are included in OC. Discounting means that money spent (or saved) at some future date has less value than the same expenditures (or savings) in the immediate present.

$$LCC = PC + \sum_{t=1}^{N} \frac{OC_t}{(1+r)^t}$$

Data inputs to LCC calculations include end-user prices for ballasts and lamps, electricity rates, annual lighting operating hours, lamp/ballast system watts, labor rates, installation times, analysis period, ballast lifetimes, lamp lifetimes, and discount rates.

2.2 Importance Assessment

The first step in the LCC analysis is to evaluate the relative importance of each variable in the calculation. To determine the ranking of input variables in order of significance of their effect on the uncertainty of the outputs, we conduct an importance analysis on the input variables by calculating the rank order correlation coefficients between each of the input variables and the output variable. Rank order correlation (or Spearman correlation) is measured by computing the ranks of the probability samples, with the largest value assigned a rank of 1 and the smallest, the rank of n, and then computing their correlation. Rank order is a good measure of the strength of monotonic relations regardless of the underlying distributions of the variables involved.

By using the rank order of the samples instead of the actual samples, the measure of correlation is not affected by skewed distributions or extreme values (so-called distribution-free and outlier-resistance measure), and is therefore more robust than the simple correlation (or Pearson correlation).

Importance is then defined as the absolute rank-order correlation coefficient between the sample of output values and the sample for each uncertain input. Unlike commonly used deterministic measures of sensitivity, importance averages over the entire joint probability distribution. Therefore, it works well even for models where the sensitivity to one input depends strongly on the value of another [Lumina Decision Systems, 1996].

Figure 1 shows the relative importance of 9 variables on change in total life cycle cost (dLCC) of the two alternative technologies (Cathode Cutout and Electronic Rapid Start) with respect to Energy Efficient Magnetic. For dLCC, the four most important inputs having significantly large correlations with the output for all three technologies, are ballast life, electricity price, incremental ballast price and annual operating hours. These four inputs will have the dominant effect on change in total life cycle cost. The same analysis for payback times shows that annual operating hours becomes the most influential input, followed by electricity price and ballast consumer price relative to other inputs.

We conclude that the variables that tend to have the most significance in determining the dLCC and payback times are electricity price, annual operating hours, ballast life, and ballast consumer price. The consistently unimportant variables are times to change lamps and ballasts, labor rates for installing ballasts and lamps, and lamp price. These five variables influence dLCC for cathode cutout ballasts (not for electronic ballasts) because the lamp lifetime decreases when cathode cutout ballasts are used relative to when magnetic ballasts are used. In the case of ballasts, these results could have been predicted in a qualitative sense. However, for other products (e.g., water heaters), it can be more difficult to predict which are the most important inputs, without an importance analysis. Reducing the number of input variables for which detailed distributions are needed allows the analysts to concentrate data collection on fewer input variables. Average point values can be used for the less important input variables. To address discount rate variability, we ran these calculations for three scenarios: discount rates of 4%, 8% and 12%. This paper reports only results from the 8% scenario.

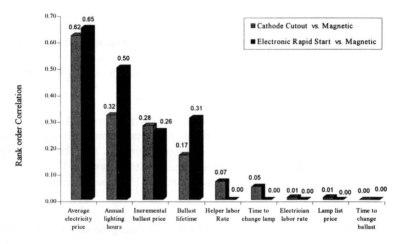

Input Distributions to LCC Change

Figure 1: Relative importance of 9 variables on change in total life cycle cost of two alternative technologies

3 Data

In this section we discuss the four most important inputs and show their distributions.

3.1 Ballast Price Distribution

The LCC analysis uses incremental price for converting from the baseline EEM T12 ballast to a more efficient design option. Since there are ballasts in the marketplace for all of the efficiency levels that were considered by the U.S. DOE, we collected price information in the marketplace. In most instances, ballasts, either in fixtures or not in fixtures, are purchased by contractors from electrical distributors. We collected selling price data from distributors around the country and applied one markup of 13% to the distributor selling price. The incremental prices are shown for one ballast type in Figure 2 below. A normal distribution with the mean equal to the incremental price and a standard deviation of 15% was used to describe the distribution of incremental ballast prices. Details of the ballast price analysis can be found in chapter 3 of the ballast TSD (www.eren.doe.gov/buildings/codes_standards/applbrf/ballast.html).

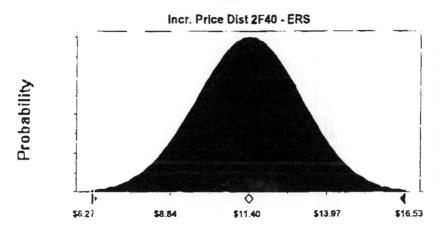

Incr. Price Dist 2F40 - ERS

| $6.27 | $8.84 | $11.40 | $13.97 | $16.53 |

Figure 2: Distribution of prices for 2F40T12 ERS ballast

3.2 Electricity Price Distribution

Marginal electricity prices in $1997/kWh are used. A detailed explanation of the process for deriving these marginal electricity rates can be found in Appendix B of the ballast TSD Technical Support Document (www.eren.doe.gov/buildings/codes_standards/applbrf/ballast.html). Electricity prices projected for the year 2003 were used because that is the earliest year in which changes to the minimum ballast efficiency standards could have occurred.

Electricity price is a function of the average price distribution obtained from the Energy Information Agency's (EIA) data on revenue and sales, for the commercial sector by utility, and the epsilons discussed in Appendix B of the TSD. Epsilon is the percent difference between the average electricity price and the marginal price calculated for each customer on each tariff (the average epsilon is 5.5%). Figures 3 presents the distribution of sales-weighted average electricity rates in the commercial ($0.0762/kWh weighted average) sector and Figure 4 presents the epsilon distribution. For each selected point on the average rate distribution, the software program Crystal Ball, selects an epsilon value. Thus, a marginal rate is calculated by the following equation:

Marginal Electricity Rate = Sales Weighted Average Rate x (1 + Epsilon)

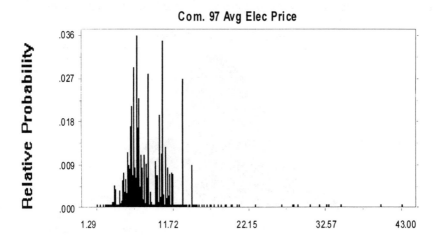

Figure 3: Distribution of 1997 Average Commercial Electricity Prices (c/kWh)

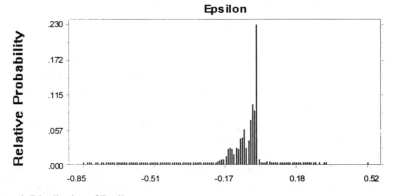

Figure 4: Distribution of Epsilons

In order to project electricity prices into the future, we used the Annual Energy Outlook (AEO) 1999 forecast for the price of electricity relative to the 1997 baseline (EIA, 1998). Using the Reference case, these ratios were used to estimate the future marginal prices. There is a forecast of a 21% decrease in average electricity price for 2020 relative to 1997.

3.3 Annual Lighting Operating Hours.

Annual lighting hours were derived from a large data set of lighting energy audits done throughout the U.S (Xenergy, 1995). Annual interior lighting hours were calculated for various commercial-sector building types: office, retail, grocery, restaurant, lodging, health, assembly, school, college, and warehouse. Interior lighting hours were also calculated for industrial buildings. Appendix A of the TSD describes the derivation of the weighted averages for each lamp type. Figure 5 shows the annual operating hours distribution for ballasts operating two four-foot 40W T12 lamps (2F40T12)

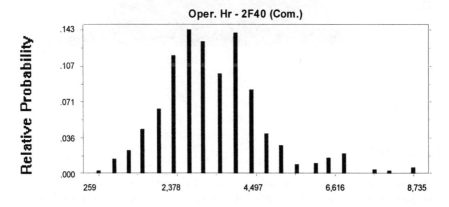

Figure 5: Annual Lighting Operation Hours Distribution for 2F40T12 in Commercial Sector

3.4 Ballast Life

The period for this analysis is the ballast service life, which is the ballast rated lifetime divided by the annual lighting hours. The value of 50,000 hours for both magnetic and electronic ballasts was agreed to by most stakeholders and used in this analysis. For a four-foot lamp with a life of 50,000 hours, operated 3,400 hours/year, the period of

374

analysis would be 50,000 hours/ 3,600 hours/year = 14.7 years. The period of analysis therefore varies according the life of each ballast and its annual operating hours. Figure 6 shows the ballast life distribution assumed for this analysis. This is a Weibull distribution starting at 18,000 hour life. Weibull distributions describe data resulting from life and fatigue tests and are commonly used to describe failure time in reliability studies.

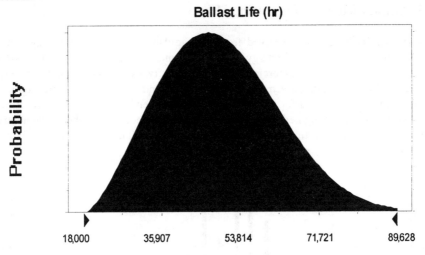

Ballast Life (hr)

| 18,000 | 35,907 | 53,814 | 71,721 | 89,628 |

Figure 6: Ballast Life Distribution in hours

4 Results

In this section, we describe probability-based life-cycle cost analyses performed to account for the full range of possible values for four of the most important LCC input variables: ballast life, ballast price, ballast operating hours, and electricity rate. LCC changes and payback periods are calculated using a probability distribution of possible values for the input variables weighted by their likelihood of occurrence and Monte Carlo simulation. The model used to calculate the LCC and payback periods can be found at the DOE Office of Codes and Standards website (URL: http://www.eren.doe.gov/buildings/codes_standards/applbrf/ballast.html). A description of the LCC model can also be found in Section A.7 of Appendix A of the ballast TSD. The probability distributions for the key variables were shown earlier.

Figure 7 presents the changes in life-cycle cost as a result of switching from the baseline energy-efficient magnetic ballast to an electronic ballast operating two T12 lamps. We calculated the change in LCC for 10,000 combinations of the four input variables. A Monte Carlo simulation is used to select from the distributions according to the frequency of occurrence of each possible value of input. In particular, the figures show the probability that a particular delta life-cycle cost will occur. This can be interpreted as the fraction of the ballast population that experiences a particular delta LCC. In this case, the LCC will be reduced about 80% of the time (represented by the values to the right of the $0 marker on the x-axis), with a mean savings of $6 over the ballast lifetime.

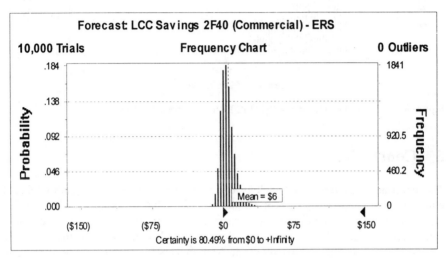

Figure 7: Life-Cycle Cost Savings for 2F40T12 ERS in Commercial Sector

4.1 Payback Period

Payback period is the amount of time needed to recover, through lower operating costs, the additional investment in increased efficiency. From Figure 8, we can see that, for the case of switching from the baseline energy-efficient magnetic ballast to an electronic ballast operating two T12 lamps, about 96 % of the time the payback period is less than the mean lifetime of the ballast (~15 yrs). The median payback period is 5.4 years

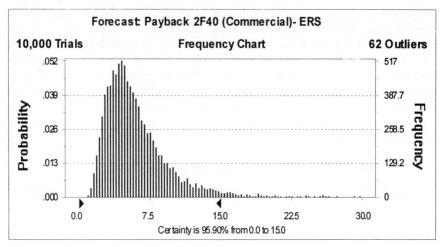

Figure 8: Payback Distribution for 2F40T12 ERS in Commercial Sector

5 Conclusion

The uncertainty analysis methodology described here has one major advantage, it eliminates the need for agreement on one specific value for point estimates of key input variables. This reduces the time spent on multiple reanalyses where the input values are changed slightly, and thus the time needed to complete the standards process. The methodology also allows decision makers to present a range of possible outcomes to interested parties, or stakeholders in the efficiency standards process.

Acknowledgment

This work was supported by the Assistant Secretary for Energy Efficiency and Renewable Energy, Office of Building Research and Standards, of the U.S. Department of Energy under Contract No. DE-AC03-76SF00098.

6 References

[1] Energy Information Administration, 1998. Annual Energy Outlook, 1999: with Projections through 2020, DOE/EIA-0383(99), December, 1998.

[2] Lumina Decision Systems, 1996. *Analytica User Guide*, Lumina Decision Systems, Los Altos, CA 94022.
[3] Xenergy, 1995. Extract of XENCAP lighting audit database showing lighting equipment and lighting hours of operation for over 24,000 buildings nationwide for years 1990-95.

A General Design for Energy Test Procedures

Alan Meier

Lawrence Berkeley National Laboratory

Abstract. Appliances are increasingly controlled by microprocessors. Unfortunately, energy test procedures have not been modified to capture the positive and negative contributions of the microprocessor to the appliance's energy use. A new test procedure is described, which captures both the mechanical and logical features present in many new appliances. We developed an energy test procedure for refrigerators that incorporates most aspects of the proposed new approach. Some of the strengths and weaknesses of the new test are described.

1 Why Are New Energy Test Procedures Needed?

The technologies employed in major appliances are undergoing a major transformation. In the past, most aspects of an appliance's operation were controlled by the user. New appliances, however, have microprocessor controls, which may adjust the appliance's operation without any action by (or even knowledge of) the user. The microprocessor can gather information through sensors or from memory of previous cycles to select an operating strategy that results in enhanced amenities or services to the user (Meier 1997; Meier 1998).

This trend has the potential to save energy, water, and other resources in many different ways. For example in washing machines, sensors may measure the weight of clothing and the extent of soiling. The microprocessor will use this information to select the minimum amount of water and detergent to achieve clean clothes. Microprocessors can also control variable-speed motors in air conditioners and refrigerators; this will allow better temperature regulation, dehumidification, often with less energy than traditional approaches.

Unfortunately the energy savings from microprocessor controls are not fully captured in the present energy test procedures. Some omissions in the present tests are that they:

- Fail to include part-load conditions (less than full loads in washing machines, cooling or heating at less than steady-state output, etc.)
- Ignore learning capability from previous cycles
- Ignore sensing special conditions for service (such as level of soiling and type of fabric)
- Fail to recognize communication between the appliance and a network (including the internet)

In some cases these omissions lead to only a small discrepancy between the laboratory measurements and actual use, but this discrepancy is likely to grow as microprocessors become more sophisticated.

At the same time, some manufacturers are programming the microprocessors in appliances to recognize when the appliances are being tested. When the unique test conditions are identified, the microprocessor modifies performance such that it uses less energy than it would under ordinary conditions.

This paper outlines a general approach to energy tests for this new generation of appliances. These tests aim to assess the integrated performance of both the appliance's mechanical features (the "hardware") and its logical features (the "software").

2 Elements of the New Approach

In the new approach, the test procedure consists of "hardware tests" and "software tests." The results of these tests are inputs to a model that simulates the appliance's energy use. The simulation model programmed to predict energy use in any conditions requested by the user.

The test conditions may be those specified by any of the major existing test procedures (ISO, IEC, DOE or JIS). This predicted energy use is the information needed to demonstrate compliance with energy efficiency regulations or to prepare an energy label. Figure 1 shows the flow of information in the proposed test procedure.

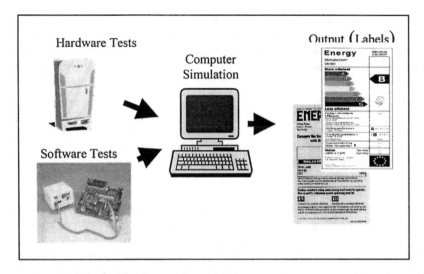

Figure 1: Elements of the new energy test procedure.

The hardware tests differ from present tests because the goal is to extract performance parameters, such as the COP, the water consumption, the heat loss coefficient, etc., which can be used to simulate the appliance's performance. The software tests are a new concept but are similar to "reverse engineering", which is often performed by electronics manufacturers and others in order to understand a competitor's product. In the following sections, each of these elements is described in detail for the refrigerator test procedure.

3 Application of General Approach to Refrigerator Test Procedures

This approach differs significantly from existing test procedures, although certain aspects are similar to specific tests. For example, the U.S. DOE test procedure for central air conditioners uses a simple hardware test to derive performance parameters that are used to calculate a seasonal performance value (Domanski 1989). The original Japanese refrigerator test (Japanese Standards Association 1979) combines results from a "winter" and a "summer" condition to derive a weighted result. This is a crude simulation model that produces energy consumption in an "average" condition. However, no test has combined all aspects of the proposed test procedure. For this reason, we applied the general approach to a specific appliance: the refrigerator. The goal was to learn what kinds of problems might arise. Some of the results are described below.

3.1 Hardware Tests

The goal of the hardware tests is to collect overall mechanical efficiency parameters for the appliance. An example of the kind of efficiency parameters collected in the hardware tests for a refrigerator is shown in Figure 2.

Four separate hardware tests are performed. The shell test seeks to measure the overall heat loss coefficient of the refrigerator's box. The door test measures the heat gain caused by door openings (and thus captures certain aspects of the refrigerator's geometry). The COP test measures the efficiency of heat extraction over a range of temperatures. Finally, the load test captures the refrigerator's response to internal sensible and latent loads. These parameters are entered in the simulation model to help predict energy consumption of the refrigerator.

The hardware tests need to satisfy two conflicting requirements. The simulation model needs detailed data in order to be properly calibrated. On the other hand, hardware tests are expensive and have limited precision. The challenge is to devise hardware tests that are both simple to perform and yield sufficient information to model the appliance's performance accurately over a wide range of conditions.

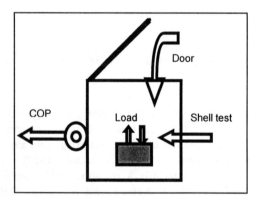

Figure 2: Major parameters in the refrigerator hardware tests

The hardware test proposed to measure the refrigerator's efficiency of heat extraction (COP test) illustrates our attempts to satisfy the need for simple tests and performance data over a wide range of conditions. The goal is to measure the COP from part-load to full-load and at different ambient temperatures. In addition, the test must recognize that future compressors are likely to operate at different speeds or capacities. The test requires the placement of a heater that can be remotely-controlled inside the refrigerator. The test consists of the following steps:

1. Place the refrigerator in a test chamber at a specified ambient temperature and lower the inside temperature until the compressor must cycle to maintain the desired temperature.
2. The test begins when the compressor on-time is much less than the compressor off-time.
3. The heater is switched on at low power, and the refrigerator's temperature is allowed to stabilize at the original temperature.
4. The power of the heater is increased in small steps, each time waiting for temperature stabilization to be achieved. Intervals between cycling become gradually shorter.
5. The power is increased until the compressor is no longer able to maintain the desired inside temperature–even when the compressor is operating constantly–and the inside temperature begins to rise.
6. Repeat steps 1–5 at a second ambient temperature.

The results of one series of tests are illustrated in Figure 3. The left chart shows the incremental increases in heater power over time.

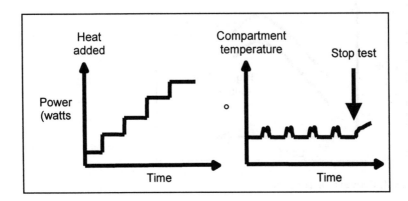

Figure 3: Hardware test to measure COPs.

The right chart shows the internal temperature at each level of heat input. The temperature fluctuates a little after each increase but then returns to the thermostat setting. When the compressor's heat removal capacity is finally overwhelmed, the inside temperature climbs above the thermostat setting.

These measurements are sufficient to develop a performance curve (as shown in Figure 4) for the compressor system. The surface captures efficiency at both part-load conditions and at two ambient temperatures. Interpolation to other temperatures is also possible without significant loss of accuracy. These performance curves will be used by the simulation model to predict energy consumption of the tested unit over a broad range of conditions.

We have developed hardware tests for measuring heat loss, door-opening, and loads, but there is not enough space here to describe them and are reported elsewhere (Wihlborg and Ernebrant 1999). Most of the tests could be easily automated, so they need not necessarily be more complicated to perform. Some have already been done and reported in the literature (Alissi, Ramadhyani et al. 1988).

3.2 Software Tests

The goal of the software test is to evaluate the "intelligence" of the microprocessor. How does it respond to different situations? Are its algorithms crude or sophisticated? Are there certain combinations of conditions where the appliance behaves erratically? Are the algorithms consistent with the appliance's mechanical parameters?

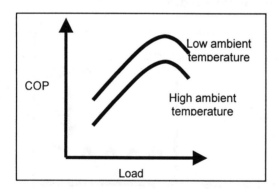

Figure 4: Derived COP parameters

We assume that the microprocessor can be interrogated, either in place via a communications cable, or extracted and tested on the bench. Even though this feature is not now generally available, some manufacturers already have installed this capability to interrogate the microprocessor in order to diagnose technical problems. (One manufacturer uses the communication to erase the microprocessor's memory of previous cycles in order to ensure consistent testing.) Microprocessors can already be removed in most cars and some dishwashers (because manufacturers expect to change the cleaning algorithms when new detergents are introduced). Manufacturers are also creating "network ready" appliances which assume communications ability.

Our approach to assessing the appliance's microprocessor relies on a form of reverse engineering. A computer creates thousands of different combinations of conditions. It submits them to the appliance microprocessor (as if these were the inputs from the controls and sensors) and records the microprocessor's responses. The computer assembles the microprocessor's replies into "response surfaces." An example of a 2-dimensional hypothetical response surface is shown in Figure 5.

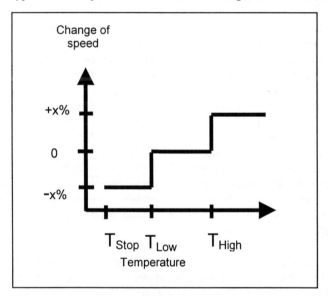

Figure 5: A hypothetical response surface for a variable speed motor.

The response surfaces are converted into equations and passed to the simulation model. These equations will be used to control the appliance's operation.

Response surfaces may be relatively simple for most of today's appliances because the microprocessor typically controls one or two aspects of the appliance. For example, microprocessors in many refrigerators control the defrost interval and data displayed on a small screen. Future response surfaces will become very complex when the microprocessor controls several functions (such as defrost interval, compressor speed, and temperature distribution in a refrigerator) and relies on more than a few sensors. Microprocessors employing fuzzy logic, where their response depends on earlier conditions, add yet another layer of complexity. Most defrost controls remember only a few defrost cycles but at least one manufacturer tracks the last ten cycles.

We anticipate that the microprocessor will need to be interrogated millions of times in order to develop smooth response surfaces. This can nevertheless be accomplished in a reasonably short time because the interrogations require very small fractions of a second.

3.3 Simulation Model and Outputs

The third element in the test procedure is a dynamic simulation model of the appliance. This model simulates the operation of the appliance and predicts its energy consumption under specified conditions. The specified conditions may be those specified by the ISO, DOE, or JIS tests or any other conditions.

Simulation models for several appliances have already been developed for refrigerators (Arthur D. Little Inc. 1982), air conditioners, furnaces, and water heaters (Hiller and Lowenstein 1992; Paul and Whitacre 1993). Most of these models assume steady-state behavior; these are not realistic when trying to simulate the decisions of the microprocessor, which are dynamic. Recently, two water heater models have been developed that simulate dynamic aspects, such as hot water draw-downs, the recovery period, and thermal stratification. These models resemble those used to simulate building energy loads, in which the materials, their physical properties, and geometry are all specified.

We also use a dynamic model but employ a different approach for specification of the appliance. Our model relies on lumped parameters instead of detailed specifications of all of the physical characteristics of the materials and components. The model is very general and relies on inputs from the hardware and software tests to give it structure and uniqueness. Complex, dynamic models are now reasonably easy to construct using commercially available software, so models for other appliances could be made. The software and hardware tests must be carefully coordinated with the simulation model. The challenge is to develop hardware and software tests that provide suitable parameters for use by the simulation model. We found that for the refrigerator model the approach to simulation depended upon the kinds of data available from the hardware and software tests. Likewise, we revised the hardware tests several times in order to provide the most useful information to the model. In other words, an iterative approach is most successful.

The model simulates the appliance's operation and estimates its energy consumption. The model requires a detailed description of the hypothetical schedule and conditions, that is, the period of measurement, the ambient temperature, inside temperatures, humidity, number of door openings, etc. For example, the DOE refrigerator test specifies that the refrigerator be tested at ambient temperature of 32°C for 24 hours, or from defrost to defrost. The JIS refrigerator test includes door openings with a specific schedule.

We successfully simulated the energy use of the same refrigerator when tested according to the ISO, DOE, and JIS test conditions. Thus, we were able to prepare energy consumption estimates (such as those needed for energy labels or regulatory purposes) for three different markets using a single test procedure.

4 Conclusions

We have presented a framework for a new energy test procedure. It addresses a major flaw in current test procedures, that is, they test only the mechanical aspects of the appliances and ignore software aspects. Microprocessor control (coupled to extensive use of sensors) is likely to save significantly more energy than mechanical improvements in the next decade, so it is important to capture those benefits.

Our approach also offers a novel solution to the harmonization problem. In the case of our approach, everyone can agree on the same hardware tests, software tests, and simulation model. Each country may select its own output from the model. This output (which might appear on its energy-use labels, or be part of its minimum efficiency regulations) would reflect the unique conditions faced there.

Our approach is considerably more complicated than current test procedures. This is not surprising, because it seeks to capture the energy impacts of complex interactions between mechanical and controls aspects of the appliance. Furthermore, we have not demonstrated that the hardware and software tests for one appliance are technically feasible and that a model can simulate all the key aspects of operation. Nevertheless, we have shown how the parts fit together, and we have shown potential benefits from the proposed test procedure. We plan to refine the test and apply it to a real refrigerator soon. In the meantime, we invite your comments and suggestions.

Acknowledgements

This work was supported by the Assistant Secretary for Energy Efficiency and Renewable Energy, Building Technologies, of the U.S. Department of Energy under Contract No. DE-AC03-76SF00098.

5   References

[1] Alissi, M.S., S. Ramadhyani and R.J. Schoenhals 1988. "Effects of Ambient Temperature, Ambient Humidity, and Door Openings on Energy Consumption of a Household Refrigerator-Freezer." *ASHRAE Transactions* **94**(2): 1713-1735.

[2] Arthur D. Little Inc. 1982. *Refrigerator and Freezer Computer Model User's Guide*. Cambridge (Mass.): Arthur D. Little Inc.

[3] Domanski, P.A. 1989. *Rating Procedure for Mixed Air-Source Unitary Air Conditioners and Heat Pumps Operating in the Cooling Mode - Revision 1*. NISTIR 89-4071. National Institute of Standards and Technology.

[4] Hiller, C. and A Lowenstein. 1992. *WATSIM User's Manual, Version 1.0: EPRI Detailed Water Heating Simulation Model User's Manual*. Palo Alto (Calif.): Electric Power Research Institute.

[5] Japanese Standards Association. 1979. *Household Electric Refrigerators, Refrigerator-Freezers and Freezers, Japanese Industrial Standard JIS C 9607*. Tokyo: Japanese Standards Association.

[6] Meier, Alan. 1998. "Energy Test Procedures for the Twenty-First Century" In the *Proceedings of* 1998 Appliance Manufacturer Conference and Expo. Nashville, Tenn.: Appliance Manufacturer Magazine.

[7] Meier, Alan K. 1997. "The Next Generation of Energy Test Procedures" In the *Proceedings of* Proc. of First International Conference on Energy Efficiency in Household Appliances. Florence, Italy: Springer Verlag.

[8] Paul, D. and G. Whitacre. 1993. *TANK Computer Program User's Manual with Diskettes: An Interactive Personal Computer Program to Aid in the Design and Analysis of Storage-Type Water Heaters*. Columbus (Ohio): Battelle Memorial Institute.

[9] Wihlborg, Mats and Stefan Ernebrant 1999. "Energy Test Procedures for Appliances". Department of Engineering. Lund (Sweden), University of Lund.

Refrigeration Energy Label Standard Measurements Linked to Energy Consumption in Daily Use

Franco Moretti

Government Affairs, Whirlpool Europe

Performance standard tests have been selected by the ISO (International Standard Organisation) with the purpose of representing the real world average use. Within climatic classification the global standard testing was customised by the European Team CEN (Comité Européen de Normalisation) TC 44 (with Italian Presidency and Secretariat at UNI - Milan - [1]) to regional conditions. Energy label directive 94/2 EC (21.1.1994) and the Energy Efficiency Directive 96/57 EC (3.9.1996) have made the refrigeration standards legally binding for the performance and energy consumption declaration. Therefore the CEN TC44 has got a mandate from EU Commission to define the testing conditions. The four standards covering the cooling appliance type ([II]) have been condensed into the single EU standard (EN 153 - 1990 revised 1995). Cooling appliances of all climate classes should be labelled in Europe with energy consumption test at 25°C ambient temperature with closed door(s).

Purpose of the tests is to emulate the real use with simple, fast, repeatable and reproducible testing procedures by all test laboratories around the globe. Testing with closed door is the best answer as long as the ambient temperature conditions are settled in the proper way compared to the real use.

The Italian delegation of the EU Team CEN TC44 decided in 1995 to assess the suitability of the standard test in relation to real use by financing a series of tests commissioned to IMQ (Italian Quality Institute, Milan - [III]) to evaluate the energy consumption with closed door and by opening the doors.

The appliance chosen was a static refrigerator freezer "four stars" with "Top Mount" freezer. As door opening sequence the Japanese standard JIS C9607-1993 was chosen to emulate the practical use

390

Table I

Item	Refrigerator-freezer	
	Door of fresh food compartment	*Door of freezer compartment*
• Rate of open and close	Every 12 min.	Every 40 min.
• Daily cycles of open and close	50	15
• Opening angle	45° (1)	
• Opening time	10 sec. (fully opened for at least 5 sec.)	

(source IMQ report)

The ambient conditions in the various tests carried out were:

testing room temperature = 16°, 25° and 32°C
with Relative Humidity RH 75%

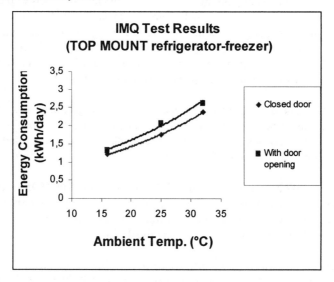

Figure 1

The door opening sequence utilised according to the 1993 Japanese standard is heavier compared to the normal use. In fact the Japanese standard was changed in 1999 reducing the daily door opening to 25 cycles for refrigerator compartment and 12 cycles for the freezer.

The results are in IMQ report *n. 55S0341 of 1/12/1995.* In graph n. 1 IMQ test on a Top Mount Refrigerator-Freezer are detailed.

By analytical calculation and /or graphical analysis as shown in next graph (*Graph 2*), it is possible to calculate the equivalent ambient temperature of the test at 25°C with closed door finding the ambient temperature giving the same amount of energy consumption by opening the door.

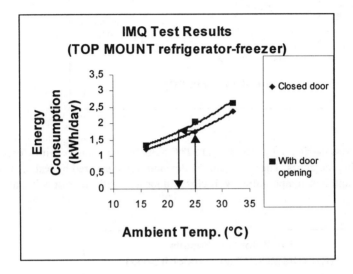

Figure 2

Using the door opening sequence of *Table I*, you can see from *Graph 2* that testing at 25°C ambient temperature with closed door, gives equivalent test result compared to the test by opening the door at 21,5-22°C ambient temperature.

As a parallel analysis Whirlpool has carried out the same comparative testing on a Combi refrigerator freezer with "Bottom Freezer", notified to CEN TC44 team in January 1996.

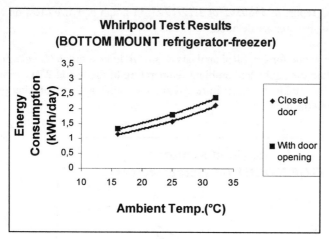

Figure 3

The test was carried out at 16, 25 and 32° C ambient temperature with closed door and by opening the doors. The same energy demand was measured by testing an appliance at 25°C ambient temperature with closed door and with door opening at 21,5-22°C (*Figure 4*).

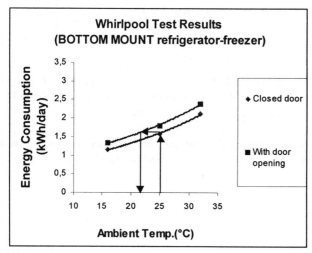

Figure 4

Considering that the European yearly average kitchen temperature in modern houses is around 18-19°C (16-17°C in older houses - *source Good Housekeeping -* [IV]), the tests confirmed that today's energy label values required to be declared by the energy label directive 94/2 according to the EN153/90 and ISO standard testing at 25°C with closed door are 10 to 12% higher than the yearly energy consumption in real use.

Testing with closed door allows an easier appliance stabilisation of cooling appliances in the test room with better accuracy of the measurement and better repeatability of test results between laboratories.

Within the SAVE program, TNO (Holland) has led a Ring Test showing a large difference even with closed door due to the performance standard interpretation .

In case of test with door opening results among the 18 laboratories (9 manufacturers' laboratories and 9 independent test laboratories) would have varied to a much greater extent!

A further activity to define an "Operative Code of Laboratory Practice" to minimise these differences according to the present standard regulation is ongoing. CECED ([V]) Manufacturers are involved.

1 Conclusions

IMQ study, financed by Italian manufacturers in the framework of the CEN TC 44 revealed and confirmed that present energy consumption declared on cooling appliances is 10 to 12% higher than the yearly energy consumption in real use also supported by Whirlpool Europe test results.

As a consequence the concept of the "maximum value in energy declaration" mentioned as one of the potential hypotheses within SAVE discussion, will impose 18-25% higher declared value compared to the yearly energy consumption in the real use of refrigerator appliances!

A further test programme carried out by Energy Ambient / Consumer Organisations may allow to collect more detailed information than the comparative test result I mentioned in my presentation. Test programme should measure appliances in laboratory before installation in customer houses monitoring also the average yearly kitchen ambient temperature.

In any event I strongly recommend taking into consideration the outcomes of the IMQ/ CEN study when dealing with potential new European energy declaration policies for cooling appliances.

Endnotes

[I] UNI - Ente Nazionale Italiano di Unificazione, Via Battistotti Sassi 11/b - Milano - Tel. +39 02 7002400; Fax +39 02 70105992

[II] Normative References:

EN 28187 /ISO 8187 freezers	Performance of household refrigerating appliances - Refrigerator-
EN 25155 /ISO 8187 Essential	Household frozen food storage cabinets and food freezers -
	Characteristics and test methods
EN 27371 /ISO 8187 with or	Performance of household refrigerating appliances - Refrigerators
	Without low temperature compartment
EN 2856 1 /ISO 8187 refrigerator-	Household frost free refrigerating appliances - Refrigerators,
	freezer- Frozen food storage cabinets and food freezers cooled by
forced air	
	circulation - Characteristics and test methods

[III] IMQ - Istituto Marchio di Qualita' - Via Quintiliano 43, 20138 Milano - Tel. +39 02 50731; Fax 02 5073271

[IV] Source "Good Housekeeping Institute" London - according to the National Energy Foundation , average room temperature in kitchen in modern house is 19-19°C while in older house was 16-17°C

[V] CECED - European Committee of Manufacturers of Domestic Equipment

Cold Appliance European Ring Test

S.M. van der Sluis

Netherlands Organisation for Applied Scientific Research (TNO)

1 Summary

The project "Inter laboratory comparison of test results on household refrigerators and freezers" is an EC SAVE project carried out in the 1998 project round.

The first objective of the project "Inter laboratory comparison of test results on household refrigerators and freezers" is to experience and evaluate potential differences in test results between laboratories for testing of refrigerators and freezers under the EN 153 standard, by means of a ring test covering a large number (18) of laboratories. Based on the test results, the objective is to formulate a "testing guide" which can be used to minimise the inter laboratory spread in measurement results and the level of disputes on data declarations.
A second objective is to study possibilities for creating a smaller error margin between energy consumption declarations and test results. The currently applicable margin of 15% constitutes de facto an obstacle in reaching agreement between industry and consumer associations on the interpretation of test results. When an agreement can be made on interpretations among all the labs involved, the level of disputes on data declarations can be reduced, and confidence in energy labels can be boosted.

Measurement results are presented from the ring test between 18 European laboratories, participating in the project. The results reported concern energy consumption and volume. Results on energy consumption measurement show a large spread between laboratories, supporting the call for a "testing guide" which details the energy consumption measurement further than the Standard EN 153.

The project has been made possible by additional contributions from CECED, NOVEM, DEA, van de Bunt and the participating laboratories.

2 Introduction

The energy consumption of refrigerating appliances can be measured according to the European Standard EN 153. Measurements are performed by industry, e.g. in order to establish the "rated" energy consumption, by consumer organisation laboratories to establish the actual energy consumption of specific products, and by independent laboratories e.g. in order to verify the rated energy consumption underlying the energy label.

The first objective of the project "Inter laboratory comparison of test results on household refrigerators and freezers" is to experience and evaluate potential differences in test results between laboratories for testing of refrigerators and freezers under the EN 153 standard, by means of a ring test covering a large number of laboratories.

3 Participating Laboratories

The project "Inter laboratory comparison of test results on household refrigerators and freezers" is carried out in the framework of the EC SAVE programme, 1998 project round. The main contractor for this project is the Netherlands Organisation for Applied Scientific Research TNO.

CECED, the association of European white goods manufacturers, participates in the project as a partner and is involved in all project phases. CECED provides advice to the project, represents the manufacturers and acts as a channel for the input to the project from white goods manufacturers.

Table 1: Participating laboratories.

Manufacturer's labs / CECED	Independent & consumer organisation labs
ARÇELIK (Turkey)	CA Research & Testing Centre (U.K)
Antonio Merloni (Italy)	ENEA ERG - QUA Icelab (Italy)
BSHG (Germany)	Forbruger Styrelsen (Denmark)
Candy elettrodomestici (Italy)	IMQ (Italy)
Electrolux (Sweden)	REGENT (The Netherlands)
Groupe Brandt (France)	TNO - MEP (The Netherlands)
Liebherr (Germany)	TTS Työtehoseura (Finland)
Merloni (Italy)	LCOE (Spain)
Vestfrost (Denmark)	
Whirlpool (Italy)	

4 Testing Programme

Measurements (energy consumption at 25 °C & volume) are made at each lab on 3 appliances: refrigerator, refrigerator/freezer and freezer. With 18 laboratories participating, the logistics plan includes three separate "rings", each with one set of 3 appliances. All appliances are tested at the start at TNO laboratory, and again at the end.

Each laboratory fills out a checklist, containing the testing procedures (installation, stabilisation, etc.) and testing details (spacers, cardboard, sensor locations etc.) as used by the laboratory. Visiting personnel from other labs fill out a "visitor checklist", on which remarks concerning the procedures & details can be made.

For the ring test, it is decided to make use of the same version of EN 153 and underlying standards, in order to eliminate any additional deviations in test results caused by differences in translations, versions and national adaptations to the standards. It is agreed that the standards to be used for measurement will be the following versions:
-	EN 153 : 1995 E.
-	for a freezer: EN/ISO 5155 : 1995 E
-	for a refrigerator: EN/ISO 7371 : 1995 E
-	for a refrigerator/freezer: EN 28187 : 1991

For the same reason, manufacturer's loading plans - for the refrigerator/freezer and for the upright freezer – have been used in testing in the ring test.

The refrigerator used in the ring test is a table-top refrigerator without frozen food compartment, model Whirlpool ART 417 /G with a declared volume of 158 litre and a declared energy consumption of 219 kWh/year (Energy labelling class C).

Figure 1a: Ring test refrigerator.

The Refrigerator / freezer selected is a two-door one compressor static appliance model Bosch KGV 3104. The fresh food compartment (declared volume 198 litre) is located above he frozen food compartment (declared volume 105 litre), and the compressor is located at the bottom of the appliance. The declared energy consumption for this appliance is 438 kWh/year (energy labelling class B).

Figure 1b: Ring test refrigerator/freezer

The (upright) freezer selected is a static appliance; model AEG Arctis 2232 GS, with a declared volume of 184 litres. The compressor is located at the bottom of the appliance. The declared energy consumption for this appliance is 420 kWh/year (energy labelling class C).

Figure 1c: Ring test freezer)

5   Test Results (Energy Consumption)

The test results from the different laboratories are presented on the basis of the "ring" in which these labs participated (North, West and South). Each ring measured a different appliance (of the same type/model), with slightly differing "real" energy consumption (caused by production spread). Test results are presented based on group averages which have been set by definition to be "100".

Table IIa: Test results for refrigerators, compared to group averages (separately for groups North, West & South). Results are listed chronologically.

		Range		91.6 - 108.7		
		Average μ		100.0		
		Standard deviation σ		4.7		
		σ/μ		0.047		
	1	2	3	4	5	6	7
North	96.2	103.4	101.9	96.8	99.2	102.6	
West	91.6	99.4			108.7		100.2
South	93.5		98.7		107.4	97.6	102.7

Table IIb: Test results for refrigerator / freezers, compared to group averages (separately for groups North, West & South). Results are listed chronologically.

Range	91.8 - 108.7	
Average μ	100.0	
Standard deviation σ	4.7	
σ/μ	0.047	

	1	2	3	4	5	6	7
North	94.0	98.3	104.3	97.7	101.7	101.7	102.4
West	96.7	108.7			96.5		98.1
South	91.8		97.7		105.8	98.0	106.7

Table IIc: Test results for freezers, compared to group averages (separately for groups North, West & South). Results are listed chronologically.

Range	92.7 - 105.2	
Average μ	100.0	
Standard deviation σ	3.8	
σ/μ	0.038	

	1	2	3	4	5	6	7
North	99.2	100.1	92.7	104.5	105.2	99.8	98.5
West	101.4	103.7			94.5		100.4
South	102.7		100.1		104.9	95.7	96.5

6 Test Results (Volume Measurements)

Table IIIa: Results for refrigerator storage volume measurements (in litres)

Range	158.0 - 162.0	
Average μ	159.7	
Standard deviation σ	1.5	
σ/μ	0.009	

	1	2	3	4	5	6	7
North		158	161	162	158	158	
West	161	159			162		161
South			160		158	160	159

Table IIIb: Results for ref. / freezer volume measurements, refrigerator volume (in litres)

Range	190.0 - 198.0	
Average μ	194.3	
Standard deviation σ	2.5	
σ/μ	0.013	

	1	2	3	4	5	6	7
North		194	194	194	198	196	
West	192	192			198		194
South			194		194	197	190

Table IIIc: Results for refr./freezer volume measurements, freezer volume (in litres)

Range   77.9 - 105.0
Average μ 96.4
Standard deviation σ 7.9
σ/μ 0.082

	1	2	3	4	5	6	7
North		99	98	78	105	104	
West	98	86			101		98
South			99		90	105	93

Table IIId: Results for freezer storage volume measurements (in litres)

Range   174.5 - 185.0
Average μ 181.3
Standard deviation σ 3.2
σ/μ 0.018

	1	2	3	4	5	6	7
North		181	181	184	184	185	
West	180	179			184		181
South			184		184	175	177

7 Analysis of Test Results

The results of the cold appliance ring test have been presented in the preceding section in an anonymous representation. In this representation it is not possible to discern the different laboratories. Still, interest has risen in the differences in results between the group of industrial laboratories and the group of independent/consumer laboratories participating in the project. The results are presented in table 4, split up for the two groups (μ = average, σ = standard deviation). In this presentation the TNO laboratory has been counted as one lab, taking the average of three tests on each appliance.

Table IV: Comparison of test results from industrial and independent/consumer association labs.

	Industrial labs	Independent/consumer labs
refrigerator test, $\mu \pm \sigma$	101.4 ± 3.8	100.2 ± 5.0
refrigerator-freezer test, $\mu \pm \sigma$	102.4 ± 2.9	98.7 ± 4.8
freezer test, $\mu \pm \sigma$	99.9 ± 4.1	100.0 ± 3.8

Further interest in the analysis extends to variables that may influence the inter laboratory spread, and can therefore contribute to the reduction of spread when these variables are better controlled.

A check has been made on the linearity of energy consumption versus compartment temperature for the ring test refrigerator, by plotting all test results in accordance with the temperature interval between measurements "A" and "B" used for the interpolation (Figure 2). The (horizontal) regression line for this plot indicates there is no relation between interpolation temperature interval and measured energy consumption for this specific refrigerator. We conclude that for this specific refrigerator non-linearity is not the cause of spread in test results.

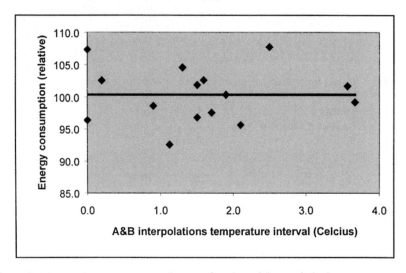

Figure 2: Measured energy consumption as a function of (interpolation) temperature interval

Based on reported use or non-use of the Cold Accumulators during the energy consumption test in the participating labs, a comparison has been made between the energy consumption measurement results for three labs not using Cold Accumulators and 7 laboratories that did use the Accumulators. The results are presented in Figure 3, showing an identical average for both groups. The use or non-use of cold accumulators in this specific freezer is therefore not a probable cause for the observed spread in energy consumption test results.

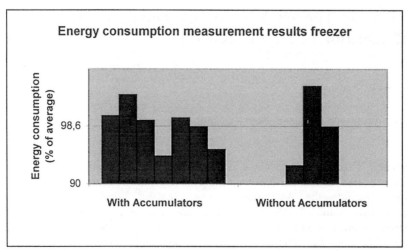

Figure 3: Analysis of energy consumption with and without Cold Accumulators.

Factors that now are the most prominent suspect to contributing to inter laboratory spread are the following:
- Apparent stability versus real stability
- Test room disturbances (airflow, lighting)

Ongoing research and testing in the framework of the project will aim on further investigation of these matters, and their influence on the inter laboratory spread in test results. Based on the outcome of this work, the objective is to formulate a "testing guide" which can be used to minimise the inter laboratory spread in measurement results and the level of disputes on data declarations.

8   Conclusion

The ring test has provided results on the inter laboratory spread in outcomes for the energy consumption test of three simple appliances (a refrigerator, a refrigerator freezer and an upright freezer). These results show that the inter laboratory spread is quite considerable. Differences between the "highest" and the "lowest" test results are observed amounting to 17% (for refrigerators and refrigerator/freezers) and 13% (for freezers). Considering the allowable error margin in the European standard EN 153 of 15%, the observed inter laboratory spread is too high (as it leaves no space for other sources of differences, such as production spread).
Ongoing research and testing in the framework of the project will aim on finding the sources of inter laboratory spread in test results. Based on the outcome of this work, the objective is to formulate a "testing guide" which can be used to minimise the inter laboratory spread in measurement results for laboratories adhering to this testing guide.

A Generic Approach to the Determination of Refrigerator Energy Efficiency

Pradeep K. Bansal

The University of Auckland, New Zealand

Abstract. This paper presents a comprehensive research study on the energy efficiency test standards for household refrigerators/freezers. The standards studied, include the Australian - New Zealand Standard (AS/NZS), the International Standard (ISO), the American National Standard (ANSI), the Japanese Industrial Standard (JIS), and the Chinese Taipei National Standard (CNS). The paper highlights the salient differences among these test standards and presents energy consumption data for some of the standards.

In order to study the realism and weaknesses of test standards, various refrigerator-freezers were tested under varying test conditions of AS/NZS and ISO with a view to quantify the effect of different variables (e.g. ambient air temperature, relative humidity, cabinet internal air temperature and cabinet size) on the energy consumption. The effect of relative humidity on energy consumption was found to be marginal whereas the energy consumption increased significantly, as expected, with cabinet size, ambient air temperature and doors openings. Further, the study examines the merits and demerits of current test standards and proposes new guidelines that should be considered on how to improve them so that they can represent realistic *"real"* world energy use.

Key Words: Refrigerator-freezers, Energy Consumption, International Standards, AS/NZS, ANSI, CNS, ISO, JIS

1 Introduction

There are a large number of energy consumption test standards for domestic refrigerators being used around the world. The energy consumption results can differ significantly from one test standard to another for the same cabinet. This occurs due to different test ambient temperatures and relative humidities, compartment internal temperatures, door openings and operational requirements away from the test conditions. Earlier testing analyses by Bansal et al.[1-2] revealed that refrigerators perform best when tested to the standard for which they were designed. Testing of a cabinet to other standards gives different energy consumption results. Due to the complexity of refrigeration systems, converting energy consumption from one test procedure to another is generally very difficult unless substantial additional information is collected during the tests.

For a fairer comparison, it is desirable to develop an algorithm, which will enable the energy consumption under one standard to be estimated when the same cabinet is tested to a different standard. A general relationship of this type would be of special interest to refrigerator manufacturers, governments, environmental groups, consumer organisations, utilities etc. Earlier, Bansal and Kruger[3] presented an analysis of test standards and performed tests on a number of cabinets (*all-refrigerators* and *refrigerator-freezers*) under five test standards, namely the Australian - New Zealand Standard (AS/NZS)[4], the International Standard (ISO)[5-6], the American National Standard (ANSI)[7], the Japanese Industrial Standard (JIS)[8] and the Chinese Taipei National Standard (CNS)[9-10]. Note that the Korean Standard is essentially identical to the Chinese Taipei National Standard - both appear to be a JIS version of the standard (at the warmer ambient) but without door openings.

They proposed an algorithm to predict the energy consumption of a cabinet from one standard to another of the following general form -

$$E_A = \left(\frac{COP_B}{COP_A} \cdot \sqrt{\frac{\Delta T_A}{\Delta T_B}} \right) \cdot E_B \qquad (1)$$

where suffixes 'A' and 'B' represent any two standards, E is the energy consumption, and ΔT is the temperature difference between the compartment and the ambient air. This formula represented the experimental results reasonably well for most of the standards except the JIS where the agreement was not that good, probably because the effect of door openings was neglected.

It would be desirable to harmonise various test standards into one standard that can be accepted internationally for reasons of uniformity and trade. Therefore, it is crucial to quantify the effect of various performance variables on the energy consumption of refrigerators as the actual impact of each of these elements may vary significantly between different countries of use. A disaggregated test procedure will therefore enable the development of a general relationship to predict energy consumption under different operating conditions. The most important variables include the impact of:

a.   the ambient air temperature,
b.   ambient air relative humidity (where door openings introduce humidity into the cabinet),
c.   size of the cabinet,
d.   food loading, and
e.   frequency of door openings (introduction of warm air and humidity).

In order to study the realism and weaknesses of test standards, various refrigerator-freezers were tested under different test conditions of Australian/New Zealand (AS/NZS), variations in AS/NZS and International test standards (ISO) with a view to quantify the effect of different stated variables on the energy consumption. Further, the study examines the merits and demerits of current test standards and proposes new guidelines and approaches that should be looked into to improve the existing standards.

2   Test Standards and their Differences

Most test standards measure energy consumption at the food compartment internal temperature of 3°C and the ambient temperature of either 32° or 30°C. The only exception is the ISO which specifies 5°C for the food compartment temperature and two different ambient temperatures (25/32°C) depending on the climate classification. However, the quoted energy consumption figures in ISO are usually based on the temperate climate classification of 25°C. Also, ISO is the only test standard that specifies food loading in the freezer compartment of the frost-free refrigerator-freezers. The CNS requires the relative humidity of the ambient air to be 75% ± 5%, while the ISO recommends between 45 and 75%. AS/NZS, JIS and the ANSI do not prescribe any humidity requirements. JIS is the only standard that prescribes door openings as shown in Table II and requires tests at a second test ambient of 15°C. It weights the two results assuming 100 days at 30°C (27%) and 265 days at 15°C (73%) to evaluate the annual energy consumption.

The JIS has been recently revised (Banse[11]) to be compatible with ISO. The new standard (Method C) prescribes the food compartment temperature to be 5°C and the frequency of door openings to be 25 times/25 minutes/day and 8 times/25 minutes/day for the food compartment and the freezer doors respectively. It appears that frequency of door openings has been reduced to half in the revised standard.

3   Cabinets Chosen for the Tests

A refrigerator-freezer (R/F) consists of two or more compartments, with at least one of the compartments designed for the refrigerated storage of food at temperatures (T_{FF}) above 0°C and with at least one of the compartments designed for the freezing (T_{FR}) and storage of frozen food. Six cabinets each of refrigerator-freezers (two models of each N169, N249 and N375 type) and all-refrigerators were tested under AS/NZS, variations in AS/NZS and ISO. However, this paper discusses results only for the refrigerator-freezer cabinets (Bansal[12]). All of these R/F cabinets were frost free and had storage volumes of respectively 46, 46 and 85 litres in the freezer and 108, 181 and 278 litres in the fresh food compartments. These cabinets were divided into two sets, as shown in Table V to test the reliability and repeatability of the test data for the cabinets of the same model and manufacturer. Three cabinets of each set were tested simultaneously in the environmental chamber to assess the effect of the following set of parameters-

1. Size of the cabinet
2. Ambient temperature, and
3. Relative humidity of ambient air.

Further, the tests were extended to assess the effect of other variables, including food loading and door opening under varying test conditions.

Table I: General testing requirements for various test standards

Cabinet Type			AS/NZS	ISO	ANSI	JIS[#]	CNS
Testing Parameters →		Ambient (T_A)	32±0.5 °C	25/32± 0.5 °C	32.2±0.6 °C	15/30 ±1 °C	30 ±1 °C
		Relative Humidity	-	45 -75 %		-	75±5%
		Door Openings	No	No	No	Yes	No
Refrigerator- Freezers	*	Fresh-Food	3 ±0.5 °C	5°C	7.2 °C	3 ±0.5 °C	-
		Freezer	-9 ±0.5 °C	-6°C	-9.4 °C	-6 ± 0.5 °C	-
	**	Fresh-Food	3 ±0.5 °C	5 °C	7.2 °C	3 ±0.5 °C	3 ±0.5 °C
		Freezer	-15±0.5 °C	-12 °C	-15 °C	-12 ± 0.5 °C	-12/-15 ±0.5°C
	***	Fresh-Food	-	5 °C	-	3 ±0.5 °C	3 ±0.5 °C
		Freezer	-	-18 °C	-	-18 ± 0.5 °C	-18 ± 0.5 °C
Energy Measurement Period			Lesser of 1 kWh or 16h operation [s]	≥ 24 h	3h< t <24h 2 or more cycles	= 24 h of testing	= 24 h of testing

For star ratings (*, **, ***), please refer Bansal and Kruger[3].

\# 73% of the consumption is weighted at an ambient of 15°C and 27% at 30°C.

[s] Note that for cyclic products, the test period consists of a whole number of compressor cycles. For frost free models, the test period consists of a whole number of defrost cycles.

Table II: Door Opening Requirement For the Japanese Industrial Standard

Type	Compartment	Rate	Number of Openings	Opening Angle	Opening Time
Refrigerator- Freezer	Fresh-Food	Every 12 Min	50	90°	10 s [#]
	Freezer	Every 40 Min	15	90°	10 s [#]

\# The door shall be fully opened for at least 5 s. This is for Method A.

4 Measurement Devices and Test Requirements

T-type thermocouples were used for temperature measurements and watt/watt-hour transducers for measuring power/energy consumption of the cabinets. As per "standards" requirements, the temperature-sensing end of each thermocouple was inserted in a brass cylinder to increase its heat capacity. A National Instruments[13-14] AT-MIO 16X data acquisition (plug-in type) board with an AMUX-64T front-end analog multiplexer was used to acquire data from the devices. A LabVIEW[15] software program was used for the data acquisition and analysis of results. Details can be found in Bansal and Kruger[1].

Table III presents the conditions used to assess the effect of ambient air temperature on energy consumption. Two tests were performed for each parameter - one above and the other below the characteristic temperature (i.e. T_{FF} for the fresh food compartment or T_{FR} for the frozen food compartment) to interpolate the result. Both the tests were performed at ambient relative humidity of 60%, except when relative humidity was a variable. The last column of Table III specifies variations in AS/NZS test conditions, where the effect of different parameters (e.g. ambient temperature and relative humidity) on the cabinets is studied by varying the base test conditions of the AS/NZS. In these tests, the fresh food compartment temperature (T_{FF}) was kept at 3°C (contrary to 5°C of ISO) so that the base conditions can be consistently maintained and compared with AS/NZS. The temperature difference, $\Delta T \{= (T_A - T_{FF})$ or $(T_A - T_{FR})\}$, is the main driving force of the heat load on the cabinet. The larger the temperature difference, the higher the heat load and hence the higher the energy consumption.

Table III: General Testing Requirements for Refrigerator-Freezers

Cabinets	Testing Parameters	AS/NZS	ISO	Variation (AS/NZS)		
				I	II	III
	Ambient (T_A)	32±0.5 °C	25 ±0.5 °C	32±1 °C	25 ±1 °C	10 ±1 °C
	Relative Humidity	-	45 -75 %	40 - 80±5%	40 - 80±5%	40 - 80±5 %
Refrigerator-	Freezer (T_{FR})	<= - 15°C	<= - 18°C	<= - 15°C	<= - 15°C	<= - 15°C
Freezer	Door Openings	No	No	No	YES[$]	No
	ΔT ($T_A - T_{FR}$)	>47°C	>43°C	>47°C	>40°C	>35° C

[$] Two door-opening schemes were used as shown in Tables II and V.

A new test scheme was formulated where the effect of both the food packs and the door openings (at ambient temperature of 25°C and cabinet temperature of 3°C - referred AS/NZS 25/3°C) could be compared with the base (closed door) test of AS/NZS with inflated ambient temperature of 32°C. The adopted scheme comprised the following steps-

a. Test the cabinets as per AS/NZS 25/3°C with no food loads and no door openings (empty),

b. Load the freezer compartment with food packs but leave the food compartment empty and measure the energy consumption as per AS/NZS 25/3°C,

c. Load the freezer compartment with food packs and the food compartment with bottles (filled with water) and measure the energy consumption as per AS/NZS 25/3°C. The volume occupied by water bottles in the food compartment was about 23 and 21% respectively for N169 and N375,

d. Both the compartments loaded as in (c) above, measure the energy consumption as per AS/NZS 25/3°C but with door openings as per JIS (see Table II), and finally

e. Both the compartments loaded as in (c) above but measure the energy consumption as per AS/NZS 25/3°C with a *"real"* scheme of door openings as suggested in Table IV.

Table IV: Real door opening scheme for the measurement of energy consumption in AS/NZS

Times during The day [$]	Time period	Door Opening Frequency Fresh Food	Freezer	Total openings in given time Fresh Food	Freezer
Breakfast	0 – 2 hours	10 min.	40 min.	12	3
Morning	2 – 4 hours	20 min.	60 min.	6	2
Lunch	4 – 6 hours	15 min.	60 min.	8	2
Afternoon	6 – 8 hours	20 min.	60 min.	6	2
Dinner	8 – 11 hours	10 min.	40 min.	18	4
Evening	11 – 14 hours	12 min.	40 min.	15	4
Total	Openings	in 24	Hours	65	17

[$] The testing was done for at least 24 hours.

It may be inferred from Table IV that the total number of door openings on an average is 82 per day in a typical home with a family of four members. This includes door openings of fresh food and freezer compartments respectively of about 65 and 17 times over a 24-hour period. Note that door openings were evenly spaced throughout each period. The period from 0 – 2 hours represent the breakfast time, when the doors of fresh food and freezer compartments are opened at every 10 minutes and 40-minute respectively, yielding 12 and 3 door openings for the two compartments. This scheme is quite different from the JIS (see Table II) where the total number of door openings for both the compartments over a 24 hour period is only 65 but the day is quite short (10 hours only) and door openings are evenly spread out.

5 Experimental Results and Discussion

5.1 Relationship Between and ISO and AS/NZS

Table V presents the test results for both sets of refrigerator-freezer cabinets, tested under ISO and variations of AS/NZS (as shown in Table III). The last column of the table compares the energy consumption values from these tests with the labeled values by the manufacturer as per AS/NZS 32/3°. As explained earlier, the energy consumption will be highest for AS/NZS 32/3°, followed by AS/NZS 25/3°, ISO25/5° and AS/NZS 10/3°. The largest cabinet N_375 displays lower energy consumptions than the labeled value by 5.5% (for Set 1) and 11% (for Set 2). The other four cabinets show higher energy consumptions than the labeled figures. However, the performance difference is consistent through the entire test regime.

Table V: Annual Energy Consumption (kWh/year) of two sets of cabinets (empty) in ISO and AS/NZS

	Cabinets	ISO 25/5° $T_{FR}<=-18°$	AS/NZS 10/3° $T_{FR}<=-15°$	AS/NZS 25/3° $T_{FR}<=-15°$	AS/NZS 32/3° $T_{FR}<=-15°$	AS/NZS 32/3° Labeled (kWh/y)	% variation with labeled
Set 1	N169_1	514.6	319.4	515.0	647.1	600	7.9%
	N249_1	537.4	-	517.8	669.5	650	3.0%
	N375_1	675.7	-	716.3	878.5	930	-5.5%
Set 2	N169_2	486.5	307.6	518.7	640.8	600	6.8%
	N249_2	569.0	348.1	551.1	705.0	650	8.5%
	N375_2	641.6	513.5	677.7	824.3	930	-11.4%

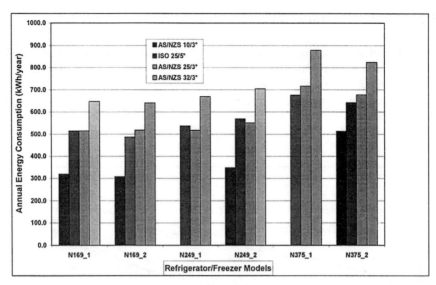

Figure 1: Annual energy consumption of different models of refrigerator-freezers with varying ambient conditions

Further comparison of ISO 25/5°, AS/NZS 25/3° and AS/NZS 32/3° is shown in Figure 1. Note that for all of these cabinets, the freezer compartment was running cold enough so that when the fresh food compartment temperature was raised to +5°C, the freezer compartment still met the −15°C requirement. For N169_2, the energy consumption in AS/NZS 32/3° is about 19% and 24% higher than AS/NZS 25/3° and ISO 25/5° respectively. For N375_2, the corresponding energy consumption is 18% and 22% higher than AS/NZS 25/3° and ISO 25/5° respectively. However, the energy consumption in ISO 25/5° is respectively 7%, +3% and 6% lower than AS/NZS 25/3°. Experimental data for cabinets N249_1 and N249_2 show higher energy consumption for ISO 25/5° than AS/NZS 25/3°. This is because the fresh food compartment temperature (T_{FF}) was less than 3°C for these cabinets in order to obtain the freezer compartment temperature (T_{FR}) less than −18°C, whereas for N169 and N375 cabinets, T_{FF} was higher than 3°C for T_{FR} = -18°C.

5.2 Effect of Ambient Temperature

The effect of ambient temperature on energy consumption is shown Table V and Figures 2 and 3. Data for N249_1 and N375_1 were not available. The energy consumption increases significantly by about 38%, 50% and 20% when the ambient air temperature is raised from 10 to 25° C, 10 to 32° C and 25 to 32° C respectively.

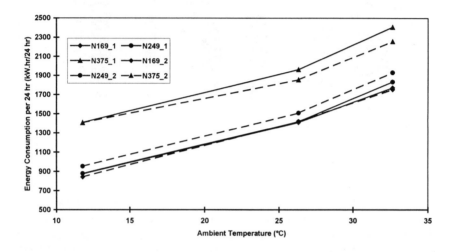

Figure 2: Annual energy consumption of different models of refrigerator-freezers with ambient air temperature (air relative humidity at 60%).

5.3 Effect of Food Packs and Doors Openings

The experimental results for these parameters are presented in Table VI and Figure 3. Only two cabinets (N169_2 and N375_2) were tested for all conditions. The second, third, and fourth columns in Table VI show experimental results of these cabinets as per AS/NZS 25/3° respectively when tested with no food load (i.e. *empty* compartments), food packs loaded only in the freezer compartment (i.e. food compartment being empty); and both freezer and food compartments being loaded with food packs and water bottles respectively. The fifth column shows the corresponding results with door openings when both compartments are loaded as in the forth column. The "Real" and "JIS" door opening schemes are shown earlier in Tables IV and II respectively.

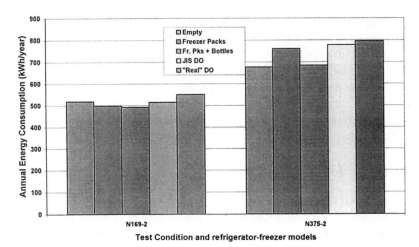

Figure 3: Variation of annual energy consumption of different models of refrigerator-freezers with food packs and different schemes of doors openings

Table VI: Variation in annual energy consumption (kWh/year) with cabinet loading and two schemes of doors openings

Cabinet	Loading Conditions				
	Empty	Freezer Packs	Freezer packs and bottles	Door Opening	
				"Real" D.O.#	JIS D.O.$
N169-2	518.7	499.3	492.8	551.2	515.3
N375-2	677.7	760.9	684.6	795.7	777.8

$ Real door opening scheme is shown in Table IV;   # JIS door opening scheme is shown in Table II.

The relative effect of these variables on energy consumption is shown in Figure 3. There does not seem to be any appreciable difference between the energy consumption results when cabinets were tested either empty or with food packs.

However, the door openings make a significant impact on the energy consumption as may be seen from the fourth and fifth bars for cabinets N169_2 and N375_2 in Figure 3. The energy consumption in the *real world test*, shown by "Real" D.O., is about 6.5% more than JIS D.O., 10.5% more than the AS/NZS 25/3°C test (with food packs and bottles) for cabinet N169_2. The corresponding values for N375_2 are 2.3% and 14% more than the AS/NZS 25/3°C test (with food packs and bottles). The *real world tests* (i.e. "Real" D.O.) for N169_2 and N375_2 consume about 16% and 4% less energy respectively than their closed door test as per AS/NZS 32/3°C. This is in line with the results of Meier and Jansky[16], who reported that the field energy use of American refrigerators is about 15% less than their laboratory (as per ANSI) test values. A recent study[17] on the energy use in New Zealand households also reported that the annual energy consumption of refrigerator/freezers in New Zealand is about 25% less than the label ratings. It may be concluded here that the closed door test with highly inflated ambient temperature (as in AS/NZS 32/3°C) is unrealistic and yield considerably higher energy consumption (15 to 25%) values than the *real world use* conditions. This may be due to the fact that the yearly average temperature in a New Zealand home kitchen may be about 20°C as against 32°C, prescribed by the AS/NZS. Therefore, it is desirable to design a test, which can simulate the *real world* usage conditions and may be truly representative of the ambient conditions of a country.

5.4 Effect of Relative Humidity

For N169_2 and N375_2, tests were performed with the "Real" door opening scheme, where both the compartments were loaded with food packs and the doors of both compartments were opened according to the scheme laid down in Table IV. At ambient temperature of 25°C, the air relative humidity was varied in stages between 40 and 80%. As may be seen from Figure 4, the relative humidity has a marginal effect on the energy consumption, when tested with door openings. This may be due to the fact both the compartments were loaded either with food packs or bottles and therefore, fresh air changes to the compartment were very limited.

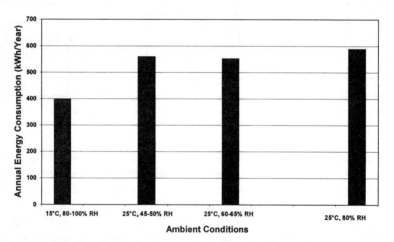

Figure 4: Variation of annual energy consumption of N169_2 with "Real" door openings and varying air humidity.

6   A Generic Approach to Model Energy Consumption

It is obvious that high ambients of ANSI or AS/NZS crudely compensate for door opening loads, since kitchens rarely exceed 20°C. This compensation becomes less realistic as cabinet insulation is improved. Therefore, well-insulated cabinets that perform well in ANSI or AS/NZS, would be less impressive in JIS or "Real" world door opening schemes. ISO is seen to underestimate the "real" world energy consumption of a refrigerator. It is also well known that tests with door openings have poor repeatability and are expensive.

There is, therefore, an obvious need for the alignment of the test standards. Current test standards need to be modified to ensure some modeling of actual user behaviour and refrigerator/freezer performance such as the impact of ambient temperature, the introduction of warm foods into the refrigerator compartment, the impact of door opening, and the introduction of humidity into the refrigerator compartments. Considering all these issues, the total energy consumption of a refrigerator (E, in kWh/year) can be defined as the summation of the following four terms-

$$E = E_{ambient} + E_{processing} + E_{defrost} + E_{other} \qquad (2)$$

where the terms $E_{ambient}$, $E_{processing}$, $E_{defrost}$ and E_{other} respectively represent the energy consumption due to the heat loads of the ambient air (on the closed door cabinet), the air-infiltration and loading of warm food load into the cabinet (with door openings), the defrost heater and other accessories such as condensation heaters (or anti-sweat heaters), lights, switches etc. Note that some of these terms may be incremental.

One way of achieving the best of both worlds is to perform a closed-door test at 20°C (or different ambient) but with an electrical heating load inside a refrigerator to emulate the processing load (i.e. door opening and/or introduced food load). In the simplest form, the electrical load will be steady but better results may be obtained with a regime that varies with time. If desired, water vapour could also be introduced into the cabinet by heating water in an electric pan in the cabinet (for example).

From Figure 2, it is possible to deduce the closed-door energy consumption of a cabinet at typical kitchen temperature of 20°C. From field surveys, it is possible to deduce the total energy consumption of an equivalent refrigerator in normal use in a kitchen. The difference between these two figures can be used to establish wattage required for processing load inside the refrigerators (and freezers). Therefore, there is a need of a new standard that has the realism of "Real" door opening scheme but the simplicity of AS/NZS or ANSI.

7   Conclusions and Recommendations

This paper presented a comprehensive research study on the energy consumption of refrigerator-freezers. Most of the current standards use closed door tests for reasons of repeatability but highly inflated ambient air temperatures to compensate for door openings. However, these standards yield up to 25% more energy consumption than the real world in-field end use data.

In order to design a realistic standard to match the laboratory energy consumption with the in-field use data, as well as testing results to be more repeatable, a number of tests were performed on six cabinets of refrigerator-freezers. The analysis revealed the following conclusions-

a. The effect of relative humidity on energy consumption is marginal,

b. Energy consumption decreases by 20%, 50% and 40% respectively as the temperature in AS/NZS decreases from 32°C to 25°C, 32 to 10°C and 25 to 10°C,

c. AS/NZS 25/3° consumes about 7% more energy than ISO 25/5° (except where $T_{FF} < 3°C$ in ISO to achieve $T_{FR} = -18°C$),

d. Tests with *"Real"* world door opening (as per AS/NZS) consume about 7% more energy than the JIS door opening scheme,

e. AS/NZS32/3° consumes about 24% and 20% more energy than ISO 25/5° and AS/NZS 25/3° respectively, and

f. AS/NZS 32/3° consumes about 15% more energy than the test with the "real" world door openings at AS/NZS 25/3°.

It may be concluded that there is a need to investigate the potential for conversion algorithms (further to equation 1) to translate the energy consumption results of a cabinet from one test standard to the other. Further, a generic approach has been proposed in the paper to modify the current test standards to enable some modeling of actual user behaviour and refrigerator/freezer performance. The modeling parameters in this approach include the impact of ambient temperature, the introduction of warm food loads into the cabinet, the impact of door opening, and the introduction of humidity into the refrigerator compartments.

Acknowledgments

The author is thankful to Messrs. Lindsey Roke and Ian McGill (Fisher & Paykel NZ Ltd.), Mr David Cogan (EECA, Ministry of Commerce, NZ), Mr Gareth Jones (The University of Auckland, NZ) and Lloyd Harrington (Energy Efficient Strategies, Australia) for their support. Thanks are also to many students who worked on this project over the years at the University of Auckland.

8   References

[1]   Bansal P K. An insight into the comparison of test standards for household freezers, *Proc. IIF – IIR conference on "Refrigeration, climate control and energy conservation"*, Melbourne (Australia), pp 172 – 180 (1996-1)

[2] Bansal P K., McGill I., Analysis of household all-refrigerators for different test standards, *ASHRAE Transactions*, 101(1), pp 1439 – 1445 (1995)

[3] Bansal P K., Kruger R. Test standards for household refrigerators and freezers I: preliminary comparisons, *Int. J Refrigeration*, 18 (1), 4 –20 (1995)

[4] Standards Association of New Zealand. *New Zealand Standard AS/NZS 4474-1997*, Performance of household electrical appliances – Refrigerating appliances, Part 1: Energy consumption and performance, Part 2: Energy labeling and minimum energy performance standard requirements, Wellington, 1997

[5] International Organisation for Standardization. *International Standard ISO 8187.3* Household refrigerating appliances - refrigerator-freezers - characteristics and test methods, 1989

[6] International Organisation for Standardization. *International Standard ISO 5155-1983*, Household frozen food storage cabinets and food freezers - essential characteristics and test methods, 1985

[7] Association of Home Appliance Manufacturers. American National Standard, Household refrigerators/household freezers, *ANSI/AHAM HRF-1-1988*, Chicago, 1988

[8] Japanese Standards Association. Japanese Industrial Standard, Household electric refrigerators, refrigerator-freezers and freezers, *JIS C 9607-1986*, Japan, 1986. Note this is known as Method A.

[9] National Bureau of Standards Chinese Taipei (Taiwan): Chinese National Standard. Electric refrigerators and freezers, *General No. CNS2062, Classified No. C4048, 1989*

[10] National Bureau of Standards Chinese Taipei (Taiwan): Chinese National Standard., Method of test for electric refrigerators and freezers, *General No. CNS9577, Classified No. C3164, 1989*

[11] Banse, T. The promotion situation of energy saving in Japanese electric refrigerators, *APEC symposium on "Domestic Refrigerator/Freezers"*, Wellington (New Zealand) during March 5 –8 (2000)

[12] Bansal, P K. Energy efficiency test standards for household refrigerators, *APEC symposium on "Domestic Refrigerator/Freezers"*, Wellington (New Zealand) during March 5 –8 (2000)

[13] National Instruments : AT-MIO-16X. *User manual, Austin*, 1992

[14] National Instruments : AMUX-64T. *User manual, Austin*, 1992

[15] LabView Graphical Programming For Instruments. National Instruments Corporation, Austin, 1992

[16] Meier, A., Jansky, R., Field performance of residential refrigerators: a comparison of laboratory and field use, *ASHRAE Trans* (1988) 94 (2)

[17] Energy use in New Zealand households, Report on the household energy end use project, Year 1, *Energy Efficiency and Conservation Authority, Wellington*, May, 1997.

A New Method for Detailed Electric Consumption of Domestic Appliances

Magali Deschizeau[1], Paul Bertrand[1], Alain Anglade[2] and Michel Grimaldi[3]

[1] Conseil en Technologies Innovantes
[2] Agence De l'Environnement et de la Maîtrise de l'Energie
[3] Laboratoire d'Optique Appliquée - Université de Toulon et du Var

1   Introduction

Electric consumption constitutes an important part of the operational expenses of housekeeping. Through the meter, the distributors of electricity supply a global information of this consumption but, this day, there are no simple means to obtain the detail of its consumption, device by device.

Method described here allows to identify starting up and extinction of the various electric appliances connected on a domestic network. Eventually, it could lead to the realization of a device intended to estimate the expense of electricity, as well as to reduce global consumption.

The method presented in this paper uses a device plugged to domestic power network as well as a wattmeter on the main meter. The device detects the signatures of every present device on the network and the wattmeter measures instantaneous global power.

2   Principle from there Method

The method is based on the measure of the power variation in the main meter when the impulse of starting up or extinction of an appliance is detected. Impulse and the variation of mean real power allow to identify without ambiguity the starting up or the extinction of the appliance.

The Figure 1 shows a characteristic signature of an electric radiator.

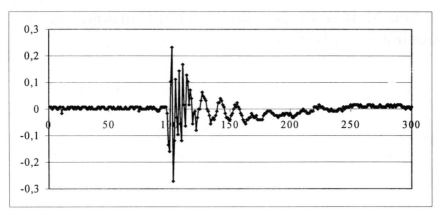

Figure 1

3 Cross Analysis of Electromagnetic Signatures

To make sure that impulses are characteristic of every device, a signatures database has been realized in laboratory, by measuring one by one, 20 electric appliances (lamps, television, hairdryer, radiator...). For each of them, about twenty impulses (ignition and/or extinction) has been recorded.

The algorithm of analysis of the electromagnetic signature includes, for every impulse of the base, 3 stages :

- Stage 1 : crosscorrelation of this signal with the other impulses of the base
- Stage 2 : for every device, search for the maximum of crosscorrelation
- Stage 3 : among these maximum, the higher corresponds to the most likely device

The success rate of this analysis is determined by the number of good identifications on the total number of signatures by device (in %).

The Figure 2 presents the results. The mean success rate is 61,5 %.

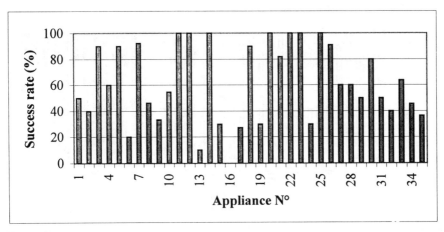

Figure 2

In most cases, the impulse is very characteristic of the appliance but not always of its ignition or its extinction. Indeed, some devices have signatures of ignition and extinction very similar and present high rates of correlation.
To improve identification, a supplementary information of power consumption is added to the algorithm to minimize the ignition-extinction ambiguity.

4 Signatures Analysis and Power Measures

The previous method is modified so that it realizes simultaneously :
- An analysis of signals by a technique of crosscorrelations,
- And a confrontation between the appliance mean powers

The mean values of real power Pr of 20 tested appliances were so measured first.
Algorithm includes previous stages 1, 2 and the following 3 stages :

- Stage 3 : the values of real powers Pr are sorted out according to the values of the maximum of decreasing crosscorrelations determined in the stage 2.
- Stage 4 : the power Pr of the tested appliance is likened to the variation of instantaneous global power if it would has been measured in the main meter.
- Stage 5 : we search for, among the values of real power Pr, the nearest power equal to the variation of mean power.

Figure 3 presents the results of this last method.

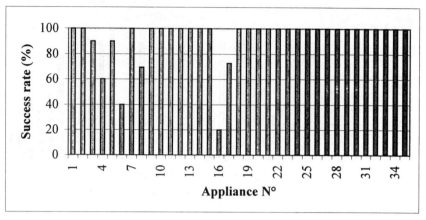

Figure 3

The signature analysis and the confrontation of mean powers, have allowed to identify 92,6 % of the appliances of the database.

To test this method of measure, a campaign of measures has been realized in a house. The results of this campaign are the object of the last part.

5 Discrimination by a Neurons Network

Concerning the signature analysis, another method has been envisaged and is being tested at present. This one is based on the use of an artificial neurons network. After a phase of learning on a complete database, this approach should allow to obtain the discrimination of the different appliances in a very brief time (propagation through the network).

The signature is decomposed into atoms, on a base of wavelets, by means of a "Matching pursuit" type adaptive algorithm. The coefficients of the most representative atoms, that is to say of higher energy, are then presented to an artificial neurons network. A multilayers network using the method of backpropagation for the learning, has been chosen. The last layer contains as many neurons as there are devices to discriminate.

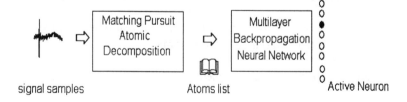

Learning is done on a base of 590 signals representing the signatures of 9 appliances (about 60 signals by appliance). The system is then tested on a second base of 89 signals representing these same devices but not belonging to the learning base.

The following table shows the statistical distribution of results obtained on the two bases simultaneously. For every line associated to an appliance (more exactly an action of ignition or extinction), columns represent the distribution of choices made by the neurons network.

The correspondence between numbers and actual appliances is :
- Coffee machine : 0 ignition, 1 extinction
- Hair dryer : 2 ignition, 3 extinction
- Lamp : 4 ignition, 5 extinction
- Toaster : 6 ignition, 7 extinction
- Hot plate : 8 ignition

N°	0	1	2	3	4	5	6	7	8
0	**68,9%**	11,1%	0,0%	8,9%	0,0%	2,2%	8,9%	0,0%	0,0%
1	4,9%	**58,0%**	4,9%	3,7%	4,9%	11,1%	12,3%	0,0%	0,0%
2	0,0%	5,0%	**75,0%**	12,5%	2,5%	0,0%	0,0%	0,0%	5,0%
3	1,0%	0,0%	0,0%	**90,2%**	0,0%	3,9%	1,0%	0,0%	3,9%
4	0,0%	2,0%	2,0%	0,0%	**79,4%**	1,0%	1,0%	0,0%	14,7%
5	0,0%	1,0%	0,0%	3,9%	1,0%	**93,1%**	1,0%	0,0%	0,0%
6	1,0%	11,8%	2,9%	3,9%	2,9%	0,0%	**72,5%**	0,0%	4,9%
7	0,0%	33,3%	0,0%	33,3%	0,0%	0,0%	0,0%	**33,3%**	0,0%
8	0,0%	1,0%	0,0%	0,0%	6,9%	0,0%	0,0%	0,0%	**92,2%**

The mean success rate of this method is 73,6 %.

The Figure 4 shows the same distribution applied only to the sample of test. Corresponding signals so have not contributed to the learning of the neurons network, and are "seen" for the first time by this one. The mean success rate of identification is 60,9 %.

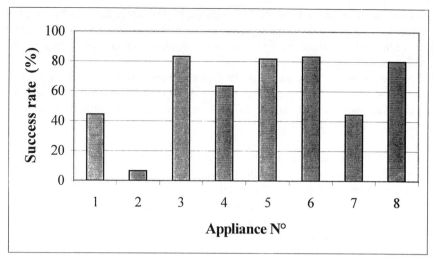

Figure 4

As for the crosscorrelation, we notice that the main source of error comes from the same device exchanging the ignition and the extinction. The Figure n° 5 shows results obtained in that case, that is to say without taking into account the notion of ignition and of extinction. The mean success rate is then 70 %.

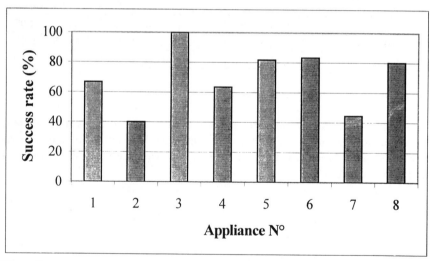

Figure 5

6 Measurement Campaign

This measurement campaign has been realized in a house in which about thirty appliances has been tested. Every appliance was connected through a wattmeter which measured and recorded every 10 seconds, the instantaneous real power of the device, the date and the hour of the measure.

At the same time, signatures dated from devices were recorded on a computer synchronized with the various wattmeters.

This measurement campaign took place during 4 days including 2 full 24 hours periods.

6.1 Exploitation

6.1.1 Power Measurement

For every instrumented appliance, the dated variations of instantaneous real power have been determined. Absolute variations superiors to 3 watts and higher than 10% of the maximum absolute variation have only been kept.

6.1.2 Electromagnetic Signatures

For every previous power variation, we have searched for, 15 seconds before and 5 seconds after the variation, all the impulses detected in this interval of time. They represent signatures of appliance susceptible to have lead the power variation. The variation sign has allowed us to differentiate the ignitions of the extinctions.
Then, a database including the signals of ignition or extinction of the various appliances of the house has been constituted.

All the signatures might not be associated to a power variation and conversely, either because an impulse might correspond to a device not connected to a wattmeter, or because a power variation might not be preceded by a signature.
Finally the same impulse signal might be associated to several power variations, if for instance the corresponding devices have been switched on or have been switched off in the same interval of 10 seconds.

6.2 Identification by Signatures Analysis

The previous method has been used, to identify signals contained in the database of the house.
For some devices used very rarely during the 4 days of measurement, only 2 or 4 signals were detected and the success rate is not significant and is not presented.

The following results are obtained on devices for which we have been able to detect a great number of signatures and corresponding to devices presenting regular phases of ignition and extinction. The following table lists some of these devices.

APPLIANCE	N °
Ignition Radiator 1	1
Extinction Radiator 1	2
Ignition Fridge	3
Extinction Fridge	4
Ignition Dehumidifier	5
Extinction Dehumidifier	6
Ignition Radiator 2	7
Extinction Radiator 2	8

The Figure 6 presents results on devices contained in the previous table. The mean success rate by signature analysis (without information of power) is 68,7 %.

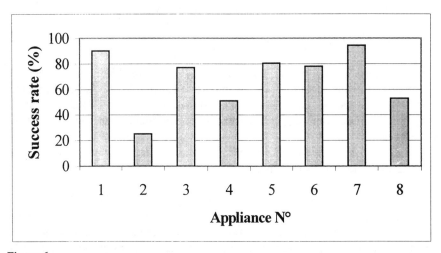

Figure 6

As in laboratory, the method identifies the good appliance but does not differentiate always the ignition of the extinction.

The Figure 7 presents the identification rates obtained on these same appliances without taking into account the notion of ignition or of extinction. The mean success rate is then 81,4 %.

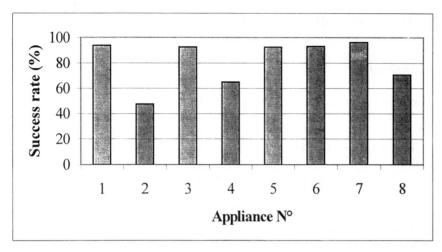

Figure 7

7 Conclusion

Results obtained in laboratory and in the house are coherent. They show that the electromagnetic signature analysis method can be used to identify ignition or extinction from 80 to 90 % devices.

The two methods tested in laboratory to discriminate signals (crosscorrelation and neurons network) give similar results and lead to the same observations. The main sources of error come from the confusion between ignition and extinction for the same device. By adding to the signature analysis, the information of power, these two methods should obtain even better performances.

Other measurement campaigns in a house as well as an extension of the method with signals that are not present in the database, are necessary and will be soon realized to confirm its performance.

Eventually, this technique will be implemented into a real time equipment near the main meter and will supply a real detailed electric consumption appliance by appliance.

A Measured Factor-4 House

Willem Groote

Department Of Fluid, Heat and Combustion Mechanics, University of Gent

Abstract. The results of three years measurements of the power consumption of the different appliances in a low-electricity house will be presented . The house is for most end-uses equipped with state-of-the-art electric household appliances, and has all comfort and appliances as in a standard house. The total annual electricity consumption is down to 850 kWh per year, which is four times less than an average 3-persons household in Belgium . The appliances used will be briefly discussed. Based on the measurements and on statistical data of households, the influence of the residents' behaviour versus the influence of the most efficient technology in the resulting energy use will be discussed.

1 Introduction and Background

The power consumption in a low electricity house has been measured monthly during 3 years. The house is inhabited by 3 persons. It has all comfort and appliances as in a standard house but is for most end-uses equipped with state-of-the-art electric household appliances. For the main appliances, 40 to 100 end-use measurements have been done with individual metering devices. The largest part of the home office consumption (PC, but not lighting, printer and modem) is excluded in the data presented, because one of the home occupants was working home full-time the first two years.

2 Total Electricity Consumption

In Figure 1, we show the evolution of the monthly power consumption in the period 1997-1999. The power consumption during the summer months July and August have been split equally between these two months to cancel the effect of sliding holiday periods.

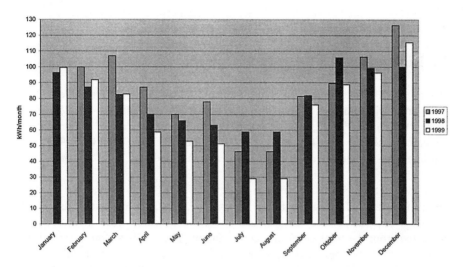

Figure 1: Evolution of the monthly power consumption in the period 1997-1999

In Figure 2, we compare the annual electricity consumption in the house with the average power consumption of a 3-persons household in Belgium (BBRI 1999).

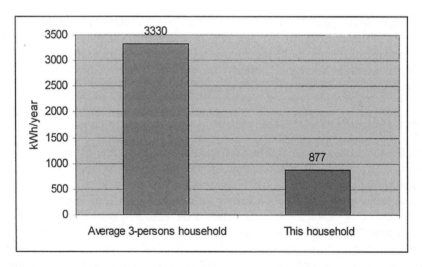

Figure 2: Comparison with average electricity consumption in Belgium

From Figures 1 and 2 can be seen that the power consumption is much lower (26,3. % of the Belgian average) and constantly declining.

End-use monitoring and information on user behaviour has allowed us to disaggregate the power consumption into end-uses, and to distinguish the technology improvement impacts from user behaviour.

3   End-Use Disaggregation

3.1 Lighting

Lighting is distributed over 20 lighting fixtures and lamps. Lamp types are 6 CFL, 7 FL and 7 incandescent. Total installed power is 790 Watt (= 4,4 W/m²), of which 19,5 % is CFL, 22,3 % is FL and 58,2 % is incandescent. Compared to an average household in Flanders, which has 1,21 CFL's (VIREG 1998), this household has 9 times more efficient light bulbs installed (11).

It has not been possible to measure power use of lighting seperately. Lighting energy is estimated using the following method. First, the electricity use of all important appliances (see further) and the leaking losses were measured and distracted from the total electricity consumption. Then, it is assumed that of the remaining fraction only the lighting load is season-dependant. Finally, it is assumed that the lighting consumption during June is between 4 and 8 kWh (average 6 kWh). Under these assumptions, we can calculate the annual energy use for lighting as 175 kWh/year. As reference value for the lighting energy use of an average household, we use 700 kWh/year (Cabinet Sidler 1996, Boardman et. Al. 1997).

Although this is hard to measure, we have no indications that the household behaves different than an average one in using their lighting. We therefore attribute the difference in the lighting energy consumption between the household and an average household entirely to the technology choice (substituting incandescent lamps with more efficient ones).

3.2 Food Storage (Refrigerator)

The household has an old 160 liter refrigerator with a 16 liter **-cabinet, but no freezer. The measured consumption of the refrigerator is 163 kWh/year. In february 2000, this appliance has been replaced with the most efficient appliance on the Belgian market without a freezer cabinet. In Figure 3, the impact is shown of the technology choice versus the behavioural choice of not having a freezer. It is assumed that the reference appliance here is the combination of a single door refrigerator and a seperate freezer (average number of appliances per Belgian household resp. 0,83 and 0,67, source VIREG 1998). Further, it is assumed that if a freezer would have been installed, it would be an efficient middle-seize freezer (205 liter, 178 kWh/year).

It can be seen that a replacement of the existing refrigerator with 2 efficient new appliances (160 liter refrigerator + 178 liter **** freezer) would cause an increase in the annual electricy consumption of 133 kWh/year. Due the decision (for the moment) of not having a freezer the consumption will decrease with 45 kWh/year (29 %) compared to the old existing refrigerator.

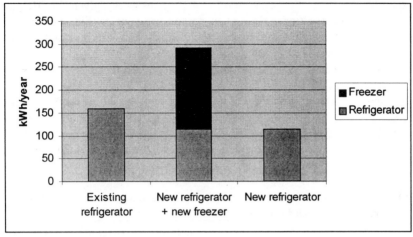

Figure 3: Technology choice versus behavioural choice of not having a freezer in the household

3.3 Washing Machine

The washing machine is a hot-fill type with an A/A/A-label. The hot water comes from an instantaneous pilotless gas water heater at a distance of 5 meters from the machine. During one year, each individual washing cycle has been recorded. In Table I, the recorded data are summarised.

Table I: Principal characteristics of the measured washing machine cycles

	Cycle					
	All cycles	Hot (≥30 °C)	Cold	30/40 °C	50/60 °C	75/90 °C
Number of cycles	113	108	5	80	26	2
Average energy consumed (Wh/cycle)	422	438	71	339	691	1094
Annual energy consumed (kWh/year)	47,70	47,35	0,35	27,18	17,99	2,19
Share of annual consumption (%)	100	99,2	0,8	56.9	37.7	4.6

In total, 113 washing cycles per year were recorded. Taking into account 4 weeks of holiday, this results in an average of 2,35 washing cycles per week. This is substantially less than the average of 4,3 cycles/week, given by Belgian respondents on the question « How often do you use your washing machine ? » (VIREG 1998), although these figures are guesses and not recorded values. However, also the recorded values of the CIEL study (Cabinet O. Sidler 1996) are much higher (average of 257 cycli per year, 1.832 cycli recorded).

We thus may conclude that because of the behaviour of the family (attitude to use their machine almost allways full-load) there are ca. 50 % less washing cycles and therefore less energy use. To quantify this effect we calculate the annual energy use of the measured machine if it had been used with the same washing frequency pattern measured in France (CIEL 1996) (Table II).

Table II: Measured annual number of cycles per family and average energy consumed per cycle in the French CIEL study (Cabinet O. Sidler 1996)

Cycle	Cold	30/40 °C	50/60 °C	75/90 °C
Number of cycles per year	85	178	63	16
Average energy consumed (Wh/cycle)	105	619	1298	2065

Regarding another well-known important washing habit, the frequency of different temperature cycles, we find no significant difference between this household and the ones monitored in France in the CIEL study.

We now compare the power consumption of the measured machine with that of the average machine measured in the CIEL study, with an AAA Cold Fill-machine and with this AAA-Hot Fill machine that are used as the average machine in the CIEL study (Table II) and with the measured AAA Hot-fill machine had it been used as in an average CIEL household (Figure 4). The assumed power consumptions per cycle for the AAA-Cold Fill machine are 0,60 kWh (30/40°), 0,95 kWh (60°) and 1,7 kWh (90°) respectively.

The influence of the users behaviour on the energy consumption, by only using the washing machine with full loads, is very clear (reduction of energy consumption with 59 %). It is even stronger than the influence of the technology choice, a hot-fill AAA machine (reduction of energy consumption with 48 %). User behaviour and technology choice together result in reduction of the power consumption of 79,6 % compared to an average washing machine in the CIEL sample. Of course also the additional heating energy of a hot-fill washing machine has to be taken into account, but using electricity for hot water supply is extremely wasteful from a physics point of view (Norgard et. Al. 1997)

Another striking conclusion from Figure 4 is that a Cold-Fill washing machine with an AAA-label is only slightly (14 %) more efficient than an average existing washing machine in France (of which more than 50 % in the sample was manufactured before 1990). It is an open question if any Cold-Fill machine deserves an A-label for energy efficiency.

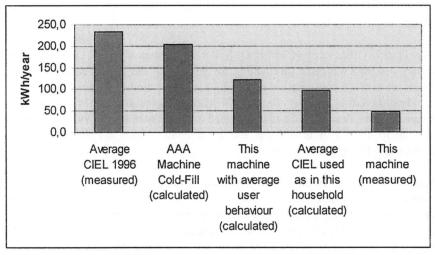

Figure 4: Power consumption of measured and calculated washing machines

3.4 Clothes Dryer

The drying tumble in the household is a room air ventilating type, with an outlet to the outside. There is an additional electric heating element in the machine, which was seldom used during the monitoring period. The advantage of this type of clothes dryers is their low energy consumption. Disadvantages are their noise, the long time they need to dry, and the bad outlet of humid air to the outside. Therefore, the machine has been replaced shortly after the monitoring period with a gas-fired clothes dryer.

Table III: Principal characteristics of the measured drying machine cycles

Cycle	All cycles	Room air vented	Electric heated
Number of cycles per year	84	81	3
Average energy consumed (Wh/cycle)	981	934	2.259
Annual energy consumption (kWh/year)	82,4	75,6	6,8

The clothes dryer was monitored during 2 periods totalling 12 months. Individual cycles were recorded during 4 months. Usage pattern and average consumption are summarised in Table III. In total, 84 drying cycles per year were recorded. Taking into account 4 weeks of holiday, this results in an average of 1,75 washing cycles per week. This is almost 50 % less than the average of 3,3 cycles/week, given by Belgian respondents on the question « How often do you use your clothes dryer ? » (VIREG 1998).

The French CIEL study (Cabinet O. Sidler 1996) measured 3,7 cycles/week for cycles > 0,90 kWh/cycle (accounting for 93,1 % of the annual energy use) and found an average consumption of 480 kWh/year for clothes dryers. However, the standard deviation was very large (374 kWh/year) because of strongly different usage patterns. A study in Scotland, quoted in (Boardman et. al. 1997), even found that up to a third of households owning a tumble dryer did not use it at all, presumably for reasons of economy.

We may conclude that due to the behaviour of the family (dry clothes as much as possible outside, attitude to use their machine almost allways full-load) there are ca. 50 % less drying cycles compared to an average household. To quantify the impact of this different behaviour we calculate the annual energy use of the measured machine if it had been used with the same frequency measured in France (Table IV). We neglect cycles < 900 Wh because they have only a small influence (less than 7 %) on the annual energy use of the measured clothes dryers.

Table IV: Measured annual number of cycles per family and average energy consumed per cycle in the French CIEL study (Cabinet O. Sidler 1996)

	Cycles > 900 Wh
Number of cycles per year	192
Average energy consumed (Wh/cycle)	2.319

We now compare the power consumption of the measured machine with that of the average machine measured in the CIEL study, with an average CIEL machine had it been used as in this household and with the measured clothes dryer had it been used as in an average CIEL household (Figure 5).

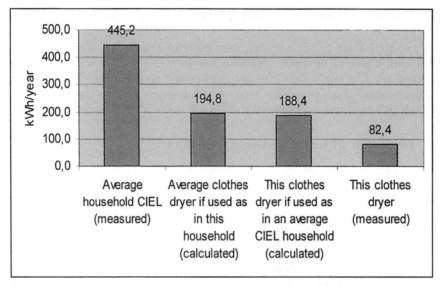

Figure 5: Power consumption of measured and calculated clothes dryers

The influence of the users behaviour on the energy consumption, by only using the washing machine with full loads, is very clear (reduction of energy consumption with 56 %). It is almost as strong as the influence of the technology choice, a room air vented clothes dryer. User behaviour and technology choice together result in reduction of the power consumption of 81 % compared to an average clothes dryer in the CIEL sample.

3.5 Television

The television is more than 10 years old. Its energy use has been measured during two periods of 5 months, resulting in an annual power consumption of 104.2 kWh of which 81,4 kWh in active mode. With a measured leaking loss of 3,0 W we calculate an average of 3 h 30 daily active mode. These values are similar to the averages measured or used in (Cabinet Sidler 1996) and (Fawcett et. Al. 2000).

3.6 Heating Auxilliary

The heating system in the house are pilotless decentral gas heaters. Electricity is needed for the ignition and mainly for the pulsing ventilator. Power use has been recorded during a full heating season, and was 44,7 kWh/year (1999). This is favourable compared to the standard house with an average power consumption of 297 kWh/year for the circulation pump of the hydronic central heating (Cabinet Sidler 1996, Norgard et. Al. 1997). We attribute this difference entirely to a technology choice.

436

3.7 Leaking Losses (Including Metering Equipment)

The biggest leaking losses measured were those of a radio (15 Watt) and of the VCR (20 Watt !). The latter appliance has been broken down and must be replaced, the first one is now only plugged in when needed. Other leaking losses are the cordless phone (6 W), the TV (3 W), the modem (2,75 W), the CD-player (3,1 W) and (not really leaking losses) the power meters (1,2 W each).

3.8 Overview of End-Use Disaggregation

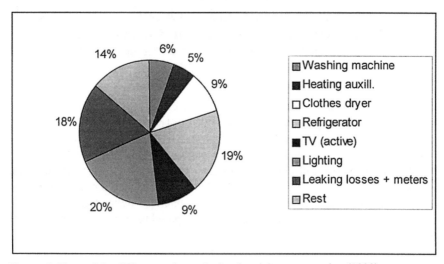

Figure 6: Share of the different end-uses in the electricity consumption (1999)

As in an average household, lighting and cold food storage remain the most important end-uses. As could be expected, in a low electricity house leaking losses become much more important than in an average house.

4   Technology and User Behaviour: Conclusions

Finally, by aggregating the different end-use consumptions under different scenario's, we find the total impact of the better electro-technology used in this household versus the more conscious user behaviour (Figure 7).

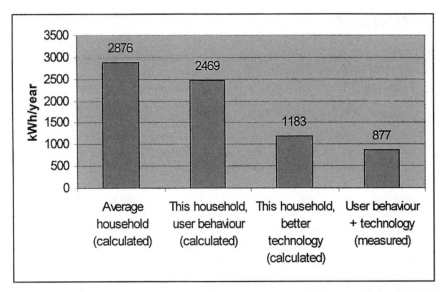

Figure 7: Annual electricity consumption of an average household versus this household under different assumptions.

For the average household, measured end-uses from (Cabinet Sidler 1996), (Boardman et. Al. 1997) and (Fawcett et. Al. 2000), as mentioned in the previous paragraphs, were compiled. For the categories « Leaking losses » and « Rest », the same measured values as in the investigated household were used, which may underestimate the consumption of an average household.

From Figure 7 , it can be seen that an energy-conscious behaviour did reduce the power consumption with 14 % compared to an average family. The main factors, each about equally contributing, are washing full-load, drying clothes outdoors when possible and in a full loaded machine otherwise, and not having a freezer.

The more efficient technology alone, if used as in an average family, would reduce the power consumption with 59 % compared to an average family.

User behaviour and technology choice together result in a reduction of the electricity consumption of 70 % compared to an average household.

We conclude that in this case technology choice is more important than user behaviour. However, technology choices themselves are a kind of user behaviour.

5 References

[1] BBRI 1999, *Study on Energy Use, Insulation and Ventilation in Dwellings (SENVIVV)*. Belgian Building Research Institute, Brussels

[2] Boardman B. et. Al. 1997, *Domestic Equipment and Carbon Dioxide Emissions (DECADE)*. Environmental Change Unit, University of Oxford

[3] Cabinet O. Sidler 1996, *Demand-Side Management End-Use Metering Campaign in the Residential Sector (*CIEL), Paris, 166 pp.

[4] Fawcett T. et Al. (2000) *Lower Carbon Futures for European Households*. Environmental Change Institute, University of Oxford, UK.

[5] Norgard J. & A. Gydesen 1994, *Energy Efficient Domestic Appliances: Analyses and Field Tests. NATO Advanced Research Workshop on Integrated Electricity Resource Planning*, July, Espinho

[6] Norgard J. & Guldbrandsen T. 1997, *The Next Generation of Appliances: Visions for Sustainability. Invited contribution to the first International Conference on Energy Efficiency in Household Appliances*, 10-12 November 1997, Firenze

[7] VIREG 1998, *Energy Enquiry Households in Flanders*. Flemish Institute for a Rational Use of Energy (VIREG), Brussels 1998 (in Dutch)

[8] Von Weiszacker E., Lovins A.B. & Lovins H., 1996, *Factor-4: Doubling Welfare and Halving Resource Use*. Earthscan, London

Metering Campaign on All Cooking End-Uses in 100 Households

Olivier Sidler

Enertech (France)

Abstract. This paper presents the findings of an experimental study performed in 100 French households on the end-use power demand and energy consumption of domestic appliances focusing on cooking appliances [1].

The study centred on the analysis of a database containing metered results for 517 cases of 32 types of domestic electrical appliance covering the main forms of electric cooking, as well as auxiliary uses, such as coffee-makers, kettles, etc... The households, all located in central France, were metered for a period of one month between January and July 1998. The power demand and energy consumption of each appliance was recorded every 10 minutes over the whole month using the DIACE monitoring system [2]. Annual energy consumption estimates were made from interpretation of this data in most cases by assuming that the usage in the second half of the year would be identical to that in the first. This enabled a hierarchy of cooking appliance energy consumption to be established wherein electric cookers (combined hob and oven) came top with 457 kWh/year, ahead of induction hobs (337 kWh/year), ceramic hobs (281 kWh/year), and ovens (224 kWh/year).

The combined cooking-related energy consumption accounted for 14 % of the total electricity-specific energy consumption of the households surveyed. The average annual energy consumption of all electric cooking appliances was 568 kWh/year. A pronounced and regular monthly variation in cooking-related electricity consumption is found, with the energy consumption in January being 1.25 times the annual average monthly value and 0.72 times in June. Some 96% of all cooking related energy consumption was attributable to loads of less than 3 kW.

Interestingly, the standby power consumption of induction hobs was found to be very significant (30% of their total annual energy use) and to offset all their energy savings in the cooking mode compared to conventional hobs. However, this conclusion is not generic and would not apply to many newer induction hobs with low standby power levels.

1 Introduction

The *ECUEL* project [1] was carried out in partnership with **ADEME** (The French National Energy and Environmental Agency), **EDF** (The French National Electricity Utility) and the **CEC** (Commission of the European Communities). The collection, analysis and processing of data were carried out by **Cabinet SIDLER** of France and **PW Consulting** of the UK. The aims of the study were, firstly, to provide an evaluation of the energy consumption levels involved in electric cooking, secondly, to gain a fuller understanding of the effect of external conditions on the operation and energy consumption of domestic cold appliances, and lastly, to ascertain whether using a tumble-dryer can reduce the energy consumption involved in ironing laundry.

The study centred around the establishment and analysis of a database **containing 517 examples of 32 types of domestic electrical appliance** covering the main forms of electric cooking, as well as auxiliary usages, such as coffee-makers, kettles, etc. Some 98 households in the Drôme and Ardèche regions, with an average of 3.2 people per household, were metered for a period of one month between January and July 1998. The measurement system used in the project, called *DIACE*, enabled specific information to be gathered about the daily and sub-daily operation of the appliances being monitored. The DIACE system is unobtrusive and reliable and uses power line carrier technology to transmit the readings from the individual meters to a collector. The stored data is remotely downloaded each day to a computer acting as a central data logger via a modem.

2 Overview of the Electric Cooking

The saturation level of the main appliances monitored in the panel was the following :

Table I : list of appliances monitored and their distribution within the household sample.

Appliance type	Saturation level in the panel	Appliance type	Saturation level in the panel
Iron	90.8%	Induction hob	9.2%
Coffee-maker	81.6%	Catalytic mini-oven	6.1%
Ceramic hob	57.1%	Condensor clothes-dryer – automatic sensing of end of cycle	6.1%
Microwave	44.9%	Evacution clothes-dryer – automatic sensing of end of cycle	5.1%
Microwave + grill (and/orfan-assisted oven)	32.7%	Steam-cooker (atmospheric pressure)	4.1%
Deep-fryer	24.5%	Condensor clothes-dryer – timer controlled	4.1%
Pyrolytic fan-assisted main oven	22.4%	Catalytic main oven using convection	3.1%
Evacuation clothes dryers – timer controlled	22.4%	Manually cleaned main oven using convection	2.1%
Electric kettle	19.4%	Washer-dryer	2.0%
Electric cooker	18.6%	Vorwerk	2.0%
Pyrolytic main oven using convection	17.3%	Portable induction hob	2.0%
Catalytic fan-assisted main oven	14.3%	Manually cleaned fan-assisted main oven	2.0%
Sealed hob	13.3%	Grill	2.0%
Manually cleaned mini-oven	12.2%		

Ranking the cooking appliances according to their average annual energy consumption enables those appliances that consume the most electricity to be highlighted. Electric cookers came top with **457 kWh/year,** ahead of induction hobs **(337 kWh/year),** ceramic hobs **(281 kWh/year),** and ovens **(224 kWh/year),** see Figure 1.

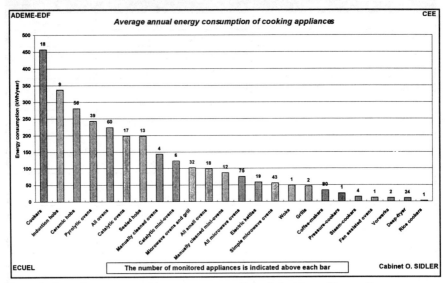

Figure 1: Annual average monthly energy consumption of electric cooking appliances

Some **50 %** of the total cooking-related energy consumption was attributable to electric hobs. All together the different types of ovens accounted for approximately **42 %** of the total cooking-related energy consumption, which accounted for **14 %** of the total household electricity-specific energy consumption i.e. of electricity demand that was not for space heating or domestic hot water. **The average annual energy consumption of all electric cooking appliances was 568 kWh/year.** The monthly fluctuation of cooking-related electricity consumption was very marked with 1.25 times the average annual monthly energy consumption in January and 0.72 times in June, see Figure 2.

Some 99 % of the power demand of electric cooking appliances was under 3 kW, Figure 3. This means that householders who want to use electricity for cooking do not necessarily need to be charged using a higher electricity tariff associated with higher peak power demand as is commonly the case in France. The most significant financial gains are to be made, not so much in the area of energy savings, but in the improvement of household demand-side management.

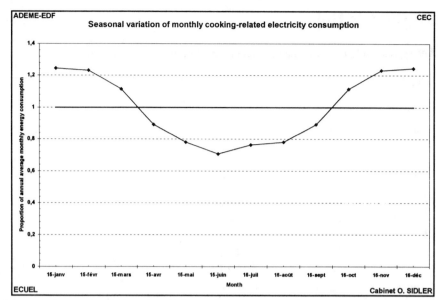

Figure 2: Seasonal variation of monthly cooking-related electricity consumption

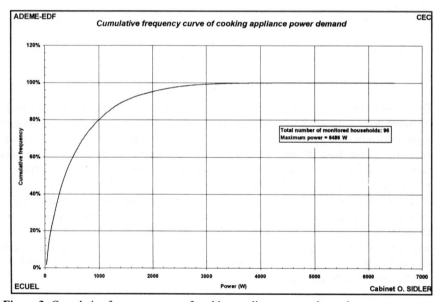

Figure 3: Cumulative frequency curve of cooking appliance power demand

3 Electric Hobs

The typical daily power demand of an electric hob is shown in Figure 4 :

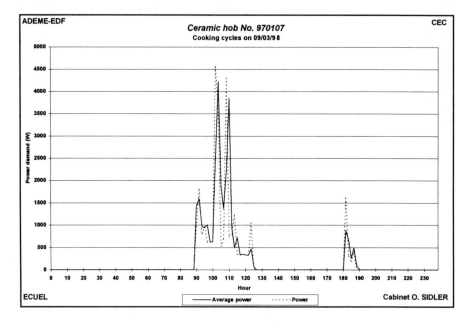

Figure 4: Daily power demand of an electric hob

Due to their standby power demand (between 8 and 18 W, **30 % of the total energy consumption**) and their heavy daily use (**58 min./day**.), induction hobs were those that used the most energy. If, however, we view the results in terms of energy consumption per hour of use, we find that the different forms of cooking technology ranked in the order we would have expected, given their relative energy efficiency. Sealed hobs (efficiency: 50 %, energy consumption: 1161 Wh/h) were less efficient than ceramic hobs (efficiency: up to 70 %, energy consumption: 999 Wh/h) or induction hobs (efficiency: 82 %, energy consumption: 588 Wh/h). From a purely financial perspective, there is no advantage in choosing an induction hob over a sealed hob (it would take 282 years to achieve payback). The criteria involved in the purchase of a hob of this type are related more to such considerations as ease of use, aesthetics, and the safety of the appliance.

4 Ovens

The typical daily power demand of an electric oven is given in Figure 5.

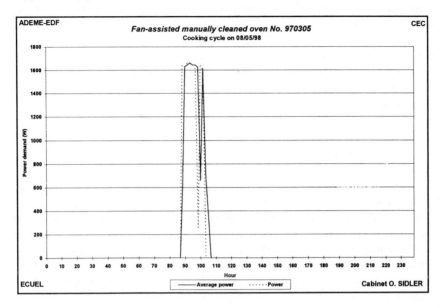

Figure 5: Daily power demand of an electric oven

The average energy consumption per household of ovens was **224 kWh/year**. The value recorded for convection ovens was 233 kWh/year and for fan-assisted ovens was 219 kWh/year. The average energy consumption of a cooking period of an oven was **889 Wh**. Catalytic-ovens used **199 kWh/year**, pyrolytic-ovens used **243 kWh/year** and manually cleaned ovens used **224 kWh/year**. The average cleaning cycle of a pyrolytic-oven used **3490 Wh**, although this facility was rarely used, and only accounted for **11 %** of the total energy consumption of ovens with this kind of cleaning system. This level of energy consumption could be reduced by improving oven insulation quality.

For electric ovens, just as for electric hobs, the most efficient form of technology, from an energy saving perspective, is also that which exhibited the highest overall energy consumption levels. This was due to the relatively heavier use of these appliances.

5   Mini-Ovens

In most cases in which the large capacity of a traditional oven is unnecessary mini-ovens are a good substitute for large ovens in terms of power demand and energy consumption savings. The maximum recorded power demand of mini-ovens was **34 %** lower (2410 W) than that of large ovens and their annual energy consumption was less than half (**99 kWh/year**). Given an equal amount of use, replacing a large oven with a mini-oven could lead to a **27 %** energy saving on average. As cooking periods for mini-ovens are usually shorter than for large ovens, the resulting savings could be greater still.

6   Microvave Ovens

The average annual energy consumption of microwaves that were monitored during the *ECUEL* study was **75 kWh/year**. The power demand of basic microwaves was around 1500 W. In the case of combination microwaves, this figure was as much as double. Basic microwave ovens used **55 kWh/year,** whereas microwave ovens with grill or combination modes used **102 kWh/year**. Most of the microwave ovens that were monitored, were used mainly for defrosting or reheating food, rather than for actual cooking as such. The average energy consumption of a cooking period was **69 Wh**. Microwave ovens do not give any energy savings compared to electric ovens when they are used for traditional cooking techniques.

7   Other Cooking Appliances

Electric kettles should not be overlooked when considering overall energy consumption levels. Their average annual energy consumption of **58 kWh/year** was greater than that of basic microwave ovens. Depending on the appliance, the power demand ranged from between **750 W** and **1750 W**. Most electric coffee-makers had a lower power demand than that of kettles (**686 W on average**). Their average annual energy consumption was also lower at **34 kWh/year**. The average power demand of deep-fryers was **1542 W**. They used **11 kWh/year** and were used mainly during the summer probably due to the French preference for using these appliances outside in order to minimise the accumulation of fatty odours in the kitchen. The average power demand of steam-cookers was **683 W**. They used an average of **15 kWh/year**.

8   Consolidated Results

The comparison of the results of the *ECUEL* measurement campaign with those from other countries, indicates that they were far more accurate than the estimates used up until now (see Table II and [4], [5]). Detailed end-use metering campaigns, such as *ECUEL,* are needed to improve understanding of the different areas of electricity use, and to provide more accurate data for forecasting with predictive models.

Table II: Table comparing electric cooking annual energy consumption values between studies in different countries (kWh/year).

Appliance	Ecuel [1]	CIEL [6]	French Guyana [3]	Swed [7]	Aust [8]	France [9]	France [10]	GE [4]	UK [4]	IT [4]	NL [4]	AU [4]	DK [4]	FI [4]
	Measured values					Estimated values							
Cooking total	568			570		1000			631		143			435
Ovens (all types)	224		169	194	233	300	111	90	277	138	46	456	153	200
Cookers	457			496				600	547	885				
Hobs (all types)	273		186	317	187		418							
Sealed Hotplate Hobs	198					600	432							
Ceramic Hobs	281					550	389							
Induction Hobs	337					400	285							
Micro-waves	75	49	36	50	67	100	60							
Mini-ovens	99	64												
Coffee-makers	34	24	34	36										

9 Conclusion

The principal interest of the ECUEL measurement campaign is to be able to precisely identify the energy consumption levels of various common cooking appliances and thereby significantly improve upon the hirtherto crude estimates of their energy use. It might be questioned whether a measurement campaign in a 100 households is really representative of the average behaviour in a large European country? If a precision of 0.1% is required then the answer is certainly no, as a sample of at least 1000 households would be required. Nonetheless a measurement campaign on a sample of 100 households would be expected to give results within ~±5% of the national average, which is far more reliable than those made without end-use metering. Furthermore, end-use measurement campaigns using very fine measurement intervals also enable typical appliance load profiles to be measured. The ECUEL campaign has shown that contrary to what was previously believed the household cooking energy consumption for all cooking appliances combined is mostly satisfied by loads of less than 3kW in France.

The study has also illustrated the continued importance of reducing standby demand and lends weight to the call to reduce such loads to less than 1W in all but the most exceptional cases.

10 References

[1] « *Electricity demand-side management: an experimental investigation of cooking appliances, domestic cold appliances and clothes-dryers in 100 households: project ECUEL* », Cabinet Sidler & PW Consulting for the Commission of the European Communities, contract N° 4.1031/Z/96-164, ADEME, and EDF, June, 1999.

[2] « *Demonstration of the DIACE end-use metering system* » Jérôme Gilbert, in *Non-intrusive and other end-use metering systems,* Amsterdam, The Netherlands, 21-23 September, Novem, Sittard, The Netherlands, 1998.

[3] « *Campagne de mesures sur les usages électriques dans le secteur résidentiel en Guyane* » - Cabinet Sidler for ADEME, Contract N° 96.45.045, Paris, France, June, 1998.

[4] « *Efficient Domestic Ovens* » - TTS Institute (Work Efficiency Institute) for the Commission of the European Communities, contract N° 4.1031/D/97-047, Helsinki, Finland, January, 2000.

[5] « *Ovens provide food for thought* », Pirkko Kasanen, *Appliance Efficiency,* Vol 4(1), 2000.

[6] « *Demand Side Management: End-Use Metering Campaign in the Residential Sector* » - Cabinet SIDLER for the Commission of the European Communities, contract N° 4.1031/Z/93-58, ADEME and EDF. Translated by PW Consulting. 1996.

[7] « *Domestic electricity in detached houses* » - R1995:30 NUTEK, Stockolm Sweden, 1995.

[8] « *The residential end-use study* » - Pacific Power and Electricity Distributors of New South Wales and the Australian Capital Territory, Australia, 1995.

[9] « *La cuisson domestique : un pas vers le tout-électrique* », CFE in *Bâtiment Relations Elec* - N°22, 1997.

[10] Values used in the MURELEC software - personal correspondence, INESTENE, Paris 1998.

Should Research Money Been Spent for Intrusive or Non-Intrusive End-Use Metering?

Harry Vreuls

Novem, the Netherlands

1 Summary

This paper describes the main elements of several end-use metering systems and an open, two way communication gateway system. It argues that information from intrusive as well as from non-intrusive end-use metering is needed. The developments in the energy distribution sector (two way communication and offering a wider range of services) are main factors in the process that determines whether intrusive metering will be in the future the better option than non-intrusive metering. As priorities has to been set, intrusive metering seems to have the best opportunities. In that case an international field experiment for a number of appliances in houses should start as soon as possible.

2 Introduction

Telemetering became the last decades more and more a correct way for collecting information on continues energy use of companies, in buildings and in houses. The meters itself are improved, and prices for digital meters are decreasing. Several international organisations are promoting the use of this technique and the standardisation process. Also conferences are organised dealing with telemetering, e.g. in the USA the bi-annual ITC and in Europe the ETSC.

In the Netherlands in the period 1992-1994 two experiments (one in Ten Hague and one in Amsterdam) were done with telemetering combined with feedback to consumers. Both experiments resulted in energy savings, but it had to be concluded that the metering system at that moment had too many technical problems and was too expensive to introduce it at a larger scale. An experiment that combined telemetering with different communication tools (written information and weekly information on the energy use and energy saving suggestions on television) in the city of Helmond in 1998/1999 proved these results. Two universities (Amsterdam and Groningen) worked in the nineties in the field of analysing load curves and splitting it up in load curves for individual appliances. This way of getting information is often referred to as 'non-intrusive end-use meting', Also some experiments where done dealing with metering the energy use of a specific appliance in a household. This way of data collecting and communication is referred to as intrusive end-use meting, as the appliance or group of appliances are metered separately inside the house.

At the moment a Dutch field experiment is researched to get proven energy data for the electricity use of appliances in households in the Netherlands. This information is needed to improve models for future energy use (scenario's), to evaluate the impact of energy efficient appliances (that are promoted by the system of European energy labels) and the influence of the behavioural aspects on this energy use, as well as to set priorities for future energy saving programmes.

In the next section the main characteristics for intrusive and non-intrusive end use metering are presented. This is followed by an overview on available systems to meter. The future of metering system depends on the advantages of two way connections and standardisation in techniques. Trends and international co operation work within the IEA in this field are described in the fifth section. In the last section it is argued that the Dutch field experiment should be done with intrusive end use metering, and that an international experiment is preferable.

3 Intrusive and Non-Intrusive End Use Metering Systems

Utilities are metering the energy use of a household. New techniques makes is possible to get frequent information on the energy use. Especial for electricity the technique to produce load curves is well developed. If additional information on the energy uses within the houses is needed, two option are available for getting these:
1. to (sub)meter the appliances within the house;
2. to filter the specific electricity consumption for an appliance from the load curve.

The first option often is referred to as intrusive end use metering: admission into the house being metered is needed to install the (sub)meters. Several technical options for meters and communication (e.g. power wire and infra red signal) are available for the intrusive metering.

The second one is non-intrusive metering; no (sub)meters are installed, but the load curve is analysed and disaggregated in individual loads. To do this, five methods are in use:
1. library method

A library with the characteristic load curves for household equipment is created. This library is used to split up the total load curve. For this splitting up, two techniques are in use:
   a. filtering: step by step the easiest recognised load curves (e.g. the continue load of the fridge) are subtracted, so the remaining curve is easier to analyse

b. regression: using the statistical technique of regression analysis to determine which specific appliance causes a specific electricity use at a specific moment. To do this, information is needed on the number of time an appliance is switched off.

Most of the time the regression technique is used additional to the filtering.

2. library with restricted metering method
The use of the library method is made less complex, as it is know whether an appliance is in use or not. In this method the on/off status is separate measured, e.g. using a resistor connected to the lead.

3. econometric method
The electricity use by the smaller electrical appliance is not filtered from the load curve for a single house, but for a group of houses using mathematical or statistical techniques

4. fuzzy set method
Fuzzy logic is used for recognition of the appliances. Although this method is capable of taking care of many imprecise criteria, at the moment no additional research is ongoing for this technique

5. neural network method
A self learning computer system is used for modelling the load curves.; a pattern recognition system based on a training process using data on energy loads.

4   Available End Use Metering Systems

A research on available (non)-intrusive end-use metering systems concluded that less than ten systems are (almost) available for metering load curves for domestic appliances in the Netherlands (Holstein en Kemna, 1999). The basic input for this research came from an international workshop on non-intrusive and other end-use metering system, held in November 1998. Five operational systems are intrusive end-use metering systems and three non-intrusive (and some are still in the developing phase). The following systems were researched:
1. DIACE
2. POEM2000,
3. EMU 10
4. Power master
5. Useload
6. NIALM
7. SPEED
8. IVAM

Ad 1. DIACE is an intrusive metering system. For each appliance that should be measured an adapter is placed before the plug. The energy use is sent to a data collector, located in the house, by the lighting system (power-line carrier technology) with a frequency between 1 and 60 minutes. The data collector sends the information by modem to a PC. The system is used in France (Enertech), and surveyed more than 400 households, accounting for a total of 2571 individual end users. It is distributed by Landis and Gyr as well as by Euro CP. The cost for hard- and software and metering for 10 appliances in 15 houses is 40-45.000 Euro.

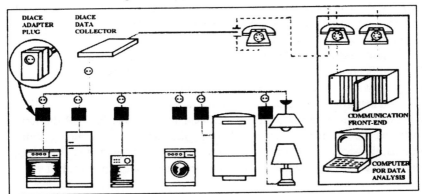

Figure 1: The Diace operating principle

Ad 2. POEM2000 is also an intrusive metering system, but the signal is sent by the adapter using radio waves. The local collector unit sends the load curves to a PC. Additional a laptop can be use as a portable commissioning unit. The system is marketed by Datum Solutions (UK). This system was used in two projects in the UK: one in 1992, dealing with 175 houses and 715 appliances and one in 1995 for domestic lighting in 100 houses (in three different regions). The cost for metering (including hard- and software) is about twice as high as with the DIACE system: about 85.000 Euro.

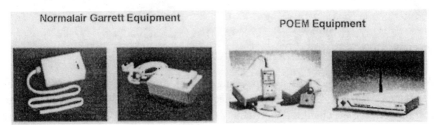

Figure 2: Elements of the POEM system

Ad 3. The EMU 10 system uses infrared for signal transfer. This system was used in a field experiment by Ademe and is distributed by EMU Elektronik AG (Switzerland) The costs for 15 houses are 90-95.000 Euro.

Ad 4 and 5. The system Powermaster (Sattler Energie Consulting, Austria) is restricted to appliances that are connected to the power line at centralised points, and so not suitable for metering houses. Also the system Useload (Sintef, Norway) is not a proper system for a field experiment, as it predicts load curves for groups of users and developed as a simulation model.

Ad 6. NIALM is a non-intrusive metering system, developed by VTT Energy (Finland), based on a standard 3 phase kWh meter. A laptop, functioning as a data recorder, is connected to this meter. The data metered are: voltage, active load and reactive load. The load curve is split in the following steps:
1) normalisation (to nominal load);
2) margin detection (finding the smallest detectable load change)
3) modification of the data (differentiate 1- and 3-phase load)
4) present appliance register (using cluster analysis)
5) load identification (remaining error tolerance 10%)
The system was tested in a field experiment in 1997 and in 1999 the test of a prototype started for 3 houses. Although the meter is a normal commercial product, the analyses software is still under construction and not yet user friendly. For the moment the system is not suitable for a lager scale experiment.

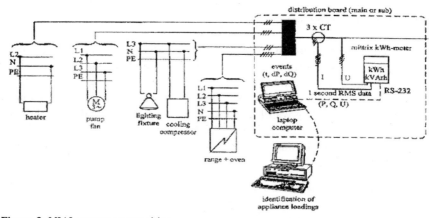

Figure 3: NIAL system composition

Ad 7. SPEED, a non-intrusive metering system, functions with a data recorder connected with an US standard connection within the kWh meter. The data recorder logs load changes, Watt and VAR, and sends data every two weeks by modem to the master station, located outside the house. A pc is used for analysis and standard reports with graphs, spreadsheets and tables. The load curve is analysis in five steps:
1) detection of load changes
2) cluster analysis
3) cluster matching (finding switch on and switch moments belonging to each other)
4) solve data deviation (e.g. switch on of 2 appliances at the same moment)
5) identification of the load of the specific appliance.

The SPEED system is on the market (Enetics INC, New York US). The hard- and software cost are bout 30.000 Euro, but the data transmission control and data analysis raises the cost with 155.000 Euro.

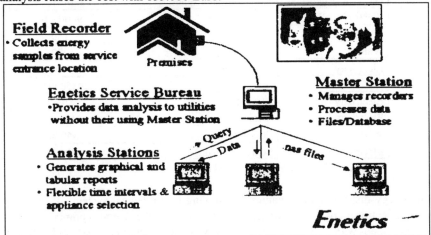

Figure 4: Elements of the SPEED system

Ad 8. IVAM is a software package under development by IVAM research international (Amsterdam). The software uses as input just electric load and not active or reactive load an cosine phi. The system functions for data collection on moment and frequency that (bigger) energy users change their electricity use. As just every two to five minutes the load is measured, it is dubious whether the system is good enough for detailed analyses.

5 Telemetering, Two Way Communaction and Trends

Some telemetering projects are mentioned ahead, indicating that utilities can start installing and using this type of meters. At the moment they are doing this too. As an example, I refer to the Metering and Automation Telemanagement System (STAM), developed by two Spanish companies, IKUSI and Iberdrola. This system was implemented first on smaller scale (60 – 300 meters), but at the moment projects up to 35.000 meters are under construction. In Table 1 is listed where the STAM is installed and operative by the end of 1998.

456

Distribution	Town (country)	Number of meters
Electricity	San Sebastián (Spain)	200
Gas / Electricity	San Sebastián (Spain)	320 / 328*
Electricity / Water	Zarauz (Spain)	9000 / 40
Water	Torrevieja (Spain)	180
Water	Alicante (Spain)	280
Electricity / Water	Hernani (Spain)	150 / 300 *
Water	(Russia)	35000*
Electricity / Water	(Brazil)	80/60
Electricity / Water	(Brazil)	214/ 100
Water	(Russia)	35000*
Electricity	Logroño (Spain)	3000 *

* these installations are being carried out

The energy distribution companies are looking to implement systems of two way communication between the domestic customer and the energy company. It will be combined with a wider range of services to customers than only energy delivery. The information exchange requirements can be summarised as:

- Improved monitoring/measurement/control/communication (e.g. load, devices, generation, export, self energy audit, auto energy audit);
- Improved information feedback (more frequent billing, bill disaggregation, energy prices, energy labelling awareness, promotion)
- Linking of energy services to other "in house" applications to improve market potential

The value added service, that can be developed related tot these information exchanges, will not only having benefit to utility businesses, but also have benefit to other actors. So a combined market push (by utilities and services companies) and a market pull (by consumers asking more services) will stimulate new metering systems.

An IEA workgroup (from national consortia in five participating countries, including utilities, manufacturers and R&D organisations) within the Implementing Agreement on DSM, researched the preferred communication options for any specific service or group of services. It took also into account that these may well change over relatively short timescales form the perspectives of cast and/or performance. An illustration of the service provider achitectures, using a flexible customer gateway, is shown in Figure 5.

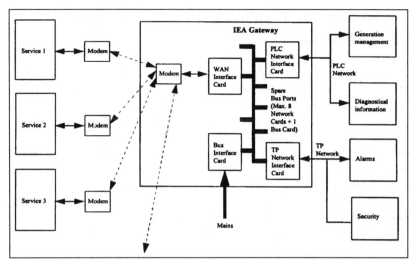

Figure 5: IEA Gateway, illustration of a complete system

Recently a prototype of this flexible gateway was produced. It will initially be used to demonstrate applications using the telephone network (PSTN) outside and LON protocol on power line and Mbus protocol on twisted pair inside customers premises. A schematic diagram of the Flexible Gateway is shown in Figure 6. This system, that is ready for demonstration, will stimulate standardisation and the implementation of intrusive metering.

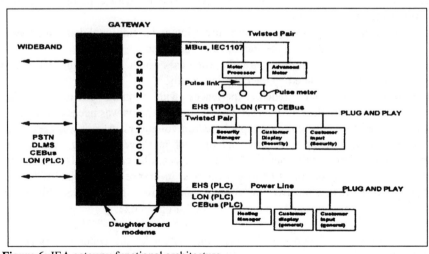

Figure 6: IEA gateway functional architecture

6 A Dutch Intrusive Metering System Field Experiment

In the Netherlands policy makers are also looking for information on specific energy uses by households. This information can be used to improve models for future energy use, to evaluate the impact of energy efficient appliances (that are promoted by the system of European energy labels) and the influence of the behavioural aspects on this energy use, and to set priorities for energy saving programmes. The needed information should be based on metered use, especially on a level of energy functions and appliances. At the moment Novem is preparing a survey for the Ministry of Economic Affairs on this issue.

As this survey should result in detailed information over the coming years, during the preparation phase it is now under discussion:

- whether one can rely on just non-intrusive metering for detailed information on energy use for appliances;
- which technology should be used for metering and data communication;
- whether a smaller field test is needed before the survey could start or not;
- how European co-operation could lower the cost and raise the quality of information.

In this matter I refer to experiences from three projects. The first is from a project in 1993/1994 by the energy Research Group, University of Sydney. They combined direct intrusive metering data for 289 households, metering on total household load plus up to 7 appliances per dwelling, with conditional demand analysis and concluded that:

- direct intrusive metering increases precision and conditional demand analysis lowers cost; so they can be successfully combined;
- consumption of high-penetration end uses such as lighting has been successfully estimated by the combined method.

The second are from French experiences with running end-use metering campaigns for more than 400 households. It showed that is was necessary to allow enough time to find the survey sample based on volunteers. Also it was concluded that even with a reliable, precise and discreet intrusive metering system, one should not underestimate that the key to a successful campaign is in the data analysis. A small experiment (8 to 9 dwellings) in France by EDF in 1998/1999, using the SPEED non intrusive metering system, showed that the USA library on domestic appliances does not fit correct to the European one.

An advise on using a non-intrusive metering system for the Dutch survey (about 300 households), concluded that it will cost circa 400.000 Euro for the first year. But after evaluating the systems mentioned ahead, it was also concluded that none non-intrusive system was good enough at the moment and that prior to and combined with this also intrusive metering should be done.

7 Conclusions

Looking to the selection of European experiences as presented ahead, and the available systems, my provisional conclusion for the moment is that priority should be set to address (research) money to upgrade intrusive metering and that European co-operation will bring a lot of advantages. This conclusion is based on two main reasons:

a. it is necessary to create a basic for good European libraries on appliances specific load curves; especially the labelled appliances;
b. the experiences with submetering should result in increasing market implementation for this systems, taken into account the growing utilities interest in two way communication.

Within one and a half year it should be possible:

a. to judge whether the created European libraries for the non-intrusive metering system will be good enough, and cost effective in combination with demand analysis;
b. to decide on how standardisation (related to the IEA gateway protocol) could be combined with available non-intrusive metering systems;
c. to use European co-operation for ongoing smaller non-intrusive metering projects to enrich the libraries.

8 References

[1] Alvarez, Eguren & Zamora, 1998, *Telemanagement system and measurement automatisation*, CEPSI conference, Thailand.
[2] Bartels, R & Fiebig, D.G. 1998, *Residential end-use electricity demand: results from a designed experiment*, Energy Research Group, University of Sydney, Australia
[3] Formbey, R. April 2000, *Customer Energy Services and Communication in competitive energy markets*, Conference on Energy efficiency, Turkey
[4] Uitzinger, M.J., IVAM, 1995, *Telemetering and monitoring,* (in Dutch)
[5] Novem, Non-intrusive and other end-use meting systems, international workshop, the Netherlands, September 1998
[6] Van Holsteijn en Kemna, 1999, *Non-intrusive end-use metering; phase 1,* report ordered by Novem, the Netherlands (in Dutch)

A Consumer Advice on Energy Efficient Use and Purchase of Household Appliances and Lighting

Annemie Loozen[1] and Catelijne van de Moosdijk[2]

[1] Novem, the Netherlands
[2] Hendriks, Novem, the Netherlands

1   Purpose

The goal of this project, which is executed by the 'Dutch Agency for Energy and the Environment (Novem)' by order of the Ministry of Economic Affairs of the Netherlands, is to save energy by achieving energy efficient behaviour towards detached household appliances and lighting. Experiments have shown that 5 % to 10 % of the domestic energy use can be saved by correct domestic behaviour. This project tries to encourage energy efficient domestic behaviour by developing a set of coherent energy saving behavioural measures, concerning both the energy efficient use of household appliances and lighting as well as the purchase of energy efficient household appliances and lighting. The selection of household appliances and lighting is based on the fact that the presence of these appliances can be easily established and that the advice therefore can be concrete and specific.

2   Results

The advice can be given in addition to an existing individual advice about the energetic constructional quality of dwellings (existing buildings), or as a separate advice (existing buildings and newly built houses). Experiences from the advice about the energetic constructional quality of dwellings learn that many consumers wonder why the advice does not contain information on household appliances and lighting. Therefore, and because the demand for the existing customised advice about the energetic constructional quality of dwellings is high, it is to be expected that many consumers will also ask for a customised advice about their household appliances and lighting and will carry out the proposed measures.

3 Prior Conditions

The conditions mentioned beneath are very important, as they are crucial to success or failure of the advice on consumer behaviour in general and on consumer behaviour as to household appliances and lighting in particular.

First of all the project aims at giving consumers advice about the energetic situation in their individual households. Because consumers ask for the advice, they appear to be interested in energy efficiency. By giving them the advice, their needs are met. But to make the advice really successful, the contents and the approach of the advice have to be such that consumers are inclined to carry out the proposed measures. Only by doing so, energy saving will actually be realised.

Secondly, Novem has made a selection of measures and has chosen for those measures that are highly energy saving as well as easy to execute by consumers. This selection was based on existing theoretical and practical information on energy saving behaviour. Examples of these measures are:

- less use of standby functions of appliances;
- right regulation of thermostats, refrigerators, washing machines, etc.;
- placing of the refrigerator away from the cooking stove;
- purchase of energy efficient white goods, etc.

Finally, earlier projects have shown that an advice customised to the specific home situation of the consumer increases the implementation of the proposed measures. Therefore these measures about household appliances will also be tailored to the consumers specific home situation.

4 Approach

4.1 January 2000 – July 2000

The project started with a survey of all possible kinds of measures as to consumer behaviour and energy saving. Based on existing theoretical and practical information on energy saving behaviour Novem decided to select a number of measures. The theoretical information was based on literature, the practical information was gathered from interviews with groups of citizens involved in environment and energy saving, so called Eco-teams. The literature and the experiences of the members of the Eco-teams made it possible to select measures: those measures which are highly energy saving as well as easy to execute by consumers.

Based on both the literature and the experiences of Eco-teams, a questionnaire was developed by order of Novem by SWOKA, an organisation specialised in analysing consumers needs. Within the selection of measures included in the advice at first a long list of questions and measures was presented. In April this list was submitted in a workshop to organisations experienced in advising households about the energetic constructional quality of dwellings. In general, these advisors were very positive on the initiative to develop an advice on household appliances and lighting: they regarded the advice as significant and feasible. During their advice-visits to households, consumers also asked for advice on energy saving as to household appliances and lighting. The advisors were unanimous about the time of the advice: the advice may not take up too much time of the consumer or of the advisor. It was also brought to our attention that the consumer is not very likely to pay much for the advice. As for the existing advice on construction of houses, households have to pay 160 Euro; whenever they execute proposed measures, the amount of 160 Euro is paid back. In addition the consumer can benefit from the subventions by the government on purchasing appliances or isolation material in order to carry out measures in the advice. Energy saving on the construction of the house however is generally bringing in a higher amount of money than energy saving on household appliances and lighting. Therefore the amount of money for an advice on household appliances and lighting may not mount up to the same extent as an advice on the construction.

These experiences led to the conclusion that the questionnaire had to be altered from a long list into a short list of measures. Preliminary to the advice, the advisor asks the consumer by means of a small questionnaire to explicit the items of energy saving that he/she is especially interested in. To enable the consumer to make a considered decision, information about the average consumption of gas and electricity in households accompanies the questionnaire. Along with some personal information (name, address, phone, number of members of the family etc.) and data on the energy bill (quantities and amounts of money on the consumption of gas and electricity) the consumer returns this questionnaire to the advisor. Thus the advisor will only inform the consumer on the subjects of his/her choice. The revised questionnaire was submitted not only to several advisors that attended the workshop, but also to consumers to check the questionnaire on usefulness and comprehensibility. Their comments – especially the need to take into account the different household situations – were included in a revised questionnaire.

4.2 July 2000 – January 2001

The questionnaire, developed as mentioned above, will be tested among a larger number of customers (advisors and consumers) in a field test. This field test is scheduled for September and October 2000. Advisors can claim subvention from Novem in order to test the questionnaire. Especially the way in which the advisor is going to check the contents and approach of the advice is interesting for Novem. Not only can advisors choose between advising on household appliances and lighting in addition to the constructional quality of the house on one hand and the separate advice on household appliances and lighting on the other hand. Moreover the advisor can decide to co-operate in advising consumers with e.g. brokers, energy consultants, municipalities. Assuming that the advice will still be thought

464

of as significant and feasible, the results of the field test will be used to further perfect the questionnaire.

The existing advice on the energetic constructional quality of houses is supported by software, which has been developed last year and is due to be revised in the next few months. The software that yet has to be developed for the advice on household appliances and lighting will have to be according to the existing software, in case the advice on household appliances and lighting will be added to the existing advice. This software will be developed in the forthcoming months.

One of the conclusions of the April-workshop was that the costs of the advice must not be for account of the consumer. Analogous to the advice on the construction of houses, it is the intention of Novem to return the costs of the advice to the consumer as soon as he/she has decided to follow up at least one of the measures proposed in the advice. Financing the advice is an item that will be considered in detail during the next months.

The problem with executing measures is how to check whether or not the consumer is actually taking measures as advised. This might make another visit necessary, because the energy bill can not be used as proof of energy saving behaviour: the effects of energy saving behaviour may well have been undone by extra energy use and/or the purchase of energy devouring appliances (e.g. tumble-drier, waterbed). The experiences of the field test maybe bring about a solution to this problem. The coming months will be used to elaborate the measurement and evaluation of the field test.

According to the primary timetable consumers will, at the beginning of 2001, be able to ask consultants for a customised energy efficiency advice about household appliances and lighting based on the set of measures as developed. Because the results of the field test will not be due until November 2000, the timetable is most likely to be adapted.

5 Contents: Measures

The questionnaire and measures to be proposed concern both the gas and electricity consumption. Gas is used in order to cook (3 %), hot water supply (19 %) and heating (78 %). The electricity consumption is due to cleaning (21 %), cooking (8 %), cooling/freezing (18 %), lighting (16 %), hot water supply and heating (15 %), audio/video/communications (14 %) and other appliances (8 %). In the Dutch households the average consumption of natural gas is about 1945 m^3 and the average electricity consumption about 3280 kWh.

Examples of proposed measures – as to the energy efficient use of household appliances and lighting as well as the purchase of energy efficient appliances and lighting – are:
- less use of standby functions of appliances;

- right regulation of thermostats, refrigerators, washing machines, etc.;
- placing of the refrigerator away from the cooking stove;
- purchase of energy efficient white goods and lighting, etc.

Examples of questions and of an advice as a result of it are presented below, in order to be more specific about the contents of the advice.

a. *Refrigerator*
   - Do you keep tomatoes, paprika's and cucumbers in the refrigerator? Advice: these vegetables can be well preserved outside the refrigerator.
   - How often do you put food in the refrigerator that is still warm? Advice: don't put warm food in the refrigerator and defrost food in the refrigerator.
   - How often do you defrost the freezing compartment of your refrigerator? Advice: defrosting results in a, more efficiently working, refrigerator.
   - Is the refrigerator placed within 50 centimetres of a heating radiator, cooking stove or oven? Advice: change the position of the refrigerator.
   - Which type of refrigerator do you intend to buy in the future? Advice 1: choose a highly efficient refrigerator of energy efficiency class A (the label of the European Community). Together with a subvention at NLG 100 of utilities your energy bill will decrease by ... kWh and thus by NLG ... a year. Advice 2: choose – depending on the number of family members – a refrigerator of the right size.

b. *Washing machine*
   - How much of your laundry do you wash at 30 degrees, how many at 40 degrees, how many at 60 degrees an how many at 90 degrees? Advice: each program is compared with other programs, both in energy use and energetic costs.
   - How many times a week do you use the washing machine for cleaning your laundry? Advice: washing laundry at 90 degrees costs twice as much energy as washing at 60 degrees, while washing at 60 degrees is twice as energy devouring as washing at 40 degrees.

c. *Drier*
   - How many times a week do you dry the laundry by using the clothes-line in stead of the drier? Advice: use the clothes-line more often, if possible in a draughty place. Using the clothes-line in one of three cases in stead of the drier decreases energy consumption and energy expenses.
   - At which spin speed do you centrifuge? Advice: use the maximum spin speed. The energetic benefit you gain from centrifuging at lower speed is less than the extra energy consumption of the drier.

d. *Dishwasher*
   - Do you wash the dishes before putting them in the dishwasher? Advice: do not wash the dishes twice and whenever you do wash the dishes before putting them in the dishwasher, use cold water.

- How many times do you use a dishwasher that is only partly filled? Advice: fill the dishwasher up to the maximum.

e. *Cooking*
   - Do you cook using gas or using electricity? Advice: use gas.

f. *Heating*
   - How many rooms are regularly heated without someone being there? Advice: don't heat rooms that you don't use.
   - What is the average time during which you ventilate the living room? Advice: it is better to ventilate 15 – 20 minutes by creating a draught, than to ventilate for a much longer period by opening only one window.
   - Is, in the living room, furniture placed within 50 centimetres of the radiator? Advice: change the position of the furniture.
   - Are, in the winter, the curtains usually drawn? Advice: draw the curtains.

g. *Hot water supply*
   - How often during one week's time do you take a bath? Advice: use the shower in stead of the bath.
   - Is your shower fit with a low flow showerhead? Advice: have a low flow shower head installed.

h. *Televisions, videos, audio equipment and computers*
   For how long do televisions, videos, audio equipment and computers function in a stand-by mode? Advice 1: install powermanagement. Advice 2: turn out the equipment whenever you don't use it.

i. *Lighting*
   Based on a survey of all types of bulbs in the house, the measure is proposed to buy Compact Fluorescent Lamp (CFL) and to turn out the light whenever you leave a room.

6   Discussion

In chapter 3 it is stated that several conditions are crucial to turn an advice on consumer behaviour concerning household appliances and lighting into a success or a failure.

To be able to discuss this subject profoundly, it is important to focus on the following items:

1. Do you believe a tailor-made advice to consumers – concerning both the energy efficient use of household appliances and lighting as well as the purchase of energy efficient household appliances and lighting – is significant and feasible?

2. Do you endorse the emphasis on measures that are highly energy saving as well as easy to execute by consumers?

3. This project tries to encourage energy efficient domestic behaviour by developing a set of coherent energy saving behavioural measures. Do you think that consumers will execute the proposed measures?

7 References

[1] *Basisonderzoek Aardgasverbruik Kleinverbruikers 1998* (BAK 98); Arnhem: EnergieNed, 1999.

[2] *Basisonderzoek Elektriciteitsverbruik Kleinverbruikers 1997* (BEK 97); Arnhem: EnergieNed, 1998.

[3] *Elektrische apparaten - milieu en budget*, Utrecht: Milieu Centraal / NIBUD, 1999.

[4] *Energielabels, instrument voor energiebesparing – afwasmachines* (brochure), Utrecht: Novem.

[5] Heijs, W., *Huishoudelijk energieverbruik: Gewoontegedrag en Interventiemogelijkheden*, Utrecht: Novem, 1999.

[6] Jorritsma et al., *Eco(no)Teams?*, 1998.

[7] Sabbé, T. & O. Praalder, *Praktische instructie gebruik energielabels*, IMK Nederland.

[8] *Wonen - milieu en budget*, Utrecht: Milieu Centraal / NIBUD, 1999.

[9] *Wij kiezen zelf - werkmap voor EcoTeams*; Den Haag: Global Action Plan Nederland, 1999.

[10] *Energiebesparingsadvies voor huishoudelijke apparaten en verlichting;* Leiden: SWOKA, 2000.

Influencing Consumer Behaviour - Danish Clothes Washing as an Example

Malene Hein[1] and Birgitta Jacobsen[2]

[1] Elkraft System
[2] Danish Energy Agency

This paper presents the successful experience of a Danish campaign regarding clothes washing, which is an area that is usually considered to be both uninteresting and tedious. The aim of the campaign was to shift the clothes washing behaviour of consumers in an energy-efficient direction.

The overall purpose was to contribute to reducing Denmark's CO_2-emissions. More specifically the campaign was aimed at lowering the energy consumption of the household sector, which is a major source of CO_2-emissions. At the campaign's outset in 1995, electricity consumed in conjunction with washing and drying clothes accounted for approximately 18% of the Danish households' total electricity consumption. Washing at 90°C or more accounted for 15% of all washing in 1997 - a very high percentage in comparison with other European countries. This combined with the fact that washing at 90°C uses approximately twice as much electricity as washing at 60°C, and modern detergents make washing at temperatures above 60°C superfluous, founded the motivation for the campaign. The aim and message of the campaign was therefore that one could lower the washing temperature, and thereby save electricity, without lowering the cleanliness or comfort of the consumers.

In 1995 a co-ordination group was formed on the initiative of the Danish Energy Agency. The group consisted of the Energy Agency, the Danish Environmental Agency, the National Consumer Agency of Denmark and the Danish electricity companies. The aim of the group was to find a campaign object within the area of household clothes washing. After examining the different possibilities it was decided to concentrate on reducing the amount of washing at above 60°C. This target was chosen in spite of the fact that more electricity is used for dry tumbling clothes, and therefor the savings possibilities within this field ought to be bigger. But it was assessed that it would be easier to change the washing temperature, and that thereby the campaign would stand a bigger chance of success. The main argument was that lowering the washing temperature does not impose extra work or other discomforts on the consumer.

1 Background for the Campaign

One or two generations ago washing at very high temperatures was the only way to get the laundry clean. In those days detergents had little effect at low water temperatures. A copper or wash boiler was a labour-saving way of getting the wash water in circulation. The alternative was doing it by hand.

This situation has changed: Modern detergents contain a number of active ingredients, such as enzymes, which, even at quite low temperatures, are able to dissolve organic dirt, to loosen inorganic dirt trapped between fibres and to prevent dirt in the water from redepositing on fabrics.

It is becoming increasingly common for people to change their clothes daily. Furthermore, the number of people whose work involves their clothes getting heavily soiled is declining. Today most laundry is thus only lightly soiled and normally free from old, dried-up stains. This means that the demands on the washing effect, including the temperature, are not the same as earlier.

Due to the developments outlined above, the percentage of all 90°C wash cycles is declining. However, as washing habits are deep-seated the population's washing behaviour seems to be lagging behind. It was thus estimated that 90°C wash cycles were used unnecessarily frequent.

It was also estimated that the biggest obstacle towards changing the washing habits in the target-group was objections that the clothes will not be completely clean, odours will not be removed or washing at lower temperatures is unhygienic.

As background and foundation for the campaign the National Consumer Agency therefore made a study together with the Danish Institute of Technology, which showed that there were no health or hygienic problems connected with washing household clothes at only 60°C.

2 Starting Point

Before starting the campaign assessments of the Danish washing habits were made. The first survey was done already in 1995, and the second was done in the summer 1997, just before the campaign start.
The survey in 1997 showed that washing at 90°C or more accounted for 15% of all washing. The earlier assessment in 1995 showed a higher percentage, but the figures cannot be compared directly, as the people who were interviewed in the first assessment where not the persons responsible for clothes washing in the

households. Nevertheless it can be assumed that already before the campaign started there was a tendency toward washing at lower temperatures.

According to the 1997-survey, the best estimate of the total number of wash cycles in private households in Denmark per year was around 318 million, corresponding to 1.2 per week per capita. While the distribution between the different temperatures was as shown below.

Washing temperature	Electricity consumption	Frequency
40°C	0.65 kWh	47%
60°C	1.20 kWh	38%
90°C	2.00 kWh	15%

The aim of the campaign was to change the population's attitudes and habits. More specifically, over a period of three years the objective was to convert approximately one fourth of all wash cycles at 90°C or more to 60°C. In addition it was presumed that the message of the campaign would have a rub-off effect, resulting in conversion of some of the 60°C wash cycles to lower temperatures. It was estimated that each time two 90°C wash cycles are converted to 60°C, one 60°C wash cycle will be converted to 40°C.

As the energy saved per wash cycle converted from 90°C to 60°C amounts to 0.8 kWh and the corresponding figure per wash cycle converted from 60°C to 40°C is 0.55 kWh, the potential energy savings were expected to amount to approximately 13-17 million kWh.

3   The Campaign Design – Clean Washing at 60°C

After these preliminary actions an advertising agency was assigned and it was decided to run a three-year campaign combining a mass media strategy with a network strategy. The budget for the three-year period was set at 8 million DKK – or 1.07 million EURO. An advertising agency was assigned to create and administer the campaign.

The overall campaign message was very simple: Clean washing at 60°C. Where clean refers to both the clothes being clean and to the environment being less polluted. As an eye-catcher a washing label with the simple message was used on all campaign material and thereby repeated again and again.

Vask rent ved 60°

Figure 1: Campaign message: Clean washing at 60°C

It was chosen that the main argument for changing washing habits should be the effects on the environment while less attention was drawn to the potential savings on the electricity bill. The reason for this priority was that both the messages that less electricity is used when lowering the temperature and that this would have effects on the environment were expected to be accepted straight away.

The main target group for the campaign was women between 25-49 years with private washing machines in the household. This group represents the largest amount of washes, as women are often the ones who are responsible for washing in the family, and women in this age group often are part of families with children still living at home. It was also concluded that changing the habits of older women would be more difficult.

4 Mass Media Campaign

The mass media strategy consisted of two parts. The first part was advertising in newspapers, magazines and local papers. This was aimed at creating interest in washing habits and getting the topic on the agenda in the households and among other parties. The other part of the mass media strategy was public relations work aimed at getting press coverage of the campaign and thereby increasing the exposure of the message.

5 Network Campaign

The aim of the network part of the campaign was to use electricity companies, NGOs and local Agenda 21-workers as ambassadors for the campaign. This was done in order to spread the message more effectively and in order to catch the consumers in situations where washing was naturally on the agenda. In addition

472

to these ambassadors also libraries and pharmacies were used for distributing campaign material.

Special education material was made for schools, with the aim of teaching the school children about environmentally friendly washing and through the children putting the message on the agenda in the families.

Collaboration with commercial parties was another element in the network part of the campaign. Contacts were made with producers of white goods and detergents, traders and electricians/service mechanics working with washing machines and retail traders. The general idea was to get the commercial parties to include the campaign message in sales information about washing machines and detergents.

This mix of activities created a very strong synergy and the fact that the message was communicated through many channels gave it much more strength and credibility. The participation from the networking organisations and companies was extremely positive as the visibility of the campaign and the possibility of combining it with their own activities inspired them.

The campaign started in the autumn of 1997 and measurements of the development have been carried out during the summers of 1998 and 1999. The objective was to lower the proportion of washes above 90°C from 15% to 11%. Towards the end of the campaign the proportion of 90°C washes was measured as being 9% which is regarded as highly satisfying. Another measurement of the effect of the campaign is the percentage of population, which say that they never or seldom wash at temperatures above 60°C. Also this percentage has increased, as has the percentage of people who agree, they do not boil laundry as often as they used to. Likewise the share of people who agree that underwear must be boiled has decreased. The figures are shown below.

	Survey 1997	Survey 1998	Survey 1999
Share of washes at 90°C or more	15%	12%	9%
"Never wash at temperatures above 60°C"	33%	46%	51%
"Underwear must be boiled"	45%	44%	27%
"Do not boil laundry as often as used to."	55%	69%	70%

The savings generated from the campaign are substantial and prove that a campaign with a relatively low budget regarding a tedious subject can be successful if the right strategy is chosen.

6   Why Did the Campaign Succeed?

On top of the surveys a more thorough evaluation of the campaign was done in 1999 by PLS Consult. The evaluators conclude that the campaign was well planned and carried out in all aspects.

The fact that the campaign message has been repeated often throughout a fairly long period of time is important for the success. Changing of habits demand prolonged and continuous activities in order to have impact and in order to maintain the new habits after the campaign is over. Especially the combination of mass media and network campaign is emphasised as a reason for the success. The many different campaign parts have set focus on the message from many different angels and supported each other.

A common campaign-identity – the washing label – has increased the attention to the campaign. The fact that the message is simple is also important – both simple to understand and simple to comply with.

Another very important aspect was the co-operation with the National Consumer Agency, which insured credibility to the message that 60°C is enough for hygienic washing.

The mass media campaign was successful as it resulted in a high exposure, which is important for putting an issue on the agenda among the consumers. Without using television advertising it was still possible to get a high exposure through magazines, newspapers and especially through local papers. The public relations activities were most successful with the local papers – resulting in 813 press notes in 1997 and 1998.

The network campaign was useful as it resulted in contact with the consumers in situations where washing was on the agenda, and as the many different ambassadors resulted in the message being repeated frequently.

In the network campaign especially the role of the electricity utilities was a success. Many utilities had a very active role as ambassadors, while it was more difficult to involve the NGOs and local agenda 21 workers in the campaign. This is partially explained by the fact that the campaign message fits naturally into the normal activities of the electricity utilities. But also the fact that the electricity utilities have theoretical and practical experience in advising consumers played a role in the efficiency.

Also the collaboration with the commercial partners was successful as a number of the largest retail and electrician chains in Denmark used the campaign in their sales activities. Commercial partners are in general positive towards taking part in

474

public information activities if they can see synergy on their own products and because a "green" profile is regarded positively both ideologically and commercially.

By combining mass media activities with local activities in electricity utilities, schools and among sellers of relevant products, it was possible with a fairly small budget to maintain focus on the issue over a period of three years and thereby possible to influence the behaviour of the Danish consumers.

Notes:
Changing the Danes' Washing Habits (http://www.ens.dk/Vask/Rapwash.htm).
Vask rent ved 60 grader effektmåling, August 1999, Jysk Analyseinstitut A/S
Evaluering af kampagnen "Vask rent ved 60°", October 1999, PLS Consult

Room Air Conditioners: Consumer Survey in Italy and Spain

Bill Mebane[1] and Milena Presutto[2]

[1] Consultant
[2] ENEA

Abstract. A consumer survey was carried out in November 1998 using CATI telephone interview methodology in Italy and Spain with the same structured questionnaire. The purpose was to study: the degree of consumer satisfaction regarding performance of installed room air conditioners; the consumers' motivation of their purchase; and finally the propensity to buy new, high efficiency, room air conditioners.

Over 2000 contacts in Italy and 3000 contacts in Spain were necessary to obtain representative interviews from 210 households owning an air conditioner with a cooling capacity below 12 kW and 90 households without air conditioning equipment, in each country. Questions were related to five different air conditioner types: multiple split units, single split mobile units, single split fixed units, packaged (window) units, and single-duct air conditioners.

A comparative analysis shows an existing market considerably larger in Italy with ten- percent penetration as compared to five percent for Spain. Both in Italy and Spain, fixed (single) split units are the most frequently installed.

Spanish consumers seem more satisfied than Italian ones with the cooling and acoustic performance (noise level) of their installed appliances. In both countries consumers owning multiple split and (mobile) split air conditioners seem the most satisfied. Owners of (fixed) split and multiple split systems would certainly re-purchase the same type of air conditioner.

In both countries, but particularly in Italy, there is a high potential propensity for purchasing efficient air conditioners, which guarantee savings on electricity bills.

1 Introduction and Methodology of the Consumer Survey

Consumers were interviewed in two countries, **Italy**[1] and **Spain,** considered as a prototype of the demand side of the room air conditioner (RAC) market. An external specialised company[2] carried out this part of the work in November 1998. The consumer survey is also presented in the more general SAVE study on "Energy Efficiency of Room Air Conditioners"[3]

The interviews were based on an *ad hoc* questionnaire with 21 questions covering the following main areas of information:
- the characteristics of the present market of installed RACs;
- the degree of satisfaction of consumers for the various performance characteristics of installed air conditioning systems;
- the factors motivating the purchase decisions;
- the consumer response to a new and improved high efficiency product.

The study on the impact on consumers included the following steps: household sample definition; preparation of the questionnaire; pilot interviews for the validation of the questionnaire; telephone interviews and household sample identification; data analysis and drafting the field survey results.

In the choice of the sample we tried to balance the relatively low penetration of RACs in households with the need to assure statistical validity to air conditioners owners results. Therefore, a 2000 (Italy) and 3000 (Spain) households sample - called "screening sample" - was chosen with about 5,000 total contacts.

Through telephone interviews a "valid output sample", of 300 questionnaires/country out of the total contacts, was chosen in such a way to include 70% of RAC owners (210 households/country) and 30% of households without room air conditioning (90 households/country). 600 valid questionnaires were finally collected.

Questions were related to five different air conditioners types:
- Multiple-split units

---

[1] Italy is covered using extra funding provided by ENEA in the framework of an agreement with Italian Ministry of Industry; results are given to the Commission free, in order to improve SAVE study results. No Commission funds coming from the SAVE/EERAC study is used for Italian survey.
[2] ASM-Analisi e Strategie di Mercato, is the Italian society of the GfK group; GfK is a market research company active in all Europe.
[3] European Commission; DG XVII-SAVE PROGRAMME; Energy Efficiency of Room Air Conditioners: Final Report",1999.

- Single split mobile units
- Single split fixed units
- Packaged (window) units
- Single-ducts air conditioners

Main survey results are described in the following paragraphs, where a comparison of the two countries is also presented.

2 The Present Market

The penetration of air conditioners for domestic use in Italy is approximately 10 percent; instead for Spain it is the half, or five percent of households

As illustrated in Figure 1, Italy and Spain have a preference for installing single split (fixed) units, which represent 40% of the total installed units in both countries. Single-ducts are preferred much more in Italy, while packaged (window) units are favoured in Spain. Multiple-split units have greater penetration in Spain.

Base = those having an air conditioning system installed; percentage values

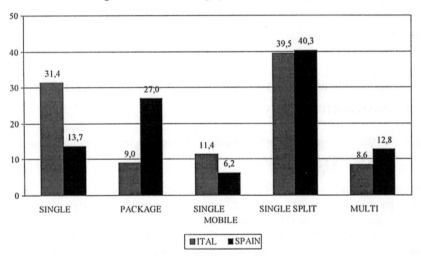

Figure 1: Installed air conditioning systems

With the respect to the utilisation of air conditioners, the mean values for Italy are 2.8 months/year and 3.2 months/year for Spain, consistent with climatic differences. Instead the mean number of hours/day is slightly higher for Italy, 6.1, as compared to 5.2 for Spain, probably reflecting differences in custom, organisation of time and possibly income differences. Italy also has a higher general penetration (10% compared with 5% for Spain) which also may impact on this.

The comparison of RAC average use in the two countries in terms of hours per year is shown in Figure 2. Despite the difference in the pattern of use the average total hours per year of RAC utilisation are similar for the two countries, even if slightly higher for Italy: 510 against 499 for Spain. RAC use depends on the RAC type: in general, fixed air conditioners more utilised than mobile units. However, while single-ducts and mobile splits are used more in Spain, fixed split and multiple-splits are used more in Italy. For packaged units the difference between the two countries is very low.

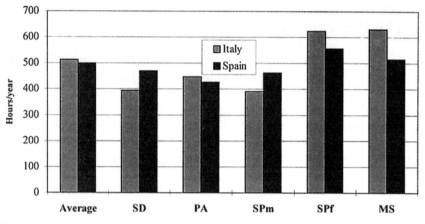

Figure 2: Comparison of average RACs pattern of use in Italy and Spain

3 Satisfaction of Consumers with the Installed Air Conditioning Systems

In general there are greater similarities than differences in the evaluation of the functional performance between Spain and Italy.

In the specific evaluation of ordinary and extraordinary maintenance (Figure 3) the percentage responses given as "good" are nearly equal: 50% (Italy) vs. 48% (Spain), whereas Italy has a lower response for "very good" (22% vs. 33%) and higher response for "sufficient" (25% vs. 12%). In general the Spanish sample has expressed a slightly more favourable evaluation of this characteristic.

Base = those having an air conditioning system installed; percentage values

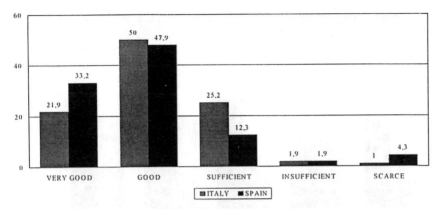

Figure 3: Evaluation of the functioning of the air conditioning system installed, considering possible ordinary and extraordinary maintenance

With regard to noiselessness, the Spanish sample gave higher marks both in the "very good" and "good" responses (Figure 4). This is particularly due to higher satisfaction with single-ducts and mobile split units in Spain, as shown in Figure 5.

Base = those having an air conditioning system installed; percentage values

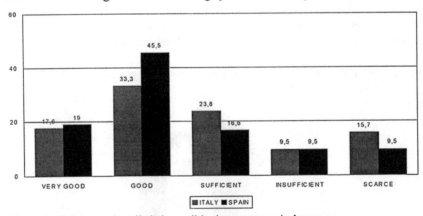

Figure 4: Opinion on installed air conditioning system noiselessness

480

Base = those having an air conditioning system installed; AVERAGE VALUES
(1=SCARSE; 5=VERY GOOD)

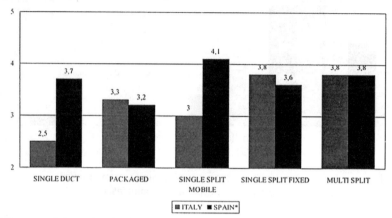

Figure 5: Opinion on installed air conditioning system noiselessness
(*) Small and statistically not significant base for Multiple-split

Concerning the time needed to reach the desired comfort level; there is a
significant difference in the perception given by the two samples. The Spanish
panel gives a general lower amount of time (Figure 6): 18% vs. 6% classified as
"very short" and 47% vs. 38% as "short". This more rapid time to reach the
comfort level on the part of Spanish sample is reflected among all models.

Base = those having an air conditioning system installed; percentage values

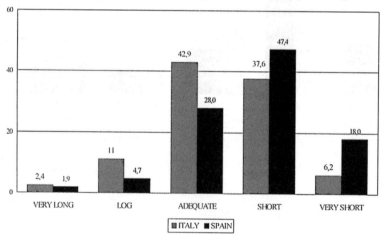

Figure 6: Opinion on time needed to reach a comfort level temperature
(*) Small and statistically not significant base for Multiple-split

With regard to the influence of the air conditioners on the electricity bill, Figure 7, Italy reveals a greater influence for all categories of response except "high". The overall greater influence of air conditioners on the electricity bill in Italy is probably due – at least in part – to Italy's higher electricity prices.

Base = those having an air conditioning system installed; SPAIN: NO ANSWER 9,5%, Percentage values

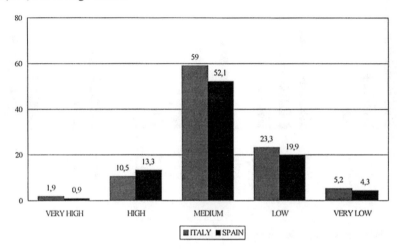

Figure 7: Influence of the air conditioning system on overall electricity bill

4   Purchase and Re-purchase of Air Conditioning Systems by Current Owners

The nature of the purchase decision is analysed from various points of view. An important dimension is who decided to install the RAC. In Italy 73% of the sample claimed it was their own decision as opposed to 58% for Spain. With respect to the brand: 66% of the Italians chose their own compared to 56% of Spanish respondents

In a more detailed analysis of what factors were given as to why they chose their air conditioning system, numerous different responses were given (Figure 8).

482

Base = those having an air conditioning system installed; percentage values

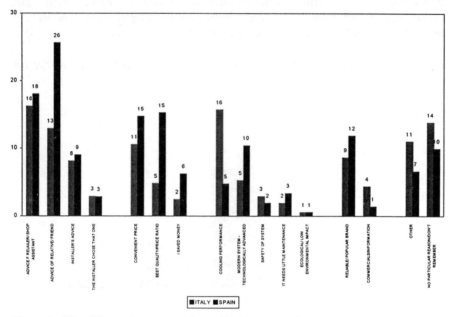

Figure 8: Why did you choose your air conditioning system?

In general the most important groups of responses were, in order of importance: advice; performance characteristics; economic factors; and marketing factors.

Within the response related to advice, the Spanish sample was in general more dependant on advice, in particular from friends or relatives. With reference to performance characteristics, Italians gave more frequently the reason of cooling performance whereas Spaniards reported the modern, technologically advanced aspects more often. With respect to economic issues such "convenient price", "best quality/price ratio" and "saving money" the Spanish panel consistently gave more importance to these factors. Marketing aspects played a less important role in both countries with brands being slightly more important in Spain and general consumer information slightly more important in Italy.

Regarding model loyalty, that is if the RAC owners would purchase the same type of model again (Figure 9) the distribution of the response is different in form between the two countries; however, taken together the "yes, certainly" and "yes, probably" responses, the percentages are very close: 57.2% (Italy) and 57.3% (Spain). There is greater uncertainty about the decision ("don't know" response) in Spain: 23.2% vs. 13%; and more certainty about the no repeat purchases in Italy, with 29% vs. 19% in the "not, probably" or "not, certainly" categories.

Analysing the model loyalty with respect to the models themselves (Figure 10), we see that in general there is greater loyalty with increasing costs and complexity of the systems. Least loyalty (lower repeat purchase) go to the single-ducts, packaged systems and mobile splits are next, followed by fixed split and multiple-splits. This may represent the general phenomenon of trading up to larger units.

Base = those having an air conditioning system installed; percentage values

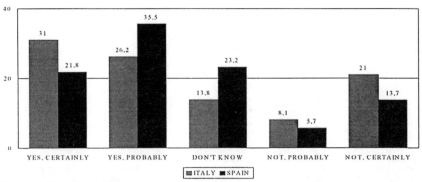

Figure 9: Should you change your present system, would you purchase the same type again?

Analysis by sub-samples: type of air conditioner installed
Base = those having an air conditioning system installed of the different categories;

AVERAGE VALUES (1=NOT, CERTAILNLY; 5=YES, CERTAINLY)

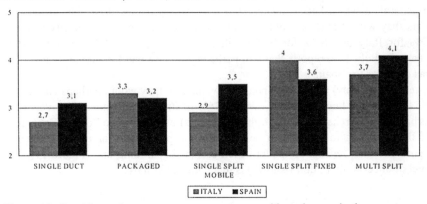

Figure 10: Should you change your present system, would you buy again the same type you have now?

Also when asked what type of air conditioner would be purchased we obtain the same general pattern favouring fixed split units and multiple-split, as illustrated in

484

Figure 11. There are some country differences between types: Spaniards favouring more multiple-splits and mobile splits whereas Italian gave greater preference to fixed single splits.

Base = those who would buy a different type of air conditioning system; percentage values

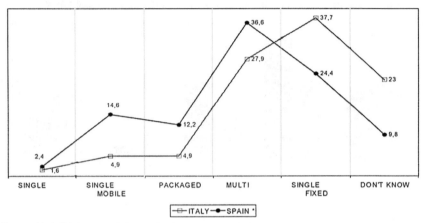

Figure 11: Which type of air conditioning system would you buy?
(*) Small and statistically not significant base for Multiple-split

5 Future Purchases

In general, it was asked to owners and non-owners if they think that in the next years they will change or buy a new one. For owners, the great majority of them, 67% for Italy and 83% for Spain will not purchase unless the present system breaks. In Italy 14% reports that will perhaps purchase one for another house. This was only 1% for Spain.

Instead 19% (Italy) and 16% (Spain) answered that they would perhaps change the one they have at home. For non-owners: 36% of Italians responded positively (yes, perhaps they purchase a RAC in the next years) as opposed to 26% of Spaniards. Of those willing to purchase a new system the preference for types is given in Figure 12. This repeats the pattern seen earlier for "trading up" (Figure 11). The level of uncertainty ("don't know") is much higher in Italy.

Base = those willing to change or purchase a new air conditioning system in the next years; percentage values

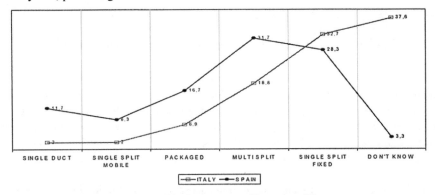

Figure 12: Which type of air conditioning system will you install?

Base = those willing to change or purchase a new air conditioning system in the next years; percentage values

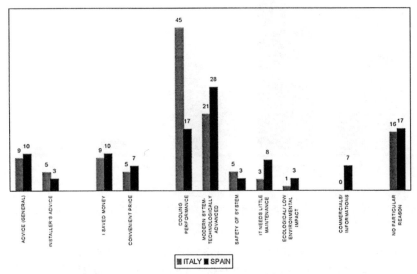

Figure 13: Reasons to choose an air conditioning system

Questioned about the reasons for purchase (or change) of a new system (Figure 13) the relative importance of the response groups changes with respect to the reasons given for the past purchase (Figure 8). For the future purchase performance characteristics are given as most important, followed by economic factors and advice, at a much lower levels of frequency. Marketing factors are given as the least important. The same relative differences between countries are shown for the performance characteristics as given in the past purchase decisions: for example

Italians greatly emphasising cooling performance and Spanish the general modernity of the systems.

6 Market Potential and Technological Innovations

Respondents were informed about technological development that allows the manufacturing of new type of air conditioning system, which guarantees an actual saving on the electricity bill; and asked if they would buy the new system instead of the one they had in mind. The positive responses (Figure 14) were higher for Italy summing the "Yes, for sure" and "Yes, probably" answers: 75% vs. 63% (in Spain). This again could be partially due to the higher electricity prices in Italy.

Base = those willing to change or purchase a new system in the next years; percentage values

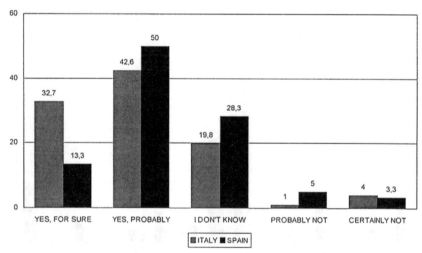

Figure 14: Would you buy the new improved system instead of the one you have in mind to choose?

Consumers answering Yes ("Yes, for sure" + "Yes, probably") are considered to have "a purchasing propensity" towards new improved RAC models.

7 Consumer Interest in Improved Models

In order to evaluate consumers' interest in improved models, telephone interviews were considered hypothesising a new technologically improved air conditioner, with an increased purchase price but giving an energy (and money) saving on the electric bill.

Cost increase of the RAC system and cost decrease on the electric bill have been selected on the basis of the actual type of air conditioning system owned by the interviewed consumers. In the question reference is made to the cost of the owners' traditional product, in addition to the increase in cost (price) of the system and the corresponding annual savings. Table I gives the three hypotheses that have been tested.

Table I: Scheme used for the measurement of consumer interest in improved models

RAC TYPE	Cost of traditional system		Increase in the cost of the system		Annual savings on the bill		CASE
	ITALY (Lire)	SPAIN (Pesetas)	ITALY (Lire)	SPAIN (Pesetas)	ITALY (Lire)	SPAIN (Pesetas)	
SINGLE-DUCT/	1,600,000	135,000	45,000	3,600	15,000	1,200	1
			66,000	5,400	22,000	1,800	2
			90,000	7,200	30,000	2,400	3
PACKAGED	1,500,000	115,000	90,000	8,000	30,000	2,700	1
			135,000	12,000	45,000	4,000	2
			180,000	16,000	60,000	5,400	3
MOBILE SPLIT	2,100,000	180,000	60,000	5,400	20,000	1,800	1
			90,000	7,800	30,000	2,600	2
			120,000	10,800	40,000	3,600	3
FIXED SPLIT	2,800,000	200,000	84,000	7,000	28,000	2,300	1
			126,00	11,000	42,000	3,500	2
			168,000	14,000	58,000	4,600	3

The general results, shown in Figure 15, indicate a considerable difference in interest in the purchase of the improved system between the two countries. The Spanish respondents give a much lower level of interest: 12% vs. 65% in the case 1 hypothesis, as shown in Figure 23. The Spanish level of interest rises significantly to 46% in the heavy saving (case 3). There is a non linear increase between case 2 and case 3, corresponding to a possible threshold of transaction costs - below case 2 performance it is not worth the effort to be informed, to find the specific energy savings model and to waste ones time and money for extra savings – above this it is. Instead the Italian response is very strong and rather linear, without such threshold effects.

The answers for the same question were divided into owners and non-owners (a 30% minority of the sample) and the results presented in Figure 16. The "Yes" response is in general lower for non-owners, significantly for Spain in the case 3 which goes from 55% of owners to 28% of non-owners. The degree of uncertainty (don't know) is very high in the case of non-owners.

Base = total sample (owners of multiple-splits excluded); 283 for Italy and 275 for Spain

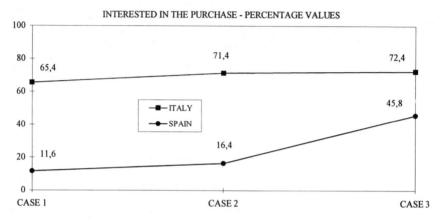

Figure 15: If there was an increase in the price of the air conditioning system of ... lire/pesetas, with an annual saving of ... lire/pesetas on the electricity bill, would you be interested in the purchase of this new type of system, instead of a similar traditional one?

Analysis by sub-samples: owners and non-owners of an air conditioning system
Base = those belonging to each category (owners of multiple-splits excluded)

	Price increase and energy saving hypothesis					
	case 1		**case 2**		**case 3**	
Consumers	ITALY	SPAIN	ITALY	SPAIN	ITALY	SPAIN
Owners						
YES	67.2	12.0	74.5	18.5	75.5	54.9
NO	32.8	88.0	25.5	81.5	24.5	45.1
Don't know		0		0		0
Non-owners						
YES	61.5	11.0	64.8	12.1	65.9	27.5
NO	38.5	16.5	35.2	15.4	34.1	0
Don't know	0	72.5	0	72.5	0	72.5

Figure 16: If there was an increase in the price of the air conditioning system of ...
lire/pesetas, with an annual saving of ... lire/pesetas on the electricity bill,
would you be interested in the purchase of this new system, instead of a similar
traditional one?

The answers to the same questions may be broken down according to other sample
characteristics, for example, according to the general propensity to buy a new
system. Clearly the higher the general propensity the higher the specific interest of
the energy saving new system. The difference in interest to this purchase was also
studied in terms of hours of use. A small difference is shown in the direction
expected: more interest, the greater the hours of use. It was also hypothesised that
larger households, in terms of more floor space, would have greater interest. This
was also confirmed for both countries (Figure 17). This may correspond to greater
need (cooling space) and possibly greater family income, having a larger home.

Analysis by sub-samples: house size
Base = those belonging to each category (owners of multiple-splits excluded)

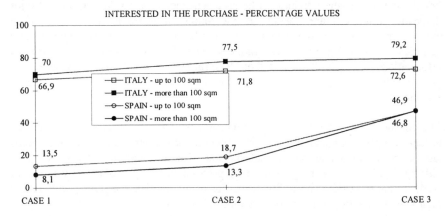

Figure 17: If there was an increase in the price of the air conditioning system of ...
lire/pesetas, with an annual saving of ... lire/pesetas on the electricity bill,
would you be interested in the purchase of this new system, instead of a similar
traditional one?

Domestic Appliance Salesperson as Consumer's Information Source

Pirkko Kasanen, Arja Rytkonen and Anneli Reisbacka

TTS-Institute

1 The Aim and Background of the Study

The aim of the study was to investigate what kind of information the consumer gets from a domestic appliances salesperson in a shopping situation and how the sales staff are able to adapt the available information to the customer's needs. Another aim was to survey the way in which the energy label was utilised by the salespersons and in what domain would extra training be needed for the staff.

For the consumer, the purchase of home electronics is not a daily routine. Major domestic appliances are bought only a couple of times in a lifetime. When entering the store, the consumer may be totally unaware of the new choices and qualities offered by the latest models. The first and foremost source of information, hence, is the salesperson. Several studies emphasise the central role of the salesperson as a source of information (Kuusela 1996, Moisander & Kasanen 1993).

The consumer buying his/her first appliance faces the most difficult situation; due to lack of previous personal experience he/she finds it hard to estimate the claims and propositions made by the salesperson. The growing population of elder citizens is also increasingly in need of professional help provided by the sales staff, so that the safety and utility properties of the appliance acquired would meet their specific needs and restricted capabilities.

Different methods are used for getting information on the appliance desired (Kuusela 1996). Some buyers acquire comparative technical information from specialised stores whereas others rely on advice given by friends, colleagues, and other consumers. Retailers, on their part, stressed the importance of specific appliance training offered by manufacturers since without it, giving technical information on the qualities of a certain appliance would be impossible, as well as discerning which qualities the seller should highlight in a selling situation (Timonen, Heiskanen, Niva 1998, 38-40). But, on the other hand, importers were complaining the difficulty of getting the staff to participate in the training offered.

It has been found that with extra training and information given to salespeople, favourable attitude can be awakened towards energy efficient domestic appliance. For example, in France, where in 1994 a training course was arranged for home electronics sellers on the energy efficiency of the appliance, the results acquired were beneficial. 85% of the salespersons felt they had received useful information utilisable in their work. Due to the positive experience of the training course, a Save II project was launched in France, in order to create a training module for home electronics salespeople (Heulard 1999). Retailers can also be given a broader responsibility in encouraging energy efficient choices. The results of the French experiment also underlined the necessity of an active participation of the retailers: Success depends on whether they anticipate the market and create demand by introducing more efficient appliances (Colombier & Menanteau 1997).

In the pro graduate paper made by Satu Routa in 1997, "The Domestic Appliance Salesperson as the Consumer's Information Source," the role of the salesman was investigated as the information source for the consumer in the purchase situation of a washing machine. The survey revealed that especially the buyer of one's first domestic appliance was getting inadequate, variable, or even misleading information from the salesperson. The energy label was not utilised and in some cases it was even misinterpreted.

2 Research Methods

The material for the survey was collected in test shopping events where a test shopper (4 shoppers in total) was investigating the possibilities available to meet his demands with the help of the sales staff. For the sake of accuracy, the information acquired was recorded on tape -- without the salesperson's knowledge -- so that it would mimick a real purchase situation. After the test shopping event, the shopper was to fill a questionnaire inquiring of the purchase situation. The data concerning each device was analysed by a researcher who was not present in the purchase situation.

There were 117 test purchase situations in total, of which 31 dealt with the purchase of a washing machine and tumble dryer, 25 of a cooker, 32 of a fridge freezer, and 29 of a television set with a VCR (video cassette recorder). The stores were situated all over the country in 13 different towns. There were 58 stores altogether, and occasionally two test buyers with different goals were visiting the same store.

3 Customer Service and the Display of Appliance in the Stores

The willingness to serve, based on a subjective estimate of the customers, was good in most of the stores (53%) or at least satisfying (34%). Only 15% of the shops were estimated to have poor service. The completeness of satisfaction on the part of the customer was diminished by the ignorance and superficiality of the salesperson, as well as his/her insufficient interest in presenting the appliance displayed and its specific features, but also by the restricted choice of device available. In some cases, the arrogant attitude of the salesperson towards the customer lowered the estimation given on the service.

The salespersons were mostly male (93%) and generally fairly young. According to customer estimates, almost half (44%) of them were under 30. There were few saleswomen (7 altogether) but the information given by them was somewhat more substantial and practical than that given by the salesmen. This was displayed also in the consumer satisfaction, since in all test shopping events, the service given by a saleswoman was estimated good.

Brochures were generally the material presented to the customer (60%), whereas user manuals were not so often used (5%). Other written material shown included the energy label (46%), but also research material was used in some stores (6 occasions).

The price was discussed in almost every shopping situation (86-100%), in addition to which the prices were clearly displayed in some of the shops. Guarantee was mostly discussed in connection with the purchase of a washing machine and tumble dryer (94%) or a fridge freezer (84%). It did not come out as often when cookers (40%), televisions and VCRs (38%) were purchased.

4 Information Given on Washing Machines and Tumble Dryers

In the shopping event of a washing machine, the location, installation, and space demands were mostly overlooked. For example, the opening direction of the lid has a great significance in what comes to space demands and flexibility of usage. The opening direction of the lid should be checked already in the purchase situation, since it is not changeable. Also the height adjustment and movability of the appliance, as well as the safety systems were neglected.

For wash systems, one of the selling principles was the spin rpm; it was referred to in almost every shopping event (94%). The case was similar with the wash load capacity of the machine (90%). According to the test buyer, the energy label was clearly or sufficiently visible in roughly half of the stores. Energy efficiency or energy consumption was referred to in 77% of the test purchases. Differences in energy consumption were generally regarded insignificant. Usually the consumer saw no great differences in energy consumption between different models. Wash result was dealt in reference to the energy label classification (70%). The use of washer controls and programmes was explained by 71% of the sellers. Cleansing and maintenance, however, were mostly overlooked.

Connector specifications were discussed more often in the case of a tumble dryer investment than with washing machines. They were mostly brought up by the customer. Tumble dryer needs a well vented place to function properly. This fact was mentioned only in one case. When stacked, the tumble dryer is attached with a special stacking kit. This was brought up in 58% of the test purchase situations, although a part of the sales staff (16%) claimed that attachment was not always necessary.

Tumble dryer capacity was discussed in almost every test case (97%). Similarly, tumble dryer programmes were generally explained (81%), as was the basis of operation; whether it was based on manual choosing of operating time or on automatic moisture level monitoring. Energy efficiency and energy consumption were mentioned in 58% of the test purchases.

Some of the salespeople were insufficiently informed on the principles of granting the energy label as well as the energy efficient use of washer and dryer.

5 Information Given on Fridge Freezers

The space and ventilation demands of fridge freezers were discussed in 86% of the cases. The free space needed for ventilation was brought up mostly by the buyer (78%). Mostly the sales staff announced the space needed to be a few centimetres, but in a couple of cases the staff misleadingly claimed that no extra space was needed. The appliance presented was of the kind where ventilation space was not included in the given measurement. The easy utilisation of a refrigeration product depends on which direction the door opens and whether it is reversible. This came up in one fifth of the test shopping events. Maintenance and the possibility of cleansing was discussed in about half of the cases.

Energy efficiency and energy consumption were discussed in almost every test shopping event. According to the test buyer, the energy label was aptly attached in half of the stores. Only 47% of the salespersons rated excellent as to giving information on the energy label. Some sellers even gave misleading information on the principles of the label.

6 Information Given on Cookers

Usually the sales event began with investigations of the customer's needs considering the capacity of the cooker. In most of the cases (68%), the installation was to be made by a professional when delivered.

The differences between hot plates were discussed in 76% of the cases as well as the differences between top-bottom heat ovens and circulating air ovens (52%).

The possibility of cleansing was mostly overlooked, it was mentioned only by 28% of the salespersons.

Safety features of cookers were treated mostly in relation to the residual heat indicator of the ceramic hob (40%). For example, exterior temperature of the oven was mostly neglected, although it is significant for families with small children.

In certain cases, the sales staff had poor knowledge of differences between cookers and of their specific features -- such as the difference between iron and ceramic hotplate, the existence or lack of circulating air in the appliance, temperature marker differences between models, and energy consumption.

7   Information Given on Television Sets and VCR

The test buyers of television sets and VCRs (2 persons) were the same as for the cookers. It was mostly the same salesperson who was in charge of all three purchase situations. Certain salespersons showed more interest in presenting televisions and VCRs than cookers. This became evident in their way of providing detailed information on the specific features of TV-sets and VCRs.

In the beginning of a television test purchase situation, the seller generally inquired the customer's wishes regarding the size of the set. The arguments given for the size varied. Some salespersons stated that nowadays the viewing distance played no significant role whereas others gave recommendations considering the distance. For the test buyers, the viewing distance was between 2 and 3 metres. The screen size of the television sets presented was between 20 and 29 inches.

Location of the TV-set was discussed in one fourth of the test purchases. The space needed for air circulation was brought up mostly by the customer. The ventilation demands given by the sales staff varied. One part of the sellers justified the space needed for ventilation on the grounds of safety, while others stressed its importance for the sound quality. Similarly, the use of the stand-by switch was unevenly substantiated.

Likewise, the salespeople started the presentation of a VCR by questions of its intended use. The taping qualities were often discussed (86%) as well as its connecting to the TV-set, interoperability of the remote, and tuning of the channels (3/4 of the test purchases).

The energy consumption of TV-sets and VCR was inquired most often by the buyer. Generally, their energy consumption was regarded low by the sales staff.

8 Things to Be Brought up in the Purchase Situation

The buyer should be advised more often to seek help from the user manual. In the cases presented, it was most often neglected. Although the manual contained necessary information for correct location, installation, use, and maintenance required by the appliance, only 5% of the salespersons referred to its utilisation.

In the purchase situation, the salesperson should make notions of
- the consumer's needs,
- the space reserved for the appliance,
- space and requirements needed for installation and use,
- fitting in of the appliance and its compatibility with other household fixtures,
- environmental impact,
- performance,
- energy consumption and related costs,
- features,
- safety,
- price,
- recycling or disposal of the old appliance,
- the guarantee and service available.

9 Recommendations

As the survey indicates, much depends on the initiative taken by the buyer himself as to what extent the location, installation, maintenance, and safety demands are treated in the shopping event. It must not be forgotten that these are the specific features according to which different models are distinguished in the market.

There are more and more elderly people in today's society, which is to be taken into consideration when safety features of domestic appliances are discussed. In the shopping event, safety features are too often overlooked, a fact maybe influenced by the young age of the salespeople. To correct the situation, we would like to propose:

- more research on how elderly or disabled people and their restricted capabilities can be taken into consideration
- more information on safety features in the use of the appliance, for both sales staff and consumers
- more comparative device-specific information especially on how structural differences of appliances influence the use, for both seller and buyer.

The study revealed that the sales staff of domestic appliance have willingness to serve but there is variation in their level of knowledge. In the shopping event, the seller treats primarily the price and the features mentioned in the energy label. But there is a lack of knowledge -- and even misinformation -- in what comes to the interpretation of the energy label.

- Adapting information on energy efficiency and appliance capacity in the real life situation, demands much more attendance to the needs of the customer and to the general features of the device from the seller's side.
- The salespeople need much more information on specific properties of the appliance: its location, installation, use, and maintenance as well as the effect of these on the durability, energy efficiency, performance, and features of the machines. Also additional information on household electricity consumption in general is required.
- Also the basic principles of granting an energy label and the measurements of performance should be treated in more detail.
- The salespeople need to be instructed on how to guide their customers into a more efficient utilisation of the user manual.

Training courses aimed at raising the general awareness of salespersons should include general knowledge on the principles of choosing a domestic appliance, such as first finding out the consumer's needs in order to match them with the appliance desired, then investigating the measurements of space reserved and required for the appliance and its installation, its compatibility with other household fixtures, its performance, environmental impacts and energy consumption, its operational features and safety, and finally the possibilities of maintenance and final disposal of an old machine.

- The immediate training should be directed to representatives of specific domestic appliance brands who, in their turn, pass on the knowledge to their own authorised outlets and personnel.
- A long-term programme is being planned also in Finland in order to develop professional training of home electronics sellers. Co-operation with adult education centres has been raised as a possible solution.

To ameliorate the quality of written information available for the consumer, product guides and user manuals need to be standardised to avoid discrepancy in contents and layout. At the same time, the consumer's possibility of choice to find an appliance suitable to his needs should be encouraged.

- It is a task for the Consumer Agency to promote the development of these domains by demanding certain standards from user manuals and product guides. Consumers need extended knowledge to base their choice on, and here co-operation with such Finnish organisations as Consumer Agency (Kuluttajavirasto), Motiva Energy Information Center, TUKES -Safety Technology Authority, Wholesale Dealers of Electronics, Home Electronics Association (Kodintekniikkaliitto) and Work Efficiency Institute (TTS-Institute) is to be proposed.

10 References

[1] Colombier, Michel & Menanteau, Philippe, 1997, Some results and propositions from a French experiment with energy labelling. In *Sustainable Energy Opportunities for a Greater Europe. The Energy Efficiency Challenge for Europe. Proceedings of the 1997 ECEEE Summer Study*. Part 1. Prague.

[2] Heulard, V. ,1999, Vocational Training. In *For An Energy Efficient Millennium. The Save Conference. Proceedings*, Volume II.

[3] Kuusela, Hannu, 1996, Factors affecting the acquisition of domestic appliances and energy saving. In *LINKKI Research program on consumer habits and energy conservation, Publication* 15/1996, pp.117-127. Helsinki.

[4] Moisander, Johanna & Kasanen, Pirkko, 1993, The choice situations and markets for household appliances from the point of view of energy conservation (in Finnish). *Helsingin yliopiston sosiaalipsykologian laitoksen energiajulkaisuja* 11/1993. Helsinki.

[5] Routa, S. , 1997, Kodinkonemyyjä kuluttajan tietolähteenä. Opinnäytetyö. Jyväskylän ammattikorkeakoulu. (Pro graduate paper, Jyväskylä Polytechnic)

[6] Timonen, P., Heiskanen, E., Kärnä, A. & Niva, M. 1998. Tuotteiden ympäristölaadun parantaminen – Tuoteketjun osapuolten näkemyksiä. *Kuluttajatutkimuskeskuksen julkaisuja* 1. (National Consumer Research Centre, Publications:1)

Changing the Electricity Customer Mentality, Under Energy Market Liberalisation Circumstances

Laurentia Predescu and Oana Popescu

S.C. ELECTRICA S.A.

Considering the present changes and the improvement of electricity supplier-customer relationship, the "shock" of changing mentality is a must, for the utility as well as for the electric energy consumers.

Romania is one of the European countries which faces specific problems in the transition from a centralised economy, highly industry oriented, towards the market economy. This situation implies a double issue: the energy intensity per product unit is high, but in the same time there is a huge capacity for energy conservation. The change of mentality means overcoming existent barriers regarding energy efficiency, in terms of political, informational, technological, financial and cultural aspects.

1 Market Economy Transition

The transition towards market economy obviously includes reorganisation of the Romanian energy sector. One of the major targets is to encourage competition on the domestic energy market, in order to ensure efficient generation, transmission, distribution and supply, considering environment protection. Another goal is to improve the quality of the services offered to customers.

The initial structure of the Romanian energy sector is specific for a vertically integrated monopoly. Thus, the transition towards a competitive market should be done step by step, in order to ensure an efficient risk control.

Due to the monopolistic structure of the energy sector, the consumers have access to a limited set of options. Competition is the most effective mechanism that can protect the consumers' interests, enabling them to choose the package of services they really need (the right combination of price, security, quality, additional services, risk, terms of delivery and payment). Competition also implies efficient allocation of resources and reorganisation of activities with a view to flexibility increase. Generally, competition reveals inefficiency and identifies areas of malfunction [5].

The first step towards market economy in Romania was to abolish the quota system and to eliminate the restrictions in the electricity supply.

Previously contracts were formal, since both supplier and consumer were state owned. The contract system started to be gradually improved [2]. So, it was decided to supply electricity only based on firm contracts concluded between the supplier and the consumer, taking into account the following:
- entirely meeting the demand;
- establishing correct supplier-customer relationships, based on economical principles.

From now on, the contract settles the relations between supplier and consumers, regarding supply, payment and use of electricity.

According to the subscribed demand and the specific of electricity use, consumers are divided in the following groups (*see* Fig. 1):
a) domestic consumers (no power limit);
b) small customers, with subscribed demand less than 100kW (including especially commercial customers, small workshops, schools, hospitals etc.)
c) big customers, with subscribed demand over 100 kW (generally industrial consumers).

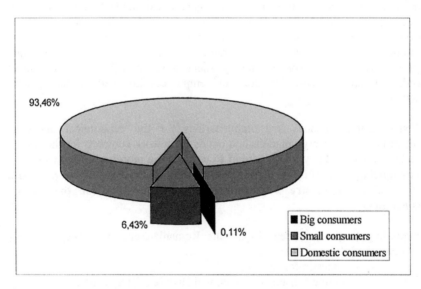

Figure 1: The structure of electricity consumers in Romania

The contract period for the domestic consumers is unlimited; for the other categories the contract is concluded over a specified period of time, for each consumption place. The contract duration can be extended upon written request from the consumer.

At any consumers category, the electricity supply contracts are drawn according to the necessary demand requested. Except for the domestic consumers, the quantities, mutually agreed upon, are stipulated in the annexes to the contract. The consumers are entitled to ask for a modification of the subscribed quantities, on a quarterly or monthly basis, as well as for the permission to exceed the maximum power over the subscribed values.

2 Restructuring the Romanian Energy Sector

Romania is determined to adhere to the European Union. This is an option clearly expressed by all government programmes. The only way to reach this goal is by lining up all components of the society with the Directive of the European Union 96/92 EC.

The National Energy Regulatory Authority (ANRE) was established in 1998, through the Emergency Government Ordinance OUG 29/1998.

The role of ANRE is to create and apply the regulatory system at national level; this will ensure an efficient, transparent and stable functioning of the electricity and heat sector and market and will **protect the interests of the consumers and investors** [1]. In its activity, ANRE keeps an equidistant position from the three poles of interests of the energy sector: the state (strategies, taxes, environment policy, social programmes), energy consumers and, last but not least, investors and employees involved in this specific activity.

ANRE grants licences and authorisations for the economical operators activating on the energy markets and issues framework contracts for the energy sector activities. **They are intended to correct the monopoly behaviour, by introducing – in this very first reform stage – a competitive pressure. They also try ensure a certain level of protection for the energy consumers**.

According to the new regulations, the Romanian energy market presently involves the following categories:
- domestic consumers;
- captive consumers, which will have regulated tariffs;
- eligible consumers; they are entitled to choose the supplier and to sign a direct energy supply contract with it, at negotiated prices, having access to the transmission and/or distribution grids.

The electricity consumption breakdown for the new categories of consumers is presented in Figure 2.

3   Regulation of Prices and Tariffs

Prices and tariffs play an important role in market liberalisation; they encourage competition and protect the interests of consumers [1]. The improvement of electricity tariffs system made possible to gradually eliminate the cross subsidy between industrial and domestic consumers, thus reflecting the energy behaviour of each category. Two new tariffs have been introduced for the domestic sector: the *one-part tariff with subscription rate* and the *one-part time-of-day tariff with subscription rate.* This new tariff structure allowed recreating a normal situation, that is cutting the electricity pay-back price while the consumption goes up. The poorer categories of population benefit from a *social tariff* (*See* Figure 3).

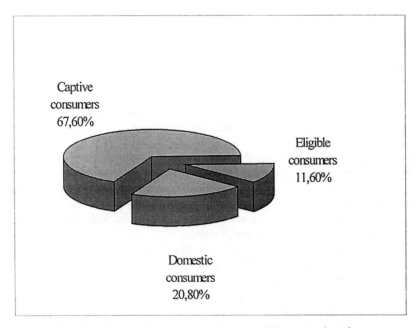

Figure 2: Electricity consumption in Romania structured by categories of consumers

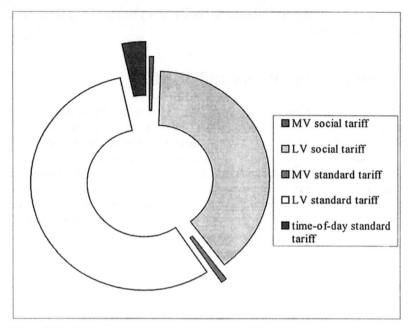

Figure 3: Breakdown of domestic consumers by tariff and voltage level

In the future, the present way of establishing tariffs based on average costs will be replaced by the marginal cost method, the only one capable of transmitting correct economical signals to the electricity consumers.

Setting up electricity prices and tariffs for the end-users is a complex process and an act of great responsibility. Romania intends to establish a Committee for Electricity Tariffs for non-eligible consumers, with a consultative role. The Committee is formed by representatives of the electricity sector, of the consumers, trade unions, as well as other organisations or institutions interested. The Tariff Committee will ensure:
- transparency at prices and tariffs setting;
- consumers understanding of the way of founding and establishing the tariff system;
- direct participation at negotiations of the consumers representatives;
- a forum for consumers and agents from electricity sector, for supporting their interests in the framework of new regulatory system;
- avoiding dissatisfaction from trade unions or population.

One of our energy policy strategies, **that contributes to influencing consumers' behaviour and changing their mentality**, is:

4 Consulting Services and Protection for the Electricity Customer

Considering the period of ongoing transformation for the energy market, one of our main concerns should be gaining the customers loyalty, which first implies keeping them satisfied with the products and services offered [2].

Unlike other products and services, the electricity price is highly dependent of WHEN AND HOW this energy is consumed. Therefore, it is most important that the supplier offers consulting services to its customers.

The first opportunity to meet and get to know the customers is given by the moment of contract signing. The customer receives all necessary information regarding the terms of supply and tariff system. Subsequently, the customer is entitled to receive free advice concerning all aspects of electricity supply, at any of the supply centres or subsidiaries.

In order to increase the electricity consumption, by attracting new customers – according to their needs and possibilities – a series of measures have been taken:

- granting certain facilities (some of them substantial) for several categories of new customers, for the electricity supply equipment;
- free of charge connection for the new domestic consumers: single-phased connection; meter mounting; actual connection of the interior installation to the exterior supply installation; concluding all contract forms;
- more operative connection and electricity supply for the new customers;
- consumer protection through increased contractual liability: the supplier is obliged to pay compensations to the domestic consumer for any damage caused to the electric appliances (except for lamps and electro-thermical receivers); this apply if the supplier is responsible for voltage values higher than those permitted by the technical norms in force, in the supplier-customer delimitation point.

5 Consumers Education for Efficient Energy Use

Our company goes further in the effort of **changing the mentality of the electricity consumer.** Beside a full range of electricity supply services, it offers to its customers **education regarding energy efficiency problems**.

We have a special task in educating the domestic customers. This sector will have a paradoxical destiny: we are trying to get an increase in the energy consumption, as a reflex of general civilisation status going up; but in the same time we have in view to eliminate the old, uneconomical appliances and to promote efficient and environmental friendly technologies.

In this period of transition, the electricity consumer should be prepared by means of a sustained information and education campaign. This is the main road to follow for CONEL – S.C. ELECTRICA S.A. in the near future. The campaign will be held on different plans, within a SAVE II project [6].

The most important aspect, the spinal column of the whole project, will be the opening of a pilot public Energy Efficiency Information "Shop", at a major intersection in central Bucharest.

The concept is new world-wide; it gathers several types of activity, easing the customer's access to both energy efficiency information and products:
- **classical information and education activity**, through written information for domestic, commercial and industrial customers (leaflets, brochures, folders, designated to address sub-sector specific energy efficiency issues, such as what is a compact fluorescent light bulb or what can schools do to reduce their energy bill);
- **specialised consulting activity**: free, objective energy efficiency advice to visitors, including directing them to energy efficiency specialists (equipment suppliers, contractors, ESCOs). A range of databases will be developed, including: local vendors of energy efficient products (for domestic/small commercial consumers); local small contractors, with experience of carrying out domestic energy efficiency projects; larger contractors, able to carry out energy efficiency projects for entire blocks and for commercial customers; highly specialised contractors (for industrial projects);
- **selling or give away of energy efficiency products** for domestic consumers, primarily to help overcome paradigms which associate energy conservation with energy deprivation. There will be 20 simple, low-cost, attractive product lines, such as: draught excluders, window insulation, combined movement/night time switchers for exterior lights, compact fluorescent light bulbs, thermostatic radiator valves, solid state dimmer switches, solar powered rechargeable AA battery units, electric timers, small "A" or "B" rated refrigerators etc.

The concept of "Energy Efficiency Information Shop" has three unique features, as future trends for this activity:
- bridging the gap between supply side and demand side activities, by building on the energy supplier – consumer relationship;

- for the first time in Romania, a permanent, free energy efficiency information facility will be available to the public;
- we believe that it is the first time in the world that an energy efficiency information centre will be involved directly in the market deployment of low-cost energy efficiency products for residential consumers.

If the project is successful, S.C. ELECTRICA may replicate it in its 42 centres throughout Romania, bringing energy efficiency awareness to 7.7 million domestic and over 500,000 commercial, tertiary and industrial customers. In addition, the project may be suitable for replication in EU and other countries where there is a need to improve public awareness of the benefits of energy efficiency.

The project will be supported by a media advertising and publicity campaign, that will include the key elements of:
- improving public awareness of energy efficiency issues;
- notification of the existence of the "energy efficiency shop";
- building a corporate image for S.C. ELECTRICA S.A. and tightening the supplier-consumer relationship.

A strong project team has been assembled to bring together: direct access to energy consumers, communication of the energy efficiency message and specific expertise in the design and implementation of energy efficiency information projects.

6   Barriers to Energy Efficiency

Romania has a huge potential for energy saving. But to achieve this there is a need for a **culture change**, involving breaking down existing **barriers to energy efficiency**. The energy efficiency information project addresses such barriers, which include:
- **Political barriers**: key efficiency messages (e.g. energy efficiency saves money; energy efficiency creates jobs; energy efficiency improves national industrial competitiveness; energy efficiency improves the environment) are not well understood in Romania. Until now, energy efficiency was not a major political option.
- **Informational barriers**: customers, especially domestic ones, didn't have access to information regarding energy efficiency; this was a field for specialists only, who could read specific literature. We will make good quality, Romanian language energy efficiency information widely available, that anyone can understand.

- **Technological barriers**: technologies which are increasingly commonplace in EU countries are often unknown in Romania. The free access to databases of specialist energy efficiency contractors will bring together technologies and their applications.
- **Financial barriers**: compact fluorescent lamps will be made widely available on a "one-per-family" basis, at low cost. We will conduct experiments with pricing policies to assess the price level at which wide-scale market deployment of the key residential energy efficiency technologies takes place, with the objective of maximising the market deployment of such technologies at minimum cost.
- **Historical barrier**: energy saving has a bad name in Romania (energy conservation went hand-in-hand with energy deprivation under the former regime). This important barrier will be addressed through the advertising and publicity campaign, and by demonstrating and selling a small range of well-designed, attractive, modern energy efficient products in the shop.

From this perspective, the "Energy Efficiency Information Shop" project may be considered as a catalyst to help the concept of energy efficiency take better hold, throughout the Romanian economy and beyond.

7 Conclusions

The way in which electricity suppliers will take care of their customers is a vital subject of growing importance. Decentralising of energy markets gives the consumer a greater options range for choosing his supplier; moreover, it contributes to the lowering of prices level, it increases the complexity degree for public service offer and it improves the quality of services [4]. The consumers will no longer be obliged to buy energy from the same monopolistic supplier, neither from the regional one. If they are not satisfied with the price or the services they receive, another supplier – either internal or external – will be ready to attract them.

The new competitive environment implies a change of orientation, from product to consumer. The utility has to know its customers: not only the energy consumption, but also their activities, life style, needs, aspirations. The customers should be treated as individuals, not just a meter at the end of a supply cable [3].

Thus, customers' loyalty can be gained. Also, the supplier gets several benefits: loyal customers offer publicity through recommendations made to others, they are more receptive to cross sales, offer valuable feedback, tend to be more profitable etc.

We agree that only the companies that succeed to adapt rapidly, professionally and without compromises to the new changes will survive the ongoing process.

8 Bibliography

[1] ANRE – *Regulatory Strategy for Supporting Liberalisation and Privatisation Processes in the Heat and Electricity Sector*, Bucharest, April, 2000.

[2] Predescu, L., Vilt, L. – *Supplier-Customer Relationship in Romania and its Diversification.* DistribuTECH Europe, Madrid, September 1999.

[3] The Boston Consulting Group – *Best Practice in Customer Relationships and Retail Marketing*, 2000.

[4] Petrovic, L., Nardoni, L. – *Switched on to Customer Demands*, The Business Management Magazine Electricity International, March 2000.

[5] Schmit, D. – *Experience of Energy Market Deregulation in the European Countries*, VGB, Electrical Power Plants, 2/1999.

[6] Predescu, L., Popescu, O., Vilt, L., Popescu, C. – *The Energy Efficiency Information "Boutique". Proposal for a SAVE Action.* Project Round 1999.

The Second Turn of the Energy Efficiency Spiral

David Cogan

Energy Efficiency and Conservation Authority

Abstract. Appliance energy efficiency programmes have generally followed basically similar tracks, and have often been justified via unsophisticated assumptions of energy use and consumption patterns. To obtain optimum results from extensions of energy efficiency programmes, it will be necessary to investigate actual energy use and adapt future energy performance requirements to suit. Factors that are presently often overlooked include baseload consumption, seasonal and daily variations, intelligence in appliances, and the changing mix of appliances owned in households. Planning for the next round of household appliance energy efficiency measures in New Zealand is using the results of detailed household energy end-use surveys in New Zealand and New South Wales, the outputs from a series of APEC workshop type events and consultation with manufacturers and importers.

1 Energy Performance Test Methods

Energy performance test methods are used for a number of purposes. They are used to check that the model as manufactured bears a resemblance to its design; to provide a comparison between different models; to allocate grades or classes to models, perhaps for energy labelling purposes; and they are sometimes used to determine whether or not the model may be sold. The test procedures may also check that the appliance performs its intended task. Freezers must freeze, refrigerators must refrigerate but not freeze, and clothes dryers must dry clothes. But problems may arise if energy performance tests are used to predict how much energy the tested appliance will use in service.

There is an uncertainty principle, attributed to Lindsey Roke[1], that states: "An energy performance test may be repeatable, or it may be realistic, but not both."

The mode of energy performance that may be measured most consistently is a steady state condition, although in the case of some products steady state may be replaced by a standard cycle. In such tests, ambient conditions are defined and kept constant. Loads are kept the same, and human influence and presence avoided as much as possible. The model under test is run until everything is stabilised, and only then are measurements taken.

In real life, things rarely stay the same for long. Not only do conditions vary enormously from place to place, even within the same country, but changes occur on several different time-scales. People open doors, thus changing local ambient conditions. Cold and wet appliances have different loads put in and taken out. Patterns of use change over the years. Temperature and humidity may vary within a day, on a diurnal nocturnal cycle, seasonally and on even longer time-scales, such as over the millennia from ice age to global warming.

These factors may or may not be of significance. For inefficient appliances, slight differences in consumption due to differences in conditions of use may well be of less importance than the energy that could be saved if a more efficient appliance were used. In such cases, inaccuracies introduced by attempts at realism may well be greater than differences between tested and real life energy consumption. The use of a steady-state method of performance measurement test would then be appropriate. This conventional type of energy test measurement method will also continue to be appropriate for certain appliances. Lighting appliances and components, for example, do generally operate under conditions that are, on average, approximately those of the test. While lamp output does vary over its life, this is either not noticed or is allowed for during the process of designing the lighting installation.

But for certain other products, particularly those that have already been subjected to a round of energy efficiency improvement, the selection of test procedures for measuring energy use now needs very careful consideration.

One major factor is the increasing use of micro-controllers within appliances and domestic systems. It has been pointed out [Meier 1998] that these devices may be programmed to make an appliance perform well under test conditions, or they may be programmed to make an appliance use less energy in service. But these two aims may be not only different; they may be incompatible with, or even in opposition to, each other.

A freezer that is well insulated could be programmed to give preference to operate during the cool part of the daily temperature cycle. Under such conditions, the refrigeration process is more efficient, and in many cases the electricity used may be at a cheaper rate, thus giving both economic and energy-saving benefits. But such a feature would never show any advantage under a test that demands an ambient temperature that is constant at 25°C. On the other hand, a refrigerator that defrosts only when there has been a build-up of frost shows up well during a test that never introduces any humidity into the cabinet, but would perform worse in real life.

Thus there is an increasing need to know how appliances behave and are used in actual service, and to devise appropriate energy performance measurement methods.

510

An initial step is to measure all that happens in actual buildings occupied by real human beings.

2 The Household Energy End-Use Project

One study of energy use in homes is New Zealand's Household Energy End-use Project (HEEP). This project is being phased-in gradually over several years, starting with a few houses and then gradually increasing the numbers. This way, lessons learnt during the early years are used to refine the measurement and analysis techniques and processes. The third year study [HEEP 1999] has a number of findings that are relevant to the development of energy efficiency standards.

2.1 Control Circuit Supply

Many modern appliances have a transformer to supply low-voltage control circuits. This transformer may draw power continuously, even though it probably supplies no useful power. The problem of this "standby power" is now well known for electronic appliances, but the HEEP study shows that it is also of significance to some "power appliances".

One washing machine tested had a standby power of 9 W, giving a profile typified by Figure1.

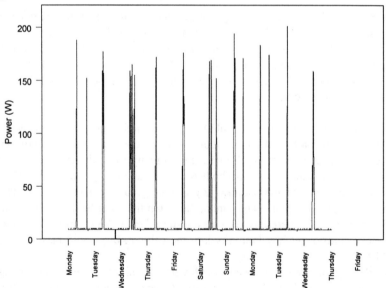

Figure 1: Energy use profile of a washing machine

The overall result in this case is that 44% of the energy used by that washing machine was for keeping an indicator lamp glowing, and only 56% was used for actually washing clothes. Clearly in such cases it is better first to address the standby loss in preference to shaving a few percent off the operational energy consumption.

Similarly, in energy consumption terms, it is possible to argue that microwave ovens ought really to be called clocks. On average, 43% of the total energy used in microwave ovens was during standby, with standby powers ranging from 1 to 7 watts. Figure 2 shows the range of standby losses for microwave ovens and other appliances in percentage terms.

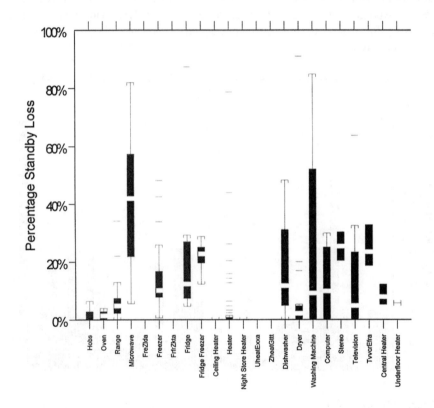

Figure 2: Appliance standby losses

(In Figure 2, the solid bars go between the 25% and the 75% quartiles. The horizontal line within each bar is the median. The horizontal lines above and below the bars indicate outlying values.)

512

It is noticeable that some cold appliances have what appear to be standby losses. Most of these will in fact be operational losses, and all would be captured by conventional measurements of energy performance. However, in a surprisingly high number of cases, freezers showed a very high "baseload" power consumption. In these cases, the freezers were in fact faulty, and were running continuously at full power. No one had noticed. Of course the freezer gets a bit colder, but then system efficiency drops off so its not so very much colder and after all, the freezer contents are still frozen .... So perhaps a campaign to make freezer owners aware of how to tell if their freezer is over-cooling would be worthwhile.[2]

Total house baseloads, that is, standby losses plus continuous operational loads, vary from around 15 watts to over 400 watts. The spread of values is fairly flat, so that an average value cannot be used as a typical value. However, there is certainly a lot of potential for relatively easy energy savings. Figure 3 shows the values measured.

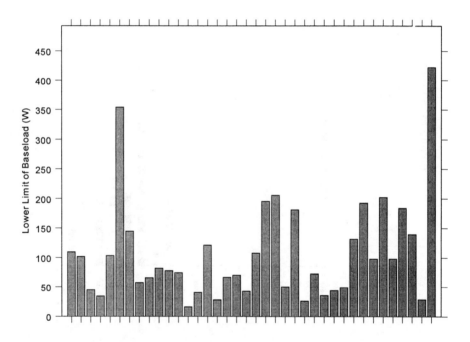

Figure 3: House baseloads

3 Future Developments

The results from the HEEP study indicate that there is no single solution covering all products. It is necessary to give separate consideration to each product group. This reinforces the conclusions reached by two recent events in the Asia Pacific region.

A colloquium on the technical aspects of minimum energy performance standards was held in October 1999 in Seoul. Two groups of experts gathered to discuss domestic air-conditioners and ballasts for fluorescent lamps. The future strategies agreed on by each group clearly pointed a preferred way forward, but were different from each other [SGES 1999].

A similar event, a symposium on energy efficiency aspects of domestic refrigerating appliances, was held in Wellington, New Zealand in March 2000. Perhaps surprisingly, the recommendations made differed from those for air-conditioners, despite the similar physical process performed by the two sorts of product [SGES 2000]. However, in both cases there was recognition of the need for test or modelling procedures to provide some idea of how the tested model will function under different operating conditions.

3.1 Possible Approaches

Suitable approaches for different appliance types could be as follows. Products within classification groups are generally as listed by Fawcett, Lane and Boardman [Fawcett, Lane and Boardman 2000].

3.2 Refrigerating Appliances

There needs to be a single comprehensive test procedure to cater for all appliance types, including kimchi stores and cellars as well as refrigerators and freezers. AS/NZS 4474.1 has such an approach. There also needs to be testing at two points to obtain an idea of how performance varies with ambient temperature, and an additional test at one of the points with an introduced humidity load. These additional tests would provide enough information for approximate modelling, which would be useful for coarse screening of appliances being exported to countries where different ambient test conditions apply. Sophisticated modelling of refrigerating appliances may not be feasible in the short to medium term. [SGES 2000]

3.3 Consumer Electronics

The main requirement for consumer electronics is to have an accurate and generally agreed method of measuring low values of standby power. Measurement of harmonic load imposed by consumer electronic appliances will also be important in some cases.

3.4 Major Cooking Appliances

Standby power is a major issue for microwave ovens, and may also be of importance for hobs and ovens.

3.5 Minor Cooking Appliances

The main requirement for minor cooking appliances will be that "OFF" must actually mean "OFF". Any power indicator light circuits should consume as little power as possible.

3.6 Lighting

There are several different ways of expressing the efficiency of ballasts for fluorescent lamps, but all are based on the same measured parameters. A single test method is feasible but does not yet exist [SGES 1999].

3.7 Miscellaneous Appliances

Electric towel rails can often be a major consumer of energy [HEEP 1999]. Problems that hinder attempts to reduce their energy consumption include their low purchase price and the consequent relatively high cost of adding controls. In addition they are used for more than one purpose. As well as warming and drying towels, a towel rail may act as an anti-condensation heater for the bathroom and provide background warmth.

Other miscellaneous appliances may have plug transformers associated with them. These often represent a constant power drain, which for much of the time is unnecessary. A study of how to reduce the wasted energy taken by plug transformers would be potentially very valuable.

3.8 Water Heating

Most effort is aimed at heat loss from electric storage water heaters. However, instantaneous water heaters can also have a degree of inefficiency, and an agreed means of measuring this will, at some stage, be necessary.

3.9 Wet Appliances

A standby loss limit requirement needs to be added to specifications for wet appliances.

3.10   Space Heating Appliances

These are mainly an issue for those regions which are neither cold enough to warrant central heating, nor warm enough to be able to dispense with heating entirely.

A solid fuel heater may in some instances be considered to have a cyclic operation rather than a steady-state one. In this context, the test methods for solid fuel heaters will at some time need to take the effects of thermal mass into account.

Electric space heaters are not all equal. It may be necessary to introduce the idea of task efficiency, with several different tasks defined, so that consumers may select an appropriate heater for the task they want it to perform.

Heat pumps and room air-conditioners will need to be tested or modelled at various operating points to enable overall performance to be assessed. To some extent this is done by the US concept of a seasonal energy efficiency ratio (SEER), but in that instance the data are hidden behind a single rating figure [Rosenquist 1997, Cogan and Harrington 1997]. Testing at several points will better demonstrate the advantages of units with variable speed compressors and intelligent controls. See also the conclusions of the Colloquium on Technical Aspects of Minimum Energy Performance Standards [SGES 1999].

4   Conclusion

The main inference that can be drawn from the recent studies is that the potential for saving energy in households is not primarily in using the current procedures for energy performance testing and making future requirements just a bit more stringent. This approach leads to an increasingly small return, like travelling down a snail or nautilus shell. Once the early gains have been made, the potential for further savings is constricted.

516

There is more energy saving potential by looking in additional dimensions, and discovering the different layers waiting to be explored. The best approach can be determined only by studies of what happens in real life and why it happens. These studies should then be followed up by workshops that consider the differences that exist between climates, cultures and economic situations and that then agree on directions to be followed. Different approaches will be indicated for the various product groups.

---

[1] Lindsey Roke is Chief Refrigeration Engineer of Fisher & Paykel Ltd, whiteware manufacturers of Auckland, New Zealand.

[2] A suitable test is to place in the freezer a two-litre tub of New Zealand ice cream. After taking out the tub and opening it, it should still be possible to scoop out the ice cream easily, and there should be no crystallisation. It is admitted, however, that Italian gelato tastes better.

5 References

[1] Camilleri, M., Isaacs, I., Pollard, A., Stoecklein, A., Tries, J., Jowett, J., Fitzgerald, G., Jamieson, R.E., and Pool, F. *"Energy Use in New Zealand Households, Report on the Household Energy End Use Project (HEEP) Year 3"* Energy Efficiency and Conservation Authority, Wellington, New Zealand June 1999 (HEEP 1999)

[2] Cogan, D.B., and Harrington, L. with Egan, K. (Ed) 1997, *"Proceedings of the Technical Colloquium on Energy Efficiency Testing Procedures for Industrial Motors, Household Refrigerators, and Air-Conditioners"*. International Institute for Energy Conservation, Bangkok

[3] Fawcett, T., Lane, K., Boardman, B. et al *"Lower Carbon Futures"*. Environmental Change Institute, University of Oxford, March 2000

[4] Harrington, L., Wilkenfeld, G., Ratandilok Na Phuket, S., and Cogan, D.B. (Eds) for the APEC Steering Group on Energy Standards. *"Proceedings of the APEC Colloquium on Technical Issues of Minimum Energy Performance Standards"* International Institution for Energy Conservation, Bangkok, 1999 (SGES 1999)

[5] Harrington, L., Ratandilok Na Phuket, S., and Cogan, D.B. (Eds) for the APEC Steering Group on Energy Standards. *"Proceedings of the APEC Symposium on Domestic Refrigeration Appliances"* International Institution for Energy Conservation, Bangkok, — in preparation at the time of writing (SGES 2000)

[6] Meier, A.K. (1998). *"Energy Test Procedures for the Twenty-First Century."* Circulated by International Organisation for Standardisation as document ISO/TC 86/SC 5 N 495.

[7] Rosenquist, G. 1997 *Personal Communication,* Lawrence Berkeley Laboratory

Lower Carbon Futures for European Households

Tina Fawcett

Environmental Change Institute, University of Oxford

Abstract. Substantial carbon savings can be made by European households from increasing the efficiency of use of gas and electricity. Although this may not be a new message, it is still a necessary one. Kyoto targets are fast approaching, and delay in agreeing EU level efficiency targets compromises the ability of member states to meet their targets. In the UK, a delay of two years in implementing suggested EU policy has resulted in the potential for carbon savings by 2010 decreasing by 25%.

Domestic energy consumption across the EU is still rising. To reduce this the main emphasis continues to be on the more efficient use of electricity, partly because this is the more polluting fuel in the EU, but also because product-level policy is the easiest to implement. The powerful approach of market transformation strategies, if supported by strong EU commitments, can improve the efficiency of products with certainty and speed.

Around 3.7 MtC could be saved by 2010 in the UK, Netherlands and Portugal through policies to increase the efficiency of gas and electricity use, and to encourage fuel switching to natural gas in lighting, appliances and water heating. These savings would be achieved without any drop in the level of service provided to consumers and are delivered through the sale of more energy efficient LAWH and more gas fired appliances. The policies depend upon a strategic approach to carbon dioxide emissions in this sector that is strongly supported both in the member states and in the European Commission. Many new and innovative policy instruments have been identified to make these savings, both for energy efficiency and fuel switching, thus offering policy makers a variety of pathways to a lower carbon future.

1 Introduction

The aim of this paper is to present policy solutions to the problem of domestic energy carbon emissions and to identify routes to lower carbon futures for European households. More particularly, for the UK, Netherlands and Portugal, it outlines policy opportunities for more efficient use of gas and electricity and fuel switching for lights, appliances and water heating, and to quantify the resultant carbon savings.

This paper summarises the results of the CADENCE study (carbon dioxide from domestic equipment: end-use efficiency and consumer education), which has been a collaboration between three partner organisations over a period of two years. The three partners are: Environmental Change Institute (ECI), Oxford University, UK (main contractor); ISR, Coimbra, Portugal; Ecofys, Utrecht, the Netherlands.

2   The Case for Further Action

The policy measures planned and implemented at EU and nation state level so far will not be sufficient to meet the Kyoto targets; further action is needed. The need for decisive action is partially the result of failure of previous policies to provide sufficient energy savings, as evidenced by rising household energy demand: "programmes dealing with energy efficiency have had an impact on reducing energy intensity but have not been in themselves sufficient to bring about changes on a scale necessary" (European Commission, DGXVII, 1998). The importance of more action on energy efficiency is stressed in the recent Action Plan (European Commission, 2000): "there is a pressing need to renew commitment both at Community and Member State level to promote energy efficiency more actively".

In 1997, calculations for electrical lights and appliances in the UK established a potential of 2.7MtC by 2010 given immediate action. Two years have passed in which firm EU action has not been taken and the savings available from this sector are now only 2MtC by 2010. Based on past performance, strong EU action can not be guaranteed. With each passing year, the required policy programme becomes even more concentrated if the European Kyoto target is to be achieved.

The targets set at Kyoto are only a stage towards a less carbon intensive future and more stringent objectives will have to be set and met beyond 2010. The overall reduction required by 2100 may be as much as 90% of current European levels of greenhouse gas emissions if we are to avoid 'dangerous climate change' (Section 6). There are policies other than energy conservation that will contribute to the Kyoto targets, such as increased use of renewable energy and combined heat and power. However, reduced domestic demand has an important contribution to make both to Kyoto targets, and to savings beyond Kyoto.

3   Expected Trends

The basis of the projections presented is a detailed end-use (bottom-up) model, of domestic electricity and gas use in lights, appliances and water heating (LAWH). The model contains information on the ownership, sales, usage and electricity and gas consumption of these household sectors. Linked to this is information about household numbers. The model has been validated wherever possible by using monitored energy consumption data, average household bills and by comparison with total, domestic sector energy consumption. Discussion and consultation with the policy community, manufacturers, retailers and other stakeholders through the UK Market Transformation Programme (www.mtprog.com) has assisted in confirming results from the model for electricity use by lights and appliances in the UK. Data limitations mean the results for the Netherlands and Portugal are less certain in some cases, however the potential savings described are still robust.

3.1 Electricity

Figure 1 presents reference case projections for household electricity use from LAWH in the UK, Netherlands and Portugal from 1970-2020. The reference case represents the base line for the domestic sector - showing where energy consumption has been in the past and where it is projected to go if the future follows expected trends. The reference case includes the expected effects of existing national and EU policies.

National domestic electricity consumption is projected to increase for all three countries. Between 1998 and 2010 this increase will be 9% for the UK and 14% and 12% for the Netherlands and Portugal respectively.

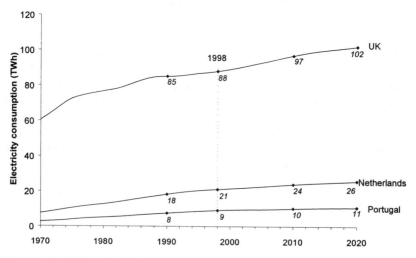

Figure 1: LAWH electricity consumption in the reference case scenario, 1970-2020

Household numbers are rising in all three countries and thus represent a major driver of increasing energy consumption. Ownership of energy consuming products is also rising, and new technologies enjoying rapid uptake, such as digital consumer electronics, represent one of the major threats to meeting energy conservation targets. Increased usage due to changing lifestyles and rising service requirements is another driver for increased energy consumption. However, this is not universally true, and some changes to usage patterns are reducing energy consumption, an example being the Europe-wide trend to lower temperature use of washing machines.

The major policy approach, which is currently being used to counter-act all these drivers, is to improve the energy efficiency of end-use. Significant carbon emission reductions can be achieved through efficiency improvements beyond those already being made (Section 4). However, these are expected to be insufficient to challenge the major sources of increasing energy demand, and may face problems in meeting longer term carbon reduction targets (see Section 6).

3.2 Gas

Despite a rise in the number of households and an increase in ownership of appliances, domestic gas consumption by LAWH is expected to level off in all three countries by 2010 as efficiency improvements in new water heating equipment continue to be introduced to the stock of appliances (Figure 2).

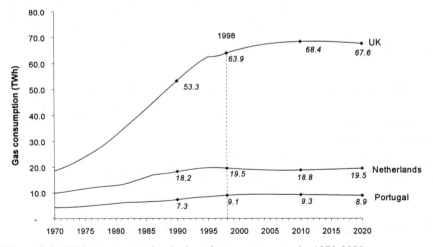

Figure 2: LAWH gas consumption in the reference case scenario, 1970-2020

In all three partner countries in this study, most gas consumption is for the production of hot water. With 81% of households connected, the majority of households in the UK have access to and use natural gas for some application. In

522

the Netherlands the connection is even higher with 96% of households while in Portugal only 4% are connected but use of liquid petroleum gas (LPG) is widespread. Other gas uses modelled are for cooking and tumble drying. Gas means natural gas for the UK and the Netherlands and refers to LPG and natural gas in Portugal, unless otherwise specified.

Half of all European households have natural gas and this is expected to grow to over two-thirds by 2020. Provided gas is used at an efficiency of 65% or more, there are carbon savings from the more extensive use of gas (at the expense of electricity) in all European countries, except France and Sweden. European policy on domestic gas use will become more important. When space heating is included, gas usage is substantial and the benefits multiply in countries with combined space and water heating boilers (UK and NL). There is a 20% carbon saving when natural gas displaces LPG and more than 40% when electricity is replaced.

3.3 Carbon Emissions

Increased energy consumption in all three countries will mean that carbon emissions are expected to increase between 1990 and 2010 in Portugal and the Netherlands. As shown in Table I this will be an increase of 6% in the Netherlands and 22% in Portugal. In the UK, this increase in energy consumption will not result in increased emissions between 1990 and 2010 due to the large decrease in the carbon content of electricity in the early 1990s. In fact household emissions from LAWH are projected to fall by 24%. When calculated by household, emissions are highest for the Netherlands with 6.45 tonnes of carbon in 1998 and lowest in Portugal. The lower emissions for Portugal are explained partially by lower appliance ownership. Carbon intensity of electricity production is also important in explaining emission levels in each country.

Table I: Carbon emissions from LAWH, reference case, 1990-2010

	UK	NL	PO
% change in emissions	-24	+6	+22
LAWH as % of total domestic gas + elec. emissions*	43	39	99
Tonnes of carbon for LAWH/household, 1998	5.52	6.45	5.05

*This figure is for 1998 in the UK, 1997 in the Netherlands and 1996 in Portugal

4 Potential for Reducing Carbon Emissions

As demonstrated in numerous previous studies, for almost all areas of energy usage there is a considerable technical potential for improving efficiency. Further, many of these technical advances are cost effective to the consumer, particularly when introduced via a market transformation strategy which operates to encourage new 'future practice' equipment on to the market, increases the market share of the best current equipment, and gradually removes the worst. The policy tools used to effect this transformation include: technology procurement, aggregated purchase, targeted rebates and subsidies, improved information to consumers, retailers and other stakeholders, energy labels, negotiated agreements and mandatory minimum standards.

The 'Kyoto+' scenario demonstrates the scale of savings that can be achieved through cost-effective technologies which are introduced via implementing a suite of policies at EU and national level. This scenario shows savings compared with the reference case of around 20% in 2010 and 30% in 2020 (Table II). The savings result from increased energy efficiency and fuel switching from electricity to gas and from LPG to natural gas in Portugal. No additional behavioural or lifestyle changes are assumed. Most savings (80%) come from more electrical efficiency, particularly cold appliances, lighting and consumer electronics. Only 5% of the savings come from more efficient gas and the remaining 15% from fuel switching. Fuel switching means that there is actually an increase in gas consumption in Kyoto+ (except for the UK in 2010).

Table II: Energy consumption, carbon emissions and savings, reference case and Kyoto+ scenarios

	UK		NL		PO	
	2010	2020	2010	2020	2010	2020
Reference case (RC)						
Electricity (TWh)	96.9	102.0	24.0	25.8	10.2	10.9
Gas (TWh)	68.4	67.6	18.8	19.5	9.3	8.9
Carbon (MtC)	14.0	15.8	4.2	4.2	1.9	1.8
Kyoto+						
Electricity (TWh)	74.3	62.0	17.8	15.4	7.3	5.5
Gas (TWh)	68.3	70.0	19.4	20.4	9.5	9.5
Carbon (MtC)	11.6	11.1	3.4	3.0	1.6	1.2
National savings (RC – Kyoto+)						
Electricity (TWh)	22.6	40.0	6.2	10.3	2.9	5.3
Gas (TWh)	0.08	-2.46	-0.61	-0.92	-0.20	-0.55
Carbon (MtC)	2.44	4.68	0.81	1.22	0.39	0.63
Carbon (% of RC)	18	30	19	29	21	34

When compared with carbon emissions in 1990, the Kyoto+ scenario could deliver savings of 44% for the UK, 12.5% for the Netherlands and growth in emissions of 2.5% for Portugal. The large savings in the UK reflect the drop in carbon intensity of electricity since 1990 in addition to the energy savings made under Kyoto+. For all three countries, the savings are greater than the national Kyoto target for reduction of greenhouse gases from 1990 to 2010 (UK target -12.5%, Netherlands -6%, Portugal +27%).

The Kyoto+ Scenario was constructed based on the following objectives, in order of importance:
- there is a high degree of certainty that the energy and carbon dioxide savings will be achieved if the policies are implemented;
- the cost to consumers is minimised;
- the importance of equity is recognised, to enable all income groups to participate.

To achieve certainty, all efficiency savings are underpinned by mandatory minimum standards. Industry agreements could, in theory, deliver the same savings (although it seems unlikely in practice, due to the high target levels) and this alternative mechanism may be preferred. Ambitious efficiency targets are set for individual products, both in terms of the technology and the time scale within which this is achieved. The time scale and sequence for actions at the EU level is based on experience of policy making to date, using the shortest realistic time scales. The cost to consumers is minimised by either setting the efficiency level at today's best, or supporting more ambitious efficiency levels by programmes to ensure the cost of the technology will be economic by the time it is mandatory. The third objective means that many of the financial incentives for efficiency serve a dual purpose: both growing the market for energy efficient products and ensuring that the subsidised products are given to low-income households.

Correctly designing policies also requires a sophisticated understanding of consumption. Consumption is not the isolated act of an individual – choices about appliances and heating systems take place in complex social, institutional and cultural settings. The policies to deliver the potential savings depend upon a society with shared environmental concerns and commitments and an institutional framework which clearly encourages sustainable choices. These issues are discussed in more detail in Banks, 2000.

4.1 EU-Wide Carbon Savings

Simple estimates of savings across the EU have been made, based on country specific knowledge and the very detailed modelling that has been carried out for the three partner countries. Across all member states, the Kyoto+ policies affect 82% of all domestic electricity (500 TWh) and 22% of natural gas (260TWh) use. The adoption of the Kyoto+ Scenario would result in 16MtC savings by 2010 and 29MtC in 2020. As there is likely to be a 14% gap between the growth in emissions and the EU Kyoto commitment, these savings in LAWH would make an important contribution. A common and co-ordinated approach to minimum standards would seem desirable and would also provide a useful contribution to post-Kyoto debates.

5 A case for Action: Digital TV

In all three countries, the fastest growing sector of electricity demand is consumer electronics, which is expected to almost double between 1998 and 2020, to 27TWh. The suggested solution in the Kyoto+ Scenario is illustrated by the case of digital TV in the UK.

Digital TV alone is set to increase UK LAWH electricity consumption in lights and appliances by 7% by 2010 – more than half of the total projected increase. At present a separate digital decoder is required for each mode of transmission: terrestrial, cable and satellite. Once digital TV is introduced fully, every TV is likely to require at least one decoder, and many will have more than one. If this remains the case, by 2020 there could be 53m TVs and 72m decoders (either integrated into the TV or as a separate set-top box). Today's digital set-top boxes use more electricity per year than the average TV (130 kWh/yr compared with 120 kWh/yr for a TV). Set-top boxes for the reception of digital TV are being given way for free, by 2010 they could be costing UK households £357m (571m Euros) every year in electricity.

Slightly more efficient set-top boxes are expected in future (this technology improvement has been included in the reference case) – but far more could be achieved by manufacturers and TV service providers. According to research done by the Consumers' Association (Harrison, pers. comm.) it would cost just £2 (3.2 Euros) per set-top box to reduce the stand-by energy demand to 1W, around one tenth of its current value.

In the Kyoto+ Scenario the policy suggestion is rather simple – to announce a 1W standby minimum standard in 2001 and to make this mandatory in 2005. The issues of on mode consumption and platform type are not addressed. This single step would reduce energy consumption from digital decoders by 60% by 2010 compared with the reference case.

This is an area where some of the energy efficiency policies used for appliances such as refrigerators and freezers are not appropriate: the consumption per appliance is too low to be deemed important by individual consumers, the technology is developing fast and there is a wide range of manufacturers. In addition, in the UK consumers are currently being given digital set-top boxes by service providers, rather than making a choice between different models in a retail environment. Thus the suggested policy intervention route does not include energy labels, rebates etc. but goes for the more simple minimum standard approach to permanently transform the market. For other products, such as light bulbs, a much more complex approach is necessary, and the Kyoto+ transformation takes place over a longer time scale and involving many more actors and policy instruments.

6 Sustainable Energy Use

Sustainable energy use involves:
1. reducing greenhouse gas concentrations to a level that will not cause longer-term global temperatures to rise unacceptably, and
2. sharing 'permission' to emit greenhouse gases fairly between nations.

Various organisations have suggested sustainability targets for energy, including the Global Commons Institute (1998), Friends of the Earth (Maclaren et al, 1998) and the European Environment Agency (1999). Using these figures, to achieve 'true' world sustainability (0.2tC/person/year) by 2100, UK carbon emissions in 2020 need to be approximately1.6tC/person/year.

For the UK, this means a 35% reduction from 1998 to 2020. These reductions will have to be achieved across all sectors, not just domestic LAWH. However, there is likely to be severe difficulty in achieving any savings at all from the transport sector, whose emissions currently show strong growth, and are expected to continue to rise by 2020 (DETR, 1997). Assuming transport emissions could be stabilised at 1998 levels, other sectors would need to reduce their carbon emissions by 43% by 2020 to reach sustainability (if transport emissions grow as projected, 63% reductions from other sectors would be required). Clearly, further savings will be required beyond 2020 to keep moving towards the target of 0.2tC per capita in 2100.

The Kyoto+ Scenario, if implemented in full and on time, would deliver 17% carbon savings by 2020 compared with emissions in 1998. This is much less than the 43% identified above: considerable further action is needed to move towards sustainability. Achieving the Sustainability Scenario will require people to change their behaviour: not only how they buy, use and dispose of their LAWH, but also how they live, work and travel. In many cases, consumer responsibility will go hand in hand with improved quality of life - lower bills, greater choice, better health – so individuals could voluntarily change their behaviour to reduce their carbon emissions. In other cases, consumer responsibility reduces individuals' amenity but benefits the environment and wider society: such measures would require potentially unpopular regulation and enforcement.

7 Conclusions and Recommendations

Considerable carbon savings can be delivered via the policy programme outlined in the Kyoto+ Scenario, an average of 18% of the reference case in 2010 and 30% by 2020. Market transformation is a powerful strategic approach for delivering significant savings, which can provide real benefits to lower income consumers. New policies that empower the consumer (web based) or the retailer (aggregated purchase) are being developed in Europe, showing a new emphasis on education and awareness in market transformation strategies.

However, in order for Kyoto+ to become a reality there would need to be strong political support at an EU and member state level for reducing carbon dioxide emissions, procurement and rebate programmes at the national level underpinned by EU-wide efficiency standards. The adoption of a framework directive, by the Commission, would simplify and speed up the process of setting these tough, industry standards and is needed to guarantee the savings in this time scale. Minimum standards deliver savings at a low cost, if they are part of a clear strategy.

In the absence of strong EU action, member states could ensure they still achieved significant carbon savings by increasing the efficiency appliances sold in their own markets, and by encouraging further fuel switching. However, it will be more difficult to ensure that the same degree of savings are made, since the majority of savings are available from strong action on product efficiency, action which is difficult to take at member state level. In addition, without the underpinning of minimum standards the certainty of savings is also harder to achieve.

A national approach which could deliver guaranteed savings would be the introduction of carbon budgets for utilities (this is discussed in more detail in Boardman, 2000). The utilities would have to ensure reducing carbon emissions by a combination of efficiency and fuel switching measures both in supply and through changing demand. To reinforce this approach, consumers would also be made aware of the impact of their own energy use in carbon terms. Increasing consumer education and awareness of energy and carbon issues is key to the success of many policies, whether within a carbon market framework or not.

Although improved technology can deliver the first steps towards sustainability, it can not guarantee sufficient savings on its own. The Sustainability Scenario demonstrates the need to move beyond product level energy efficiency on the path to a lower carbon future.

8 References

[1] Banks, N. (2000) *Socio-technical networks and the sad case of the condensing boiler*, in proceedings from the 2000 ACEEE Summer Study on Energy Efficiency in Buildings

[2] Boardman, B. (2000) *Creating a carbon market,* in proceedings from the Energy Efficiency in Household Appliances and Lighting conference, Naples.

[3] DETR (1997) *Press Release 203.* Environment ministerial speech at WWF conference, 4 June 1997, London, UK

[4] European Environment Agency (1999) *Environment in the European Union at the Turn of the Century*, Office for Official Publications of the European Communities, Luxembourg

[5] European Commission, DG XVII (1998) COM (98) 246. *Energy Efficiency in the European Community – Towards a Strategy for the Rational Use of Energy*

[6] European Commission (2000) TREN D1 17/399 *Action Plan to Improve Energy Efficiency in the European Community*

[7] Global Commons Institute (1998) http://www.gn.apc.org/gci

[8] McLaren, D., S. Bullock and N. Yousuf (1998) *Tomorrow's World: Britain's Share in a Sustainable Future*, Earthscan, London Global Commons Institute (1999)

Acknowledgements

This paper is based on work of many people:
ECI – Kevin Lane, Brenda Boardman, Nick Banks, Harriet Griffin, Judith Lipp, Pernille Schiellerup, Riki Therivel
Ecofys - Kornelis Blok, Margreet van Brummelen, Koen Eising, Frank Zegers, Edith Molenbroek
ISR – Anibal T. de Almeida, Catarina Nunes, Jorge da Silva Mariano

The full findings are published in:
Fawcett, T,. Lane, K., Boardman, B. et al (2000) *Lower carbon futures* Environmental Change Institute, University of Oxford, UK.

The work was funded by the European Commission, DG XVII, the UK Department of the Environment, Transport and the Regions, PowerGen, EnergieNed, Novem, Gasunie N.V, the University of Coimbra and the Polytechnic Institute of Coimbra.

High Efficiency Household Appliances and Low Income Families in Italy

Stefano Faberi[1], William Mebane[2] and Milena Presutto[3]

[1] ISIS
[2] Consultant
[3] ENEA

Abstract. With increasing globalisation of the economy, income distribution and consumer preferences are changing. This may impact upon the possibility for low-income families to purchase and utilise energy efficient home appliances.

The aim of the study is to access the present degree of difficulty that low income families in Italy have in purchasing high efficiency refrigerator/freezers and clothes washers.

Results show that, given their substantially lower level of expenditure for household appliances, the higher prices of energy efficient models represent an obstacle for these families in Italy to overcome energy inefficiency and energy poverty. an important input to energy policy. Over two million 600 thousand families, 11.9 percent of the total families in 1999 are estimated to be below the poverty threshold based upon consumption. An independent survey shows that these families possess less technologically advanced models, keeping them for a much longer period of 36 years, more than double the recommended lifetime of 15 years, and four times the national average of nine years. This technical disadvantage is aggravated by the additional operating costs of the more inefficient models. The higher prices of the more efficient models make it difficult for the poor families to afford them. Even at the annualised price of the average refrigerator - belonging to the energy efficiency C class - the annual expenditure of the poor families in Italy is less than half that necessary to purchase this model, consistent with the longer turnover rates. According to latest statistics in 1999, the B model is priced 70% above the C model in Italy, making it very difficult for poor families to purchase the more efficient models. Extrapolating the relative consumption data to other European countries and comparing it to local prices indicates that the situation may be similar in other EU Member States.

The implications for energy and social policy are clear: without some form of public incentive it is unlikely that poor families in Italy will purchase the energy efficient models, remaining in an energy poverty trap for many years to come.

1 Introduction

The social, economic and institutional transformation in act in the last decade, coupled with globalisation, has had two main implications for the Italian labour market: the increased rate of unemployment and the reduction of the related social benefits of employment, for example the participation in pension programmes. This not only affects the capacity of the family to produce income but may impact also on the scale of preferences and propensity for consumption, which determines the quantity and quality of operative choices. In this context it is not simple to determine the potential for purchase of new higher efficiency, higher priced household appliances, particularly by low income and low consumption families. The present paper draws upon the previous study presented in the 1998 Second National Conference on Energy and Environment [1] in attempting to address these questions.

2 Definition of the Poverty Threshold

Poverty can be measured with respect to income or level of consumption, and may be given in relative or absolute terms, The Italian Commission on Poverty and Emargination has defined poor families as those whose consumption per member is one half that of the average pro capita expenditure at a national level. A two-member family is used as a reference. This definition mirrors that of the method used in for the *International Standard of Poverty Line*.

For 1997 the poverty line is estimated at 1.233.829 Lire/month expenditures for a two-person family. In terms of pro-capita expenditure the threshold is 616.914 lire per month as shown in Table I[2]. With regard to the extent of the poverty, in 1997, 11.2 percent of the total families, or 2.2 million families are considered to be living below this poverty threshold. The historic trend of the percentage of families below the poverty line has gone from a minimum of 8.3 percent in 1980 to a maximum of 14.8 percent in 1988 to around 11 percent in the early 1990's and subsequently increasing to 11.9 in 1999.

Table I: Extent of Poverty in Italy: 1980-1997

Year	Poverty threshold*	"poor" Families	
	Liras/month	(%)	(number)
1980	267.257	8,3	
1981	314.408	9,6	
1982	367.952	9,5	
1983	422.266	10,6	
1984	470.426	11,3	
1985	549.270	11,6	
1986	623.878	12,6	
1987	691.788	14,4	
1988	749.385	14,8	
1989	838.260	14,4	
1990	914.266	11,7	
1991	1.010.336	11,8	
1992	1.041.651	11,7	
1993	1.024.973 (512.487)	10,7	
1994	1.094.296 (547.148)	10,2	2.038.000
1995	1.143.355	10,6	2.128.000
1996	1.190.273	10,3	2.079.000
1997	1.233.829	11,2	2.245000
1999	(**)	11,9	2.600.000

(*) For a two-person family (**) Revised method used resulting in poverty level about 10% above previous method. No revised historic trend based on new method is available.

As stated previously, the results of poverty evaluation depend on adopted methodology. The poverty threshold in Europe [3] in 1993 and 1994 as proportion of persons below a certain income is shown in Table II. In this case poverty line is "50% of the arithmetic mean of equivalent[1] net income" of each country. The European average involves 17 and 16% of the households (and total population) respectively. In absolute terms in 1994, 56 million persons or 23 million families were living in a state of poverty in the European Union. For Italy, the incidence of poverty according to this measure is slightly over the average, with 18% of households involved in 1993 and barely below the average with 15% of the households in 1994. The poverty threshold adopted in the EU based upon income is considerably higher than the level adopted in Italy using consumption expenditures, involving more families as shown for year 1994: 15 % vs.10.2%.

---

[1] Equivalent net monetary income is derived dividing the total monetary income of the household by the number of "adult equivalents". The equivalence scale used is the modified OECD scale, i.e. 1,0 for the first adult, 0,5 for every other adult in the household and 0,3 for every child younger than 14.

3   Expenditure for Household Appliances

The Italian Institute of Statistics publish annual reports covering family expenditures [4]. For the two most important appliances, the level of penetration amounts to 98.0 percent of all households for refrigerators and 92.8 percent for washing machines in 1996. Dishwashers lag considerably with only 23.3 percent penetration.

In Table III the expenditure for household appliances and number of household appliance purchased is estimated for families classified according to monthly expenditure [4, 5]. The expenditure for household appliances shown in column five includes only the most important ones: all types of refrigerators and freezers and clothes washers, while in the sixth column the estimated expenses for all cooling and washing appliances (i.e. including also air conditioning systems and dishwashers) were estimated. Finally, the total number of purchased household appliances given in the last column of Table III was estimated dividing the total annual expenses for household appliances by the average price of the appliances (calculated as market turnover divided by annual unit sales).

The families have been grouped into intervals of expenditure as shown, the first intervals corresponding approximately to the families below the poverty line according to the Italian Commission on Poverty and Emargination. It is striking that the low expenditure families, 10,3 percent of the total families, constitute only 2.9 percent of the annual purchases (both for the most important one and all cooling and washing appliances). The poorest 30 percent spend 11.9 percent. Instead the highest expenditure group shown, with monthly expenditure over three million lire, representing 45 percent of the population purchased 72 percent of the total.

Table II: Poverty in the EU in 1993 and 1994 as Equivalent Net Income below 50% of the National Average

Country	B	DK	D	GR	ES	FR	IR	IT	L	NL	A	P	UK	EU
1994														
per-capita equivalent net income poverty threshold (NI/year)	286.208	68.064	14.931	939.642	598.331	48.451	3.967	8.142.000	440.820	14.275	101.741	525.148	4.660	n.a.
per capita (NI/month)	23.851	5.672	1.244	78.304	49.861	4.038	331	678.000	36.735	1.190	8.478	43.762	388	n.d.
persons (%)	15	7	14	20	18	14	22	16	14	8	14	25	20	16
persons (10³)	1.516	365	11.310	2.048	6.993	7.980	788	9.026	56	1.214	1.107	2.462	11.505	56.487
per-household equivalent net income poverty threshold (NI/year)	483.890	102.604	23.710	1.698.083	1.190.114	80.875	7.998	15.279.000	782.172	23.267	179.768.500	1.011.841	7.765	n.d.
per household (NI/month)	40.324	8.550	1.976	141.507	99.176	6.740	667	1.273.000	65.181	1.939	14.980.708	84.320	647	n.d.
household (%)	15	10	14	23	17	15	23	15	13	9	14	28	20	16
household (10³)	618	236	5.072	864	2.063	3.457	264	3.054	19	586	440	923	4.890	22.512
1993														
per-capita equivalent net income poverty threshold (NI/year)	267.960	63.936	14.976	815.280	577.080	44.592	3.528	8.011.000	465.000	14.100		510.960	4.536	n.d.
per capita (NI/month)	22.330	5.328	1.248	67.940	48.090	3.716	294	667.600	38.750	1.175		42.580	378	n.d
persons (%)	13	6	11	22	20	14	21	20	15	13		26	22	17
persons (10³)	1.289	318	9.099	2.258	7.631	7.591	759.000	10.895	60	1.919		2.537	12.805	57.162
household (%)	13	9	13	24	19	16	21	18	14	14		29	23	17
household 10³	508	216	4.515	872	2.272	3.523	238.000	3.429	22	842		915	5.474	22.825

NI = Net Income in National Currency

Table III: Estimate of the family expenditure and number of household appliances (hh-appliances.) purchased by Italian families in 1996

Class of monthly expenditure (lire/family)	Families (number)	Expenditure per family member (lire/y)	Av. expenditure for main hh-appliances* (lire/y family)	Total expenditure for main hh-appliances* (lire/y)	Estimated total expenditure for domestic appliances (lire/y)	Estimated n° of purchased main hh-appliances* (number)
At least 400.000	40.176	315.696	0	0	0	0
400.001 - 600.000	200.880	425.220	0	0	0	0
600.001 - 800.000	421.848	542.851	7.452	3.143.611.296	4.554.511.132	11.353
800.001 - 1.000.000	602.640	624.139	10.956	6.602.523.840	9.565.835.436	23.845
1.000.001 - 1.200.000	803.640	649.456	27.648	22.219.038.720	32.186.469.287	80.232
Total low expenditure	*2.069.184*	*592.099*	*15.448*	*31.965.173.856*	*46.306.815.855*	*115.430*
1.200.001 - 1.400.000	883.872	682.134	45.648	40.346.989.056	58.455.322.085	145.713
1.400.001 - 1.600.000	1.024.488	690.487	10.000	10.244.880.000	14.961.679.082	37.295
1.600.001 - 1.800.000	1.024.488	713.284	17.988	18.428.490.144	26.699.472.553	66.554
1.800.001 - 2.000.000	1.104.840	773.063	26.304	29.061.711.360	42.105.042.721	104.956
Total medium expenditure	*4.037.688*	*717.038*	*24.292*	*98.082.070.560*	*142.221.516.441*	*354.518*
2.000.001 - 2.200.000	1.124.928	777.467	49.956	56.196.903.168	81.418.915.059	202.955
2.200.001 - 2.400.000	1.024.488	830.749	26.880	27.538.237.440	39.897.810.887	99.454
2.400.001 - 2.600.000	984.312	857.868	28.896	28.442.679.552	41.208.180.166	102.721
2.600.001 - 2.800.000	944.136	920.218	32.220	30.420.061.920	44.073.041.359	109.862
2.800.001 - 3.000.000	883.872	950.109	33.360	29.485.969.920	42.719.714.878	106.489
Greater than 3.000.000	9.019.512	1.546.419	87.480	789.026.909.760	1.143.154.005.356	2.849.570
Total high expenditure	*13.981.248*	*1.303.648*	*68.743*	*961.110.761.760*	*1.392.471.667.705*	*3.471.051*
TOTAL	20.088.000	1.112.447	54.319	1.091.158.006.176	1.581.000.000.000	3.941.000

Note: (*) household appliances = refrigerators, freezers, and clothes washers

The low level of consumption of the poor families is confirmed in the estimates of the rate of renewal of the existing stock of main household appliances shown in Table IV.

Table IV: Rate of substitution of household appliances in Italy 1995-1996 by class of Family expenditure

Class of monthly expenditure (lire/family)	Families (number)	Stock of refrigerators and washing machines (A) (number)	Refrigerators and washing machines purchased (B) (No., 1995-1996)	Annual rate of substitution (A/B)*2 (years)
At least 400.000	40.176	77.258	≅ 0	
400.001 - 600.000	200.880	386.292	672	
600.001 - 800.000	421.848	811.214	25.095	64,6
800.001 - 1.000.000	602.640	1.158.877	76.517	30,2
1.000.001 - 1.200.000	803.640	1.545.169	116.734	26,4
Total low expenditure	*2.069.184*	*3.978.810*	*219.018*	*36,3*
1.200.001 - 1.400.000	883.872	1.699.686	225.839	15
1.400.001 - 1.600.000	1.024.488	1.970.090	140.347	28
1.600.001 - 1.800.000	1.024.488	1.970.090	174.160	22,6
1.800.001 - 2.000.000	1.104.840	2.124.607	314.994	13,4
Total medium expenditure	*4.037.688*	*7.764.473*	*855.340*	*18,1*
2.000.001 - 2.200.000	1.124.928	2.163.237	323.778	13,4
2.200.001 - 2.400.000	1.024.488	1.970.090	374.138	10,6
2.400.001 - 2.600.000	984.312	1.892.832	239.711	15,8
2.600.001 - 2.800.000	944.136	1.815.574	328.244	11
2.800.001 - 3.000.000	883.872	1.699.686	306.705	11
Greater than 3.000.000	9.019.512	17.344.522	5.876.067	6
Total high expenditure	*13.981.248*	*26.885.940*	*7.448.643*	*7,2*
TOTAL	20.088.000	38.629.224	8.523.001	9

Note: refrigerators and washing machines stock comes from ISTAT, representing 96,5% of the total 40.176.000 installed household appliances.

Dividing the stock of refrigerators, freezers and clothes washers by the number of their estimated purchases calculated for 1995 and 1996, we obtain the biannual rate of substitution, in terms of number of years to renew the stock. To get the annual rate we multiply by two. Thus for the low expenditure group, with consumption of 1.2 million lire/month and less, we have an annual rate of renewal of 36 years. This compares to a technologically sound value of 15 years. The

European average is 12-15 years for refrigerators and 7-10 years for clothes washers. The Italian national average is less, nine years as shown.

This implies that the families living below the poverty threshold have inefficient household appliances with a turnover rate 2.4 times slower than that recommended and four times the national average. They are inefficient due to increased maintenance problems and due to the fact that the older models themselves were designed with more lenient energy criteria. For the poorest 30 percent of Italian families the annual turnover is 22 years.

4 Expenditure for Household Appliances and Income Levels

Utilising an approximate function of income versus consumption expenditure, the expenditure in household appliances can be estimated for different intervals of family income as shown in Figure 1 for years 1993-1996.

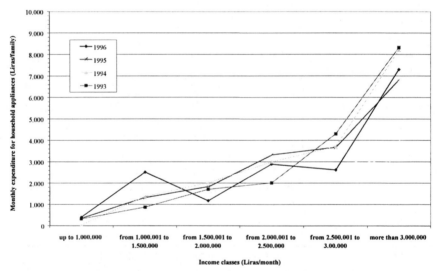

Figure 1: Monthly expenditure in household appliances of Italian families by class of Income in 1993-1996.

Unfortunately this conversion may not be homogeneous for all years as evidenced by some anomalies in 1996. However, the general trend for the four years is consistent. Within the family income interval up to 3.0 million lire per month the relationship is rather linear going from about 1.000 lire/month (12.000 lire/year) to 3.500 lire/month (42.000 lire/year). Instead for the category with more than 3.0

million per month the average rises to about 7.500 lire/month or 90.000 lire per year.

5   Technical Characteristics of the Household Appliances of Poor Families

In 1994, within the framework of a study financed by the SAVE Programme [6] of the European Union, a sample of 1000 families resident in all the national territory was interview to understand the technical characteristic and modality of use of their household appliances. Through analysis of the demographic data it was possible to identify families with high probability of being poor. These family members had a low level of education, older head of family, and greater number of family members etc. Two groups were distinguished: those with a high risk of poverty and those without such risk.

Two technical characteristics stand out as significantly different as shown in Table V. The more advanced four and three star models are less frequent in poor families as compared to the others, 22 % compared to 34%. The less advanced 1 and 2 star models instead are more frequent. Another measure of technological sophistication is the degree of automation of the anti-frost system. The old systems are manual that is with no anti-frost system. Others are automatic with different kind of subsystems. In the survey 71 percent of the poor families still had manual system as compared to 64 percent for the rest. This finding is also consistent with the slower turnover of the stock in poor families. Older models are less likely to have these more modern features

Table V: Technical features of installed refrigerators in Italian families

Technical Feature	Frequency: Poor Families	Frequency: Other Families
3 and 4 stars	22 %	34%
1 and 2 Stars	6%	3%
Manual Frost	71%	64%
Automatic Frost	29%	36%

6   Prices of Household Appliances with Respect to Efficiency Class

According to a study conducted by ADEME [7] on refrigerators prices within Europe, the average difference between an A model and a C model is 127 Euro and the difference between a B model and a C model is 40 Euro. The year is 1996 covering all EU member states except Portugal and Greece This is not far from the Italian average difference between the A and C models found in same study, as

538

127 Euro. Instead, for Italy the difference between B and C model is much less, 10 Euro in year 1996 This could be part of a strategy to force an easy upgrade from models C to B, however, in the long term price should reflect underlying costs.
In fact in a more recent analysis realised by ENEA in 1999[2] [8], a more distinct price distribution is revealed as illustrated in Figure 2.

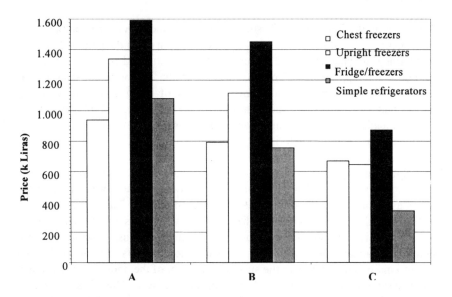

Figure 2: Average price of refrigerators by energy efficiency class in Italy in 1999.

Now there are substantial price differences between each class. For example for refrigerator-freezers the difference between A and C is about 730.000 lire (85% increase over class C) and the difference between B and C is about 600.000 lire (70% increase over class C). Even hypothesising a strong policy of discounts the difference in prices between the higher efficiency classes A and B is noteworthy, also at the expense on the poor families.

7   The Annual Expenditure for Household Appliances in Relation to their Prices

Dividing the average price of the refrigerators by the average lifetime, considered 15 years one obtains an annualised price. This has been confronted with the annual family expenditures for the main household appliances for year 1996 in Italy for

[2] Only the classes A, B e C are permitted in 1999, with the exception of horizontal freezers for which class D and E are also permitted in commerce.

539

the low expenditure families as shown in Figure 3. It is alarming that for the poor families with monthly expenditures of 1,200,000 lire and less, their budget for the main appliances does not reach even half the cost of model C. Only families with high annual expenditure for household appliances can think to reach A class models even if with some extra effort compared to "usual" annual appliance expenditure.

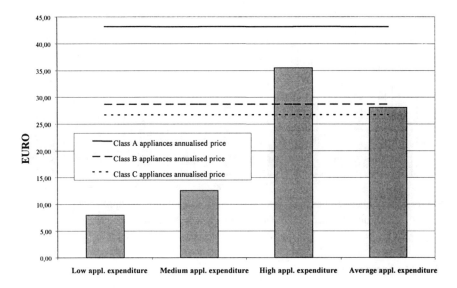

Figure 3: Annual expenditure for main household appliances by families and annualised prices of class A, B and C refrigerators in 1996 in Italy

8   Extension to Europe

Not having detailed data on appliance expenditures in the other Member States we may explore the hypothesis that the relative proportions are similar and compare that to Member State prices for appliances. In particular, if we assume that the relative income spent by the poor families is between 0.22% and 0.30%, the case for Italy, and apply these proportions to the other Member States we obtain the result reported in Figure 4 for year 1995. Prices are relative to those from each Member State and again the prices are annualised by dividing by 15.

Again we are alarmed by the distance from the level consumption for household appliance – specifically refrigerators - by the poor families and the price of the more efficient appliances. Even if the Italian relative consumption parameters are underestimated for the other Member States by a factor of two (e.g. the relative

540

consumption is double that of Italy), almost no country's consumption arrive to the level of the class A models.

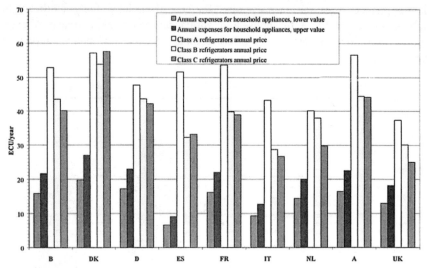

Figure 4: Hypothesised families' annual appliance expenditure and appliance prices according to energy efficiency classes in selected Member States in 1995

9 References

[1] ENEA, Studio di fattibilità per la Costituzione di un Fondo per l'Acquisto di Elettrodomestici ad Elevata Efficienza da parte di Categorie Sociali a Basso Reddito, November 1998.
[2] Istituto Poligrafico e Zecca dello Stato, "Poverty in Italy 1980-1995", 1997.
[3] EUROSTAT, European Community Household Panel (ECHP), First Wave, 1994.
[4] 3a EUROSTAT, European Community Household Panel (ECHP), Selected Indicators from the 1995 Wave, Edition 1999.
[5] ISTAT, Consumi delle Famiglie, 1997.
[6] ANIE, L'industria Italiana degli Apparecchi Domestici nel 1997.
[7] ANIE, ENEA, Monitoring of Electrical Appliance Consumer Habits and Definition of Potential Energy Saving by Optimising Behaviour", SAVE contract N° XVII/4.1031/S/94-001, 1994.
[8] ADEME, Monitoring of Energy Efficiency Trends of European Domestic Refrigeration Appliances, SAVE contract N° XVII/4.1031/D/97-021, 1999.
[9] IFR, Frigoriferi e Congelatori, Caratteristiche Tecniche e Prezzi Informativi (Refrigerators and Freezers, Technical Characteristics and Informative Prices), Apparecchi Elettrodomestici, N° 11, December 1999.

Present Status and Perspective of Energy Efficiency and Conservation Policies in Japan Residential and Commercial Sector

Hidetoshi Nakagami and Yoshiaki Shibata

Jyukankyo Research Institute, Japan

1 Introduction

This paper presents the status quo of the energy efficiency and conservation policies adopted by the government of Japan, and also addresses some controversial issues on these policies. The major policies concerned in this paper are the followings. (1) Efficiency improvement of appliances by the Law Concerning Rational Use of Energy into which new Appliance Energy Efficiency Standard, commonly known as Top-Runner approach has been incorporated in 1998. (2) Drastic reform of life style guided by introduction of the labeling scheme. (3) Technology development such as stand-by electricity curtailing technology.

The effect that these policies, which target not only suppliers (appliance manufacturers) but also consumers, have on energy conservation should be evaluated hereafter from the viewpoint of the both sides.

2 The Law Concerning Rational Use of Energy

The Law Concerning Rational Use of Energy, commonly known as the Energy Conservation Law, was promulgated in June 1979 in the wake of the oil crisis. With a view to promoting further energy conservation to cope with global warming issues, this law has been revised twice in 1993 and 1998.

This law, which is administered mainly by Ministry of International Trade and Industry (MITI), established general objectives, policy principles, reporting requirements, financial incentives and sanctions on manufacturers to achieve energy efficiency standards. Within the extent of the law's application are factory, building and energy consuming equipment, and the energy efficiency standards are set for energy consuming equipment.

Energy Efficiency Standards under pre-revised system had been covering gasoline-fired passenger cars and trucks, air conditioners, fluorescent lamps, television sets, copying machines, electronic computers, magnetic disk drive units and video cassette recorders. The revised law extended the list to include refrigerators, diesel-powered passenger cars and diesel-powered trucks. At the

542

same time, the top-runner approach was introduced as a method to specify the appliance energy efficiency standards. On top of that, sanction was strengthened (see Table I).

Table I: Revision of The Law Concerning Rational Use of Energy
(revised in May 1998, came into effect in April 1999)

Before revision	After revision
<Product items applied> air conditioners, fluorescent lamps, television sets, copying machines, electronic computers, magnetic disk drive units, video cassette recorders, gasoline-fueled passenger cars and gasoline-fueled trucks	<Additional product items > refrigerators, diesel-powered passenger cars and diesel-powered trucks were added.
<Energy Efficiency Standard> The standards are set at slightly higher level than the average energy efficiency of the currently commercialized products.	<Energy Efficiency Standard> The standards are set at the levels higher than the most efficient products among those currently commercialized.
<Sanction> Recommendation to improve efficiency of the product	<Sanction> Recommendation followed by negative publicity, orders to comply and fines if recommendation is disregarded.

3 The Status Quo of Energy Efficiency and Conservation Policies

At the Advisory Committee for Energy set up by the Agency of Natural Resources and Energy (ANRE) within the MITI in April 2000, the status quo of the energy conservation policies was presented along with the explanations of energy supply-demand and energy trend in Japan and abroad. Major policies are described below.

3.1 Top-Runner Approach

Top-runner approach of Japan is the strictest standard for appliance efficiency in the world. Under this scheme, appliance energy efficiency standards are set at the levels higher than the most efficient products among those, already commercialized in the current market.

Sanctions on failure to comply with the Government's standards by the target year come to negative publicity, orders to comply and fines. The manufacturers are expected to comply by recognizing that more efficient products will have better market appeal.

3.2 Labeling Scheme

It was approved at the Subcommittee for Energy Conservation under the Advisory Committee for Energy in December 1998 that with a view to disseminating and penetrating energy efficiency appliances, discussion should be initiated focusing on the optimal labeling scheme under the peculiar circumstance in Japan. Since the top runner system has been already introduced, the labeling scheme should be studied carefully.

In response to this, at the Subcommittee for Energy Conservation in January 2000, the followings were approved through discussion by the representatives of both manufacturers and consumers:

- The labels should be standardized to contain common information for every appliance.
- A symbol mark and the energy efficiency standard achievement percentage should be shown on the label. (This standard is that identified by the top runner system.)
- It is desirable that the label also provides an estimate of the annual energy consumption of the appliance.
- The color of the symbol mark can be changed when the appliances achieve the energy efficiency standard.
- Among the household appliances applied in the Energy Conservation Law, air conditioners, fluorescent lamps, television sets, refrigerators and freezers are to be labeled.
- An applicable location where the label is to be fixed can be chosen among price tag, main body of the appliance or catalogue, depending on each sale arrangement.

In order to build concrete institutional structure of the energy efficiency labeling scheme, the Committee on Preparation for Draft Plan of Energy Efficiency Labeling was set up within the Energy Conservation Center, Japan (ECCJ). In April 2000, this committee selected the design of the label through public participation. It is currently deliberating the method to evaluate the energy saving achievement rate. The draft plan is to be applied to Japanese Industrial Standards (JIS) and the labeling scheme is supposed to be set forth from this summer.

3.3 Technology Development

In order to guarantee the effectiveness of energy conservation with the aid of technology, collaboration of the government, industry and academic organization is currently promoting technology development regarding energy efficiency.

Topics are high performance lighting device using light emitting diode that aims to realize 50% energy consumption reduction, stand-by electricity curtailing technology and liquid crystal display screen at ultra low energy consumption.

4 Discussion on Policies

The intention of the energy efficiency and conservation policies presented above is worthy of attention. However, several controversial issues on those schemes are raised below.

4.1 Top-Runner and Labeling

Conspiracy (manufacturer side issue)
Continuation of top runner scheme beyond the target year may not encourage further efficiency improvement, even if it is effective until manufacturers could achieve the standard. This is because manufacturers are likely to conspire to agree on refraining from development of appliances with efficiency higher than certain level (e.g. the standard) to spare further strict effort for efficiency improvement. A new target by the government should be set, apart from the top-runner approach, on and after the target year.

Rebound effect (consumer side issue)
Efficiency improvement led by top runner scheme might cause rebound effect. Against the original purpose, the information indicating efficiency improvement which leads to reduction in electricity bill may encourage people extension of operation hour of appliances.

Introduction Order (market transformation issue)
As explained above, the top runner scheme was introduced prior to enforcement of labeling scheme in Japan.
Comparative analysis from theoretical viewpoint between the case of top runner followed by labeling (Figure 1) and the case of several years of labeling followed by top runner (Figure 2) reveals that less high efficiency appliance penetrate into market in the former case than the latter case.

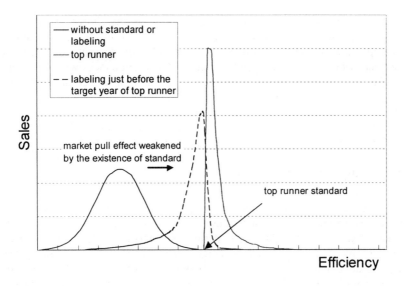

Figure 1: Top runner followed by labelling

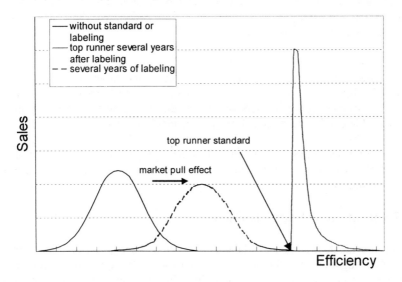

Figure 2: Several years of labeling followed by top runner

In the absence of standard (e.g. top runner scheme) or labeling, the efficiency of the appliances follows the solid black curve, with most models having only medium efficiency (see Figure 1 and 2). In general, market-pull effect by labeling causes market transformation towards higher efficiency, as shown in Figure 2. However, if top runner is introduced prior to labeling, the existence of standard would hamper the market transformation (like a tall wall damming wave), since manufacturers would be reluctant to produce (and sell) appliances with efficiency much higher than standard, being concerned about the new higher standards that might be set even after the target year (see Figure 1).

Therefore, it seems more effective for market transformation towards high efficiency to introduce top runner several years after introduction of labeling, as shown Figure 2 rather than the case of top runner followed by labeling.

4.2 Technology Development

Energy efficiency technology usually requires large amount of cost for fundamental research, practical application, production and market introduction. Since users expect shorter pay back period by energy efficiency investment, suppliers are likely to be forced to curtail the investment in the technology research and development. Besides, lack of common energy efficiency evaluation standard and information infrastructure for fundamental technology also impedes further development of energy efficiency technology.

To resolve this situation, well balanced use of enhancement of incentive to introduce energy efficiency technology through preferential tax treatment and lending mechanism, enlightenment of people's awareness, deregulation and regulation would be the effective measure.

In parallel with putting energy efficiency technology for each equipment into practical use, there is a great hope in it to create innovative energy efficiency technologies by the help of information technology. Innovative technologies are expected to encourage regional development through generating new business, ensuring stable employment and achieving sustainable society.

Last but not least, exporting energy efficiency technologies to the developing countries would make a great contribution as one of the counter measures of global warming issues.

4.3 Evaluation of Policy Impact

In order to evaluate the impact of labeling scheme on energy conservation, market survey (investigation) should be conducted to keep track on the evolution of distribution of each appliance shipments by efficiency category before and after the introduction of energy labeling in August 2000. Jyukankyo Research Institute will take in charge of this survey as a project sponsored by ECCJ. The results of this survey are supposed to be presented on a certain occasion forthcoming.

The study concerning comparison of energy consumption units between prior to and posterior to the introduction of the top runner scheme should be also conducted to analyze the rebound effect.

Impacts of U.S. Appliance Standards to Date

James E. McMahon, Peter Chan and Stuart Chaitkin

Lawrence Berkeley National Laboratory

1   Introduction

In 1975 the U.S. federal government established its role in improving appliance and lighting energy efficiency by setting voluntary labeling and efficiency guidelines for residential appliances and lighting products under the Energy Policy and Conservation Act (EPCA, P.L. 94-163). In 1987 EPCA and subsequent legislation was amended and updated by the National Appliance Energy Conservation Act (NAECA, P.L. 100-12). NAECA superceded requirements established by some individual states and set the first national energy efficiency standards for home appliances. A schedule for regular updates, currently specified to 2012, was also established. NAECA standards now influence appliances and equipment comprising about 80% of the source energy in the U.S. residential sector.

Most research during the past 20 years on appliance standards has involved prospective estimates of impacts. (Bertoldi *et al.* 1997, McMahon and Turiel 1997). This paper utilizes publicly available information to report the observed impacts over almost a decade of experience with the actual residential appliance standards that were adopted and implemented in the U.S. The energy savings, the energy cost savings, and the carbon emissions reductions that have already occurred in response to appliance standards in the U.S. are large and significant. In fact, in 1997 alone, appliance standards were responsible for reducing total U.S. residential energy consumption by approximately 2.5%, thus saving US$3.5 billion in annual energy costs to residential consumers, and reducing associated carbon emissions in the residential sector by 2.5%. Benefits to consumers, the economy, and the environment will continue to flow from these standards and their updates for years into the future.

2   Methodology

Table I lists the appliances for which NAECA standards and updated standards have been adopted and the effective date for each standard.   First updates for water heaters and central air conditioners and heat pumps and a second update for clothes washers are in process. Additional updates may occur in future.

548

Table I: Effective Dates of U.S. Residential Appliance Standards

Appliance	Original Standard Effective Date	First Update Effective Date	Second Update Effective Date
Refrigerators and Freezers	1990	1993	2001
Room Air Conditioners	1990	2000	Future
Central Air Conditioners	1992	In process	Future
Clothes Dryers	1988	1994	Future
Clothes Washers	1988	1994	In process 2004/2007
Dishwashers	1988	1994	Future
Water Heaters	1990	Proposed 2003	Future
Gas/Oil Furnaces	1990	Future	Future
Ranges and Ovens	1990	Electric (No update) Gas (Future)	Future
Showerheads and Faucets	1994	Future	Future

As explained below, we gathered the responses to the imposition of NAECA standards over the past decade (and subsequent updates to some of those standards) from a variety of publicly available data sources. For example, Figure 1 shows the changes in energy consumption for new refrigerators. The average refrigerator in 1961 had approximately 12 cubic feet of capacity including freezer, used fiberglass insulation, and consumed 1015 kWh/annum. In 1972, when the first oil price shocks occurred, average energy consumption for new refrigerators was 1726 kWh/annum. In 1980, the average new refrigerator in the U.S. had 19.6 cubic feet, used CFC-blown insulation, and consumed 1278 kWh/annum. By 2001, after the 1990 standard established by NAECA and two updates (1993 and 2001), a typical new refrigerator in the U.S. is expected to have 20 cubic feet, have more features such as through-the-door services like ice and water, use ozone-friendly foam insulation, and use about 63% less energy (476 kWh/annum) than the typical 1980 model.

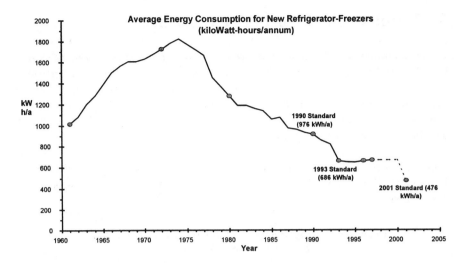

Figure 1: Average Energy Consumption for New Refrigerator-Freezers

The energy consumed each year by the nation's appliances is determined by the number of existing appliances, the product mix for each appliance, and their energy consumption per unit. Because the efficiency of units purchased generally increases over the years, the energy consumption per unit of stock varies according to vintage. The number of appliances is determined from historical data on annual shipments (from the industry trade associations), and the appliance lifetimes in years are calculated by applying a derived rate of retirement specific to each appliance. The total energy usage of the appliances in each year is calculated for two scenarios: with and without standards. In the latter case, the unit energy consumption is based upon historical (pre-standards) trends in unit energy consumption. Energy savings are calculated as the difference between the annual energy consumption in the base case and that of the standards case.

Energy cost savings are calculated based on the annual energy savings, multiplied by national average energy price in that year. A time series of energy prices is available from the U.S Department of Energy's (DOE) Energy Information Administration (U.S. DOE 1998).

Reductions in carbon emissions are calculated by multiplying the annual energy saved by carbon emissions coefficients for the various fuel types. DOE's Office of Building Technology, State and Community Programs provides these coefficients for source energy (U.S. DOE 1999).

3   Results – the 1990-1997 Experience

Using the inputs and the method described above, we designed spreadsheets that calculate and display energy savings, energy cost savings, and carbon emissions reductions for the years 1990-1997 based on the historic data. Figure 2 shows cumulative impacts from 1990 to 1997 (most recent data) of approximately 2 exajoules (EJ), US$22 billion (discounted present value at 7% to the year 2000), and 30 million metric tons of carbon.

Figure 2 graphically depicts the cumulative contribution (from 1990-1997) of the impacts of all of the appliance standards on energy savings, energy cost savings, and carbon emissions.

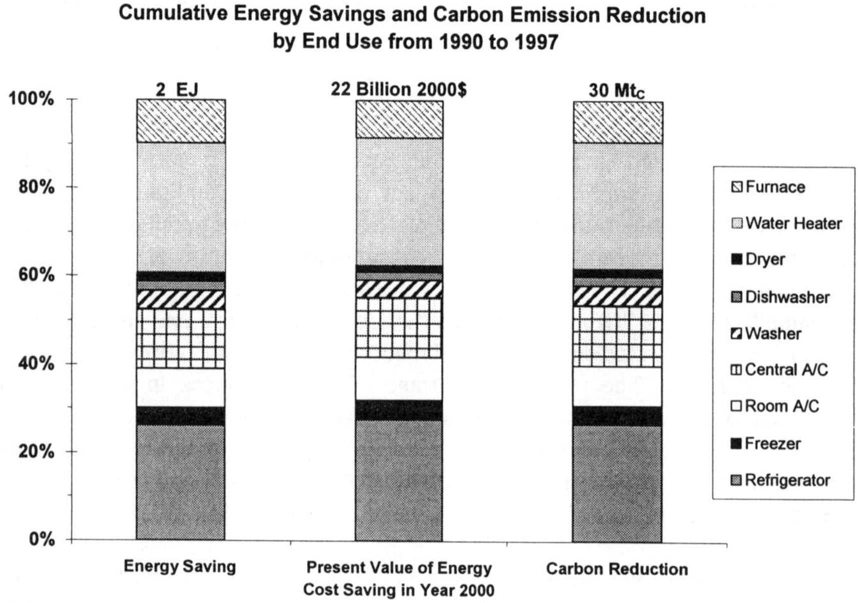

Figure 2: Impacts of Standards, 1990-1997

Tables II, III, and IV show achieved and projected impacts. The standards data are first shown individually by appliance. Then, to assemble an aggregate estimate of the national energy savings, energy costs savings, and carbon emission reductions to date, the impacts of all of the standards are added together. Projections for future years are also shown.

Table II: Annual Source Energy Savings (in EJ)*

	Refrigerator & Freezer			Room Air Conditioner		Central A/C	Clothes Washer	Dish-Washer	Clothes Dryer	Water Heater	Gas/Oil Furnace	Total All Appliances			
	NAECA 1990	Update 1993	Update 2001	NAECA 1990	Update 2000	NAECA 1992	Update 1994	Update 1994	Update 1994	NAECA 1990	NAECA 1990	NAECA	Updates	Total	Cumulative
1990	0.02	0.00	0.00	0.01	0.00	0.00	0.00	0.00	0.00	0.03	0.00	0.06	0.00	0.06	0.1
1991	0.03	0.00	0.00	0.01	0.00	0.00	0.00	0.00	0.00	0.04	0.00	0.08	0.00	0.08	0.1
1992	0.04	0.00	0.00	0.02	0.00	0.01	0.00	0.00	0.00	0.06	0.01	0.13	0.00	0.13	0.3
1993	0.04	0.02	0.00	0.02	0.00	0.02	0.00	0.00	0.00	0.07	0.02	0.17	0.02	0.19	0.5
1994	0.05	0.04	0.00	0.02	0.00	0.04	0.01	0.00	0.00	0.08	0.03	0.22	0.06	0.28	0.7
1995	0.05	0.06	0.00	0.03	0.00	0.05	0.02	0.01	0.01	0.09	0.04	0.26	0.09	0.35	1.1
1996	0.05	0.07	0.00	0.03	0.00	0.07	0.03	0.01	0.01	0.11	0.05	0.31	0.13	0.43	1.5
1997	0.05	0.09	0.00	0.04	0.00	0.08	0.03	0.02	0.02	0.12	0.06	0.34	0.16	0.50	**2.0**
2000	0.06	0.15	0.00	0.05	0.00	0.11	0.05	0.03	0.03	0.13	0.07	**0.41**	0.26	0.68	3.9
2005	0.05	0.23	0.09	0.06	0.01	0.12	0.08	0.04	0.05	0.11	0.07	0.41	0.49	0.90	8.0
2010	0.03	0.27	0.18	0.05	0.02	0.07	0.06	0.05	0.06	0.07	0.05	0.28	0.65	**0.92**	12.7
2015	0.01	0.26	0.27	0.05	0.02	0.03	0.03	0.06	0.07	0.04	0.02	0.14	0.71	0.85	17.1
2020	0.00	0.24	0.33	0.04	0.02	0.00	0.01	0.06	0.07	0.01	0.00	0.06	0.74	0.80	21.2
2025	0.00	0.23	0.35	0.03	0.03	0.00	0.00	0.06	0.08	0.00	0.00	0.04	0.74	0.78	25.1
2030	0.00	0.23	0.36	0.03	0.03	0.00	0.00	0.06	0.08	0.00	0.00	0.03	0.75	0.78	29.0

*1 EJ = 10^{18} Joules. 1 EJ = 0.9478×10^{15} Btu.

Includes electricity generation and transmission losses and household consumption of electricity, gas and oil.

Table III: Energy Cost Savings (in billion 2000 US$)

	Refrigerator & Freezer			Room Air Conditioner		Central A/C	Clothes Washer	Dish-Washer	Clothes Dryer	Water Heater	Gas/Oil Furnace	Total All Appliances				
	NAECA	Update		NAECA	Update	NAECA	Update	Update	Update	NAECA	NAECA	NAECA	Updates	Total	Cumulative	PV 2000 @7%
	1990	1993	2001	1990	2000	1992	1994	1994	1994	1990	1990					
1990	0.18	0.00	0.00	0.09	0.00	0.00	0.00	0.00	0.00	0.23	0.00	0.51	0.00	0.51	0.5	1.0
1991	0.24	0.00	0.00	0.11	0.00	0.00	0.00	0.00	0.00	0.31	0.00	0.66	0.00	0.66	1.2	2.2
1992	0.29	0.00	0.00	0.13	0.00	0.09	0.00	0.00	0.00	0.40	0.06	0.98	0.00	0.98	2.2	3.9
1993	0.33	0.15	0.00	0.16	0.00	0.19	0.00	0.00	0.00	0.50	0.13	1.32	0.15	1.47	3.6	6.3
1994	0.37	0.30	0.00	0.19	0.00	0.31	0.07	0.03	0.03	0.60	0.20	1.68	0.43	2.11	5.7	9.4
1995	0.39	0.44	0.00	0.22	0.00	0.41	0.13	0.06	0.06	0.65	0.25	1.92	0.69	2.62	8.3	13
1996	0.40	0.59	0.00	0.26	0.00	0.52	0.19	0.09	0.09	0.74	0.31	2.23	0.96	3.19	11.5	17.3
1997	0.41	0.71	0.00	0.29	0.00	0.60	0.25	0.12	0.12	0.80	0.36	2.47	1.20	3.67	15.2	**22**
2000	0.41	1.10	0.00	0.36	0.01	0.82	0.38	0.20	0.20	0.82	0.42	**2.84**	1.88	4.72	28.4	36
2005	0.34	1.64	0.66	0.40	0.07	0.90	0.52	0.30	0.33	0.67	0.42	2.73	3.52	6.26	57.0	59
2010	0.19	1.95	1.32	0.38	0.13	0.54	0.42	0.36	0.43	0.43	0.30	1.84	4.61	**6.45**	89.3	78
2015	0.06	1.89	1.92	0.34	0.16	0.20	0.20	0.37	0.48	0.21	0.11	0.92	5.02	5.94	120.0	91
2020	0.01	1.71	2.38	0.30	0.17	0.04	0.06	0.39	0.52	0.06	0.03	0.43	5.23	5.66	148.8	99
2025	0.00	1.63	2.53	0.25	0.18	0.00	0.01	0.40	0.54	0.01	0.02	0.27	5.28	5.55	176.7	105
2030	0.00	1.63	2.56	0.19	0.19	0.00	0.00	0.41	0.55	0.00	0.02	0.21	5.34	5.55	204.5	110

Table IV: Carbon Emission Reductions (in MtC/a)

	Refrigerator & Freezer			Room Air Conditioner		Central A/C	Clothes Washer	Dish-Washer	Clothes Dryer	Water Heater	Gas/Oil Furnace	Total All Appliances			
	NAECA 1990	Update 1993	Update 2001	NAECA 1990	Update 2000	NAECA 1992	Update 1994	Update 1994	Update 1994	NAECA 1990	NAECA 1990	NAECA	Updates	Total	Cumulative
1990	0.32	0.00	0.00	0.15	0.00	0.00	0.00	0.00	0.00	0.44	0.00	0.92	0.00	0.9	0.9
1991	0.43	0.00	0.00	0.19	0.00	0.00	0.00	0.00	0.00	0.60	0.01	1.23	0.00	1.2	2.2
1992	0.53	0.00	0.00	0.24	0.00	0.17	0.00	0.00	0.00	0.78	0.13	1.85	0.00	1.9	4.0
1993	0.61	0.27	0.00	0.29	0.00	0.36	0.00	0.00	0.00	0.97	0.28	2.51	0.27	2.8	6.8
1994	0.70	0.56	0.00	0.36	0.00	0.58	0.13	0.06	0.06	1.17	0.42	3.22	0.81	4.0	10.8
1995	0.72	0.83	0.00	0.42	0.00	0.77	0.26	0.12	0.11	1.33	0.54	3.77	1.32	5.1	15.9
1996	0.76	1.11	0.00	0.50	0.00	0.99	0.38	0.18	0.17	1.50	0.66	4.42	1.86	6.3	22.2
1997	0.80	1.39	0.00	0.57	0.01	1.18	0.50	0.24	0.23	1.64	0.76	4.95	2.37	7.3	**29.5**
2000	0.83	2.23	0.00	0.73	0.01	1.66	0.79	0.42	0.41	1.79	0.97	**5.99**	3.86	9.8	56.7
2005	0.68	3.35	1.34	0.82	0.14	1.83	1.12	0.64	0.68	1.51	1.02	5.87	7.28	13.2	116.9
2010	0.39	4.00	2.71	0.79	0.27	1.10	0.91	0.77	0.89	1.00	0.74	4.01	9.55	**13.6**	184.9
2015	0.12	3.93	4.01	0.72	0.34	0.41	0.44	0.81	1.02	0.50	0.27	2.03	10.55	12.6	249.9
2020	0.01	3.56	4.96	0.63	0.36	0.07	0.14	0.84	1.10	0.15	0.06	0.93	10.95	11.9	310.5
2025	0.00	3.38	5.25	0.52	0.37	0.00	0.01	0.86	1.14	0.02	0.04	0.58	11.02	11.6	368.9
2030	0.00	3.39	5.31	0.40	0.39	0.00	0.00	0.88	1.17	0.00	0.04	0.44	11.14	11.6	426.8

The present value (discounted at 7% to 2000) of cumulative energy savings from 1990 to 1997 is US$22 billion. Savings from the original NAECA legislation have increased over time and are reaching their maximum (0.41 EJ and US$2.8 billion per annum) in year 2000. A decade of standards updates effective from 1993 to 2001 shifted the maximum for total expected savings from the existing standards to 2010 (0.92 EJ and US$6.45 billion per annum). Additional standards currently under development (not shown) are expected to add to these savings.

The pattern of savings in carbon emissions follows from energy savings. The cumulative reductions in carbon emissions are nearly 30 MtC by 1997. In 2000, the original standards are expected to reduce carbon emissions by about 6 MtC (million metric tons of carbon) per annum. The updates result in an expected maximum total savings of about 13 MtC per annum in 2010.

Using refrigerators as an example, Figure 3 shows how the (undiscounted) energy cost savings and the increased equipment costs associated with new standards combine to yield positive net savings over time for the nation's consumers. For all products, the payback period required to offset increased equipment prices by energy bill savings averages 2 to 4 years, although individual consumers may experience a larger range reflecting the variability in equipment costs, energy prices, and consumer usage behaviors.

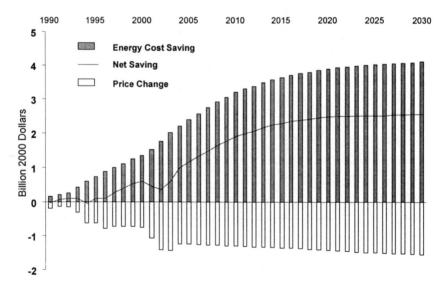

Figure 3: Annual Economic Impacts of U.S. Refrigerator Standards

Governmental administrative costs for the entire appliance, lighting, and equipment standards program represent an additional $5-10 million annually for developing test procedures, analyzing, and implementing regulations. This amount is too small to appear on the chart of national impacts.

4 Future Impacts of Adopted Standards

4.1 Existing Standards

Because of the long useful lifetimes of appliances, up to 30 years, the existing standards will continue to provide net benefits many years into the future.

4.2 Other Adopted Standards

Additional net benefits will accrue from updates to several NAECA standards that have either just taken effect (room air conditioners in 2000) or are scheduled to take effect in 2001 (refrigerators and freezers).

Table V shows a summary of the source energy savings (as per Table II, but split into electricity and natural gas) associated with all of the existing residential appliance standards. The electricity savings are also converted to show that generation corresponding to about fourteen 500-megawatt (MW) power plants has been avoided by year 2000.

Table VI shows a summary of the total energy cost savings and carbon emissions reductions from existing residential appliance standards. By 1997, annual carbon emissions associated with the residential sector are reduced by about 3% of the 1990 residential sector emissions of 253 MtC per annum. By comparing the carbon emission reductions to typical carbon emissions from automobiles, the last column shows that in year 2010 the projected savings are equivalent to removing 10 million cars from the road.

Table V: Energy Savings for Existing Residential Appliance Standards

	Source Energy Savings (EJ)				Electricity Savings (EJ Primary)					Natural Gas Savings (EJ)			
	Annual Savings			Cumu-	Annual Savings			Cumu-	Power Plants	Annual Savings			Cumu-
Year	NAECA	Update	Total	lative	NAECA	Update	Total	lative	Avoided *	NAECA	Update	Total	lative
1990	0.06	0.00	0.06	0.1	0.05	0.00	0.05	0.0	1	0.02	0.00	0.02	0.0
1991	0.08	0.00	0.08	0.1	0.06	0.00	0.06	0.1	2	0.02	0.00	0.02	0.0
1992	0.13	0.00	0.13	0.3	0.09	0.00	0.09	0.2	2	0.04	0.00	0.04	0.1
1993	0.17	0.02	0.19	0.5	0.11	0.02	0.13	0.3	4	0.06	0.00	0.06	0.1
1994	0.22	0.06	0.28	0.7	0.14	0.05	0.19	0.5	5	0.08	0.00	0.08	0.2
1995	0.26	0.09	0.35	1.1	0.17	0.08	0.25	0.8	7	0.09	0.01	0.10	0.3
1996	0.31	0.13	0.43	1.5	0.19	0.11	0.31	1.1	8	0.11	0.01	0.12	0.5
1997	0.34	0.16	0.50	2.0	0.22	0.14	0.36	1.4	10	0.12	0.02	0.14	0.6
2000	0.41	0.26	0.68	3.9	0.27	0.23	0.50	2.8	14	0.15	0.03	0.17	1.1
2005	0.41	0.49	0.90	8.0	0.26	0.45	0.71	6.0	19	0.14	0.04	0.18	2.0
2010	0.28	0.65	0.92	12.7	0.17	0.60	0.77	9.8	21	0.10	0.04	0.14	2.8
2015	0.14	0.71	0.85	17.1	0.09	0.68	0.77	13.6	21	0.05	0.03	0.08	3.4
2020	0.06	0.74	0.80	21.2	0.05	0.71	0.76	17.5	21	0.01	0.02	0.04	3.6
2025	0.04	0.74	0.78	25.1	0.03	0.72	0.75	21.2	20	0.00	0.02	0.02	3.7
2030	0.03	0.75	0.78	29.0	0.03	0.73	0.75	25.0	20	0.00	0.02	0.02	3.9

* 1 EJ of source energy (electricity) equals 27 coal-fired 500 MW power plant

Table VI: Energy Cost Savings and Carbon Emission Reductions for Adopted Residential Appliance Standards

	Energy Cost Savings (billion 2000$)				Carbon Reduction (million metric tons)			
	Annual Savings			Cumu-		% of Residential	Cumu-	million cars
Year	NAECA	Update	Total	lative	Annual	Total in 1990	lative	off the road
1990	$0.5	$0.0	$0.5	$0.5	0.9	0.4%	0.9	0.7
1991	$0.7	$0.0	$0.7	$1.2	1.2	0.5%	2.2	1.0
1992	$1.0	$0.0	$1.0	$2.2	1.9	0.7%	4.0	1.4
1993	$1.3	$0.1	$1.5	$3.6	2.8	1.1%	6.8	2.2
1994	$1.7	$0.4	$2.1	$5.7	4.0	1.6%	10.8	3.1
1995	$1.9	$0.7	$2.6	$8.3	5.1	2.0%	15.9	4.0
1996	$2.2	$1.0	$3.2	$11.5	6.3	2.5%	22.2	4.9
1997	$2.5	$1.2	$3.7	$15.2	7.3	**2.9%**	29.5	5.7
2000	$2.8	$1.9	$4.7	$28.4	9.8	3.9%	56.7	7.7
2005	$2.7	$3.5	$6.3	$57.0	13.2	5.2%	116.9	10.2
2010	$1.8	$4.6	$6.5	$89.3	13.6	5.4%	184.9	**10.6**
2015	$0.9	$5.0	$5.9	$120.0	12.6	5.0%	249.9	9.8
2020	$0.4	$5.2	$5.7	$148.8	11.9	4.7%	310.5	9.2
2025	$0.3	$5.3	$5.6	$176.7	11.6	4.6%	368.9	9.0
2030	$0.2	$5.3	$5.6	$204.5	11.6	4.6%	426.8	9.0

Carbon emission coefficient for electricity: 14.87 MtC/EJ
Carbon emission coefficient for natural gas: 13.65 MtC/EJ
Carbon emission from driving a car in a year: 1.28 metric ton of carbon

5 Standards Currently under Development

The U.S. Department of Energy is currently well along in its process to adopt updated standards for three residential appliances: water heaters, clothes washers, and central air conditioner and heat pumps. As with the previously adopted standards, these standards are expected to yield substantial savings to consumers.

Detailed results of the most current analyses of appliance standards under development in the U.S. are available from the Department of Energy's web site (U.S. DOE 2000). The current activities of the Energy Efficiency Standards Group at Lawrence Berkeley National Laboratory in support of the U.S. Department of Energy's work on appliance standards are available at the Group's web site (LBNL 2000).

6 Other Standards Activities in the U.S.

In addition to the residential appliance standards mentioned in this paper, the U.S. has also adopted or is considering energy conservation standards for such diverse products as lighting equipment (e.g., fluorescent lamp ballasts); commercial heating, cooling, and water heating equipment; distribution transformers and water conservation standards for various plumbing products.

7 Conclusions

Examining the actual impacts of U.S. appliance standards put in place during the past decade allows us to see the significant contribution they have already made in saving energy, lowering energy costs, and reducing emissions. These standards will continue to provide net benefits to U.S. consumers well into the future. In addition, the already-adopted standards that will soon take effect (for room air conditioners and refrigerators) are expected to yield substantial benefits. When the benefits from these already-adopted appliance standards are coupled with the benefits expected from standard rulemakings currently in progress and the benefits from already-adopted commercial sector standards, we can see that the U.S. has used appliance standards to reduce energy consumption, lower energy and total expenditures, and protect the environment.

Acknowledgment

This work was supported by the Office of Building Research and Standards of the U.S. Department of Energy under Contract No. DE-AC03-76SF00098.

8 References

[1] Bertoldi, Paolo, Andrea Ricci, and Boudewijn Huenges Wajer, Editors, 1997. *Proceedings of the First International Conference on Energy Efficiency in Household Appliances.* November 10-12, 1997, Florence, Italy.

[2] Lawrence Berkeley National Laboratory, Environmental Energy Technologies Division, Energy Analysis Department, Energy Efficiency Standards Group, 2000. http://eappc76.lbl.gov/tmacal/tmahome.cfm Berkeley, CA.

[3] McMahon, James E. and Isaac Turiel, Guest Editors, 1997. *Energy and Buildings.* Volume 26, Number 1, July 30, 1997. Special Issue devoted to Energy Efficiency Standards for Appliances.

[4] U.S. DOE (Department of Energy), Energy Information Administration, 1998. *Annual Energy Outlook 1999*, Washington, DC. December. DOE/EIA-0383(99).

[5] U.S. DOE, Energy Information Administration, 1997. *Emissions of Greenhouse Gases in the United States.* Washington, D.C. DOE/EIA-0573(97).

[6] U.S. DOE, Office of Building Technology, State and Community Programs, 1999. *BTS Core Databook*, June 18, 1999, Washington, D.C.

[7] U.S. DOE, Office of Codes and Standards, 2000. Appliance Specific Information http://www.eren.doe.gov/buildings/codes_standards/stkappl.htm Washington, D.C.

Standards and Labels of Household Appliances as an Opportunity to Reduce CO_2 Emissions

Mario Contaldi, Rino Caporali and Domenico Gaudioso

ANPA INT-CLIMA, Roma

Abstract. The Italian commitment to reduce GHG emissions according to the Kyoto protocol requirements includes measures in the industry – domestic sectors aiming to increase energy saving using as well standards and voluntary agreements. In particular, domestic appliances could present a consistent potential for energy saving. The EC energy label could represent a way both for pushing the market towards more energy efficient domestic appliances and for the monitoring of adopted or proposed measure. This paper summarise the work done to construct a preliminary model of the Italian stock of refrigerators from 1989 to 2000, with projections to 2010. Data since the year 1980 are needed, according to the adopted methodology. The stock is characterised according to the parameters of the EC energy label, i.e. refrigerator category, related adjusted volume and energy efficiency class. Data on refrigerators sold each year are used to construct the model, which, of course, defines also the overall number of refrigerators in use each year. Because the energy label is in use since 1995, previous years data are not directly based on label parameters, so that, besides retrieving the data, also some elaboration work is needed. This hopefully will be refined in the next future, with the cooperation of the organisations which the data belongs to.

1 Objective

A national plan to fulfil the EU commitments established by the Kyoto Protocol has been approved through the CIPE (interministerial committee for economic planning) deliberation of November 19, 1998 (# 137/98). EU is committed to reduce its greenhouse gas emissions by 8% with respect to the levels of 1990 within the time period from 2008 to 2012. The Italian commitment within the EU burden sharing agreement is to reduce national emissions by 6.5%, corresponding to an actual emissions reduction of a hundred million tons of carbon dioxide equivalent.

Those commitments refers to the main six greenhouse gases (GHG) not controlled by the Montreal Protocol for the protection of the ozone layer – namely Carbon Dioxide (CO_2), Methane (CH_4) ,

Nitrous oxide (N_2O), Hydro fluorocarbons (HFCs), Perfluorocarbons (PFCs), Sulphur hexafluoride (SF_6).

The Cipe deliberation has considered that the Kyoto protocol:

1) Has indicated that the greenhouse gas emissions have to be reduced;
2) Has identified the following actions to be implemented by Annex I Countries (Developed Countries and Countries with Economy in transition) for the reduction of emissions:
    ➤ promotion of energy efficiency in all sectors;
    ➤ development of energy production from renewable sources and of innovative technologies for emissions reduction;
    ➤ protection and extension of forests for carbon removal;
    ➤ promotion of sustainable agriculture;
    ➤ limitation and reduction of methane emissions from landfills and other energy sectors;
    ➤ fiscal measures to discourage greenhouse gas emissions, as appropriate;
3) Has created three flexible mechanisms, supplemental to the domestic actions, to contribute to the implementation of the commitments through joint actions among several Annex I Countries (Joint Implementation) or co-operation with non-Annex I Countries (Developing Countries or New Industrialised Countries) for clean development (Clean Development Mechanism), or through the international trade of emission permits (Emission Trading);
4) Has indicated the carbon removal by afforestation and reforestation, accomplished since 1990, as supplemental measure for emissions reduction;

The Cipe has approved the emission reduction targets of 95/112 Mton CO_2; they include reduction achievable through the flexible mechanisms established by the Kyoto Protocol and the domestic interventions listed in the document "Guidelines for domestic policies and measure s for the reduction of the greenhouse gas emissions". This document foresees the possible measures in the industry – domestic sectors (measures oriented to increase energy saving using as well standards as voluntary agreements) and the emission reduction potential of 24/29 Mton CO_2.

The detailed list of measures is published in the Second National Communication. In this list there are two specific points quantifying the reductions of emissions through more efficient equipment in the tertiary and residential sectors, with an emission reduction potential of 2.5 and 5.5 Mton CO_2.

In particular, as pointed out in previous works (Ref. 1), domestic appliances could present a consistent potential for energy saving. It is consequently felt the necessity of a tool enabling the analyst to evaluate in detail such potential and the impact of policies and measures aimed at improving the average energy efficiency of the Italian stock of appliances. Such a tool could be a data base of the overall stock of domestic appliances in each year, including also a characterisation of possibility to recalculate past years consumption and also to make projections for future year through the construction of an expected stock in the year of reference in correspondence of planned measures and policies. It has been judged correct to start with the refrigerators, which have the largest spread of parameters needed for their classification. A successful attempt with this typology of domestic appliances will give confidence on the feasibility of the planned stock model, with the aim to complete the work also for the other domestic appliances once this first attempt is completed.

The evaluated energy savings are finally translated in the expected reduction of GHG emissions reduction in order to compare the results with the Kyoto protocol targets.

2   Methodology

The basic idea is to characterise the Italian stock of appliances in terms of class of energy efficiency, and to combine these data with relevant statistic data in order to be able to reproduce past data for energy consumption and stock composition and to evaluate data for the years of interest, to be considered as milestones in the achievement of Kyoto protocol requirements for Italy, under different scenarios of stock evolution in terms of quantity, composition and energy efficiency.

The EU energy efficiency label is used to characterise the stock in terms of energy efficiency, and also in terms of typology by using category and adjusted volume as introduced by the European directive on energy label.

Of course such a characterisation has to be demonstrated as suitable for the designed use, which means basically that it has to reproduce both stock composition and energy consumption data which are independently available in order to give enough confidence on the results for future years.

On the other hand, there is the advantage that market data are directly available in terms of category and energy class since the year 1995, when the energy label was first introduced: the principal work to be done is then to translate market data before 1995 in the same terms, which means a classification of the appliances sold each year by evaluating the models, their volumes and declared energy consumption, which is expected as the most time consuming task.

Basing on market data for each year, on the above classification of the model brought to the market also in the years before 1995, and on a reasonable assumption on cold appliances average lifetime, it is possible to construct stock model suitable for the calculation related to the total annual energy consumption, both for past year data and for the future, under conceivable scenarios based on the implementation of emission reduction policies and measures.

3   Data Sources

Up to now global data on the appliances stock evolution from the year 1970 to the year 1998 can be found in ref. 2, which are the results of the elaboration made by ENEA and ISTAT experts of data essentially from ISTAT itself, the Italian national institute for statistic analysis, and ENEL, by far the most important Italian utility for electricity in terms of market coverage.

Data on the overall energy consumption by cold appliances and by year are given too, so that it is possible to check them with the results of the refrigerator stock simulation model.

For the time being, collaboration is being launched with the national association of appliances producers, ANIE, which should make available data on the models brought to the market in the past years to make possible the stock classification according to the energy label parameters also for the years before 1995. Up to now, some global data from ANIE have been used for the years 1986-1992 (ref. 5).

Also, some generic data have been retrieved from the literature, essentially from studies for Eco-label and for Energy label, to make a preliminary stock classification by categories and energy efficiency classes for the years 1994-1997 and to control energy consumption results for the same years.

4   Model Construction

As a first approximation the simulation model is structured to give results for the years from 1989 to 2010. According to a mean lifetime of 14 years (ref. 3), are needed data at least from the year 1980: such data are to be combined with a "decay" law in order to estimate the overall stock in each year of interest, which is a combination of the new appliances sold in the year and of the number of appliances inherited from the previous 14 years.

Enough reliable data are available only for the new models sold in the years 91-97, which are already based on categories and energy classes (ref. 3), and have been used to structure the stock of new appliances in each year.

For the previous years, it has to be taken into account that categories and energy efficiency classes have been introduced in 1995 by the EC Energy Label, so that there are no data already structured as the ones for years 91-97.

The data from ANIE, for years from 86 to 90, have been used to define the number of new appliances sold in the market, with some extrapolation to take into account that ANIE data refer to the appliances brought to the market in the year of reference, not to the sales of the year. The expected difference has been estimated from the data for years 91 and 92, where also ANIE data are available. Categories and classes of energy efficiency have been assigned through an extrapolation of the similar data for years after 1990, made by maintaining constant the number of all categories but category 1 and 7, which have the most impact on the total. The energy classes have been assigned by first estimating a specific consumption, assuming from 1986 to 1990 a technologic improvement leading linearly each class to the class immediately over.

In this way it has been possible to maintain, in each year from 1986 to 1990, the classification by categories and energy efficiency classes.

Before 1986, the average specific consumption has been maintained constant, as a first guess. The overall stock of sales have been estimated basing on the global number of installed cold appliances retrievable from the year 1970 to 1985 in ref. 2.

Also, in the end, the number of appliances not continuously used has been estimated from ISTAT-ENEL data for the total number of users connected to the electricity net, the total number of inhabited houses and the percentage of diffusion for refrigerators.

A better characterisation of the stock will be made in the future, when the necessary data will be retrieved. In any case, with the mean lifetime of 14 years and the correlated "decay " law, the errors generated by wrong data in the early years should have a little impact on the evaluation made for the nineties, in the first cut.

The specific average energy consumption for each category and each class as been evaluated by taking the higher value allowable in each class with respect to the standard energy consumption calculated according to the European directive.

Such a calculation needs the definition of the adjusted volume by each class which have been retrieved only for the years 94-97 and extrapolated for the other.

As it is easily seen, at this stage the model construction is based on a number of assumptions and extrapolations that have to be removed, as far as possible, in the next future by retrieving the needed data.

5   Model Validation

So far, a couple of checks has been made, one on the construction of the total stock for each year, and another on the overall energy consumption pertaining to the refrigerators, both data being already retrievable in ref. 2.

The first check has been successful, differences between the calculated and the real data at most are a few percent.

Different results came out from the check on the energy consumption, where in some cases the difference between calculated and actual data summed up to a 30%. Up to a 20% of the difference could come from the real life use of the appliances, as prospected in ref. 3, even if taking into consideration the Italian climate one could expect something less. The explanation of the remaining difference needs some more thinking because is not readily available, even if there are some hypothesis still to be checked: for instance in large cities, as Rome and

Milan, there are also other utilities than ENEL so that it has to be checked if the data for these utilities are included in the overall consumption or not. In any case the EC data for the Italian newly sold cold appliances average consumption are differing from ref. 2 data as well, so that it seems that this problem does not invalidate the suitability of the model for the planned evaluations.

Taking this into account it has been considered useful to run a first set of cases, whose results of course are to be considered just an indication.

6 First Results

Data in the model have been used for a simplified life cycle analysis of GHG emissions of 5 different "average" refrigerators and to evaluate the effects of two market evolution scenarios in 2010.

The effect of a combined set of measures on the life-cycle GHG emissions of an average fridge, has been computed in five different configurations:

- average 1998-marketed fridge, of an overall weight of 35 kg, being built using HFCs for expanding insulating foams and for the refrigerating circuit; categories and energy classes are the average 1998 one's (this hipotesys do not correspond to the market average in 1998 that see HFC used only for a small part of the market);
- improved 1998 fridge, analogous to the previous one except for the fact that no HFCs has been used for the insulating foams;
- no HFCs 1998 fridge, similar to the first one but without any use of HFCs either for insulating foams or for refrigerating circuit;
- 2010 fridge with market share of scenario 1 (BAU): 0% super A class, 40% A class, 30% B and C class;
- 2010 fridge with market share of scenario 2: 30% super A class, 40% A class, 30% B class;

With reference to the attached figure 1 and Table I it can be pointed out:

- the diffusion of more efficient fridges has the main impact in reducing GHGs emissions;
- an increased diffusion of existing efficient fridges and the introduction of a "super A" class with a market share of 30% in 2010 have significative improvement potential;
- the elimination of HFCs use in production of the insulating foams has a major impact in GHGs emissions (this measures is already being implemented by the industry);
- the elimination of HFCs in the refrigerating circuit has a significative impact, analogous to the energy efficiency improvements taken into consideration, the share of no-HFC in cooling circuit for domestic appliances was about 30% in 1998;

Two market evolution scenarios have been evaluated for the year 2010, which is a milestone in the fulfilment of the Kyoto protocol targets. The two scenarios point out to energy consumption of refrigerators and do not consider the inpact of HFCs on total emissions.

In all scenarios the number of households and the new appliance's sales volumes are kept constant in the 2000 – 2010 period. Moreover in all scenarios the calculations consider constant the average carbon content of the electricity distributed in Italy between 1999 and 2010. This last assumption is necessary for obvious reasons, the carbon content should be reduced of about 6% in the same period. An overall stock of refrigerators have been defined for the years 2000-2010, by maintaining constant the number of new appliances sold each year: the rule of a mean lifetime of 14 years takes into account the percentage of the stock of previous years going out of the stock and automatically define the new stock for each year.

First of all, a BAU scenario which takes into account the disappearing of energy efficiency classes below C and the continuation of actual policies that promote the diffusion of more energy efficient classes. The classes from D to G are simulated to go out of the market since the year 2000. The subdivision of the stock in categories is maintained constant. Class A appliances is simulated to reach the 20% of the new sold in 2005, and the 40% in 2010; the class B are simulated as having the 40% share of the market in 2005 and the 30% in 2010; the class C is represented as having the 40% share of the market in 2005 and the 30% in 2010. Adjusted volumes are maintained constant in each category. The overall result is an expected reduction of around 18% in 2010 with respect to 1999 of both energy consumption and "direct" (related to the average CO_2 content of the "delivered" electricity in the Italian grid) CO_2 emissions. This huge reduction is due in large part (15%) to the diffusion in the existing stock of the already efficient fridges marketed in 2000 and in a smaller portion (3%) to the diffusion of the more efficient mix of fridges.

A second scenario foresees the introduction in 2001 of a "super A" class, with 30% less energy consumption then class A. At the same time the C class is lead to an end in 2010. The "super A" class is modelled to reach 10% of the new market in 2005 and 30% in 2010: this is done by changing the classes' share of the new market from 2000 to 2010. The assumption for the construction of the overall stock of domestic appliances in 2005 and in 2010 are the same as in the case above.

The energy savings in this case is expected to be in 2010 around 24% with respect to 1999 with an increase of 6% above the first scenario.

Detailed analytical data on energy consumption and emissions for both the above described scenarios are reported in the attached Table II.

7 Conclusions

It has to be said that this calculation has been made by maintaining constant from 1999 to 2000 the overall quantity of new sold appliances in each year, the same market share in terms of refrigerator category and the same adjusted volume by category. The last assumption means that the standard energy consumption, evaluated for each energy class according to ref. 4, remain constant.

The real market data shows that the assumption of a constant market of new sold refrigerators per year has to be considered not far from the reality, which is no wonder is it is considered that the new market is substantially based on the substitution of the oldest appliances, being the market completely saturated.

On the other and the same data show that there is a movement towards more energy consuming categories, including refrigerator-freezers and no frost appliances. Also, the trend is for an increase of the adjusted volume in each refrigerator category, at least looking at the available data from 1994 to 1997. This means that the evaluated reduction of energy consumption by the 2010 can be considered as an upper boundary of the result that could be reasonably achieved. The other assumption too, on the bulk of appliances sold in each year are to be deeply checked in order to have numbers of better consistence.

In any case, the tool look as feasible and suitable for useful evaluation: for instance in the already run cases, it has been possible to check how the energy consumption could develop in the BAU scenario by evaluating the impact of the EC directive pushing out of the market classes at highest energy consumption standard and, may be more interesting in the second case, what the impact could be of the introduction of the super A energy class with the contemporaneous exclusion from the market of class C refrigerators. Also, all policies aimed at changing the market share of each class can be evaluated in this way.

Even being at this stage the results just an indication, they demonstrate in any case the usefulness of the model and the interest of having sound data for the past years in order to have reasonable data for the future. Such an objective will be hopefully pursued with the collaboration of involved organisations.

8 References

[1] F. Krause: "La risorsa efficienza" , Serie Documenti ANPA, n° 11/99
[2] C. Ardi, G. Perrella: "Dati ed analisi energetica del settore residenziale in Italia (1970-1998)
[3] ENEA report RT/ERG/2000/01 (in Italian)
[4] Revision European Eco-label criteria for refrigerators, Draft Final Report 1999

[5]   Dir. 96/57/CE, September 3 1996
[6]   L'industria italiana degli apparecchi domestici, Doc. ANIE, various years (in Italian)
[7]   6) R. Scialdoni, M. Presutto, G. Jacaz: "Ecolabel awarding criteria for refrigerators and freezers", Dec. 95

Table I: Technical potential of emission reduction of P&M in fridges
(em. reductions in 2010 with reference to 1998)

base fridge, overall CO_2 emissions, kg CO_2 eq.	3449
fridge without HFC in insulating foam	-10.5%
no HFC fridge	-5.0%
market share in 2010 = 1999	-12.6%
scenario 1, bau evolution of market share	-3.8%
scenario 2, evolution of market share with "super A"	-6.9%

Table II: Evolution of total energy consumption and emissions

	average year consumption "new"	average year consumption stock	"direct" CO_2 emissions	change from 1998
	KWh	GWh	Mt	%
market average, 1998	485	9346	5.1	-
scenario 2010 = 2000	485	7989	4.1	-
scenario "bau", 2010 (*)	398	7661	4.0	-
scenario "super A", 2010 (*)	312	7098	3.7	-

(*) see text for scenario definition

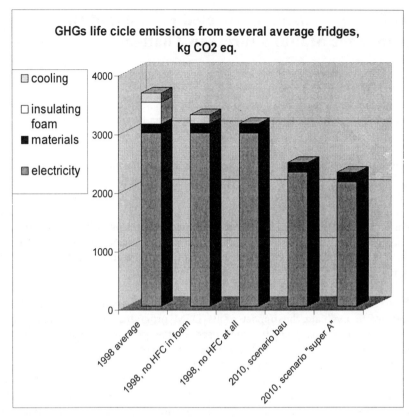

Figure 1 : GHGs life cycle emissions from several Kg CO$_2$

How Energy Labelling Affected Production Decisions of Appliance Manufacturers in Thailand

S. R. Na Phuket and C. Prijyanonda

International Institute for Energy Conservation – Asia Regional Office

Abstract. Recent experience in Thailand has shown the potential for appliance energy labelling programs to affect manufacturer production decisions. Thailand is one of the few Asian countries to have implemented a successful nation-wide energy labelling program for household appliances. The Electricity Generating Authority of Thailand (EGAT), the state-owned generating utility, implemented labelling programs as part of its national demand-side management (DSM) plan. The voluntary labelling of refrigerators and air conditioners started in 1995 and 1996, respectively.

After six years of DSM implementation, EGAT commissioned a Canadian consortium to evaluate the impact of its DSM programs in 1999. This paper describes the methods and strategies used in the process of interviewing the manufacturers. The results and our hands-on experience allowed us to draw important findings about the impact of energy labelling on production decision-making of Thai manufacturers. The Thai labelling programs are a successful example of a voluntary energy labelling effort in a developing country. The lessons learned from Thailand can provide useful guidance to policymakers in both developing and industrialised countries that are pursuing or revamping energy labelling programs.

1 Introduction

Thailand became the first country in Asia to implement a national demand-side management (DSM) program when the DSM Master Plan was developed and put into effect in 1992. By resolution of the Cabinet of the Royal Thai Government, the DSM Office (DSMO) was created within the Electricity Generating Authority of Thailand (EGAT) and given the authority to manage and implement DSM programs. The purpose of the initiative was to reduce peak energy demand while maintaining system quality and to instil an energy efficiency-oriented attitude within Thai consumers. Five DSM programs have been fully implemented and evaluated:

- Energy-Efficient Fluorescent Lamp (Thin-Tube) Program
- Compact Fluorescent Lamp (CFL) Program
- Street Lighting Pilot Project

- High Efficiency Air Conditioner Program
- High Efficiency Refrigerator Program

1.1 EGAT's Energy Labelling Programs

The High Efficiency Refrigerator and High Efficiency Air Conditioner programs are two effective labelling programs implemented under the framework of the DSM Master Plan. EGAT started implementation of energy efficiency labelling for refrigerators and air conditioners in November 1994 and February 1996, respectively. Both programs aim to promote the use of energy efficient appliances and increase their average energy efficiency.

Figure 1: Energy Label

EGAT's energy labels for refrigerators and air conditioners rank the products on a scale of #1 to #5, where a rating of #5 is the highest efficiency level and #3 is average. The label also shows consumers the average energy consumption per year (kWh/year) and the average electricity price per year (Baht/year). Since these programs are voluntary, manufacturers/distributors choose not to label their products if tests reveal that their products are less efficient than average (#3). Thus, no product in the market show #1 or #2 label.

To obtain an energy label, one sample unit must be randomly selected from a pool of at least 30 units of the same model (same size and features) and sent to the Thailand Industrial Standards Institute (TISI) laboratory for energy performance testing. Once the model has been tested, it is issued a label with a ranking between #1 to #5 according to its efficiency value compared to the average efficiency value within its size category (see Table I). Then, the manufacturer/distributor may choose whether or not to use the assigned label on their product.

572

Table I: Rating Criteria for Energy Labels

Ranking	The model will receive...
#1	...if its efficiency value is *at least 30% less than* the average efficiency value.
#2	...if its efficiency value is *15 to 30% less than* the average efficiency value.
#3	...if its efficiency value is *between –15% and +10%* of the average efficiency value.
#4	...if its efficiency value is *10 to 25% greater than* the average efficiency value.
#5	...if its efficiency value is *at least 25% greater than* the average efficiency value.

2   Methodology

In 1999, after six years of DSM implementation, the DSMO commissioned the AGRA Monenco consortium of consultants to perform process, market, and impact evaluation of its five DSM programs mentioned above. The evaluation focussed on assessing the actual reductions in electricity demand, energy use, and greenhouse gas (GHG) emissions realised by each of the programs. The evaluated savings were used to verify whether the savings reported by EGAT from rough market estimates were accurate.

The evaluation consortium was comprised of consultants from Canada (AGRA Monenco, BC Hydro International, Bureau d'études Zariffa, and Ference Weicker & Company) and Thailand (EEC-Energetics, the International Institute for Energy Conservation, and BERA). During April-July 1999, EEC-Energetics and the International Institute for Energy Conservation (IIEC) worked closely together to collect the necessary data from residential, commercial, and industrial consumers as well as from the manufacturers and distributors of fluorescent tube lamps (FTLs), compact fluorescent lamps (CFLs), air conditioners (ACs), and refrigerators. The data was collected through the use of survey instruments (mail and in-person interviews) and on-site data logging of appliance use pattern.

2.1 Survey of Manufacturers & Distributors

The manufacturer and distributor (M&D) survey required intensive in-person interviews where the questionnaires were designed to collect information on the manufacturer's production, market share, level of participation and satisfaction with the labelling programs. The questions were mostly open-ended; comments were encouraged after most questions to allow manufacturers to express their opinions and provide reasons for their responses. To cover the whole implementation period, manufacturers were requested to provide production and market data from 1993-1998.

In addition to the survey, label distribution information from EGAT was collected for the same period. We also gathered supplementary statistics and data from government agencies and public entities to fill in data gaps that could not be obtained from the survey.

2.1.1 Refrigerator Sample Size and Characteristics

All 10 refrigerator manufacturers/distributors/importers (M&D) were surveyed by in-person interviews. Out of the 10 M&Ds, there were 7 manufacturers with local plants in Thailand, 2 distributors whose brands are produced by one or more of the 7 manufacturers, and 1 importer. This sample size represented 100% of the refrigerator market.

2.1.2 Air Conditioner Sample Size and Characteristics

There are more than 200 AC manufacturers/distributors/importers/assemblers (M&D) in Thailand. Most of them are small assembling companies who use AC parts (compressors, heat exchanger, etc.) from the same parts distributors in Thailand. We selected a total of 32 M&Ds that included all major producers, a good number of medium-size producers, and several small producers. The selection strategy aimed to attain the highest possible market coverage.

2.1.3 Data Processing

Since the sample sizes for surveys are not large, the responses were compiled into simple Excel spreadsheets for data processing and analysis. For each interview, a separate data spreadsheet containing details of market and production data, as well as the summary of comments, was provided in addition to the main data sheet.

The data was formatted for simplicity in analysis and was given to the Canadian consultants to process. In this paper, we draw some results from the consortium's Final Report to EGAT (AGRA Monenco 2000), as well as key findings from further analysis of survey data by the authors.

3 Impact on Refrigerator Industry

This section describes the impact of the High Efficiency Refrigerator or "Refrigerator" program on the refrigerator industry. First, we took a look at how much the energy labels have contributed to improving the average energy efficiency of refrigerators in the Thai market. If so, did it influence the manufacturing process and costs? From our interviews, we closely examined the manufacturers' views on the program and its impact on their refrigerator production decision-making.

3.1 Improvement in Energy Efficiency

Six out of 10 M&Ds agreed that the Refrigerator program contributed significantly to increasing the average energy efficiency level of refrigerators in the Thai market. Two M&Ds indicated that the program only contributed a little to the improvement. Another two M&Ds said the program was not a contributing factor due to lack of accurate validation and their contention that the decrease in energy consumption per unit also led to a decrease in unit cooling capacity.

To verify the improvement of energy efficiency, we examined the increase or decrease of #3, #4, and #5 rated units between 1995 to 1998[1]. The data from EGAT showed that the order of #5 labels increased from 11.6% to 96.8% of the total labels ordered by the M&Ds. The number of #4 and #3 labels ordered decreased from 74.6% to 2.8% and 13.8% to 0.4%, respectively. Further, the collected data indicated that the percentage of labelled units sold compared to the total units sold ranges from 85-92% in 1996-1998. (AGRA Monenco 2000) This infers a dramatic increase of energy efficient (#5 and #4) units and elimination of average efficiency (#3) units from the market.

From their experience with consumers, eight of the 10 M&Ds said that the program has positively changed the attitude and buying habits of consumers. They believe consumers were influenced by EGAT's highly visible advertising campaign and recognise that the #5 product is energy efficient. However, one M&D who disagreed made the point that many consumers blindly look for a #5 label, but do not know what it really means.

3.2 Influence on Production Decisions

Table II summarises the reasons why the manufacturers participated in the Refrigerator program. Gaining customer trust through the program received the highest vote, receiving 32 out of 100 points. The reasons ranked 2nd to 4th are all market-oriented reasons, showing that M&Ds see the label as an influential factor on the market demand for their products. To further illustrate the importance of the label, all 10 M&Ds indicated that they used the label in their promotional campaigns and brochures to advertise their products.

Table II: Reasons for Participating in the Refrigerator Program[2]

Rank	Reasons	Points (out of 100)
1	To gain customer trust through program	32
2	Expected an increase in sales	17
3	To remain competitive in the market	13
4	Free market support (advertisements) from EGAT	12
5	To support national energy conservation efforts	10
6	No cost for testing and labels	7 (6.5)
	Participation process is easy	7 (6.5)
7	Minor or no production line modification	2
8	Minor or no production cost increase	1

Half (5 out of 10) of the M&Ds indicated that they had to modify their production line to achieve higher label rating, while 4 M&Ds did not have to. One interviewee, a distributor, did not know the answer. All five manufacturers who modified their production line said that, as a result, their production costs increased. Three of the five indicated small increases (1-8%) in production costs, but the other two claimed a 20% increase.

This explains the low score for the last two reasons in Table II. Even though five manufacturers had to modify their production lines to achieve higher efficiency rating, which also led to increases in production cost, they still chose to participate in the program. This move shows that production decisions are based more on market considerations rather than difficulties in production line modification – which is good news for the consumers.

3.3 Outlook for the Program

Seven out of 10 M&Ds supported the future enforcement of mandatory labelling and minimum energy performance standards, while the remaining three supported mandatory labelling but not minimum efficiency standards. For energy labelling, eight M&Ds felt that the efficiency requirement for each rating level should be adjusted higher. The two M&Ds that did not agree felt that the action would lead to higher production costs, which would directly become a burden on consumers.

For most aspects of the program (design of label, management, and advertising campaign), the majority of the M&Ds indicated that they were either "satisfied" or "more or less satisfied". The only one aspect of the program with which the majority (60%) of the M&Ds were dissatisfied was the testing process. They pointed out that the process takes too long and does not produce accurate and repeatable results.

Overall, six of the M&Ds are satisfied and four were "more or less satisfied" with the refrigerator labelling program. All M&Ds plan to keep labelling their refrigerators in the future (mostly due to market demand for #5 units). Six M&Ds confirmed that they would *increase* the use of labels for models that are not currently labelled and new models.

3.4 Key Comments and Suggestions from M&Ds

Below, we summarise the comments and suggestions concerning the refrigerator labelling program that appeared most frequently during the interviews.
- Increase energy efficiency testing capacity and accuracy;
- Continue the marketing support (advertisements, commercials, etc.) and education of consumers;
- Increase energy efficiency level of each label rating; the market has become saturated with refrigerators of the highest efficiency rating;

- Provide financial support to manufacturers for the development of more energy-efficient refrigerators at lower production costs.

4 Impact on Air Conditioner Industry

This section summarises the impact of the High Efficiency Air Conditioner or "AC" program on the air conditioner industry, in the same format as the refrigerator section above.

4.1 Improvement in Energy Efficiency

Although 31 out of 32 M&Ds (97%) said that the AC program contributed to increasing the average energy efficiency level of ACs, only 19 indicated that they thought the program had made a significant contribution. One M&D said the program was not a contributing factor due to the fact that there are relatively few #5 ACs on the market compared to the number of unlabelled units that are untested, and are likely to have low energy efficiency.

We examined the data from EGAT to estimate the increase or decrease of #3, #4, and #5 rated units between 1996 to 1998[3]. Results showed that the percentage of #5 labels increased from 82.8% to 91.5% of the total labels ordered by the M&Ds. The percentage #4 labels decreased from 17.2% to 8.5%. There were no #3 labels ordered. Even though more #5 labels were ordered, percentage-wise, one can not fully conclude that average energy efficiency of ACs in the total market has improved. Statistically, the percentage of labelled units sold compared to the number of total units sold only increased from 19% to 38% from 1996-1998. Therefore, more than 60% of the units sold are unlabelled and most probably are of lower efficiency. There is also another important market characteristic that we need to consider.

The Thai government requires that every AC unit sold in Thailand must be subjected to excise tax. In our survey of M&Ds, nearly every interviewee noted that there is a large "illegal" AC market, which may account for 30% of the total AC market. This "illegal" market is made up of small AC assemblers who avoid payment of excise tax and are able to sell their units at very low prices, relative to those subjected to excise tax. These units usually offer high cooling ability, but low energy efficiency and reliability. The existence of these "illegal" ACs and the lack of government enforcement causes unfair competition and discourages the "good" manufacturers from investing in production of high efficiency products.

Nevertheless, 94% of the M&Ds agreed that the program has positively changed the attitude and buying habits of consumers toward energy efficient air conditioners. However, two M&Ds who disagreed indicated that consumers still buy products based on price because the sale of unlabelled and "illegal" ACs did not seem to decline.

4.2 Influence on Production Decisions

Similar to the refrigerator industry, air conditioner M&Ds also indicated market-oriented reasons as the most important reasons for participating in the AC program (see Table III). Most M&Ds anticipated an increase in sales and expected to gain customer trust by joining the program. Remaining competitive in the market is also one of their main priorities. This reasoning and the fact that 28 out of 32 M&Ds (88%) use the label in their advertisements confirm that M&Ds recognised the energy label's strong influence on market demand.

Table III: Reasons for Participating in the AC Program[4]

Rank	Reasons	Points (out of 100)
1	Expected an increase in sales	24
2	To gain customer trust through program	23
3	To remain competitive in the market	18
4	Free market support (advertisements) from EGAT	11
5	Participation process is easy	6
6	No cost for testing and labels	5
	Minor or no production cost increase	5
7	Minor or no production line modification	4
8	To support national energy conservation efforts	4

Twenty-eight out of 32 M&Ds (88%) said they modified their AC units (design of parts, higher efficiency parts, etc.) to improve the energy efficiency ratio (EER) and rating of the units. Modification of their units caused an increase in production costs for 86% of M&Ds. The average increase in costs was estimated to be about 15%. In monetary value, to improve from #3 to #4 rating, the production cost would be about 3,800 Baht/unit (US$100), and from #4 to #5 would cost about 5,000 Baht/unit (US$132). These substantial increases in production cost are the main reasons why M&Ds still chose to produce and market unlabelled (low- and medium-efficiency) units.

All but one M&D produce at least some AC units that are unlabelled. When asked why they do not label some AC models, the most significant reason was that there was not enough demand for labelled units. Other main reasons were because the models would not achieve a #5 rating, to offer low-priced options for consumers, and to avoid testing delays. Unlabelled, low-priced units were offered to consumers to take away market share from the "illegal" AC units on the market.

4.3 Outlook for the Program

Fifty-three percent of M&Ds supported the future enforcement of mandatory labelling and minimum energy performance standards, while 25% were in favour of having only mandatory labelling and 13% were for MEPS only. Three M&Ds did not want any mandatory regulations by the government; one did not think

EGAT enforces the regulations properly and one said that local manufacturers would not be able to compete with multi-national manufacturers if standards and labels become mandatory.

In contrast to the Refrigerator program, only 6 out of 32 of M&Ds (19%) agreed that the efficiency requirement for each rating level should be adjusted higher. Most M&Ds (75%) disagreed because the current efficiency requirements for #4 and #5 ratings are already high enough and in-line with international standards. They contend that it would be too difficult for manufacturers to keep the costs down while providing the same cooling capacity. Furthermore, there is only a single set of requirements covering all sizes of ACs at present; this needs revision. Efficiency requirements should be specified for each product size category. It is relatively easy for small capacity (9,000-13,000 BTU) units to achieve high ratings; however, it is very difficult for high capacity (>13,000 BTU) units to achieve #5 ratings due to technical limitations.

Market demand and characteristics are important to consider when making revisions and changes to program requirements. If efficient AC unit prices increase due to higher requirements, several M&Ds warned that the market demand might split towards the low-end and "illegal" units and the high-end units, leaving the medium and medium-high efficiency units out of the market share. This will negatively affect the average Thai AC manufacturers, who can not compete with multi-national companies due to lack of funds for technology improvement and mass-production capability, nor compete with the "illegal" manufacturers. A majority of the M&Ds suggested that the program should focus more on enforcing the excise tax for all units sold in Thailand to provide a fair level of competition, instead of increasing efficiency requirements.

4.4 Key Comments and Suggestions from M&Ds

For the AC program, M&Ds provided more input on how to improve the program than did M&Ds of the Refrigerator program. This is because the AC market in Thailand is more complicated, with a large number of manufacturers and a large "illegal" market. Below are the comments and suggestions for the air conditioner labelling program:

- Manufacturers' decisions to participate in a voluntary labelling program are driven by consumer demand for labelled air conditioners. Since a major purchase criteria is price, consumers tend to purchase unlabelled models rather than labelled ones and this causes manufacturers to produce less of the high efficiency, labelled models;
- Encourage the government towards stricter enforcement of excise tax collection in order to decrease the market share of illegal, inefficient air conditioners.
- A mandatory labelling program will help eliminate the market distortion caused by illegal air conditioners, but will require strict monitoring and enforcement if the program is to be successful;

- Increase energy efficiency testing capacity and accuracy;
- Continue the marketing support (advertisements, commercials, etc.) and education of consumers so that they know why high-efficiency air conditioners cost more initially but can save them significant amounts of money over the air conditioners' life-cycle;
- Provide technical and financial support to manufacturers for the development of more energy-efficient air conditioners at lower production costs.

5 Conclusions

The Thai energy labelling programs are a successful example of a voluntary energy labelling effort in a developing country. The lessons learned from Thailand can provide useful guidance to policymakers in both developing and industrialised countries that are pursuing or revamping energy labelling programs. Important lessons learned from the Thai experience are presented below.

- A similar labelling strategy for the Refrigerator and AC programs yielded much different results. Programs must be designed, customised, revised, and improved specifically for each particular product market. Program design should differ not only on technical requirements, but also market transformation strategy. Each product market has different characteristics that will determine the success or failure of a labelling program.
- Voluntary labelling was effective in transforming the refrigerator market because there was not a significant spread in efficiency level and small number of manufacturers. However, for the AC market, there is a bimodal distribution of efficiency and high number of manufacturers. Stricter enforcement and stronger market intervention strategies were needed.
- Enforcement of regulations is critical for avoiding market distortion and unfair competition. The lack of excise tax enforcement allowed the "illegal" AC market to diminish the positive impact of the AC labelling program. Loopholes in mandatory regulations cause unfair competition, which will become an unyielding barrier to convincing M&Ds to participate in a labelling program.
- M&Ds weigh market demand and trends more heavily than production and testing difficulties encountered in improving energy efficiency. Therefore, the labelling programs must focus on influencing consumers to control market demand. However, the market must ensure fair competition as mentioned above.
- Each country and each product market is different. The success of labelling programs depends on both consumers and M&Ds. A survey of M&Ds and consumers should be conducted during the design stage of the program in order to obtain their views and suggestions.

Acknowledgements

The authors would like to thank EGAT and the AGRA Monenco consortium for their kind support during DSM program evaluation. We would like to also thank our reviewers, namely Ms. Iris Sulyma (BC-Hydro), Mr. Greg Wikler (EGAT DSM Advisor's Office), and Dr. Peter du Pont (IIEC-Asia) for their valuable input and review of this paper.

Endnotes

[1] Source data from EGAT – showing the number of #3, #4, and #5 labels ordered by manufacturers for refrigerator units during 1995 to 1998.

[2] Point system: Manufacturers were asked to rank 4 most important reasons. The most important reason receives 4 points, while linearly, the 4th most important reason received 1 point. The points were totalled up for each reason and normalised to 100.

[3] Source data from EGAT – showing the number of #3, #4, and #5 labels ordered by manufacturers for AC units during 1996 to 1998.

[4] Point system: same as described in Endnote #2

6 References

[1] AGRA Monenco. 2000. *DSM Program Evaluation – Conservation Program: Final Report*, Volumes 1, 4, and 5. AGRA Monenco.

[2] Egan, K. and Peter du Pont. 1998. *Asia's New Standard for Success: Energy Efficiency Standards and Labelling in 12 Asian Countries*. IIEC.

[3] Sulyma, I.M. et al. *Taking the Pulse of Thailand's DSM Market Transformation Programs*. Presented at ACEEE Summer Study on Energy Efficiency in Buildings. August 20-25, 2000.

Evaluating the Impact of Appliance Efficiency Labeling Programs and Standards

Edward Vine[1], Peter du Pont[2] and Paul Waide[3]

[1] Lawrence Berkeley National Laboratory
[2] International Institute for Energy Conservation
[3] PW Consulting

Energy-efficiency labels and standards for appliances, equipment and lighting are being implemented in many countries around the world as a cornerstone of energy policy portfolios. They have a potential for very large energy savings and are very cost effective. Once appliance labeling and standards programs have been implemented, however, it is still necessary to evaluate their effectiveness. Evaluation is important to: (1) identify areas of weakness in the program design and implementation so that these can be strengthened; and (2) measure the program impacts on product efficiency, energy consumption, environmental impact, operating costs, and manufacturing/retailing. The latter is important to justify allocating resources to the project and to ensure that it receives sufficient funding to be effective. Policy makers will find evaluation results useful during internal governmental resource allocation battles where they may be asked to prove that a program is saving an appropriate level of resources. An evaluation can be designed at almost any level of resources to meet prioritized needs of time, cost, or accuracy.

Unfortunately, there has been very little post-implementation evaluation of appliance labeling programs, although this is beginning to change. In the United States, most impact assessments of efficiency standards have taken place in the period just prior to adoption of new efficiency standards, during the standards adoption process using forecast information about product shipments and customer use. These evaluations rarely make use of any field measurements, nor do they attempt to systematically examine what would have happened if standards had not been adopted. Similarly, many of the past evaluations of appliance labeling programs have focused on consumer awareness of the label and stated intentions and have not explicitly linked the label to actual behavior (i.e., to the efficiency of the purchased appliance). Some more recent evaluations of appliance labeling programs include data on actual sales and behavior. Examples include evaluations of the European Commission's labeling program and the labeling programs in Austria, Thailand, and the U.S.

Future evaluations of labeling and standards programs are likely to be even more comprehensive as appliance labeling and standards programs are designed as market transformation strategies. As labeling and standards programs are increasingly implemented in developing countries, evaluation is expected to play a critical role in enhancing their effectiveness. In this paper, we describe the types of activities that occur in the evaluation of appliance labeling and standards programs. This paper is based on a more extensive publication that contains references to many of the activities and issues described in this paper (Vine et al. 2000).

1   Planning the Evaluation and Setting Objectives

The evaluation process should begin at the time the process of establishing the labeling and standards programs begins, so that: programs can be designed effectively, data collection can be conducted efficiently, and key stakeholders can be made aware of the importance of the evaluation and become more receptive to the evaluation's findings.

1.1 Steps in the Evaluation

Five steps need to be taken in evaluating appliance labeling programs and standards: (1) define the objectives of the evaluation, (2) determine the resources and data sources needed for conducting the evaluation, (3) collect the data, (4) analyze the data, and (5) disseminate the findings of the evaluation for use in program design. Some of the steps are interactive and, as noted above, the conceptualization of these steps should be incorporated into an evaluation research plan early in the process of designing and implementing energy efficiency labeling and standard-setting programs.

The rest of this paper discusses these steps in more detail.

1.2 Evaluating Labeling Programs vs. Standards

For both labeling and standard-setting programs, it is important to evaluate the process of the program as well as its energy and economic impacts. For appliance standards, an evaluation should focus on the decisions of manufacturers and changes in the efficiency of models sold in the marketplace. While manufacturer decisions are also affected by energy labels, an evaluation of a labeling program should place more emphasis on understanding the sales and purchase process in order to evaluate the impact of labeling on both retailer and consumer decisions. An evaluation of a labeling program will generally involve more qualitative research to understand the process of consumer decision-making and the actions of multiple stakeholders involved in the manufacture, sale, and distribution of appliances. In addition, the impacts of labeling programs occur over a longer period of time and are often more subtle than the impacts of standards which are more abrupt and can occur over a very short period of time.

1.3 The Objectives of Evaluation

An evaluation can focus on the process of a program, or on its energy and demand impact. The best evaluations should contain both process and impact components.

Process Evaluation: Process evaluation is often qualitative in nature and measures how well the program is functioning. Unfortunately, process elements are sometimes seen as less important by policymakers. In reality, these elements are critical to the implementation and success of a program. Process elements include:

- assessing consumer priorities in purchasing an appliance;
- tracking consumer awareness levels;
- monitoring correct display of labels in retailers;
- measurement of administrative efficiency (e.g. registration times etc.); and
- checking and verifying of manufacturer claims (maintaining program credibility).

Impact Evaluation: Impact evaluation is used to determine the energy and environmental impacts of a labeling program. The impact data can be used to determine cost-effectiveness as well. Impact evaluations can also assist in stock modeling and end-use (bottom up) forecasting of future trends. Impact elements include: (1) influence of the label on purchase decisions; (2) tracking of sales weighted efficiency trends; and (3) energy and demand savings.

Impacts can be very difficult to accurately determine, especially for a labeling program. One of the fundamental problems is that once a program such as energy labeling has been in place for some period, it becomes increasingly difficult and hypothetical to determine a "base case" against which to compare the program impact.

Both types of evaluation should occur regularly over the life of a labeling and standards program, especially during the initial implementation. Process and impact evaluations of labels and standards can be conducted under either "resource acquisition" or "market transformation" objectives. Under a *resource acquisition perspective*, the primary objective of evaluation is the calculation of energy and demand savings and greenhouse gas (GHG) emissions reductions from labeling programs and standards (i.e., as an alternative to purchasing energy from a power plant). Under a *market transformation perspective*, the primary objective of evaluation is to see whether sustainable changes in the marketplace have occurred as a result of labels and standards programs.

Under market transformation, program designers are increasingly using theories that contain hypotheses about how the program might effect the market players (*Theory Evaluation*). These program designers benefit from evaluations that test their hypotheses both through interview and by tracking market indicators, which can then be translated to impacts. In addition, there are short-term theories of how a market will evolve such that private actors might shift toward promoting more efficient products in the absence of a program. A theory-based approach, similar to a process evaluation, would test many of the hypotheses presented in this paper such as: "most/some/all consumers will use labels as part of their purchase decisions" or "labels will encourage manufacturers to improve the energy performance of their products."

1.4 Resources Needed for Evaluation

The cost of evaluating appliance labeling and standards programs will vary, depending on a number of factors, such as: the quantity and type of available data and whether energy savings are calculated by engineering estimates or with end-use metering of a sample of products. Most comprehensive evaluations rely on the collection of survey data, sales data, and billing data. The use of end-use monitoring equipment to measure energy consumption for specific appliances will increase the cost of evaluation, as would the purchase of commercially available market research data on model sales. While most evaluation costs occur after a program has been implemented, it is important to allocate some of the evaluation budget for up-front costs when the labeling and standard-setting programs are being discussed and the evaluation research plan is being developed.

2   Collect Data

There are many types of data useful for evaluating the impact of labeling and standard-setting programs and many methods available for collecting such data. The data requirements for evaluating the impacts of labeling programs are similar to those for standard-setting programs in many ways and different in others. For example, label impact evaluations are likely to rely more heavily on consumer surveys although some individual consumer assessment is useful in standard-setting evaluations as well. Where possible, secondary data sources (e.g., industry and government reports) should be analyzed first, since it is the most cost-effective source of information. Afterwards, primary data collection should begin through interviewing and surveys, focusing first on the country's most important data needs.

2.1 Types of Data

A first step in evaluation is to collect model-specific data for establishing a national appliance database. This database will contain information on what models are made, annual sales of each, price and technology characteristics. It can be used to monitor appliance efficiency trends at the national level. Where energy use is analyzed, utility bill data or, in some cases, end-use metered energy data, are collected. Other types of data needed include the attitudes and behavior of key market players and market characteristics (e.g., number of manufacturers and retailers, percent of appliances in stock that are energy efficient, etc.). Finally, it is important to note that it is always possible to carry out some level of evaluation, no matter how crude the data sources and how limited the resources. Evaluators should not be discouraged if they cannot gather data of the highest quality; in those cases, evaluators need to make explicit the compromise in accuracy made to limit cost.

2.2 Data Collection Methods

It is very important to collect data at the beginning of the process of designing and implementing standards and appliance labeling programs. Wherever possible, cooperative agreements with industry should be encouraged to provide data on sales and efficiency levels. Sales data can be obtained from surveys of manufacturers, retailers, and/or contractors. Visual inspection of products in stores can be conducted to assess compliance with labeling programs and to collect information on stocking practices (sometimes by a "mystery shopper" who visits stores unannounced and unidentified). Laboratory testing of appliances can be conducted to measure energy use for specific appliances and for assessing the accuracy of labels. Finally, interviews with consumers, retailers, manufacturers, and contractors often plays a central role in assessing the extent of market transformation.

3 Analyze Data

In order to evaluate resource acquisition and market transformation, a comprehensive analysis should be conducted. While the emphasis of such an analysis is usually on appliance labeling programs, this type of analysis can also be used for the evaluation of appliance standards.

3.1 Baseline

There is a critical need to establish a realistic and credible baseline, i.e. a description of what would have happened to energy use had labels and/or standards not been implemented. Determining a baseline is inherently problematic because it requires answering a hypothetical question: What would have happened in the absence of labels and/or standards? To accurately evaluate energy savings, one needs to analyze energy use for a sample of households/facilities before and after the installation of a product. For example, energy use might be measured for at least a full year before the date of the installation of the appliance and for each of several years after the installation. However, some types of appliances may not require a full year of monitoring. If the loads and operating conditions are constant over time short-term (e.g., one-week) measurements may be sufficient to estimate equipment performance and efficiency. These data would then be used for calibrating engineering estimates for generalizing to the population of energy-efficient products. Frequently load research data is available for establishing baselines for products.

Market characterization studies are also necessary for developing a baseline of existing technologies and practices. Such studies provide detailed data on end users (consumers), including estimates of market size, decision-making analyses, identification of market segments, and analysis of market share by market events (retrofit, renovation, remodeling, replacement). Market characterization studies also provide detailed data on the supply-side -- manufacturers, retailers and contractors (e.g., designers and installers) -- including relationships between supply-side actors, development of segments, business models of each entity, distribution channels, stocking/selling practices, and trade ally reaction to labeling programs.

Baseline development is often highly contentious and, at best, a good guess of what might have been. In many cases, it is as important to quantify the level of efficiency improvement from the time of the program start-up (and preferably from before) in order to demonstrate that progress is continuing to be made.

3.2 Impacts on Consumers

A key point in the evaluation of appliance labeling programs on consumers is the degree to which the label's presence affects consumer purchasing decisions in favor of more efficient appliances. In addition to observing actual consumer purchasing and sales trends, consumer evaluations should also focus on (1) their level of awareness and understanding of energy; and (2) the factors that affect their purchase of the energy-efficient appliances. Specific types of questions to address in this evaluation include the following:

- What is the level of awareness among buyers and potential buyers of the energy label, related product material, and advertising?

- What is the level of importance given to the energy label, related product material, and advertising in the buyer's choice of appliance?
- How well does the customer understand the label, related product material, and advertising?
- What is the customer's perception of usefulness of the label, related product material, and advertising?
- What sorts of changes do consumers propose to the label, related product material, and advertising in order to make them more effective?
- What is the importance of energy or fuel efficiency in the buyer's choice of the appliance? How does this relate to other customer purchase priorities?
- How does the customer use the appliance?
- What are the life cycle cost impacts, accounting for possible changes in the price of the equipment, operating expenses and installation or maintenance expenses?

Socioeconomic data can also be analyzed to help understand the effectiveness of labeling programs and standards for different sociocultural situations: e.g., low-income households versus high-income households, recent purchasers versus general public, etc. Market segmentation can be used to develop education, information and advertising programs that complement the use of labeling programs and standards.

There is an array of econometric and statistical models for analyzing the attribution of many factors to the impacts of programs on consumers. These are generally considered to be advanced evaluation tools and range widely in cost depending on many factors, especially the desired accuracy.

3.3 Impacts on Manufacturers and Retailers

Evaluators assess the impact of labeling and standards programs on appliance manufacturers by examining the following issues:
- Impact on private-sector advertising in support of labeling programs
- Impact on sales (and market share)
- Compliance with labeling and standards programs
- Promotion of labels to retailers (e.g., direct promotion, print advertising, in-house product presentations and training, trade fairs, product catalogues, help desks)
- Direct and indirect costs to manufacturers (increased cost of production, R&D efforts to improve appliance efficiency, distribution of labels, promoting and supporting labeling programs)
- Changes in the production process to produce more efficient models

588

- Similar questions to those posed to consumers (see Section 3.2)
- Distribution of energy labels on appliances in retail outlets

3.4 Impacts from a Policymaker Perspective

Policymakers, typically government and utility companies, are responsible for ensuring that suppliers and dealers comply with labeling programs and standards legislation. Accordingly, evaluation studies assess the current level of manufacturer compliance and the level of remedial enforcement activity. They may also examine the use of formal legal processes to impose penalties on persistent rule-breakers. In many cases, policymakers are responsible for implementing education and information programs that accompany the use of labels or standards. Hence, the depth and breadth of these programs are also evaluated.

3.5 Sales

As noted above, one of the two key "lagging indicators" for evaluation is sales. Market share is considered a lagging indicator because it lags behind those changes that actually cause purchase habits to change. Market share information is critical for the final analysis of a program's effects, but it is often not immediately available during program implementation. Nevertheless, by comparing the sales-weighted trends in appliance efficiencies both before the introduction of the label and after the labeling program was implemented, one can evaluate the impact of an appliance labeling program.

3.6 Energy Savings and Greenhouse Gas Emissions Reductions

At the household or facility level, it is impossible to measure energy savings directly since one would need to know how much energy would have been used if a specific appliance had not been purchased (a counter-factual question). Nevertheless, one can use any of a number of evaluation methodologies for estimating energy savings, especially for a large sample. These include engineering methods, statistical models, end-use metering, short-term monitoring and combinations of these methodologies.

For example, one can estimate changes in market share of energy-efficiency products (sales) and multiply these sales by the amount of energy saved (e.g., on average, or by type of product). Tracking change over time in product and market characteristics gives a good initial indication of the type of market shift that takes place in the early stages of labeling or in the lead-up to a new standard coming into force. Detecting trends in consumer preference toward the more efficient products on the market is a more subtle exercise. Here, both "sales-weighted" trends and consumer sentiment trends need to be monitored. To maximize the accuracy of the energy savings from shifting between any two models, one can meter a sample of products in situ to determine the actual amount of energy used.

At the national level, energy savings can be calculated using simple calculations (e.g., spreadsheets) or detailed energy end-use models. The assumptions used in the engineering analyses are adjusted to account for real-world data (e.g., actual consumption in the field, fraction of households owning a particular appliance, usage (hours per year)) from surveys and end-use monitoring.

Once the net energy savings have been calculated by subtracting baseline energy use from measured energy use, net GHG emissions reductions can be calculated in one of two ways: (1) average emissions factors can be used, based on utility or non-utility estimates; or (2) emissions factors can be calculated based on specific generation data. In both methods, emissions factors translate consumption of energy into GHG emission levels. Normally, the use of average emission factors is accurate enough for evaluating the impact of energy efficiency labels and standards. In cases where the other impact analyses are highly sophisticated and regional variations are important, using plant specific factors may be warranted.

In contrast to using average emission factors (method #1), the advantage of using calculated factors (method #2) is that they can be specifically tailored to match the characteristics of the activities being implemented by time of day or season of the year. For example, if an appliance labeling program affects electricity demand at night, then baseload power plants and emissions will probably be affected. Since different fuels are typically used for baseload and peak capacity plants, then baseload emission reductions will also differ from the average.

The calculations become more complex and more realistic if one decides to multiply the emission rate of the marginal generating plant by the energy saved for each hour of the year, rather than multiply the average emission rate for the entire system (i.e., total emissions divided by total sales) by the total energy saved. For the more detailed analysis, one must analyze the utility's existing system dispatch and expansion plans to determine the generating resources that would be replaced by saved electricity and the emissions from these electricity-supply resources.

One would have to determine if the planned energy-efficiency measures would reduce peak demand sufficiently and with enough reliability to defer or eliminate planned capacity expansion. If so, the deferred or replaced source would be the marginal expansion resource to be used as a baseline. This type of analysis may result in more accurate estimates of GHG reductions, but this method will be more costly and require expertise in utility system modeling. In addition, this type of analysis is becoming more difficult in those regions where the utility industry is being restructured. In restructured markets, the supply of energy may come from multiple energy suppliers, either within or outside the utility service area and the marginal source of power is more difficult to forecast.

3.7 Compliance

In many labeling programs, it is the responsibility of manufacturers to ensure that the information they supply is correct. Often, there is no automatic system of independent testing. Occasionally, third-party testing agencies are used. Generally, manufacturers test their own products in certified test laboratories and report the testing results on the label. Such a system can work well by relying on challenges by manufacturers, who can question the veracity of a competing manufacturer's claim. This system of self certification and challenges is used in the U.S. Under the European Union (EU) legislation, it is the responsibility of each member state to ensure that EU law is applied and enforced in their state. Since the labeling scheme was introduced in the European Union, some serious cases of inaccurate energy consumption reporting have occurred for refrigerators and freezers. Hence, there is a need to evaluate how the accuracy of manufacturer-reported energy consumption compares to third-party laboratory testing, as well as to field monitoring of the energy usage in order to determine if the appliance rating and label should be changed. Appliance labeling programs also depend on the efforts of retailers to make sure appliance labels are attached to appliances for consumers to read. Hence, it is imperative for evaluators to also assess compliance at the retailer level.

4 Apply Evaluation Results

The use of evaluation results is a critical component of the evaluation process. If a technically sound evaluation produces significant results, it is imperative that these results be used, where appropriate, to (1) refine the design, implementation and evaluation of appliance labeling programs and standards, (2) support other energy programs and policies, and (3) support accurate forecasting of energy demand for strategic planning.

4.1 Refining the Labeling and Standards Programs

The results from evaluations can be used to improve the design, implementation and future evaluations of labeling and standards programs. For example, the evaluation results can be used for reexamining the accuracy of the inputs used in designing the program. In addition, one can assess if the labeling and standards programs can (or should) be extended to other appliances that are not currently in the program.

4.2 Supporting Other Energy Programs and Policies

The evaluation of labeling programs and standards can be used for designing appliance rebate programs, appliance standards or negotiated agreements (if none exist), procurement actions, and labeling programs for other appliances. Ideally, the program designers become the clients of the evaluation department, and the evaluation results feed directly into the next round of program design or improvement.

4.3 Forecasting Energy Use and Strategic Planning

Evaluation results can be used to support forecasting and resource planning, but with caution. In particular, the following elements of an evaluation should be considered prior to attempting to use the evaluation results: (1) the representativeness of the study sample to the population of interest to planners; (2) the accuracy and precision of energy and demand impact results; and (3) the appropriate use of control samples. Nonetheless, if as part of the evaluation process, comprehensive data on market energy efficiency trends, sales volumes and usage patterns have been established, these data can be used as inputs to an end-use stock model to make long-range energy consumption and emissions forecasts. This kind of forecasting is useful to guide policy development, as it enables the estimated impact of various policy and implementation changes to be simulated in advance.

Standards and labeling program planners have a strong interest in the evaluation process. Achieving evaluation results by defining objectives, identifying the necessary resources, monitoring the program performance, and assessing the impacts is a valuable output of a standards and labeling program. The results can be used to revise an existing program's objectives or as building blocks in establishing a new program. But, the difficulty in measuring program's performance and impacts is ever present. In some cases it is due to a lack of data or a lack of resources to obtain that data. In others, it may be that the program's direct results are masked by the effects of other complementary programs that are taking place at the same time.

5 Reference

[1] Vine, E., P. du Pont, and Waide, P. 2000. "Evaluating the Impact of Appliance Efficiency Labeling Programs and Standards," in S. Wiel and J. McMahon (eds.), 2000. *Energy Efficiency Labels and Standards: A Guidebook for Appliances, Equipment and Lighting.* Washington, D.C.: Collaborative Labeling and Appliance Standards Program.

New for Old: Redesigning the Australian Energy Label

George Wilkenfeld

George Wilkenfeld & Associates, Energy Policy and Planning Consultants

Abstract. Energy labelling for household appliances was originally introduced in Australia in 1986. The program now covers refrigerators, freezers, dishwashers, clothes washers, clothes dryers and air conditioners, all with a similar "star" label design. Over 90% of appliance buyers recognise the label, and studies have found that average appliance energy efficiency has increased significantly as a result of labelling.

Even so, the program is losing its impetus. Appliance efficiency increased so rapidly that many products rate 5 stars, and even the maximum rating of 6 stars. The commercial incentive for suppliers to introduce more efficient products is now much less, since they know that buyers are satisfied with 4 or 5 star products.

For these reasons, Australian federal and State governments are relaunching the labelling program with a redesigned label, on which the most efficient models, which currently rate 5 or 6 stars, will rate no more than 3 stars. The changeover is taking place in 2000, following several years of consultation with industry and market testing of new label designs. This is the first known example of a successful program undergoing change of this magnitude, in order to maintain its effectiveness.

This paper describes the reasons for the redesign of the label, the complex and lengthy process necessary to bring it about, and the forecast costs and benefits.

1   The Development of Labelling in Australia

In Australia, steps towards the energy labelling of household electrical appliances were initiated by the New South Wales (NSW) State government in the early 1980s. In 1983 the following objectives for labelling were formally endorsed by all nine energy ministers of the Commonwealth, State and Territory governments:

- "to enable the consumer to make an informed choice between energy consuming products (a higher initial purchase price may be offset by accumulated energy cost savings over the appliance's lifetime);

- to provide an incentive for manufacturers in the medium term to design and market appliances with improved energy performance, and consequently better tailored to consumers' requirements;

- to promote energy conservation on a national scale and to retard growth in energy demand" (NECP 1983).

Informing consumer choice and providing incentive for manufacturers remain central objectives of energy labelling. The third objective has been progressively expanded to include other measures, such as minimum energy performance standards (MEPS) and the aim of reducing greenhouse gas emissions, which was not a public policy issue in 1983.

In 1985, after unsuccessful attempts to introduce a nationwide voluntary scheme, the NSW and Victorian governments regulated for the mandatory energy labelling of refrigerators, freezers, dishwashers and air conditioners, to be phased in from December 1986. Since the two States have about 60% of Australia's population, all suppliers had to participate and the program was effectively national from the beginning. In 1990 Victoria introduced labelling for clothes washers and dryers.

In the following years more States and Territories legislated for labelling, and it became national in a formal as well as an effective sense. However, it relies to this day on complementary legislation in each of the six States and two Territories. The Commonwealth government works with the State and Territory energy agencies to coordinate the program but has no regulatory powers in this area, because under the Constitution functions such as energy supply are State responsibilities.

When labelling first began, the energy tests and the form of the label (Figure 1) were all described in detail in State regulations, either because there were no appropriate Australian Standards, or because the regulators did not want to refer to Standards which could change without their control.

As each successive State adopted mandatory labelling, it tried to replicate the content of the other State's regulations, but there were inconsistencies because of different regulatory frameworks and styles. The program worked reasonably well as it was, but change was almost impossible since all the States would have had to change their technical provisions at the same time and in a coordinated manner.

2   The Need for Change

The need for change to the energy labelling program was identified as early as 1991, when the first program evaluation was carried out (GWA et al 1991). The main finding was that energy labelling has been successful because suppliers perceived a commercial value in having 5 star products, and had adopted the policy of engineering their new models to reach 5 stars where possible.

Efficiency had increased more rapidly than was expected when the label scales were set in 1985, and by 1991 there were 5 or 6 star products in nearly every category. Once products reached 5 stars, the incentive for suppliers to introduce still more efficient models was much reduced. When most products are "crowded" near the top of the scale, the effective product choice facing buyers is narrowed and the impact of the label on purchase decisions declines. Market research showed that most buyers expected 5 stars to be the best, and were often surprised when they found 6 star products in the showrooms. This was understandable, because of quality associations with "5 star hotels" and "5 star service".

The 1991 evaluation also found many technical areas where the energy tests and the "rating algorithms" (the mathematical rules for translating energy test results into star ratings) could be improved. However, it concluded that the necessary changes could be made without first setting up a new national coordination framework. In the event, the process took nine years.

3   Details of the Proposed Changes

A National Appliance Energy Labelling Co-ordinating Committee of officials, and an advisory committee of industry and consumer representatives, were established in early 1992. The revision of the label design and algorithms still took a further eight years, because of the need to develop a consensus amongst the many jurisdictions and stakeholders involved.

The following stages were necessary:

1   Revision of the regulatory and administrative structure of the program: ie agreement by the States and Territories to enact common regulations and to transfer the technical content from regulations to Australian Standards;

2   Review of the test procedures in the Australian Standards, and correction of identified anomalies;

3   Development of new label designs and market testing with consumer focus groups. A number of options were tested including extending the scale to 10 stars, and going to a 5 star rather than a 6 star scale. A streamlined 6 star design with distinct half star steps emerged as the clear preference (Figure 2);

4   Consulting with interested parties on the proposed changes;

5   Revision of the Australian Standards to include the new algorithms (ie star rating scales) and the new label designs;

6   Planning a transition process to minimise disruption and cost to manufacturers and retailers, and to minimise consumer confusion.

Much of this program development work was carried out during 1998 and 1999, following the establishment of the Australian Greenhouse Office.

The new label (Figure 2) is similar to the existing design in order to build on the already high level of consumer recognition and acceptance, but with sufficient changes in shape and text to ensure that consumers do not confuse new with old.

The new label retains a 6 star scale but with different rating algorithms, so that a product which currently rates, say, 5 stars will rate only 3 stars on the new scale. This is intended to renew the commercial incentive for suppliers to introduce new 4, 5 and 6 star products, which will be significantly more energy efficient than those which rate highly on the present scale. During the transition period the new labels will also state what the rating would have been on the previous scale. This will appear in the green band, which will eventually be discontinued.

Information based on assumptions that have been found to be difficult to sustain (eg that the annual energy consumed by an air conditioner purchased anywhere in Australia can be reasonably approximated as 500 hours of operation at full load) is to be removed, or presented in a different way (eg the energy value on the air conditioner label will be "kWh consumed per hour of operation at full load").

The guidelines for presentation of partial stars on the present label are unclear, and some suppliers have printed labels in a way that arguably exaggerates the comparative star rating of the product. The new label will allow only full stars or half stars, and so should reduce the prospect of consumers being confused or misled (the example in Figure 2 is a 2.5 star product).

The revision of the label scales gives an opportunity to make changes to the energy consumption tests on which the labels are based. These changes will generally improve the repeatability of the test, and in the main are small enough so that models previously tested need not be retested – it will be possible to produce the new label from the original test results. The exception is dishwashers, where there are major problems of test repeatability and of ambiguity in the specification of the cycle selected for the label data. This reduces the ability of the label (whether in the existing or the new format) to fairly indicate comparative energy performance to dishwasher buyers. A new test standard to address these issues is planned in early 2001. All dishwasher models remaining on the market will need to be retested to the new standard.

The proposed changes are being included in the Australian Standards which now describe energy testing and labelling requirements for each product type, and which – since the regulatory framework was streamlined - are referenced in each State and Territory's labelling regulations. These Standards are prepared with industry and public input, but cannot be changed without government agreement.

It is intended that the transition from the existing to the new label designs will be managed through the timing of Australian Standard revisions and through the product registration process.

4   The Transition Process

Under the State and Territory regulations, no refrigerator, freezer, dishwasher, clothes washer, clothes dryer or air conditioner can be sold unless a label has been registered for that model. All existing label registrations will expire on 1 October 2000. The only mechanism previously available for retiring registrations was voluntary notification, and it is estimated that about 70% of the models currently registered are no longer on the market (see Table I). In future, all registrations of "new" labels will expire after a period of 5 years from the date of registration, unless the suppliers renew.

Between 1 April 2000 and 30 September 2000 (the "overlap period") suppliers will have the option of registering new models with either the old or the new label, but if the old label is used the registration expires on 1 October 2000. After 30 September 2000 only registrations of the new label will be accepted.

Table I: Appliance Models On Labelling Registers, April 1999

	Models on register at 30 Sept 1999	Models on sale, April 99	Registered but not on sale, April 99	Current/all registered
Refrigerators and Freezers	1341	322	1019	24%
Dishwashers	400	139	261	35%
Air-conditioners	1957	681	1276	35%
Clothes Washers	642	152	490	24%
Clothes Dryers	220	53	167	24%
All registered models	4560	1347	3213	30%

Source: Energy Efficient Strategies, Personal Communication

The major potential source of buyer confusion is the possibility of seeing models with old labels next to models with new labels in the same showroom. At first glance the new label models could appear to be less efficient than the old label models because they will display fewer stars for the same level of energy efficiency. In order to minimise the possibility there will be a "display transition period" between 1 July and 30 September 2000. Government will work with appliance suppliers and retailers to try to ensure that:

- Where a unit remains on the showroom floor after 1 July 2000, the retailer will stick the new label over the old label. This will require coordination between retailers and suppliers, who will have to provide the new labels;

- Whenever a model is put on display after 1 July 2000, the retailer selects a new labelled unit from the packaged stock in preference to an old labelled unit. This will require retail staff to take more care in selecting floor stock (and would be greatly assisted if the supplier marks the cartons in some way).

It remains to be seen how successful the transition process will be. There will inevitably be some showrooms where both label types are on display for significant periods, and some where old labels remain on display after 30 September 2000. There will also be some errors when labels are over-stuck. These occurrences could disrupt the use of the label in the product selection process, but to a limited degree and for a limited period.

Furthermore, some consumers may select a model on the basis of a new label in the showroom but have a unit with the old label delivered from the warehouse. This would not impact on the selection process, but may generate some follow-up inquiries to the retailer, the product supplier or government authorities. If the reverse should occur (ie selection on the basis of the old label but having a unit with a new label delivered) the chance of confusion would be less, because of the explanations and the additional contact information (government freecall number and Internet site address) printed on every new label.

It is impossible to estimate a monetary cost for these temporary disruptions to consumers. On the one hand, consumers who visit showrooms where the display transition is not well managed may find it more difficult to take energy efficiency into account in their purchase decision, and may purchase a somewhat less efficient model than otherwise. On the other hand, noticing a new label could increase customer interest in energy efficiency, even if there are many old labels in the same showroom. For customers who use leaflets or the Internet to compare product energy efficiency, the task will be made much easier by the removal of obsolete registrations.

5 Projected Costs and Benefits

The cost of the label changeover is estimated as the sum of the following factors (all values are Australian dollars):

1. Costs to suppliers of re-registration of models (two re-registrations in the case of some dishwasher models). This is estimated at $250 per model ($150 registration fee and $100 internal administrative costs).

2. Costs to suppliers and/or retailers of verifying labels, re-labelling units or selecting new-label units for display during the transition period (and second transition period for some dishwasher models). This is estimated at A$10 for every unit displayed in appliance showrooms during the transition period.

3. Costs to suppliers of retesting dishwashers. This is estimated at $5,000 per model tested (three tests are needed per model).

4. The costs to government of supporting the new label development ($100,000, already expended) and the display transition period ($250,000, budgeted).

The Australian Greenhouse Office commissioned the author to estimate these costs from the perspective of appliance buyers (GWA 1990a). Business compliance costs and administrative costs were distinctly identified. It was assumed they would be passed on to appliance buyers, and marked up in the same proportion as the ratio of retail to wholesale price, ie a factor of 2. Therefore the total cost to consumers of elements 1 to 3 above is estimated as twice the calculated amount. Government costs are not marked up.

In order to calculate the number of models subject to re-registration, it was necessary to estimate the average number of new models registered each month, the rate at which models become obsolete, and the number of units which will incur re-labelling (or label selection) costs during the three month display transition period. It was estimated that 10% of units sold pass through showrooms and incur label replacement costs, while 90% are delivered direct from warehouse to buyer and need no label replacement.

Table II summarises the changeover costs from the perspective of appliance buyers. Nearly 90% of the total estimate of $3.2 million (undiscounted) is supplier costs passed on to buyers, and the rest government administration costs. The costs equate to about $1.63 per appliance sold if spread over one year, and $0.14 per appliance sold if spread over 10 years.

The label design changes in 2000 are the first in the 15 years of the program, and it would be reasonable to assume that there would not be another complete label redesign for another decade at least. However, there may be a need to change certain aspects of the label when there are changes in the MEPS levels affecting labelled appliances, as may occur every 4 to 5 years, and this may involve some re-registration or re-testing.

Table II: Estimated Changeover Costs from Consumer Perspective

	Total costs to suppliers	With retail markup	Share of total costs	$/unit sold in one year	$/unit sold in 10 years
Refrigerators & Freezers	$ 247,000	$ 494,000		$ 0.66	$ 0.05
Dishwashers	$ 625,300	$ 1,250,600		$ 6.95	$ 0.58
Air-conditioners	$ 256,167	$ 512,333		$ 2.05	$ 0.17
Clothes Washers	$ 175,000	$ 350,000		$ 0.70	$ 0.06
Clothes Dryers	$ 119,583	$ 239,167		$ 0.85	$ 0.07
All types	$ 1,423,050	$ 2,846,100	89.1%	$ 1.45	$ 0.12
Label design and market research		$ 100,000	3.1%	$ 0.05	$ 0.00
Publicity during transition		$ 250,000	7.8%	$ 0.13	$ 0.01
Grand total		$ 3,196,100	100.0%	$ 1.63	$ 0.14

Because of the timing of the necessary legislative changes, the major modelling of the costs and benefits of enhancing the labelling program had to be undertaken before the transition to the new label had been planned and costed in detail (GWA 1990). Therefore a preliminary cost assumption was made, which in the event was about 8 times as great as the final cost estimate in Table II.

Of the many scenarios modelled, the two relevant ones were:

A. The baseline, which assumed a continuing decline in the ability of labelling to drive efficiency improvements, due to factors such as "crowding" at the top of the star rating scale.

B. An "enhanced labelling" scenario, in which the labelling program again drives the rate of energy efficiency improvement at the rate it did when originally introduced.

The differences in the energy costs, average product prices, administrative costs and greenhouse gas emissions between the two scenarios, which together define the costs and benefits of label enhancement, are summarised in Table III. The values relate to the projected costs and lifetime energy consumption of all labelled appliances to be sold in Australia over the period 1999-2015 inclusive.

The preliminary estimate was that the costs of testing, labelling and program administration in the enhanced labelling scenario would be about $ 27 M higher than in the baseline scenario. As Table II shows, the estimate of the changeover cost is about $ 3.2 M. This means that a further $ 24 M could be spent on other supporting measures such as advertising and promotion, without exceeding the enhancement cost estimates in the model.

Table III: Modelled Benefits and Costs of Labelling Enhancement

		A. Baseline: declining labelling effectiveness	B. Enhanced labelling effectiveness (with new label)	Difference (B – A)(a)
Lifetime energy	GWh	256,004	228,003	-28,001 (b)
Lifetime CO$_2$	kt	237,990	211,962	-26,028 (b)
NPV, Energy costs	$M (1998)	25,947	23,183	-2,764 (b)
NPV, Purchase costs	$M (1998)	35,990	36,990	1,001 (c)
NPV, Labelling costs	$M (1998)	83	110	27 (c)
Total NPV (0% discount)	$M (1998)	62,020	60,284	-1,737 (b)

Source: GWA 1999. (a) Both scenarios include the impact of mandatory MEPS for refrigerators, freezers and electric water heaters, so the difference between the two scenarios is due solely to the increased effectiveness of energy labelling. (b) Negative value indicates greater benefit from enhanced labelling (c) Positive value indicates greater cost from enhanced labelling.

It should be noted that the additional administration costs are in fact only a minor part of the projected costs of the enhanced labelling scenario. By far the larger part is the estimated $ 1 billion extra that consumers are expected to voluntarily spend on appliances, as a result of greater preference for more efficient models due to the greater impact of labelling on product choice. Even so, this equates to less than 3% increase in the average purchase price of appliances.

From the modelling, the benefit of label enhancement is the net present value (NPV) of the projected energy saving over the period 1999-2015 through the purchase of more efficient appliances than otherwise. The cost is the NPV of the additional appliance cost plus and the additional labelling administration costs The undiscounted benefit/cost ratio is 2,764/(27+1,001) = 2.7. The ratio at 4% discount rate is 2.2, and at 8% discount rate, 1.8.

Because the administration cost is such a small proportion of the additional cost of the enhanced labelling scenario, the benefit/cost ratio is insensitive to changes in these cost assumptions. For example, if the cost estimated for the label changeover ($ 3.2 M) were used instead of $ 27 M, the benefit/cost ratio would only change from 2.7 to 2.8 (2,764/1,004). However, it is doubtful whether the full benefit of the changeover can be obtained without additional promotional and publicity effort – the cost of which has not been included in Table II - so some additional costs should be anticipated.

If a monetary value were assigned to the CO$_2$ emitted in generating the electricity consumed by appliances, the benefit/cost ratios of enhancing the effectiveness of labelling would increase further.

6   Conclusions

Energy labelling is a dynamic program, which changes the appliance market and is in turn affected by those changes. The Australian energy labelling program has been very successful, but signs of decline in its impact were detected as early as 1991, the 6th year of its operation. In 2000, the 15th year of operation, changes are being made which could restore the impact of the program, for some years at least, before it declines again for the same reasons as before. The costs of these changes are estimated to be modest in comparison with the projected benefits.

Energy label designs have been changed before in other countries, but usually in response to buyer confusion or a lack of impact on the market. This is the first known example of a successful program undergoing change of this magnitude, in order to maintain its effectiveness.

7   References

[1]   NECP 1993, *Consolidated Papers on Energy Labelling*, National Energy Conservation Program, Australian Minerals and Energy Council, Canberra
[2]   GWA et al 1991, *Review of Residential Appliance Energy Labelling*, George Wilkenfeld and Associates, with Test Research and Artcraft Research. Prepared for the State Electricity Commission of Victoria, September 1991
[3]   GWA 1999, *Regulatory Impact Statement: Energy Labelling and Minimum Energy Performance Standards for Household electrical appliances in Australia.* George Wilkenfeld and Associates, with assistance from Energy Efficient Strategies, February 1999.
[4]   GWA 1999a, *Regulatory Impact Assessment: Energy Labelling and Minimum Energy Performance Standards for Household electrical appliances in Australia. Supplementary cost-benefit analysis on transition to a revised label*, George Wilkenfeld and Associates, November 1999.

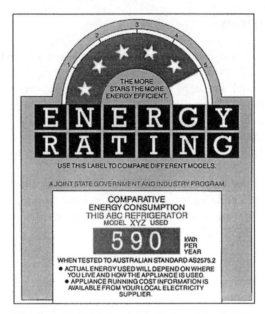

Figure 1: Original Australian Energy Label, 1986

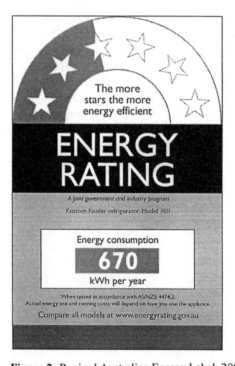

Figure 2: Revised Australian Energy Label, 2000

Testing of International Appliance Labeling Approaches with U.S. Consumers

Christine Egan

American Council for an Energy Efficient Economy

Abstract. U.S. efficiency advocates are increasingly reaching consensus that the U.S. EnergyGuide label is confusing to consumers and has little impact on purchase decisions. This paper discusses the results of primary research with U.S. consumers to test labeling approaches used or proposed elsewhere in the world. In focus groups and semi-structured interviews, five graphical designs were tested with U.S. consumers, including: a European-style letter-based graphic; an Australian-style star-based graphic; a speedometer-style graphic; a thermometer-style graphic; and the current U.S. bar-based label. This paper focuses on the comparative comprehension and preferences of American consumers regarding each of these labeling approaches and the discrete informational elements. Overall findings and cross-task analysis will be presented along with a discussion of task-by-task progress and findings. The graph modifications suggested by study participants to improve the labels are of particular interest.

1 Background

Labels depicting the energy use of home appliances are in use in many countries around the globe as part of national demand-side management and market transformation programs targeted toward reductions in energy consumption. Appliance labels typically fall into one of two categories regarding their approach to information organization—categorical or continuous. A categorical label divides the range of comparative models into distinct groups or segments. A continuous label marks the low and high end of the range of comparative models without explicitly grouping anything in between. Categorical labels are in use in Europe, Australia, Brazil, Thailand, and a few other Asian countries. Continuous labels are currently in use in the United States and Canada. Over the past several years, the trend internationally has been toward categorical labels.

Appliances in the United States have been labeled with an EnergyGuide label since 1980. The current U.S. label is shown in Figure 1. Prior research on the U.S. label suggests comprehension problems and a low level of use (du Pont 1998). In addition,

604

over the last five years alternative approaches to appliance labeling have been developed and implemented elsewhere in the world with impressive results in terms of consumer awareness, market impacts, and energy savings (Sulyma et al. 2000, du Pont 1998, Harrington 1998, Boardman et al. 1997, Waide 1997, and Wilkenfield 1997). In this context, ACEEE, with input from other organizations, decided it would be useful to evaluate the efficacy of the current EnergyGuide label and determine the best label format and graphical element for U.S. consumers

Figure 1: U.S. EnergyGuide Label

2   Introduction

The primary goal of ACEEE's appliance labeling project is to determine the best label format (e.g., bars versus letters, etc.) and critical informational elements for U.S. consumers (e.g., operating cost and/or annual kWh). A secondary goal is to uncover the opinions of other market actors (e.g., retail sales staff, manufacturers, and contractors) who come into contact with the label regarding the program efficacy and optimal label format. This paper will discuss only the results of the research with consumers.[1]

Different research methods have been used at various points in the project depending upon the specific task objectives. Overall, consumers were the highest priority as they are the primary audience and end-user of the label. Thus, a multi-method and sequential design was constructed to elicit their feedback. An initial round of consumer focus groups was conducted to gather "broad-brush" and directional feedback on the current label in a side-by-side comparison with the alternate displays. Overall, label preferences and opinions of various informational elements were emphasized. The groups led to improved graphical designs that were then tested in semi-structured interviews, which focused upon testing comprehension and interpretation of the various labels and specific informational elements along with the reasons behind preference-related statements. Various interpretive enhancements to the labels emerged from the interviews and were incorporated in the graphics used in another round of focus group testing. This second round of focus groups was intended to select the optimal designs for testing in quantitative research. All of the tasks mentioned thus far have been completed as of the writing of this paper and results are discussed in each task's section. The remaining tasks are discussed in the section on Conclusions and Next Steps.

3   Consumer Focus Groups—Round One

ACEEE contracted for an initial round of six consumer focus groups (four with white-good shoppers and two with larger household equipment shoppers) to examine consumer perceptions of the current EnergyGuide label and responses to alternative label designs. Five graphical designs were tested, including: the current style; a European-style letter-based graphic; an Australian-style star-based graphic; a speedometer-style graphic; and a thermometer-style graphic. The latter two styles had tested well with consumers in a Canadian research effort (Patterson 1991). Pairs of labels were shown side-by-side, one representing a low energy-using case and one a high energy-using case. The labels used in the focus groups were left very close to their original format. For example, the background of the letters-based European-style label was left white and the Australian-style stars-based label was left in it's dial-shaped format and color.

The focus groups began with an introductory conversation about appliance purchasing and the importance of energy efficiency. The results verified other studies in finding that energy efficiency is not reported as a major factor in U.S. consumer purchasing of appliances. They further indicated that U.S. consumers believe: (1) energy-efficient equipment is too expensive and it takes too long to recoup energy savings; (2) everything made today is energy efficient, particularly as compared to older products being replaced; (3) the differences in the energy efficiency of the models available on the market are small within product categories; and (4) white goods (versus HVAC and larger equipment) are not seen as using very much energy, with refrigerators being the most common of white appliances for which energy efficiency is relevant.

The groups found that the current label, though well-known and recognized as a source of information on energy use, was not always read or used in appliance purchase decisions. Strengths of the current label were its eye-catching and familiar bright yellow color; its official-looking appearance, and its informativeness. However, respondents also found the label to be cluttered, poorly organized and overly technical or complex. In particular, respondents found the bar graph difficult was not obvious or eye-catching and moreover difficult to understand. Improvements suggested by interviewees included: (1) reducing the amount of unnecessary text; (2)

improving the graph so that it more clearly conveys the model in question's annual kWh usage; and (3) more clearly labeling and highlighting the operating cost figure, which was considered to be among the most important informational elements on the label.

The letters label was most preferred overall. Participants liked that the label for the following reasons: it was eye-catching because of the variety of colors used; related information was grouped and blocked off into outlined spaces; the graphic was easy to recall at a later date (e.g., that model I liked was an "A" and this one was a "B"); and for some participants the graphic was easy to understand because letters are the standard system of grading in school. A number or respondents, however, were confused by the letters graphic. These respondents found it difficult to make sense of the fact that the shorter bar (labeled "A") was on top with a label "most efficient," while the longer bar (labeled "F") on the bottom was labeled "least efficient." Longer bars relating to less and shorter bars relating to more was confusing and counter-intuitive and only exacerbated an existing problem in comprehending the inverse relationship between energy use and energy efficiency. Also, some of the interviewees felt that the colors, though eye-catching, were also distracting from the main message of the graph (i.e., the model's relative energy consumption on a scale from least to most efficient). Finally, some felt that this presentation was less scientific or official-looking than the current label because of the use of multiple colors, leading to questions about the perceived credibility of this approach.

The least preferred label was the stars label. An important reason for its low ranking was that it was the only one of the alternatives not to include an operating cost figure. Operating cost appears to be one of the two critical piece of information that respondents feel should be on the label (annual kWh is the other factor). The participants were adamant that this information needed to be added to the label. Also, the color combination of the label (muted yellow and orange) was poorly received. Respondents felt it wasn't very eye-catching and looked dated. As in the other labels with color changes, it was recommended the label be returned to the expected bright yellow color. Finally, some respondents felt that stars could be misinterpreted, with more stars meaning more energy use and therefore being a lower-rated model. For others, the star rating system was a strength because of its intuitive nature and relationship to movie or hotel ratings. To avoid misinterpretation, participants recommended that the stars label needed a clearer statement that the scale was based on energy efficiency, not energy usage.

The thermometer and speedometer label fell in the middle in the analysis of respondents' label preferences. The strengths of the thermometer were its use of bright yellow (which respondents already associate with the EnergyGuide), the bold lettering for the annual kWh information, and its representation of a familiar device (i.e., a thermometer) for measuring differences. The strengths of the speedometer were its use of bold lettering and (for a minority of respondents) it's blue color. Both the thermometer and speedometer needed graphical design improvements. In the case of the thermometer, the EnergyGuide logo, which had been running alongside the label vertically, needed to be repositioned horizontally as in the current label. More importantly, the thermometer needed to be redesigned to look more like a thermometer to ensure that customers focused on the filled-in black space (not the white space) as the indicator. The speedometer needed to be redesigned to make the indicator look more like an arrow and to more equally distribute the interim tick marks along the semi-circle that formed the foundation of the speedometer. Also, some of the respondents recommended that the bright yellow of the existing label replace the blue color.

Taking into account these label-specific as well as other overall comments, the groups suggested that an ideal EnergyGuide label would:

* be formatted and outlined to communicate its messages using blocked-off spaces and relationally grouped information;
* reduce the amount of unnecessary text;
* include and highlight the estimated annual operating cost and also the annual kWh so they can be easily seen;
* use a visually appealing graphic that simply and clearly communicates the kWh usage; and
* use the color yellow as a background, as this is associated with energy information.

4 Consumer Interviews

ACEEE completed a total of 54 semi-structured customer intercept interviews in three cities: Boston (28 interviews), Denver (18 interviews), and Dallas (8 interviews). The interviews in Boston and Denver were with customers shopping for white-good appliances (refrigerators, freezers, dishwashers, clothes washers, or room air conditioners). The interviews in Dallas were with customers shopping for water heaters. In each of these interviews, five improved label designs were tested (based on the prior focus groups) depicting a high energy consuming model. Examples of the alternative graphics shown are found in Figure 2. The purpose of this set of consumer interviews was to evaluate the current label in-depth and side-by-side with alternative labeling approaches to draw out comprehension and information processing-related issues as well as to examine the reasons behind reported label preferences.

The majority of the interviewees (roughly 75 percent) correctly interpreted the single graph that was presented to them first as a test of comprehension (i.e., they deduced that the model depicted was a high energy using model). Conversely, this means that approximately one quarter of respondents did not understand the graph that was presented initially. The current label had the highest rate of misunderstanding while the speedometer graph had the lowest. The star, thermometer, and letters graph fell in the middle. Also, while the majority of participants were able to deduce that the model depicted was not very energy efficient, fewer could articulate or use the graph's comparative element. Several of the consumers interviewed saw or used only the label's individual model information. In some cases, this meant that the consumer was unable to determine that the model shown was relatively inefficient (this was categorized as non-comprehension). In other cases, the interviewee was able to make that judgement without a clear awareness of the relative nature of the graph (this was categorized as incomplete comprehension). There were also cases where the interviewee who experienced complete understanding expressed a desire for external comparisons (i.e., to walk from model to model and compare label data) to verify their interpretations and the labels themselves. The problem with understanding the comparative nature of the graph appeared to be more common with the current label format than any of the other options. Comprehension also seemed to be complicated by the interviewees' perception that an annual operating cost of ($63) was simply not that much money and therefore

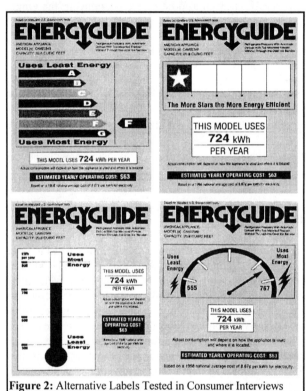

Figure 2: Alternative Labels Tested in Consumer Interviews

could not be associated with an energy-inefficient model. Finally, a few of the respondents mistook the operating cost figure for a savings number. However, this previously identified comprehension problem (du Pont 1998) was less frequent than expected, perhaps because modifications were made to all the designs except the current label to more clearly identify the operating cost number.[2]

No clear winners emerged from the analysis of the interviewees' preference-related statements. Given the small sample and qualitative nature of the questioning, the reasons behind interviewees' opinions of the labels are perhaps more interesting and significant than their selected preferences. The current label received high marks for being informative and familiar but low marks for being wordy, busy and graphically unappealing. Interestingly, the most common comment about the current label (that it contained a lot of information) was seen as both a strength and a weakness. However, far more negative comments were made regarding this aspect of the label than positive. The stars graph received high marks for being motivating and quick in conveying its message, but a subset of interviewees gave it low marks for ease of understanding. Often these interviewees had made the mistake of interpreting the stars as an energy-consumption scale (less stars equals less energy use, more stars equals more energy use) rather than as an energy efficiency rating system. For some participants, another perceived weakness of the star label was its lack of numeric kWh range information (from low to high). The letters graph received high marks for being colorful and attention-grabbing but low marks for being busy and, for some, difficult to understand. For example, a subset found confusing the presence of multiple scales for measuring energy use (varying bar lengths, colors, and letters). Others commented that the graph was inverted and that higher energy consumption should be on top, not on bottom. As with the stars label, some interviewees felt that a weakness of the letters label was its lack of kWh range information. The thermometer received high marks for the clarity of its visual element but low marks from some interviewees for ease of understanding. Many of those who had difficulty understanding the graph felt that the scale was inverted and that the top of the thermometer should be the most efficient product, not the least. The speedometer received high marks for its clarity as a graphical indicator but low marks for its visual attractiveness. Several interviewees indicated a grouped preference for the thermometer and speedometer labels. The grouping of these two labels in interviewees' preference-related statements was so strong that in some cases these labels almost seemed interchangeable and it appeared that some interviewees would be equally satisfied with either option.

As in the consumer focus groups, the interviewees made various suggestions for improving the label designs. Among the overarching comments were a range of suggestions regarding the use and presentation of operating cost data. For example, a few interviewees felt that operating cost should be a part of the comparative graphics. These comments emphasized the importance of operating costs to the participants. Another area of overarching comment was in the expression of the basis for the comparison. Several participants suggested that at a minimum all of the labels should indicate clearly that the comparison was based upon a range of similar models.

Graph-specific suggestions were made as well. Suggestions for improving the current label included: using less words, using color, using a "real" arrow rather than an upside down triangle, and marking the bar graph to somehow indicate a progression from left to right (e.g., with tick marks). Suggestions for improving the stars label included: including kWh range information and making the star rating box (and in particular the fact that it contains an efficiency-based rating) more noticeable. Suggestions for improving the letters included: flipping the graph so most energy is on top and least energy is on bottom, including a key defining the meaning of each of the bars, making the arrow stand out more, including kWh range information, including the model's actual kWh in the graphic (not just in a box below the graphic), and decreasing the number of categories. Suggestions for improving the thermometer included: using color, including the model's actual kWh in the graphic (not just in a box below the graphic), and using an arrow or some other indicator on the graphic to indicate where along the thermometer the model depicted actually falls. Suggestions for improving the speedometer included: marking each of the ticks on the speedometer with intermediate kWh amounts, including the model's actual kWh in the graphic (not just in a box below the graphic), and making the whole graph, but in particular the arrow, more visible (e.g., with thicker lines or color).

In summary, although all the label formats were comprehensible to a majority of the interviewees, the current label appeared to be most difficult for the interviewees to interpret. Furthermore, the interviews support the conclusion that from a consumer perspective improvements over the current label are possible. This is evident in the relatively high incidence of comprehension problems with the current label and the relatively low incidence of preferences for the current label over all of the other options. The thermometer and speedometer were promising label options, although the similarity in interviewees' perceptions of these two designs suggested that they were not different enough to warrant continuing to test them both. The stars label appeared promising because of its intuitive scale as well as its strong motivational potential. The letters label appeared to be the least refined of all the graph designs and needed improvement and continued testing. An overall comprehension problem for all of the formats was that many participants did not immediately grasp that the model in question was being compared to other similar models.

5   Consumer Focus Groups—Round Two

ACEEE contracted for a second round of six focus groups to examine multiple executions of each of the leading label designs with single-family homeowners in the market for household appliances and equipment sold through retail stores. The groups were expected to lead to final designs for use in quantitative testing. Four basic graph alternatives were tested[3]— the current label, the star-based label, the thermometer-based label, and the letter-based label including many of the improvements and suggestions drawn from the earlier consumer tasks. In addition, variations on these basic graphs were tested to incorporate additional informational or visual elements. For example, leading from the consumer interviews, versions of the star- and letters-based graphs were produced with kWh range information. Versions of the letter label were tested with variations on the amount of color used. For all the designs, a version was tested with a high amount of explanatory text (referred to as the high-verbiage case) and a version was produced with a low amount of text (the low-verbiage case).

Overall respondents viewed the star label most favorably. The star graphic was considered consumer-friendly because it was simple to interpret and most consumers were already familiar with the concept of using stars to connote performance. Many respondents noted that the star graphic easily and effectively communicated the energy efficiency concept to consumers. However, although the majority of consumers found the star graphic highly effective at communicating the intended message, many noted that the basic version was not very informative. Thus, most group members preferred versions that increased the amount of information available on the label. Specifically, respondents noted that the kWh range end-points were important pieces of information because the scale anchors gave consumers a context in which to evaluate the meaning of the stars. It seemed that the most desirable star graph would include the kWh range end-points along with most of the information contained in the high-verbiage versions.

Participants indicated that they liked the level of information contained in the current EnergyGuide, in spite of the fact that the graphic is relatively ineffective. Further, they indicated that they were familiar with the current label and believe it is easily recognized by consumers. This suggested the current label has considerable equity with shoppers. However, it is important to acknowledge that while consumers reported that they liked having the maximum amount of information, they also said that they did not like its cluttered appearance (in this and other tasks). Some noted that they often ignored the current EnergyGuide altogether because there is too much text. Participants appeared to have conflicting, and perhaps mutually exclusive, demands of the label. Specifically, on one hand they indicated the energy label should be high verbiage and informative, while on the other hand they indicated that they were less likely to read or use a label that appeared cluttered or busy.

The participants initially indicated that the letters label had more "stopping power" than the other concepts and was both appealing and easy to read. However, the group discussion of the letters label revealed significant confusion about the interpretation of the letters label. For example, most participants continued to find the length of the bars (longer bars means less efficient) misleading or counterintuitive. This was particularly interesting because based on earlier tasks the graph's endpoint labels were changed from efficiency-based to consumption-based to alleviate this problem. Some also found the scaling counter-intuitive, indicating that graphs typically have the lowest point of the scale (i.e., uses less energy) at the bottom of the scale and the highest point (i.e., uses most energy) at the top. Finally, some interviewees felt that although the colors drew attention to the label, they also distracted from the interpreting the label's overall meaning. This feeling is most clearly articulated in a quote from a group participant who said:

> The colors make an attractive poster but...it confuses the
> issue...There is no reason for those colors to be different...It makes
> the label less straightforward.

As a result, consumers' end-of group opinions of the letters label were much less favorable and only the thermometer label performed worse in a final written evaluation of the four label concepts. Ultimately, participants found the letters concept too busy and difficult to interpret.

Respondents' reactions to the thermometer label were unambiguously negative. In particular, consumers found the scaling counterintuitive (better energy performance at the bottom of the scale). Group members were moderately favorable to an execution done for air conditioners and based upon EER because this version was inverted (better energy performance represented at the top of the scale) and seemed more logical. However, participants indicated they were unfamiliar with EER and preferred that kWh be used because that was at least a term they associated with their electric bill. In short, the negatives of this label far outweighed the modestly positive evaluation of some specific executions.

In summary, the participants preferred the star label over the other options, with the current label being the second most preferred. The stars were strong visually and from an information-processing perspective were very clear and quick. The current label was strong because of its familiarity and depth of information. Overall, the participants wanted a label that incorporated both a strong graphic and detailed information. Designs that combined these features were most preferred.

6 Conclusions and Next Steps

Taking into account all of the tasks where the various labels were tested, there is strong evidence that improvements to the current U.S. label are possible. Many consumers found the current label complex or overly technical and a significant proportion of consumers had difficulty understanding it. Although some consumers liked the detailed and familiar nature of the label, many indicated that they don't usually read all of the text or use the label at all.

A change in the U.S. EnergyGuide label would have to offer substantial enough savings to outweigh the impacts on the various supply-side actors in implementing the changed program as well as to overcome the equity the current label has in its familiarity to consumers. In combination with some of the additional information (e.g., kWh range endpoints) that consumers suggested and that the current label already contains, a categorical system based upon stars may well meet these requirements. Another option is a re-design of the current label to enhance its visual appeal, message communication, and information organization. The remaining designs (letters, thermometer, and speedometer) do not appear to warrant continued testing.

Another interesting overall finding is that inverse nature of the relationship between energy use and efficiency makes it challenging to devise a graph that clearly and quickly explains the issue to everyone. The problem is that more of something (energy efficiency) is caused by less of something (energy use). In particular, it seemed that the vertical scales such as the thermometer and the letters were difficult. Interestingly, the scales on these two labels were reversed. In one, the best product was on top while in the other the worst product was on top. Yet in both cases we received feedback from some consumers that the scales should be reversed.

Three research tasks remain in this project: a third round of consumer focus groups, a survey, and a field test. The focus groups will address concerns raised by some of the project advisors over how a categorical system will interact with the ENERGY STAR ® logo. The next phase of the research will quantitatively test consumer comprehension of the lead designs along with the impact of those lead designs versus the current label on attention to energy use and purchasing of energy-efficient equipment. A survey will be implemented to determine (with statistical certainty) which among the lead label concepts has the highest rate of comprehension including participants' ability to decipher the main label message (i.e., that a model is or isn't energy efficient) and the label's secondary messages (e.g., that the comparison is based on similar models within certain product categories). Finally, a field test will be conducted as a pilot study of the impacts of the lead label design(s) on the sale of appliances. Once all of the research has been completed (likely in January 2001), a petition will be drafted to request that the Federal Trade Commission incorporate the project findings.

Endnotes

1. For a more detailed discussion of all aspects of the project, see Egan, C. (2000.) *An*

Evaluation of the Federal Trade Commission's Energy Guide Appliance Label: An Interim Summary of Findings. Report Number A003. Washington, D.C.: American Council for an Energy-Efficient Economy.

2.   Specifically, we shortened the explanatory text to a simple identifying label (versus two sentences as in the current U.S. label) and incorporated this label in a box along with the actual operating cost number. This is consistent with the first focus group findings that related information should be group and blocked off.

3.   The speedometer label was dropped due to its poor testing in the initial round of focus groups and seeming overlap with the thermometer label, based upon the consumer interviews. Also, several manufacturer representatives who had seen this version before as part of a Canadian study (Patterson 1991) indicated that it would be very difficult to implement.

4.   The scale ranged from one to ten where one is the lowest possible rating and ten is the highest possible rating for the attribute.

7   References

[1]   Boardman, B., K. Lane, M. Hinnells, N. Banks, G. Milne, A. Goodwin and T. Fawcett. (1997). *Transforming the UK Cold Market.* University of Oxford.

[2]   du Pont, P. (1998). "Energy Policy and Consumer Reality: The Role of Energy in the Purchase of Household Appliances in the U.S. and Thailand." Ph.D. Dissertation. University of Delaware.

[3]   Harrington, L. (1999*) Analysis of Appliance Sales Data: An Analysis to Assess Energy Consumption and Performance Trends of Major Household Appliances in the Australian Market from 1993 to 1997.* Australia: Energy Efficient Strategies.

[4]   Patterson, J. (1991). *Focus Testing of Alternative Energuide Labels: Final Integrated Report, Phases I, II, and III.* Montreal, Canada: Createc+ Recherche-Marketing.

[5]   Sulyma, I., F.K.H. Chin, P. T. du Pont, D. Ference, J. Martin, N. Phumaraphand, H. Tiedemann, G. Wikler, and S. Zariffa. (2000). "Taking the Pulse of Thailand's DSM Market Transformation Programs (Draft)." In *Proceedings of the 2000 ACEEE Summer Study on Buildings.* Washington, D.C.: American Council for an Energy-Efficient Economy.

[6]   Waide, P. (1997). "Refrigerators: Developments in the EU Market." PW Consulting. Paper presented at The First International Conference on Energy Efficiency in Household Appliances, 10-12 November, Florence, Italy.

[7]   Wilkenfeld, G. (1997). "Evaluating the Impact of the Australian Household Appliance Energy Efficiency Program." Paper presented at The First International Conference on Energy Efficiency in Household Appliances, 10-12 November, Florence, Italy.

Standardization and Labeling of Electrical Appliances in Malaysia

Mohammad Yusri Hassan, Hasimah Ab Rahman, Md Shah Majid and K.S. Kannan

Faculty of Electrical Engineering, Universiti Teknologi Malaysia

Abstract. Electricity demand is growing rapidly in the South East Asian region. For this reason several countries have implemented energy efficiency programs while others are still developing them. The Philippines and Thailand are among the South East Asia nations leading the way in developing energy efficiency programs such as standardisation and labeling programs. These two countries have well established and functioning programs for improving the efficiency of the household appliances. Realizing the importance, the Government of Malaysia also place great emphasis on the development of energy efficiency programs. Establishment of the Malaysia Energy Center and the Working Group Energy Efficiency in 1998 is a clear commitment by the Government towards energy efficiency issues.

This paper reviews research activities that have been carried out in order to formulate regulation on standardisation and labeling for household appliances. Analysis to determine the standards using statistical approach on several electrical appliances available in Malaysia market are also presented.

Keywords: energy efficiency programs, standardisation and labeling, statistical approach.

1 Introduction

Electricity demand in Malaysia has increased at a rate of over 10% for the last decade. It is expected to increase approximately 10,500 MW today to 14,500 MW in 2005, even the down turn of economy faced by the country recently. The demand is forecasted to reach 30,000 MW in the year 2020. With the existing installed capacity at 13,500 MW, this means that about 1000 MW of new capacity have to be installed by the year 2000 and over 20,000 MW before 2020.

The growth in electricity consumption has been particularly rapid in the industrial and commercial sector, but consumption in the residential sector has also continued to grow and accounts for approximately 20 to 25 percent of the electricity use.

Due to the rapid increase of energy consumption, the government should implement new regulations, policies and standard on certain electrical appliances to ensure efficient use of electrical energy.

To ensure that the activities for energy efficiency can be carried out effectively, the Department of Electricity and Gas supply, Malaysia has formulated energy efficiency regulations. Although until now the policy is yet to be implemented, many promotion activities such as conducting seminars, workshop and exhibition has been carried out.

A number of electrical appliances use in the domestic sector has been identified to meet certain standards on energy efficiency. The preliminary studies include ballast for fluorescent lamps, fan, refrigerators and air conditioners.

2   Energy Efficient Working Group

A working group of energy efficiency committee initiated by SIRIM has been setup to develop the standards and policies for Malaysia household electrical appliances. It comprises various government bodies, universities, manufactures and energy related consultant companies. Participation in the committee is voluntary and SIRIM acts as a co-ordinating secretariat. The group began working in late 1998 and currently is concentrating on the development of standards and labeling policy for refrigerators while fan and air conditioner will follow later.

3   Energy Efficiency Regulations [1]

The Energy Efficiency Regulations is prepared under the Electricity Supply Act 1990. The main areas covered under the regulations are as follow.

3.1 Specific Installations

A Specific Installations is any installation at an operating voltage of 11kV or above with an average monthly electricity consumption, taken one a period of twelve months exceeding 500,000 kWh. One of the duty of the owner or management of a specific installation is to appoint at least one Energy Efficiency Officer to take full responsibility with regards to energy efficiency programs in the installation and inform the department about the appointment.

3.2 Energy Efficiency Officer

An Energy Efficiency Officer may or may not be an employee of the owner or management of an installation but should have related certificates or degree with working experience in energy efficiency.

3.3 Scheduled Products

The product complies with the prescribed energy efficiency standard and label and is issued with a valid certificate of compliance by the Director General of Department of Electricity and Gas Supply, Malaysia. At initial stage the list of Scheduled Products includes, ballast for fluorescent lamps, fans, refrigerators not exceeding 750 litters and room air conditioners not exceeding 3kW.

3.4 Energy Using Products

The list of Energy Using Products includes lamps, clothes washers not exceeding 7kg storage water heaters, television, video monitors and vacuum cleaners not exceeding 2kW. The manufacturers, importers, advertises or sells any Energy using products Should label, advertise, indicate and describe in any manner causing any person to believe that the product is consuming less energy when compared with a product similar in function, usage type and capacity when used under similar conditions and environment.

The standards to be used for the approved and testing of the scheduled products and energy using products will be the relevant Malaysian standards or in its abrense the equivalent IEC or ISO standards or any other standards which is approved by the department.

4 Methodologies in Developing Energy Efficiency Standards [2,3]

There are two widely approaches to setting energy efficiency standards, these are statistical or engineering/economic in nature. In addition to these methods, there are other arrangements whereby the standard selection is based on stakeholder input.

4.1 Statistical Approach

The statistical approach is based on the desired energy saving and designates percentage of models to be eliminated. This approach requires (I) data that may be easier to obtain, not extensive and (ii) analysis is not very complicated than the engineering/economic approach. A significant advantage of the statistical approach is that the costs of achieving those energy savings are not explicitly determined. However it's disadvantages is, difficult to predict impact on the manufacturers. The statistical approach has been utilized in the European Union (EU) and in Australia.

4.2 Engineering and Economic Approach

The second approach to standards setting by engineering/economic; its selection is based on comprehensive evaluation of costs and benefits. The analysis seeks to determine the cost of efficiency improvements and the impact on the consumer through economic analysis, including life-cycle cost and payback period calculations. In contrast to the statistical approach, one significant advantage of the engineering/economic approach is that it allows for consideration of new designs that are not already included in existing model or of some combination of designs that result in higher efficiency than is found in any existing models. A potential disadvantage is that, the analysis is complicated and requires more time and is expensive. This approach has been used by the Lawrence Berkeley National Laboratory (LBNL) for the US Department of energy (DOE).

4.3 Consensus Approach

In a consensus approach, two or more groups get together and decide on the standards through a joint process. These groups could be some combination of a government regulatory agency, environmental/consumer groups and appliance manufacturers. This approach was used in the United States in establishing the first national efficiency standards, in 1987. The consensus approach was also used to some extent in finalizing the Australian efficiency standards for refrigerators and water heaters and for some Japanese standards.

5   Analysis on the Investigation of Electrical Appliances

A recent study conducted by Energy Research Group, UTM focused into the current status of Malaysian electrical appliances industry and the energy efficiency of its product.

In this study, the statistical approach has been used to analyze Malaysian electrical appliances based on Malaysian Standards.

5.1 Fan

The energy efficiency measurement or service value can be defined as the ratio of the capacity air delivery (m^3/min) to the power consumption (watt) of a unit with similar category. This measurement is based on the MS 141 part 2 (Malaysian Standards) adopted from IEC 342 Part Standards.

The higher the ratio, the most efficient the product. Figure 1.0, 2.0. 3.0 and 4.0 show the graph of energy efficiency measurement (service value) against modal of ceiling fans, pedestal fans, table fans and wall fan respectively.

5.2 Ballast

Based on the results of the surveys, discussion and agreement with the manufactures and importers on the fluorescent lamp ballast, the Department of Electricity and Gas Supply Malaysia has recently issued a directive on the minimum energy efficiency standards to be complied before the ballast can be approved by the department. The maximum allowable power loss (watt) for the ballast is 8W. The maximum standards is effective by 1 January 2000. Figure 5.0 and 6.0 show the graph of power loss against model of ballast for 18/20W and 36/40W fluorescent lamps.

6 Conclusions

Standardization and labeling are critical policy tools for saving energy. Standards and labeling combine "market pull" with "market push" to increase the supply and demand for energy efficient products. Minimum energy performance standards eliminate the least efficient products from the market. Label on the other hand, communicates directly with consumer, encouraging them to purchase the efficient products on the market. The overriding principle here is that a consumer who is adequately educated about energy consumption is a better consumer.

The findings show the upper and lower limit for fans and ballasts. The limit provides guidelines to establish the standard and label of different types of fans and ballast respectively.

7 References

[1] C.C. Yin, "Energy Efficiency Programs in Malaysia – Frame work and Activities", Seminar on Energy Efficiency and Appliance Standards and Labeling, Malaysia, 8 – 9 Dec, 1998.
[2] I. Turiel, T. Chan and J. E Mc Mahon, "Theory and Methodology of Appliance Standards", Energy and Building 26 (1997) pp 35 – 44.
[3] P. du Pont, "Energy Efficiency Standards. Case Study : Limited States, "Seminar on Efficiency and Appliance Standards and Labeling, Malaysia, 8 – 9 Dec, 1998.

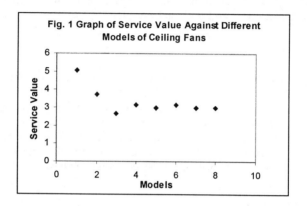

Fig. 1 Graph of Service Value Against Different Models of Ceiling Fans

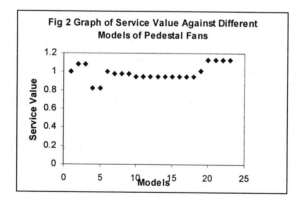

Fig 2 Graph of Service Value Against Different Models of Pedestal Fans

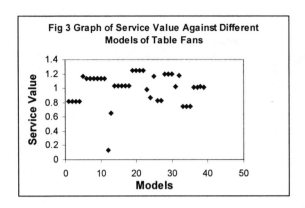

Fig 3 Graph of Service Value Against Different Models of Table Fans

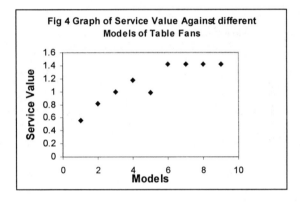

Fig 4 Graph of Service Value Against different Models of Table Fans

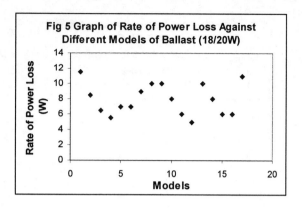

Fig 5 Graph of Rate of Power Loss Against Different Models of Ballast (18/20W)

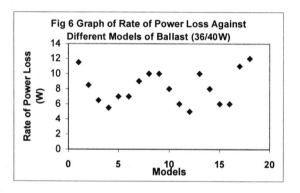

Fig 6 Graph of Rate of Power Loss Against Different Models of Ballast (36/40W)

Energy Efficiency of Electric Appliances and the Energy Label: Review of the Special Situation for Ovens

Dr. Ernst-Jürgen Breford

CECED

1 Introduction

The energy labelling of household appliances was initiated by the first EU energy labelling directive in 1992. Since than „daughter directives" covering mains operated electric appliances have been implemented for refrigerators and freezers, washing machines, tumble driers and dishwashers.

The positive effects on the reduction of energy consumption through the implementation of label is wellknown. CECED would like to point out that manufacturers are committed to improve the rational use of energy in all applications and to enhance the energy efficiency of their products to the highest possible extent.

The author of this paper is the convenor of the joint working group of CECED and CEFACD, dealing with electric and gas fired ovens. This working group has been working on several aspects of the energy consumption of ovens in especially in context with the study of efficient domestic ovens worked out within the Save project. CECED provided data on the consumption for electric ovens as well as contributions to several aspects dealt with in the save report i. e. on the design options. (See footnotes and CECED statement in that report.)

In my paper I would like to review the discussion on the energy labelling of ovens and highlight some of the special aspects that make ovens different from other household appliances.

2 Procedure

The procedure for the development of an energy consumption label is well known from the above mentioned labels for washers, dryers etc. that have been developed several years ago. The procedure consists roughly of the following steps:

First a measuring standard for the energy consumption is needed. Then a market survey is performed to get data about the existing distribution of the appliance type under consideration and the use of these appliances as well as their energy consumption. A technical study gives indications about the reduction potential of the energy consumption. The information out of the for mentioned steps and other sources finally lead to the proposal of a labelling scheme itself.

3   Problem Areas

In the following I would like to comment on some details in these main steps to highlight some aspects special to the electric and gas fired ovens:

1. **Measuring standard**

   The standard for the measurement of the energy consumption for electric ovens (EN 50304) just has been finished at the beginning of this year. The adoption of the standard to gas fired ovens is still on the way. The main point here is that the definition of a standard load for an oven is very difficult. The load has to be a compromise between a technical feasible, especially reproducible, load and a load as close as possible to real food loads cooked in an oven. The result is the well known „wet brick". Please keep in mind that this load is a good representation of a big piece of meat to be roasted in the oven, but not so good for a cake or pastry. The grilling function cannot be covered. A quick survey of the way electric ovens are used throughout the European community shows, that such a test load can be representative only for a part of the applications in the real life. Results for the assessment of total energy consumption in a country based on this standard have to be considered very carefully.

   The application to gas fired ovens is not clear get. Even bigger problems show up with the application of the „wet brick" test load to smaller ovens and ovens using microwave power. Here the fundamental question raises whether this type of load is appropriate for these appliances and the way they are used in the households, that is mostly with small loads only.

2. **Market survey**

   Even a quick survey of the ovens on the European market show a large variety of oven types varying in size and power and also in intended usage. We have everything from a small oven with a cavity less than 35 litres up to big ovens with cavities up to 90 litres, free-standing and build-in and even table-top versions exist. Not only the different sizes are used very differently, in terms of frequence and loads, but also the local way of cooking shows big differences. For instance UK, France, Sweden, and Finland households use the oven more than three times a week while others like the Netherlands or Italy are far below that. The loads differ from everything needed for a full meal in Great Britain to once in a while a Pizza in Italy. Even for the same food, big differences occur: As an example in some parts of Europe roasting of meat is done in closed pot inside the oven while in other areas roasting open on a wire rack is the normal way. Of course both procedures show different results not only in terms of the resulting roasted meat but also in terms of the overall energy consumption.

This leads to a mayor point in the actual use of household appliances. For all appliances but ovens we have automatic controls available. That means, as soon as the dishwasher, drier or washer has been filled with its load and the chosen program has been started, the machine follows automatically the preprogrammed procedure. This procedure therefore can be optimized also in terms of energy consumption through the designer. This is not true for the usage of electric ovens. Main factors in the energy consumptions are the temperature setting and the duration and exactly these things are varied widely by the user. A consideration as, „it is not brown enough for me, I will give it an extra 10 min" gives a big change in the energy consumed - bigger probably than anything that was saved by a design optimisation through the manufacturer. Other things, like the use of different assessories, (open grills, open pots or closed pots etc.) have also influence on the energy consumption.

The different settings in temperature and duration stem from different expectations on the cooking result. And these expectations are due to regional differences, but of course also exist from individual to individual. No influence through design is possible on this part of the cooking process. Even if the manufacturer provides informations for the rational use of energy, through the users manual or a cook book, suggesting for instance lower temperatures for some cooking process, these suggestions may not be accepted by the consumer because his ideas are different. This background makes it difficult to use a common approach for the usage pattern of ovens in Europe. But these datas are needed to develop a picture of the total consumption of ovens and to estimate energy reduction potentials. Indications for the problems with the poor data existing in this field can also be found in the Save report (S. 18, Tb. 2), where different reports on the energy consumptions on ovens are shown to be partly contradictuary. This means to me that a lot of further investigation is needed in this field before final conclusions on potentials and appropriate policies can be drawn!

3. Analysis of design options for reduction of energy consumption

Especially in this point the opinions of the authors of the Save report and the CECED working group diverge. Due to a lot of improvements initiated by normal market forces throughout the last ten years a good part of the improvements, discussed in the report, are already incorporated in the actual ovens. According to the manufacturers opinion ovens of today therefore do not show a huge potential for reduction of energy consumption through improvements of the design.

The SAVE study considers these potentials to be bigger than JWG experts believe. Nevertheless the special situation for ovens has led to the proposal within the Save report to set as a goal not the „strict economic and technical potential" but the „technical and economically acceptable potential". For those who are not aware of this vocabulary the „strict economic and the technical potential" approach supports the technical innovations up to the point where the total life cyle cost for the consumer is minimized and no significant increase in appliance cost can be found. The „technical and economical acceptable potential" is related to those improvements that lead to 10 % higher total life cyle cost than the present base case.

With this approach the experience with the existing label schemes for other major appliances is left behind and a completely new approach is chosen. It would mean that in practice the policy goes beyond the goal of minimizing life cycle cost and requires consumers to pay for some social costs which will never be paid back during the life time of the appliance considered. Such an approach obviously has to be studied very carefully. Manufacturers stand ready to participate constructively in a more general debate on this policy guide lines.

I have to mention that the considerations on design improvements for electric and gas ovens have to be considered separately. Furthermore the approach to transformation policies for ovens from gas to electric energy in different countries in the European community should be discussed in detail as it will be not very meaningful to put pressure on electricly heated ovens in a country where a good part of the electricity is produced through reproducible energy sources like water power. In this countries market transformation from electricity to gas as main energy source for ovens could even increase and not reduce the overall CO_2 amission.

4. Labelling scheme

The labelling scheme for ovens finally has to be a balanced conclusion out of the marked informations (What is?), the technical potentials (What could be?) and the reduction goals (What should be?). Of course requirements in terms of energy consumption that are easily met will not bring much of an improvement to the over all energy consumption. On the other hand unrealistic expectations may lead to very strong requirements that can not be met because of technical restraints. This would not give an improvement in terms of the overall effect in reduction the total energy consumption, but could give damage to the economy as the consumer would be led to the impression that only „bad" quality appliances exist on the market.

5. Conclusions

Like with other appliances the introduction of an energy label for ovens is a promising approach to the reduction of energy consumption. But one has to take into account some special aspects that make the oven case different from other appliances. The regional distribution gives a lot of differences in terms of the all day use of the ovens stemming from different ways of cooking and different expectations on the result of the preparation of the food, leading to different usage pattern and energy consumption even with the same appliance. The manufacturer in this case has no possibilities to influence the actual use in the household through the implementation of a standard „automatic-program". Up to now there is no comprehensive study on this aspects of the usage of household ovens and for the result on the overall energy consumption also the Save report gives only scarce information.

I would therefore suggest to start such a study in order to get more input data on the actual use of ovens. These datas than could be used to address directly the unrational use of energy in cooking processes. I feel that this information should be distributed through national activities according to the experience that educating the consumer through the users manuals and other informations provided by the manufacturers have been not very effective. Everybody knows how to cook and therefore very often the users manuals especially for ovens are not considered by the user.

The oven is perhaps the only appliance in the household where a good design documented in a „A" rating on the label is not enough for achieving good progress in reduction of the energy consumption. The user side for this appliances is very important. Therefore additional measures have to be taken to be really successful.

Labelling Domestic Ovens
(Save Study 4.1031/D/97-047)

Kevin Lane

Environmental Change Institute, University of Oxford

Abstract. The European Union, through its SAVE II programme, aims at promoting the rational use of energy within the community. SAVE II is a non-technological programme with many elements, among others *labelling, standardisation and other actions in the area of energy-using equipment.* This report summarises the results of the Working Group on Efficient Domestic Ovens. The purpose of this study is to identify possible EU Commission actions to improve the efficiency of domestic electric and gas ovens and to develop a common basis for possible national actions. The study focuses on conventional domestic ovens only. Other ways of cooking, such as hobs, microwave ovens, grills and electric kettles are not taken into consideration. Possibilities for and impacts of fuel switching are also outside the scope of this study. However, a wide range of possibilities to improve energy efficiency, and related policy options, are discussed in a qualitative way, to give a more complete picture of ovens in the context of climate policies.

The study, financed by the SAVE II Programme (with supplementary financing from NOVEM for the Dutch partner) was carried out in 1998-1999 by research teams from eight countries. Representatives of NOVEM and CECED as well as those of the relevant working groups of CENELEC (CLC/TC 59X/WG3) and CEN (CEN/TC 49/WG2) have participated in the meetings. CECED has also provided data for a number of tasks.

The work was organised in six tasks. The aim of Task 1 was to establish the *real life energy consumption* of the ovens concerned, as opposed to the energy consumption according to the test standard. Task 2 was to *provide a statistical analysis of the European gas and electric ovens market.* The purpose of Task 3 was to *review, identify, and analyse energy-saving design options for domestic ovens.* Task 4 was to *assess the environmental impacts* of scenarios concerning the efficiency improvements of domestic ovens. A stock model methodology was used. Task 5 aimed at analysing the *different impacts of efficiency policies on manufacturers.*

The aim of Task 6 was to give the information necessary to develop an appropriate mix of policies. The focus is on actions at EU level. The following policy options were analysed:

- EU energy labels which are warranted for domestic ovens through directive no 92/75/EEC;
- mandatory minimum energy efficiency standards;
- as an alternative to the latter, the acceptance by the EU of a voluntary initiative of the European White Goods industry to phase out ovens with a consumption higher than a certain value and/or to reach a certain average energy consumption of the appliances sold;
- co-operative procurement (technology or market procurement);
- national policies which could supplement EU policy are also analysed qualitatively.

1 Introduction

The European Union, through its SAVE II programme, aims at promoting the rational use of energy within the community. SAVE II is a non-technological programme with many elements, among others labelling, standardisation and other actions in the area of energy-using equipment.

Domestic ovens, both gas and electric, are among the last appliances to labelled as part of the European framework labelling directive of 1992 (CEC, 1992). An implementing directive is expected, which will draw on the findings of the Working Group on Efficient Domestic Ovens summarised in this paper. The purpose of this 'ovens' study conducted by the Working Group was to identify possible EU Commission actions to improve the efficiency of domestic electric and gas ovens and to develop a common basis for possible national actions (Kasanen, 2000).

The study focuses on conventional domestic ovens during their use phase only; energy used by standing pilot lights and cleaning cycles were not considered. Other ways of cooking, such as hobs, microwave ovens, grills and electric kettles, are not taken into consideration by this study. Possibilities for, and the potential impact of fuel switching, are also beyond the scope of this study. However, a wide range of possibilities to improve energy efficiency, and related policy options, are discussed in a qualitative way, to give a more complete picture of ovens in the context of climate policies.

The study, financed by the SAVE II Programme (with supplementary financing from NOVEM for the Dutch partner) was carried out in 1998-1999 by research teams from eight countries. Representatives of NOVEM and CECED as well as those of the relevant working groups of CENELEC (CLC/TC 59X/WG3) and

CEN (CEN/TC 49/WG2) have participated in the meetings. CECED has also provided data for a number of tasks.

The work was organised in six tasks. The aim of Task 1 was to establish the real life energy consumption of the ovens concerned, as opposed to the energy consumption according to the test standard. Task 2 was to provide a statistical analysis of the European gas and electric ovens market. The purpose of Task 3 was to review, identify, and analyse energy-saving design options for domestic ovens. Task 4 was to assess the environmental impacts of scenarios simulating efficiency improvements of domestic ovens, where a stock model methodology was employed. Task 5 aimed at analysing the potential impact of different efficiency policies on manufacturers. The aim of the final task was to provide the information necessary to develop an appropriate mix of policy actions, where the focus is on actions at EU level.

2 Current Situation

2.1 Ownership and Usage of Domestic Ovens

Most households in the EU own an oven, of which approximately 61% are fuelled by electricity and 38% by gas. There are inter-country variations, with some countries like Finland predominantly using electricity and others like Spain favouring gas (Table I).

Table I: Input data for ovens stock model, 1998

	House (m)	Electric ownership (%)	Gas ownership (%)	Use (times pa)
AU	3.2	79.9	19.1	110
BE	3.8	50.3	62.1	157
DK	2.4	88.0	10.4	150
FI	2.3	99.0	1.0	200
FR	23.2	51.1	48.3	224
GE	37.7	79.9	19.1	87
IR	1.2	44.4	67.8	157
IT	21.8	60.9	38.5	124
NL	6.7	63.5	20.3	45
SP	15.1	12.0	85.0	124
SW	4.2	97.3	3.0	136
UK	24.3	56.7	41.2	157
EU	145.9	60.9	37.8	135

Household use of ovens also varies quite considerably across the EU – for instance, Dutch households will, on average, use their ovens only 45 times per year, while the average French household will use its oven more than four times as often. The data shown in Table I are best estimates and change through time.

2.2 Ovens on the Market

The construction of a mass-produced oven's cooking compartment is essentially a pressed-steel cavity wrapped in thermal insulation, a hinged and often glazed door, and a vent or a flue. The oven temperature is regulated by thermostatic control of the gas burner or electric supply. The heating elements are located in the upper and lower faces of the cavity for electric ovens and in the lower face for gas ovens. The lower heating element is usually concealed, whilst the upper one can be exposed in an electric oven and double as a grill.

Usually, ovens are sold either as free-standing cookers (where the oven is combined with a hob) or as separate 'built-in' appliances. An increasing share of the market consists of fan-assisted ovens, where air is forced around the oven, or air is distributed by natural convection. The market can be segmented further, distinguishing between those appliances possessing or not possessing additional features such as an in-built grill or microwave, or self-cleaning ability. The latter – based on catalytic, pyrolitic and hydrolytic cleaning processes – is increasingly found in modern, especially electric, ovens.

2.3 Efficiency of New Ovens (and Testing Procedures)

If efficiency improvements are to be made for domestic ovens, the first step is to have a reproducible and representative test procedure. Until recently, the efficiency of ovens was compared with reference to consumption measured in the so-called 'empty' oven test. For electric ovens the test (CENLEC HD-376S2-1998) was based on measuring the energy needed to preheat the oven to 200°C (or 175°C for forced convection ovens) and maintain it at that temperature for one hour. This is now being replaced by the 'chilled wet brick' test, prEN 50304, which is considered more representative of actual oven usage. In this test procedure a saturated chilled wet brick is raised in temperature from 5°C to 60°C – the energy used to perform this task being the important measure. From a sample within the study there was little correlation between the energy values measured by the two different testing procedures. Therefore, future policy will have to aim at improving the test data under the new test.

The empty oven test (EN 30 2.1) for gas ovens is also being replaced by a wet brick test. The new procedure, very similar to the electric oven test, is still being developed. Round robin tests, where the same oven is tested in different laboratories, still need to be carried out for gas ovens.

CECED, the European manufacturers' association, provided a database on electric ovens which revealed the average natural convection oven to consume 1224Wh under the brick test, with values ranging from 710 to 1907 Wh). Data for the forced convection ovens had consumption values ranging from 736 to 1920 Wh with an average of 1269Wh.

Since the new test had not been completed for the gas oven, the detailed analysis could not be repeated for these types of ovens. However, 13 gas ovens were tested under the provisional test procedure to gain an insight into their performance. The average consumption of these ovens was 1.5 kWh (or 5.45 MJ), while the minimum was 1.14kWh (4.1 MJ).

None of the test cycles take into account the energy used by additional modes or facilities of the oven – eg clocks, grills, cleaning cycles – due to a lack of data. These may, however, consume significant amounts of energy, depending on use.

3 Potential for Improvement

The potential for improvement will be different for each oven, and will depend on its current design. For both electric and gas ovens, the energy consumption could potentially be decreased by up to 55%, under test conditions, using known design improvements. In the ovens study, two electric ovens considered representative were used as 'reference' ovens, for which an estimation of the potential energy savings to be achieved through design improvements could be made, and compared to the additional unit-manufacturing cost (Table II).

Combining design improvements is not always possible and the resulting energy savings are not always cumulative. Some of the design options were considered to be unacceptable to the consumer and these were not considered in the final estimate of potential energy savings. Thus, the single design option with the largest potential, low-emissivity, was not finally taken into account.

Table II: Estimated electric oven efficiency gains with potential improved design options

	Design option [a]	Range	Midpoint	Additional unit-manufacturing cost (Euro)	Consumer response [b]
1a	Improve thermal insulation	0-11	6	8	Acceptable
1b		0-11	6	37	Acceptable
2	Improve thermal isolation of cavity	7-8	8	6	Acceptable
3	Reduce thermal mass of even structure	10-18	14	6	Acceptable
4	Unglazed door	7-25	16	1	Unacceptable
5	Optimise glazed door design	4-12	8	20	Acceptable
6	Passive cooling for glazed door/facia	0-8	4	29	Unacceptable
7	Forced convection	0-8	4	10	Acceptable
8a	Optimise vent flow	8	8	4	Acceptable
8b		12	12	14	Acceptable
9	Low-emissivity oven design	35	35	35	Unacceptable
10	Uncover lower element	8	8	2	Unacceptable
11	Reduce auxiliary energy	1-4	3	10	Acceptable
12	Fit reflector above upper heating element	1	1	2	Acceptable

a) Design options 1 and 8 are subdivided to reflect two distinct implementation options. Option 1 may be achieved through using an extra layer of standard low-cost thermal insulation. If, owing to space constraints, a single layer of high quality insulant must replace the insulation, then the implementation cost will be greater. Attention to vent sizing may achieve small energy saving at a relatively low cost (option 8a). A more sophisticated vent controls system may achieve greater energy savings but would be more expensive (option 8b).

b) Possible consumer acceptance to design options. This is the expected initial response by consumers, which may or may not change or be influenced by fashion or through advertising.

Previous SAVE appliance studies estimated an economically justified and technically feasible (ETP) potential. At this level the life cycle cost of the oven would be minimised, which means that, for the consumer, the increased purchase price is offset by savings due to higher efficiency. Using the information in Table II, as well as data on average usage and energy prices, it is possible to estimate the point beyond which improvements in efficiency are not cost effective to the consumer.

For electric ovens, consumption at ETP level was estimated to amount to 0.87kWh, a reduction of 28% compared to the reference case of 1.2 kWh. Due to the low price of gas none of the design options would pay back within the ovens lifetime for an average gas oven. The study also examined an 'economic package' including additional design options as long as they would not raise the appliance's life cycle cost for the consumer above the present level. Under this scenario, energy savings would be higher than at the ETP level, which aims at achieving minimum life cycle cost for the consumer. In this case the potential for electric ovens is a 33% reduction compared to the reference case. The potential for gas ovens is approximately 9%.

4 Policies for Market Transformation

A high proportion of product policies is at present aimed at increasing the average efficiency of products being sold. A range of policy options is available, and is being employed, to effect this change and shift the market in favour of more efficient products. One common procedure is to rank products according to test performance (regarding energy efficiency), and to make the information available to the customer through labels or databases. Providing information on running costs is usually enough to move the market towards more efficient products to a certain extent, though this tendency can be encouraged further through rebate and procurement schemes specifically promoting these products. Moreover, the development of new, efficient products can be assisted through research, development and procurement programmes, and the resulting, highly efficient appliances increasingly enter the market. Finally, the most inefficient products can be removed from the market. For more information on market transformation see, for example, DECADE (1997).

4.1 Labelling

The main reason for conducting the 'ovens' study was to provide relevant background information to the EU Commission so domestic ovens could be labelled in accordance with the framework-labelling directive. Drawing on the information contained in the database on electric ovens provided by manufacturers, it is possible to rate the efficiency of electric ovens. A corresponding rating for gas fuelled ovens would, at present, have to be based on a smaller set of data, obtained from a provisional test procedure. Since there are two types of fuel under consideration, the study also examined the options for placing both types of oven on the same rating scale, where appliances would be rated according to carbon emissions produced. The resulting distribution of gas and electric ovens would be very similar to the one resulting from a comparison based on usage of primary energy. Table III shows the efficiency ratings on an A to G scale of the ovens on which information was available in the database, and indicates the number of appliances that would fall into each category.

Table III: Potential combined electric and gas label for EU domestic ovens

Indirect CO$_2$ emissions (g)	Efficiency class	All (n=86)	Electric (n=73)	Gas (n=13)
<200	A	0	0	0
200-299	B	9	0	62
300-399	C	7	3	31
400-499	D	8	8	8
500-599	E	62	73	0
600-699	F	12	14	0
>699	G	2	3	0

NB: data based on EU sample available to study.

It can be seen that most of the gas appliances pertain to the top categories, while electric ovens are concentrated towards the bottom of the scale. However, with further improvements it would be possible for electric ovens to gain a higher classification. This scale of ratings gives a fair reflection of carbon emissions caused by appliance use. However, it may be argued that a common rating system does not sufficiently emphasise the differences in efficiency between ovens using the same type of fuel, thus potentially reducing the effect of labelling.

So the choice between establishing a common rating system for appliances using different fuel types or instituting two separate scales (one for each fuel) has to be made according to whether the primary aim is to show the correct environmental impact of different appliances or to maximise the potential effect on sales. For further discussion on labels and carbon dioxide see Boardman (2000) and Fawcett *et al* (2000).

4.2 National Policies (Rebates, Procurement)

To complement EU level policies, national agencies still have the scope to implement measures aimed at improving the efficiency of products. Usually, this will be done in the form of providing rebates for more efficient products. Other options include procurement programmes to bring together purchasers to bulk-buy efficient products.

4.3 Minimum Standards

Removing the most inefficient products from the market by implementing minimum standards would reduce the environmental impact of ovens. The 'ovens' study recommends two phases for the removal of less efficient electric and gas ovens from the market, based on a proposed labelling scheme. In the first phase appliances rated as EFG under the proposed scheme would be eliminated, followed by the removal of products pertaining to the D category. The combined effect of this two-stage process would be to reduce electric oven consumption by 40% and gas consumption by 25%.

Table IV: Average EU test consumption figures under different scenarios

	ELECTRIC		GAS		
	kWh	(% of RC)	kWh	MJ	(% of RC)
Reference Case	1.201	100.0	1.515	5.454	100.0
ETP	0.874	72.8	1.515	5.454	100.0
'Economic Package'	0.755	62.9	1.379	4.963	91.0
Policy 1 (remove EFG)	1.117	93.0	1.348	4.854	89.0
Policy 2 (remove (DEFG)	0.721	60.0	0.894	3.218	59.0
Best on market	0.700	58.3	1.140	4.104	75.2

Note: the A-G scales are different for gas and electric in this case.

At present there is a preference on the part of the Commission for implementing voluntary measures (CEC, 2000), based on negotiation with industry associations. However, if a voluntary agreement cannot be reached, mandatory efficiency standards at EU level would be appropriate.

4.4 Carbon Emissions

An estimate of carbon emissions resulting from energy consumption of household ovens shows that emissions caused by EU electric ovens have been relatively stable since the 1980s (Figure 1). Earlier emissions are estimated at a higher level, due to the higher carbon content of EU generated electricity in the past. Accordingly, the conversion factor employed to calculate historical CO_2 emissions caused by electric ovens progressively declines to 0.4 $kgCO_2/kWh$ (EU average for 1998). Emissions are projected to remain relatively constant in the future, as the growth in the total number of ovens offsets changes in the fuel mix as well as the assumed increase in appliance efficiency.

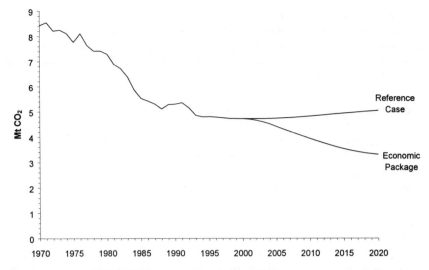

Figure 1: Estimated CO_2 emissions by EU electric ovens

Emissions by gas ovens have declined over the period under consideration, due to reduced usage of ovens as well as increasing consumer preference of electric over gas ovens, a trend which is projected to continue into the future.

5   Conclusions

EU electric ovens currently consume approximately 12TWh of delivered electricity and over 8TWh of delivered gas per annum. At present, this costs EU domestic oven users 1,200 MEuro worth of electricity and 350 MEuro worth of gas in running costs per year.

Table V: EU summary of RC, EP scenario and potential savings

		RC, 1998	RC, 2020	EP, 2020	RC-EP, 2020
ELEC	**Delivered GWh**	11,757	12,530	8,215	4,315
	CO_2 (MtCO_2)	4.70	5.01	3.29	1.73
	Cost Elec (Meuro)	1,176	1,253	821	432
GAS	**Delivered GWh**	8,765	7,548	6,909	639
	Delivered TJ	31,554	27,173	24,874	2,300
	CO_2 (MtCO_2)	1.67	1.43	1.31	0.12
	Cost gas (MEuro)	351	302	276	26

Note: LU, GR, PO excluded from analysis.

Significant energy and running cost savings could be achieved by 2020 if manufacturers produce, retailers supply and consumers purchase cost effective ovens (as under the Economic Package scenario). Projected electricity consumption could be reduced from 12.5TWh to 8.2TWh, and gas consumption could be decreased from 7.5 to 6.9 TWh (ie 27.1 to 24.9 PJ), by the year 2020.

Carbon emissions caused by electric ovens have fallen over the last 30 years due to changes in the electricity generating fuel mix away from carbon intensive modes of electricity production and to the improvements in oven efficiency. They should remain relatively constant into the future, from 4.7MtCO$_2$ to approximately 5MtCO$_2$ by 2020 under the Reference Case scenario. If the 'Economic package' potential is realised, emissions could decline by around 1.7MtCO$_2$ by the year 2020.

The purchase of more efficient gas ovens to achieve the 'economic package' savings potential would reduce associated emissions by approximately 0.1MtCO$_2$. In total, improving the efficiency of ovens, both electric and gas-fired, can potentially reduce emissions by over 2MtCO$_2$ by the year 2020.

To realise these potential reductions in carbon emissions several measures need to be enacted at the European level. Firstly, as a prerequisite for correctly labelling both gas and electric ovens, the 'brick' test for gas appliances needs to perfected. At present, this still requires round robin tests to be funded and carried out. The Commission will then be in a position to prescribe labelling procedures – whether on a common scale or on different scales according to fuel type. If the Commission decides on separate labels for gas and electric ovens, then labelling for electric ovens could be implemented before test procedures for gas ovens have been perfected. To complement labelling, measures aimed at eliminating the most inefficient ovens should also be implemented, whether on a voluntary level through negotiations with industry associations, or by mandatory regulations at EU level.

6   References

[1]   Boardman, B (2000) Creating a carbon market. *Proceedings of the AIEE appliance and lighting conference*, September 2000, Naples, Italy.
[2]   CEC (1992) Directive 95/92 Energy labelling and standard product information. *Official journal of the European Communities*. No. L297.
[3]   CEC (2000) *Energy efficiency action plan*. European Commission, May 2000, Brussels, Belgium.
[4]   DECADE (1997) *2MtC – 2 million tonnes of carbon*. Environmental Change Institute, University of Oxford, UK.

[5] Fawcett, T., Lane, K. and Boardman, B. (2000) *Lower carbon futures*. Environmental Change Institute, University of Oxford, UK.

[6] Kasananen (2000) *Efficient domestic ovens – final report*. TTS Institute, Helsinki, Finland.

Acknowledgements

This paper is a summary of the research carried out by eight European research organisations for the EU study on domestic ovens. A warm thank you should go to Pirkko Kasanen for leading the 'ovens' project and also for submitting this abstract. I should also like to add my appreciation for the task leaders of the ovens study: Marget Groot and Rene Kemna (of vhk), Andrew Pindar and Anne Rhiale Marcus Newborough and Brian Shaunnasy (at Cranfield University), Stefan Thomas (at Wuppertal), and also the other participants on the study. The study was funded by the SAVE II programme. The current author is part-funded by the UK's DETR's Market Transformation Programme (www.mtprog.com).

Variability of Consumer Impacts from Energy Efficiency Standards

James E. McMahon and Xiaomin Liu

Lawrence Berkeley National Laboratory

Abstract. A typical prospective analysis of the expected impact of energy efficiency standards on consumers is based on average economic conditions (e.g., energy price) and operating characteristics. In fact, different consumers face different economic conditions and exhibit different behaviors when using an appliance. A method has been developed to characterize the variability among individual households and to calculate the life-cycle cost of appliances taking into account those differences. Using survey data, this method is applied to a distribution of consumers representing the U.S. Examples of clothes washer standards are shown for which 70-90% of the population benefit, compared to 10-30% who are expected to bear increased costs due to new standards. In some cases, sufficient data exist to distinguish among demographic subgroups (for example, low income or elderly households) who are impacted differently from the general population.

Rank order correlations between the sampled input distributions and the sampled output distributions are calculated to determine which variability inputs are main factors. This "importance analysis" identifies the key drivers contributing to the range of results. Conversely, the importance analysis identifies variables that, while uncertain, make so little difference as to be irrelevant in deciding a particular policy. Examples will be given from analysis of water heaters to illustrate the dominance of the policy implications by a few key variables.

1 Introduction

This paper describes the method for analyzing the economic impacts on individual households of energy efficiency standards for residential appliances. We use the Life-Cycle Cost (LCC) of an appliance as a criterion to determine the effect of standards on individual households. LCC captures the tradeoff between the projected post-standard purchase price and operating expense over the life of the appliance.

The method involves replacing the point estimates of average life-cycle cost of appliances with distributions reflecting the whole spectrum of possible costs and the assessed probability associated with each value. The probabilistic modeling approach has emerged as an important and practical tool for risk assessment, mostly driven by the desire to understand the risk level that individuals face in a certain hazardous environmental condition (Finkel 1990). Since each individual's exposure to health hazards varies, risk management schemes have to be designed to reflect the inter-individual variability, that is, what is the likelihood of an individual facing a specific level of risk in a certain circumstance. Similarly, the impact of a national energy efficiency standard on consumers will not be the same for all consumers due to differences in their household characteristics, appliance use patterns, energy prices, etc. Therefore the economic impact needs to be assessed at the individual consumers' level so that policymakers are able to gain an insightful view of how the proposed energy efficiency standards would economically affect certain demographic subgroups such as low-income or elderly households.

Since 1996 researchers in the Energy Efficiency Standards Group at LBNL have applied Monte Carlo simulation to the study of energy efficiency standards of appliances in U.S. residential and commercial sectors, such as fluorescent lamp ballasts, clothes washers, water heaters, and central air-conditioning. The Monte Carlo approach has greatly improved research quality and has been adopted as normal practice for the analysis of U.S. energy efficiency standards. As a summary of the on-going research effort, this paper focuses on illustrating two techniques for using input distributions to determine the distribution of impacts on U.S. consumers from proposed standards and to identify the most relevant factors that would affect the outcome. The examples are drawn from residential clothes washer and electric water heater studies, respectively.

2 Methodology

2.1 Inter-Individual Variability

Inter-individual variability is the key to an assessment of the economic impact of energy efficiency standards on individual consumers or a group of consumers. It represents diversity or heterogeneity in a population (people or events) that is irreducible by additional measurements and refers to, in this paper, varying values of consumers' characteristics. The 1993 and 1997 *Residential Energy Consumption Surveys (RECS)* (DOE 1995, 1999) provide nationally representative samples for most variables related to appliance usage at the individual household level. In addition, RECS provides a framework to implement a probabilistic simulation such that once a household sample is selected according to its weighting, the values of all associated characteristics are used in the calculation so that consistency among variables is maintained.

The RECS data consists of a sample of more than 7,000 households from the population of all primary, occupied residential housing units in the U.S. Each sample household has a weighting factor that accounts for how often a specific household configuration occurs in the general U.S. population based upon the 1990 U.S. Census. The ratio of the weighting factor of a household to the sum of all weighting factors gives a relative frequency for that specific household, on which a probabilistic sampling process is based. The weighted sample is assumed to represent all actual households in the U.S.

The RECS household records contain rich demographic data and energy use information relevant to various residential appliances. Some of these variables illustrate a great deal of variation and ultimately will create the considerable inter-individual variability among the households in terms of energy use of household appliances. For instance, household water heater usage has much to do with the number of occupants and their age distribution. RECS provides detailed counts of number of people (from 1 to 12 or more) in each household and their ages.

When a specific variable cannot be obtained directly from RECS, data can be imputed using related RECS variables and external data sources. Based on a relationship between family size and loads of laundry washed per household per year from a survey study (DOE 1998), a new field is created to represent the variability of household clothes washer use.

2.2 Determination of Consumers' Net Benefit or Cost

Economic impacts on individual consumers from possible revisions to U.S. residential appliance energy-efficiency standards are examined using an LCC analysis. LCC is the total consumer expense over the life of an individual appliance, including purchase expense and operating expenses (which includes expenditures for energy and water). Future operating expenses are discounted to the time of purchase, and summed over the life of the appliance. The impact of standards is a combined effect due to a change in the operating expense (usually decreased) and a change in the purchase price (usually increased). The net result, either benefits or costs to the consumer, is the net change in LCC when comparing alternative efficiency levels corresponding to possible new standards to the current base. LCC and change in LCC can be defined by the following equations:

$$LCC = P + \sum_{t=1}^{n} \frac{O_t}{(1+r)^t}$$

$$\Delta LCC = LCC_{standard} - LCC_{base}$$

where: P = Purchase expense (\$),
        Σ = Sum over year t (t = 1, 2, ..., n; lifetime of appliance),
        O_t = Annual operating expense (\$),
        r = Discount rate.

If the change in LCC (ΔLCC) is negative, then there is a benefit (net savings) to the consumer; if positive, there is a net cost to the consumer. Based on this criterion, households benefiting from new efficiency standards can be distinguished from those that are bearing a loss. A Monte Carlo simulation is implemented with 10,000 trials using Microsoft Excel™ in Windows 98™, combined with Crystal Ball™ (a commercially available add-in program). During the sampling process, household records are selected based on their relative frequency derived from weightings. Those variables from sources other than RECS are simultaneously sampled according to the statistical distributions by which they are defined. Sampled values of the variables are then used in calculating LCC for various efficiency levels and the baseline. The resulting ΔLCC distribution presents an impact profile of the efficiency standards on U.S. consumers.

The primary results are depicted in two types of charts: 1) a *frequency chart* (Figure 1) showing the range of ΔLCC values with their corresponding probabilities of occurrence and 2) a *cumulative chart* (Figure 2), an integral of the frequency distribution, showing the probability that ΔLCC will be less than or equal to a certain given value. As illustrated in Figure 2, Prob(ΔLCC # 0) = 0.79 indicates that for the 35% efficiency improvement level of clothes washers, 79% of household samples will have reduced LCC compared to the baseline.

For those household samples with net savings (negative ΔLCC - shown as the shaded area in Figure 1), the 35% efficiency improvement level of clothes washers provides them reduced operating expense—energy and water—greater than the increased purchased expense. The average LCC reduction for all 10,000 samples is \$242. The chart shows a range of change in LCC from net savings of $1250 to net cost of $750 depending on household characteristics such as number of occupants in a family. (The extreme minimum and maximum values are off-scale outliers, and are provided in Table I).

Since the level of impact of energy efficiency standards varies with different households, certain groups of consumers, such as households with lower income levels, may be disproportionately affected. To evaluate the impacts on any identifiable groups, the simulations are conducted for two subgroups of the population derived from RECS: low- income and senior. Low-income households are defined as at or below 100% of poverty level and senior households are those whose household head is over 65 years old. Table I shows that low-income households have a slight higher fraction (80.7%) that benefits from the 35% efficiency improvement level than the general population (79.1%), while the senior-headed households have a lower (70.6%) percentage of beneficiaries under the new standard.

Table I: ΔLCC (1997 US$) by Percentiles between the General Population and Subgroups for 35% Efficiency Improvement Level

Groups	0%	10%	25%	50%	75%	90%	100%	Mean	Percent having net savings (□LCC<0)
General	(2,341)	(663)	(406)	(194)	(33)	111	616	(242)	79.1%
Low Income	(2,695)	(773)	(484)	(231)	(43)	100	632	(289)	80.7%
Senior	(2,541)	(462)	(263)	(104)	(22)	165	640	(132)	70.6%

2.3 Identification of Most Important Inputs

The output distribution in Figures 1 and 2 resulting from simulation modeling reflects the variability contributed by all the inputs combined. One needs to understand which of these inputs are key factors in a complicated model that involves many variables. For example, the residential electric water heater life-cycle cost model involves more than 118 input distributions, all represented by statistical distributions (triangular, normal, or used-defined discrete distributions) in five sequential modules (LCC depends on Equipment Cost and Operating Cost. Operating Cost depends on Energy Consumption, which depends on Hot Water Use). Calculations in the five modules are carried out simultaneously in a simulation run. The statistical method can help identify which factors play a dominant role in each of the five modules and consequently act as the driving forces influencing the ultimate result, the impact on consumers.

Furthermore, in the case where uncertainty (due to incomplete or insufficient data) is being explicitly addressed in the model, we want to determine which uncertainties merit an investment in additional data collection or analysis so that we can reduce that uncertainty and enhance the model accuracy. Determining which variables are important and which are not will enable us to simplify or improve the model with less effort by focusing only on those few inputs that are most relevant.

To identify the key variables, we use rank order correlation (or Spearman correlation) between each of the input distribution samples and the output distribution. It is measured by using the ranks of the samples, with the largest value assigned a rank of 1 and the smallest, the rank of n, to calculate their correlation. By using the ranks of samples, rather than the data values, the correlation is less dependent upon the specific distribution shape (Morgan and Henrion 1990). This method is more robust and more widely applicable to the cases where many different types of distributions are involved.

The importance analyses are conducted for each of the five modules in the electric water heater model with HFC-245fa blowing agent. The results of the importance analysis for the design option of 2.5" insulation, the proposed standard level, are depicted in five charts (Figure 3 to Figure 7, one for each module). The horizontal bars in the charts show the magnitude and direction of each input contributing (positively or negatively) to the output in descending order from the top. It is unambiguously shown that the operating expense—an intermediate result—has the dominant role in affecting the economic impact measured by ΔLCC (Figure 3). The operating expense itself is dependent more on the amount of energy consumed than energy prices (Figure 4). Figures 5 and 7 trace back through the model and identify the key primary variables: household composition including number of occupants and their age distribution as well as ownership of other appliances that use hot water such as clothes washers and dishwashers. In this example, the net benefit (or cost) is more dependent on the variability in operating expense than on the variability in equipment costs. On the cost side, Figure 6 shows that markup is the most dominant factor contributing to the variability in consumer equipment costs.

3 Discussion

There are two limitations in this study. One is that uncertainty (errors caused by measurement, subjective inference, and model estimation) is not explicitly addressed. For such a large-scale model as that of a water heater, it is very difficult to separate uncertainty from the inter-individual variability for all variables, let alone to quantify them. Therefore, no effort was made to simulate the effect of uncertainty unless we have sufficient information about the error term of a variable. For instance, lifetime of an appliance (e.g., water heaters) is defined as a triangular distribution using the data from a industry magazine (DOE 2000). Since no information regarding the accuracy is supplied, the distribution itself is considered to represent only the variability of the appliance's lifetime in U.S. households. When the information about errors is available, we have incorporated error terms into the simulation by using appropriately chosen distributions (e.g., normal distributions representing errors in the estimated hot water draw models by EPRI). Although this mixed treatment of variability and uncertainty could exaggerate the outcome variation (flattening the output distribution), their effects are relatively insignificant compared to those representing the variability. The comparison clearly shows in Figure 7 that water heater inlet temperature (Tin), air temperature (Tair), and thermostat setting (Ttank) have smaller correlation with the output (hot water use) than other variables. And the effects of error terms could be further reduced using a larger number of trials because of the nature of their distributions.

The second limitation is that correlations between some variables have been ignored because of lack of information. As an example, one could infer that, typically, larger families would use more hot water and consequently reduce the water heater lifetime. Since we do not have any information regarding the correlation between household size and water heater life, a lifetime sample is simply randomly matched with a sampled household record.

Acknowledgment

This work was supported by the Office of Building Research and Standards of the U.S. Department of Energy under Contract No. DE-AC03-76SF00098.

4   References

[1] Finkel, Adam M., 1990. *Confronting Uncertainty in Risk Management: A Guide for Decision-Makers,* Resources for the Future, Washington, DC.
[2] Morgan, M. and M. Henrion, 1990. *Uncertainty: A Guide to Dealing with Uncertainty in Quantitative Risk and Policy Analysis*, Cambridge University Press.

[3]   U.S. Department of Energy, Energy Information Administration, 1995. *Residential Energy Consumption Survey: Household Energy Consumption and Expenditures 1993*. Washington, DC. Report No. DOE/EIA-0321(93).

[4]   U.S. Department of Energy, Energy Information Administration, 1999. *Residential Energy Consumption Survey: Household Energy Consumption and Expenditures 1997.* http://www.eia.doe.gov/emeu/recs/recs97/publicusefiles.html

[5]   U.S. Department of Energy, Office of Codes and Standards, 1998. *Preliminary Technical Support Document: Energy Efficiency Standards for Consumer Products – Clothes Washers*. Washington, DC. http://www.eren.doe.gov/buildings/ codes_standards/reports/cwtsd/index.htm

[6]   U.S. Department of Energy, Office of Building Research and Standards, 2000. *Technical Support Document: Energy Efficiency Standards for Consumer Products: Residential Water Heaters.*<http://erendev.nrel.gov/buildings/codes_standards /reports/waterheater/index.html>

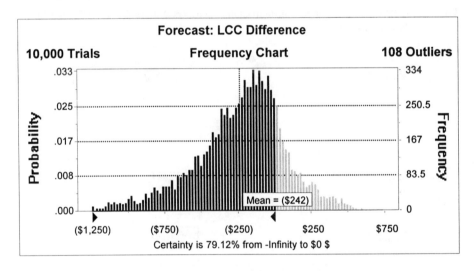

Figure 1: A 35% Efficiency Improvement in Clothes Washers Will Result in an Average Lifetime Savings of $242 Per Washer

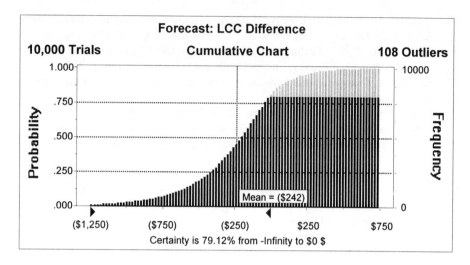

Figure 2: A 35% Efficiency Improvement in Clothes Washers Will Result in Net Savings for 79% of Washers

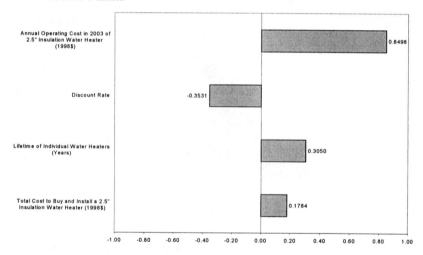

Figure 3: Annual Operating Cost is the Most Important Input Variable to Life-Cycle Cost (for Increasing Insulation to 2.5" on an Electric Water Heater with HFC-245fa)

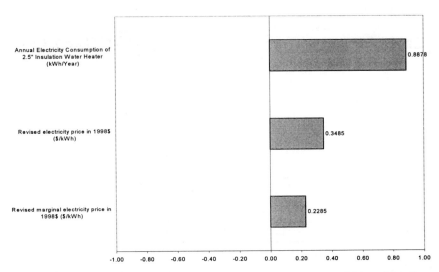

Figure 4: Annual Electricity Consumption is the Most Important Input Variable to Operating Cost (for Increasing Insulation to 2.5" on an Electric Water Heater with HFC-245fa)

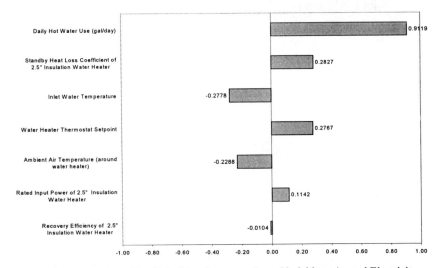

Figure 5: Daily Hot Water Use is the Most Important Input Variable to Annual Electricity Consumption (for Increasing Insulation to 2.5" on an Electric Water Heater with HFC-245fa)

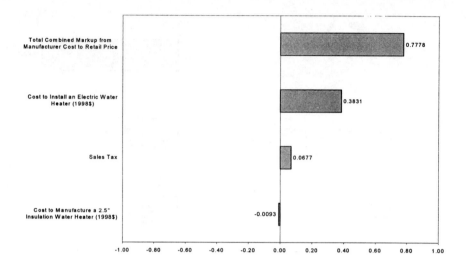

Figure 6: Markup from Manufacturer Cost to Retail Price is the Most Important Input Variable to Equipment Cost (for Increasing Insulation to 2.5" on an Electric Water Heater with HFC-245fa)

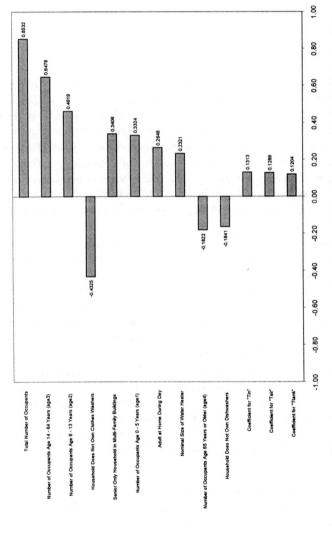

Figure 7: Total Number of Occupants is the Important Input Variable to Hot Water (for Increasing Insulation to 2.5" on an Electric Water Heater with HFC-245fa)

Note: Those inputs with correlation coefficients smaller than 0.1 are not shown on the chart.

Market Transformation in CEEs: Results from the SACHA Projects

Andrea Ricci

ISIS[1], Italy

1 Background and Introduction

Since 1995 the European Commission (Directorate General for Energy), the EU SAVE Programme and the Economic Commission for Europe of the UN (Energy Efficiency 2000 Project) jointly sponsored the so called "SACHA" projects in Central and Eastern European Countries on domestic and tertiary electric end uses. SACHA is the acronym from the title of the first project: "*State of the Art of Cooling Household Appliances Standards, Market and Technology in Central and Eastern European Countries for Energy Efficiency Programmes Implementation in ECE Member State*", completed in October 1997. SACHA-2[2] and SACHA-2.1[3] were then respectively initiated in 1998 and 1999, and both finalised in the course of 2000.

The ultimate goal is to identify and recommend policies, programmes and measures to improve and accelerate harmonistation of energy efficiency policies in the electric end-use sectors, in the framework of the on-going process of EU enlargement. The main objective of the first project was to analyse and interpret the situation of Cooling Household Appliances (CHA) in four ECE Countries (Belarus, Bulgaria, Hungary and Ukraine), in order to: i) increase current knowledge, and ii) identify possible scenarios of improvement with regard to the issues of energy efficiency and environmental friendliness, and therefore iii) identify and propose to National Authorities possible policies and measures in the electric domestic end use sector. Due to the great success of the first exercise, in SACHA-2 and 2.1 the number of investigated Countries was increased to seven, with the Czech Republic, Romania and Slovenia joining the Working Group. Moreover, Washing Household Appliances (WHA) and Domestic/Tertiary Lighting (DTL) were included in addition to CHA.

---

[1] Istituto di Studi per l'Integrazione dei Sistemi - Roma

[2] SACHA-2 project: "State of the Art of Cooling Household and Other Major Appliances Standards, Market and Technology in Central and Eastern European Countries for Energy Efficiency Improvement in ECE Member States".

[3] SACHA-2.1 project: "State of the Art of Lighting Systems and Components Standards, Market and Technology in Central and Eastern European Countries for Energy Efficiency Improvement in ECE Member States".

All projects are led by ISIS (Istituto di Studi per l'Integrazione dei Sistemi) in collaboration with ENEA (Italian National Agency for Energy, New Technology and the Environment), ANIE (Italian manufacturers Association) and the German market research firm GfK. A major asset of the projects is the strong and active participation of CEEC Energy Boards, namely: for Belarus the *Committee on Energy Saving and Energy Supervision*, for Bulgaria the *Bulgarian Foundation for Energy Efficiency* (EnEffect), for the Czech Republic the *Centre for Energy Efficiency* (SEVEN), for Hungary the *Energy Information Agency*, for Romania the *Romanian Agency for Energy Conservation* (ARCE), for Slovenia the *Agency for Efficient Use of Energy* and for Ukraine the *Ukrainian National Academy of Sciences - Institute of Energy Saving Problems*.

2 Project Implementation and Outputs Achieved

2.1 Market Structure

Original data and information to portray the market structure were collected directly on the field through *ad hoc* formats. Among the most significant descriptors is indeed the so called *apparent market*, calculated as "national production + import - export". This, in fact, corresponds to the number of units available for purchasing on the national markets. Sample results are shown in Figures 1-4 (1997 data by product type). The representation of the market structure is further completed by the description of the distribution and sales mechanisms, through the identification of the dominant models and distribution channels, and the density and diversification of the sales networks.

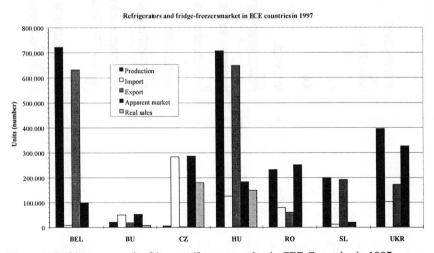

Figure 1: Refrigerators and refrigerator/freezers market in CEE Countries in 1997

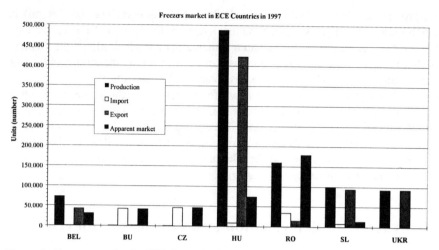

Figure 2: Freezers market in CEE Countries in 1997

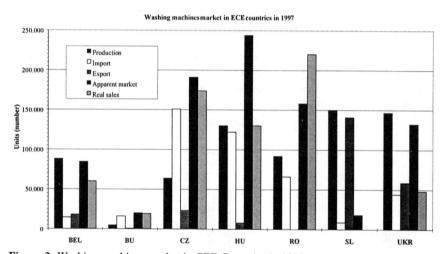

Figure 3: Washing machines market in CEE Countries in 1997

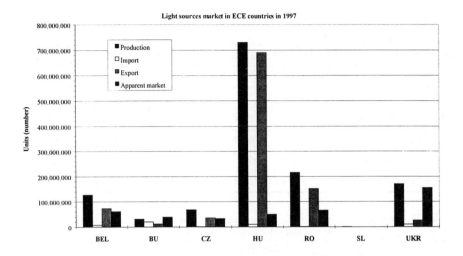

Figure 4: Light sources market in CEE Countries in 1997

2.2 Legal and Regulatory Framework

The state-of-the-art of national regulations was then analysed, based on information collected in each Country through an *ad hoc* format, so as to appraise both the nature and contents of laws and standards currently in force, as well as the organisation and functional structure operating in each national context for the enforcement of the latter, including the identification of standardisation bodies, national programmes and certification laboratories.

2.3 Consumers Habits and Purchase Behaviour

An extensive survey was conducted on the field: no less than 1000 households were interviewed in each Country, providing an original, widely documented picture of the currently installed stocks, as well as of the consumers' behaviour and habits. Examples of the data thus gathered and the output elaborated therefrom are shown in the figures presented hereafter, illustrating the present ownership levels for refrigerators, freezers and washing machines (Table I) and comparing their installed stock (Tables II and III) with population and households number. A description of consumers' purchasing attitude and products selection criteria was also attempted.

Table I: Cooling household appliance and washing machine ownership level in CEE
Countries in 1997

Equipped households (%)	BEL	BU	CZ	HU	RO	SL	UKR
Refrigerators	95.7	96.0	99.1	96.1	92.5	98.7	91.0
Freezers	9.6	31.2	40.4	58.9	35.7	70.7	4.0
Washing Machine, of which:	69.1	77.3	94.8	97.8	79.5	99.1	77.0
Automatic W. M.	**6.5**	**66.9**	**83.4**	**54.8**	**29.4**	**82.5**	**5.3**
Semi-automatic W.M.	5.7	5.8	5.4	3.9	0.5	16.8	12.3
Non-automatic W.M.	87.8	27.3	11.2	41.3	70.2	0.7	82.3

Table II: Estimated number of installed cooling household appliances and washing
machines in CEE Countries households in 1997

Country	Population	Households	Refrigerator + refrig./freezer	Freezer	Washing machine (all models)	TOTAL Appliances
Belarus	10.203.800	3.505.000	4.098.326	339.985	2.594.135	7.032.446
Bulgaria	8.948.388	2.954.577	3.409.972	938.421	2.409.839	6.758.232
Czech Rep.	10.315.000	3.984.000	5.043.628	1.868.742	5.009.492	11.921.832
Hungary	10.092.000	4.050.000	4.485.926	2.508.590	4.452.898	11.447.414
Romania	22.549.925	7.595.194	7.371.106	2.836.665	6.360.600	16.586.371
Slovenia	1.984.923	694.679	703.368	510.235	721.405	1.935.008
Ukraine	51.452.034	19.928.000	20.841.599	797.120	16.177.038	37.815.757
TOTAL	115.496.070	42.935.256	45.935.925	9.799.758	37.725..377	**93.479.060**

Table III: Estimated number of installed light sources in CEE Countries households in
1997

Country	Incandescent (No)	(%)	Halogen (No)	(%)	Dicroic (No)	(%)	Fluorescent (No)	(%)	TOTAL (No)
Belarus	46.967.000	12.4	476.540	8.2	4.346	4.2	921.045	2.4	48.368.931
Bulgaria	50.314.275	13.3	303.849	5.2	975	1.0	2.527.847	6.7	53.146.946
Czech Rep.	61.337.500	16.2	840.624	14.5	38.246	37.4	13.560.261	36.0	75.776.631
Hungary	58.642.155	15.5	1.979.235	34.0	24.057	23.5	8.209.674	21.8	68.855.121
Romania	58.183.000	15.4	444.015	7.6	4.177	4.1	8.680.394	23.1	67.311.586
Slovenia	60.454.240	16.0	394.973	6.8	30.597	29.9	1.821.002	4.8	62.700.812
Ukraine	42.761.000	11.3	1.375.032	23.6	-	-	1.937.002	5.1	46.073.034
TOTAL	378.659.170	100	5.814.268	100	102.398	100	37.657.225	100	**422.233.061**

2.4 Market Shares

The market shares of the leading brands in each Country were assessed. A detailed
analysis of the technical features of the stock was then carried out, taking into
account for CHA, the star rating and its distribution according to the age of

appliances, but also such features as insulation type, defrosting system, number of doors and thermostats, as well as positioning aspects (share of built-in, position of freezer compartments). For washing machines the stock was broken down along machine type and age, horizontal or vertical axis, top or front load, maximum load, washing temperatures, spinning characteristics etc. while for lighting products the type of light source in the different rooms and luminaires were investigated. This allowed in particular to draw a full outline of the characteristics of models installed in each examined Country. Sample outputs are presented in Tables IV-V and Figure 5. In Table VI the contribution of installed cooling household appliances, washing machines and domestic lighting to total national electric energy consumption of investigated Countries is given. Italian figures are also shown for reference purposes.

Table IV: Installed refrigerators and freezers age distribution per Country in 1997

Age class	BEL '95 (%)	BU '95 (%)	CZ (%)	HU '95 (%)	RO (%)	SL (%)	UKR '95 (%)
REFRIGERATORS							
< 5 years	23.6	36.2	32.9	37.9	15.1	27.1	12
5 - 10 years	22.6	16.2	26.1	21.7	18.3	30.9	19.2
10 - 15 years	53.8	47.6	22.6	40.4	25.4	24.1	68.8
> 15 years			18.4		41.3	17.8	
FREEZERS							
< 5 years	94.3	85.7	30.6	34	33.7	25.4	70
5 - 10 years	5.7	11.3	40.4	50.3	28.3	38.3	20
10 - 15 years	0	3	21.1	15.7	27.5	23.1	10
> 15 years			7.8		10.5	13.2	

Table V: Installed washing machines age distribution per Country in 1997

Age class	BEL (%)	BU (%)	CZ (%)	HU (%)	RO (%)	SL (%)	UKR (%)
< 5 years	23.8	26.6	40.7	26.6	23.4	36.5	12.4
5 - 10 years	29.3	26.2	24.6	30.9	17.3	37.2	29.7
10 - 15 years	19.7	22.7	16.5	21.1	23.9	16.6	24.5
> 15 years	27.2	24.5	18.2	21.3	35.4	9.7	33.4

658

Table VI: Cooling household appliances, washing machines and domestic lighting contribution (in %) to total national electric energy consumption in CEE Countries and in Italy in 1997

Domestic End Use	BEL	BU	CZ	HU	RO	SL	UKR	Italy
Cooling Household Appliances	6,2	3,8	4,9	9,4	5,7	4,4	2,9	4,8
Washing machines	1,5	1,4	4,2	3,5	1,9	5,2	0,4	2,6
Domestic Lighting	4,7	2,3	2,3	3,0	7,7	2,5	3,4	3,0
Total	*12,4*	*7,5*	*11,4*	*15,9*	*15,3*	*10,9*	*6,7*	*10,4*
Total national electric energy consumption (TWh/y)	33,5	39,3	47,6	29,8	40,8	10,7	177,8	253,7

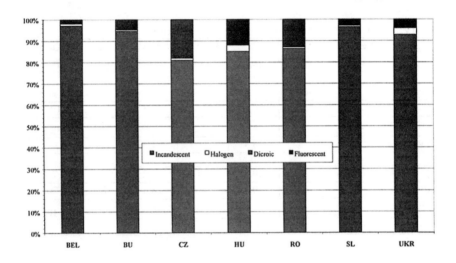

Figure 5: Light source distribution in CEE Countries households in 1997

2.5 Technical Characteristics of New Models Sold on the Market

A detailed description of the new products available on the examined national markets from the point of view of their technical characteristics and performances was carried out. Again, the information used to this end was gathered and organised by means of an *ad hoc* format, documented in close co-operation with the national institutions involved. The basis for this description is the classification of the models sold in each Country, by geographic origin and by product category.

Models classification was carried out according to existing EU Directives or internationally accepted standards[4] when existing.

All major technical characteristics of cooling and washing appliances and lighting sources and their occurrence and distribution in the models sold in 1995/1997 were analysed for the seven involved Countries. In particular, for cooling appliances, gross volume distribution by product category (defined in EU Directive 94/2) was assessed and compared according to star rating and origin of production; refrigerating and insulating fluids were examined in the context of the CFC phase-out and other significant environment oriented international agreements, their distribution highlighted according to country of origin; volume ratio (i.e. External/Total Net) was evaluated as a measure of insulation thickness; distribution along climatic classes was computed; the incidence of no-frost devices was highlighted, and that of other positioning and design characteristics. For washing machines average maximum load, washing performances and spin drying[5] efficiency were also analysed and compared across production countries (in the framework of EU Directive 95/12). Finally, energy consumption was estimated for each national market in terms of the average value (Eaver) for sold models, and for each of the 10 categories identified in the above mentioned EU Directives. Both non-weighted and weighted (with market shares) averages were established. A sample output is shown in Figures 6-10 for the most common appliance categories: refrigerator/freezers and washing machines, comparing national figures with EU models average energy consumption (EUaver). Energy efficiency of models was then calculated, to define the energy profile of the new appliance sold on the market in each Country.

---

[4] For CHA Directive 94/2/CE; for washing machines Directive 95/12/CE; for light sources ILCOS code.

[5] - Washing performance index = washing performance of the model for standard 60°C cotton cycle/ washing performance of the reference washing machines for standard 60°C cotton cycle;

- Spinning efficiency = water remaining after spin as a proportion of dry weight of wash, expressed as percentage

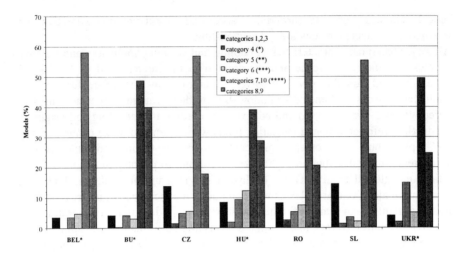

Figure 6: CHA models distribution by category/star rating in CEEC (1995* and 1997)

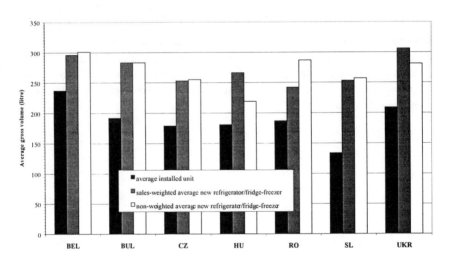

Figure 7: Average gross volume for installed and new refrigerators

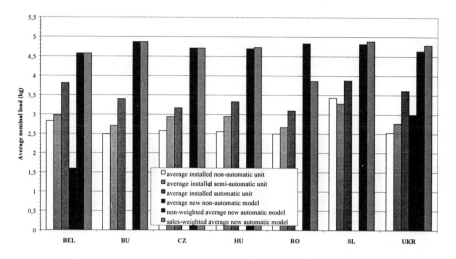

Figure 8: Average maximum load for installed and new washing machines

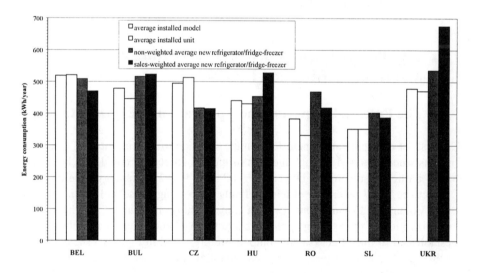

Figure 9: Average annual energy consumption for installed and new refrigerators

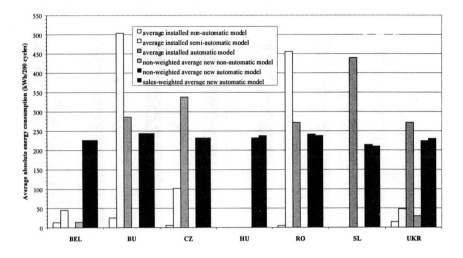

Figure 10: Average annual energy consumption for installed and new W.M.
(200 washing cycles/year)

3 The Policy Impact: Energy Labelling and Minimum Efficiency Limits

3.1 The Potential Impact of the Application of EU Directives

A first set of significant conclusions on the situation of the current CEE Countries stocks when analysed through the lens of the EU Directives could then be drawn, with particular regard to Directives 94/2 and 95/12 on energy labelling and Directives 96/57 on energy efficiency limits for cooling household appliances and the Voluntary Commitment Criteria for washing machines. A (theoretical) classification of the existing appliances into the A-G energy efficiency classes was established to this effect (sample outputs are presented in Figures 11 and 12) and subsequently with respect to the consumption limits set for each of the 10 categories (defined, as already mentioned, in Directive 96/57 and in the Voluntary Commitment). The conclusions amount to a hypothetical Pass-or-Fail verdict for each appliance, should the EU Directives be applied here and now to the examined Countries (Tables VII and VIII).

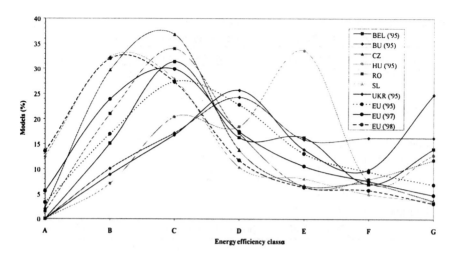

Figure 11: Cooling household appliance distribution in efficiency classes A-G in CEEC and EU in 1995-1997

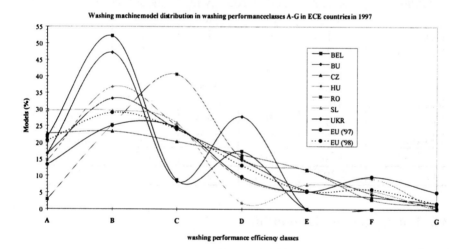

Figure 12: Washing Machines appliance distribution in efficiency classes A-G in CEEC and EU in 1997

Table VII: Cooling household appliance models complying with efficiency limits (Directive 96/57) in EU and in CEE Countries in 1995*-1997

Country	CHA models Refrig. (No)	CHA models Freez. (No)	Refrigerators (No)	Refrigerators (%)	Freezers (No)	Freezers (%)	TOT. (No)	TOT. (%)
			models complying					
Belarus*	60	26	39	65,0	2	7,7	41	47,7
Bulgaria*	178	118	68	38,2	12	10,2	80	30,1
Czech R.	222	47	172	77,5	31	66,0	203	75,5
Hungary*	151	61	51	33,8	20	32,8	71	33,5
Romania	217	55	135	62,2	16	29,1	151	57,6
Slovenia	168	54	137	81,5	30	55,6	167	75,2
Ukraine*	76	25	27	35,5	2	8,0	29	28,7
EU	5.395	2.401	5.068	93,9	1.108	46,1	6.176	79,2

Table VIII: Washing M. appliance models complying with efficiency limits (V.A. Criteria) in EU and in CEE Countries 1997

Country	WHA models (No)	I step models complying (No)	I step models complying (%)	II step models complying (No)	II step models complying (%)
Belarus	31	24	77,4	24	77,4
Bulgaria	57	55	96,5	47	82,5
Czech Rep.	137	131	95,6	123	89,8
Hungary	82	77	93,9	74	90,2
Romania	197	180	91,4	153	77,7
Slovenia	90	89	98,9	89	98,9
Ukraine	42	39	92,9	37	88,1
TOT.	636	595	93,6	547	86,0
EU	4.392	4.361	99,3	4.008	91,2

3.2 Energy Saving Evaluation

Finally the energy saving potential due to the application of the labelling/minimum efficiency requirements measures was calculated (along the methodology developed by TNO). Possible savings due to the application of the minimum efficiency requirements to the cooling appliances are indeed noteworthy. In absolute terms the estimated energy saving potential (for both appliances) is greater than 3.000 GWh/year for all 7 CEECs countries. In terms of CO_2 emissions, considering an average primary electric energy production efficiency of 31% and a power plant emission coefficient of 3,2 tons of CO_2/toe (275 ton/GWh), the mentioned savings could lead to an abatement of 2,7 Mtons/year of CO_2.

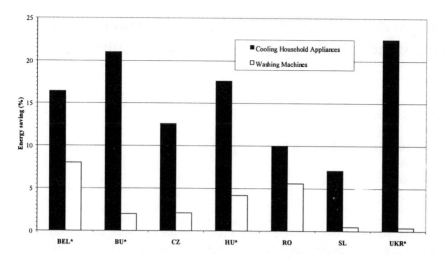

4 References

[1] Commission of the European Communities, SAVE, Directorate General for Energy, SACHA PROJECT – FINAL REPORT, Contract N° XVII/4.1031/Z/95-033, Project developed within the framework of the UN/ECE project ENERGY EFFICIENCY 2000, November 1997.

[2] Commission of the European Communities, SAVE, Directorate General for Energy, SACHA 2 PROJECT – FINAL REPORT, Contract N° 4.1031/Z/97-149, Project developed within the framework of the UN/ECE project ENERGY EFFICIENCY 2000, April 2000.

[3] Ricci, M. Presutto, Results of SACHA Project, First International Conference on Energy Efficiency in Household Appliances, Florence, 10-12 November 1997.

[4] Economic Commission for Europe, The ECE Energy Series, "East-West Energy Efficiency Standards and Labels", ECE Energy Series, Geneva, May 1998.

[5] M. Presutto, "ENEA, ANIE, SAVE-EE 2000/SACHA-II Project, Preliminary Results, May 1999, presented at the UN/ECE Steering Committee of the Energy Efficiency 2000 project, tenth session, Geneva, 31 May - 2 June 1999.

Supporting a Network for Energy Efficiency Labels and Standards Programs in Developing Countries

Mirka F. Della Cava [1], Stephen Wiel [1], Peter du Pont [2], Sood R. Na Phuket [2], Sachu Constantine [3] and James E. McMahon [1]

[1] Lawrence Berkeley National Laboratory
[2] IIEC, Asia
[3] Alliance to Save Energy

Abstract. Much of the developed world has gained experience and success with product energy efficiency standards and labeling programs over the last 20 years, and some developing countries have followed suit. Yet, many developing countries still have little or no experience in the field. Recently, a number of those countries recognized the potential economic and environmental benefits of standards and labeling and have begun to plan or develop such programs.

This paper summarizes the history of standards and labeling programs internationally and notes the current status of such programs. The paper goes on to describe the creation of a network for sharing international experience on the implementation of energy efficiency standards and labels and accessing the financial, technical, and information resources that are available globally for policymakers. A list of countries that have expressed interest in developing and implementing standards and labeling programs is also provided.[1]

1 The State of Standards and Labeling Programs Worldwide

The first mandatory minimum energy efficiency standards in modern times are widely believed to have been introduced in Poland for a range of industrial appliances as early as 1962. The French government set standards for refrigerators in 1966 and for freezers in 1978. Other European governments, including the Soviet Union, collectively introduced legislation mandating one or both of either efficiency information labels or performance standards throughout the 1960s and 1970s. However, much of this early legislation was weak and poorly implemented, had little impact on appliance energy consumption, and was repealed in the late 1970s and early 1980s under pressure to harmonize European trading conditions (Waide, Lebot 1997).

The first energy efficiency standards that dramatically affected manufacturers and significantly reduced energy consumption were mandated in the United States by the State of California in 1974 and became effective in 1977 and were followed later by national standards that became effective in 1990. Now 15 governments around the world have adopted mandatory energy-efficiency standards. Some of these are developing countries that have only recently adopted the standards. A number of other developing countries have taken initial steps in the standards-setting process, and still others are aware the opportunity exists but are searching for the means – financial, technical, and human infrastructure –to make standards happen in their country.

Mandatory labeling programs have developed in parallel with standards. In 1976 France introduced mandatory energy labeling of heating appliances, boilers, refrigerators, clothes washers, televisions, ranges and ventilation equipment, allowing consumers to compare energy performance among similar product models. Germany, Canada and the U.S. shortly followed suit. The U.S. Energy Guide labels for major household appliances became mandatory in 1980, five years after the enabling legislation was passed. Worldwide, no new labeling programs were undertaken after that until Australia implemented its labeling program in 1987. The Australian program, like seven subsequent labeling programs that were created throughout the 1990s around the world, also covers major household appliances (Duffy 1996). Table I shows a history over the past three decades of the introduction of standards and labeling programs. It indicates the order that countries first adopted some element of such programs and the frequency of application of such programs to each product by the various countries. Since the initiation dates shown for each country, some of the countries have vastly expanded and updated their programs. The initial standard levels that were set for products have varied by country. For countries using standards for long-term impact, the intent is for the stringency to be gradually increased over time as part of a basic strategy for coaxing newly emerging energy efficiency technology into the marketplace.

2 Building a Network for Program Assistance

To date there is sufficient experience worldwide with successfully establishing and running standards and labels programs that any new initiatives can benefit by drawing on previous experience and existing examples. The Collaborative Labeling and Appliance Standards Program (CLASP) is a U.S.-based organization formed in 1999 whose mission is to facilitate the design, implementation, and renforcement of energy efficiency standards and labels for appliances, equipment, and lighting products in developing and transitional countries throughout the world. In an attempt to draw on existing experiences and the wide variety of models, CLASP has developed a set of support tools for energy policymakers that includes: examples of country program experiences; sample presentations and materials that demonstrate the benefits and policy arguments in support of labeling and standards; interactive policy analysis tools; and sample labels and standards. These tools are free and available on the newly launched website, www.CLASPonline.org. While it originated in the U.S., CLASP is intended to become a worldwide collaboration open to all organizations engaged in supporting standard and label development.

One of the key tools that will be posted to the website starting October 2000, is the practical *Energy Efficiency Labels and Standards: A Guidebook for Appliances, Equipment and Lighting* designed specifically for developing country policymakers (Wiel, McMahon, et al. 2000). It discusses in step-by-step detail the rationale for standards and labeling programs and the analytical, technical, policy, legal, and regulatory actions necessary to establish successful national standards and labeling programs. It also addresses how to assess the resources needed for successful program development and how to manage the political and strategic issues that arise throughout the process of recognizing and realizing the benefits of such programs. The major steps in this process, as identified in the *Guidebook,* are outlined below in Figure 1

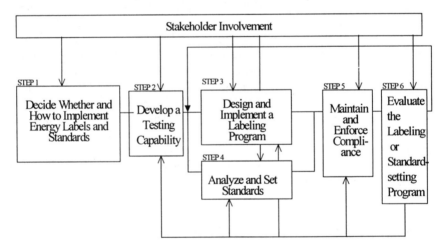

Figure 1: Typical Steps in Developing Product Energy Efficiency Standards and Labeling Programs.

Table I: The History of Energy Efficiency Labels and Standards

Product	France (1966)	U.S. (1976)	Germany (1976)	Russia (1978)	Canada (1978)	Japan (1979)	Taiwan (1981)	Australia (1981)	Brazil (1985)	N. Zealand (1986)	China (1989)	Malaysia (1989)	India (1991)	Korea (1992)	Sweden (1993)	Philippines (1993)	Thailand (1994)	Hungary (1994)	Switzerland (1994)	Mexico (1995)	Hong Kong (1995)	EU (1995)	Singapore (1996)	Norway (1999)	Indonesia (1999)	Poland (1999)	Turkey
Refrigerators	LS	LS	–	s	LS	LS	IS	LS	sL	–	S		s	LS	L	L	–	L	Ls	LS	–	SL	–	L	–	S	
Room AC	L	LS		S	LS	s	IS	L	L	–	S		s	LS		LS	–			LS	–		S				
C. Washers	S	LS	–	S	LS	Ls	–	L		–	S			S	L			L	Ls	S	–	LS	–	L		S	
Freezers		LS		S	LS	s		LS	s	–								L	Ls			SL		L			
Ballasts	L	LS			IS	s	S	L	s	–	S	S		LS		S	–	L	Ls	L		LS	–	L			
C. Dryers	L	S	–	S	LS		S	IS		–						S						S	–				
Water Heaters	L	LS	–	S	IS		S	L		–				L									–				
Dishwashers		LS	–	S	LS					–				LS					Ls					L			
Range/Ovens		S	–	S	LS	s	S			–		S		L				L	s	S		L					S
Lamps	L	LS			IS	s	S				S			LS		LS			L	S			–	L			
Central AC		LS			IS		S	L			S			LS						LS			–			S	
Motors	L	S		S	IS						S	S		L						LS			–				
Televisions		–													–				s								
Monitors	L	–				–		–						L					Ls				–				
Space Heat		S		S														L								S	
Computers		–				ls													Ls				–				

Table I: The History of Energy Efficiency Labels and Standards

Product	France (1966)	U.S. (1976)	Germany (1976)	Russia (1978)	Canada (1978)	Japan (1979)	Taiwan (1981)	Australia (1981)	Brazil (1985)	N. Zealand (1986)	China (1989)	Malaysia (1989)	India (1991)	Korea (1992)	Sweden (1993)	Philippines (1993)	Thailand (1994)	Hungary (1994)	Switzerland (1994)	Mexico (1995)	Hong Kong (1995)	EU (1995)	Singapore (1996)	Norway (1999)	Indonesia (1999)	Poland (1999)	Turkey
Printers		l		S	l	l													L s								
Heat Pumps		LS			lS	s								LS													
Boilers	L	LS												L								S					S
Furnaces		l			S														S								
Copiers		l				ls													L s								
Radio/Stereo		l		S							S																
Fax Machines		l				l																					
Automobiles		L				s								LS					L s								
VCRs		l				s					S								s								
Fans							S				S																
Irons				s																							
Showerheads		LS			l																						
Faucets		LS			l																						
Windows		l			l		S																			S	
Range Hoods																											
Scanners				S		l																					
Rice Cookers											S																
Elec. Kettles				s																							

Table I: The History of Energy Efficiency Labels and Standards

Product	France (1966)	U.S. (1976)	Germany (1976)	Russia (1978)	Canada (1978)	Japan (1979)	Taiwan (1981)	Australia (1981)	Brazil (1985)	N. Zealand (1986)	China (1989)	Malaysia (1989)	India (1991)	Korea (1992)	Sweden (1993)	Philippines (1993)	Thailand (1994)	Hungary (1994)	Switzerland (1994)	Mexico (1995)	Hong Kong (1995)	EU (1995)	Singapore (1996)	Norway (1999)	Indonesia (1999)	Poland (1999)	Turkey
Vac. Cleaners				s																							
Hard Drives						s																					
Skylights		1																									
Doors		1																									
MW Ovens			1																								
Dehumidifiers					S																						
Transformers																				S							
Pumps																				LS							
Pool heaters				S																							
Icemakers					S																						

Symbols: l= voluntary labels; L= mandatory labels; s= voluntary standards; S= mandatory standards

Sources: General data from J. McMahon & I. Turiel, editors, Energy & Buildings special issue, Vol. 26, Num. 1, 1997.

J. Duffy, 1996. Energy Labeling, Standards and Building Codes: A Global Survey and Assessment for Selected Developing Countries

K. Egan and P. du Pont, 1998. Asia's New Standard for Success: Energy Efficiency Standards and Labeling Programs in 12 Asian Countries

C. Murakoshi (Director of the Jyukankyo Research Institute), Appliance Efficiency, Issue 3, Volume 3, 1999.

J. Adnot and M. Orphelin, Appliance Efficiency, Issue 3, Volume 3, 1999.

International Energy Agency, Energy Efficiency Update, Number 22, May 1999.

Asia-Pacific Economic Cooperation, Review of Energy Efficiency Test Standards and Regulations in APEC Member Economies, November 1999

672

The *Guidebook*, information materials, other support tools, and access to work produced by other policymakers, researchers, regulators, advocates and manufacturers in the field worldwide puts the website at the center of this global network. Experiences in program development are shared through CLASP-led regional workshops where experts gather and train together. Often, receiving guidance from a regional leader in standards and labels can speed a country's progress because of possible similarities in the institutional frameworks or even data collection. One example of this is with Mexico, where standards and labeling programs have been in place since 1995. Mexico's Comision Nacional Para El Ahorro De Energia (CONAE) is being funded directly by U. S. Agency for International Development (USAID) to work with CLASP to provide outreach to Latin American nations to promote the development and adoption of efficiency standards and labels for energy consuming products.

Another example of regional leadership is the successful energy labeling program in Thailand. Countries from across Asia have recognized the success of the Thai program, which relies on a voluntary agreement between the electric utility and manufacturers. In early 1994, the Electricity Generating Authority of Thailand (EGAT) approached the five major household refrigerators manufacturers in Thailand and quickly gained their cooperation for a voluntary energy labeling program. Testing began during the fall of 1994, and labels first appeared on store models in early 1995. The remaining refrigerator manufacturers followed suit. The program was accompanied by a massive (~$8 million) national television advertising campaign. A comprehensive process and impact evaluation of the program was conducted in 1999 and found that it was highly cost effective, with a benefit cost ratio to the customer of 2.2 and to the utility of 9.8. From a total resource perspective, the benefit-cost ratio was 2.8. (Agra-Monenco 2000) This type of program shows the potential market transformation impact of labeling, in combination with a strong consumer awareness campaign. A number of countries have sent their energy officials and program managers to Thailand to study the design and operation of the Thai program so that they can adapt the experience in their own country.

2.1 Assessing the Resources Needed for Developing an Energy Efficiency Standards and Labeling Program

Access to information and experience from other countries is a start to developing energy efficiency standards and labeling programs. Actual implementation requires legal, financial, human, physical, and institutional resources. Each of these will already exist to some degree in every country and each is likely to need at least a little, if not major, bolstering to facilitate an effective standards or labeling program. Each country is unique and will need to assess its resources at home before getting started. In the U.S., where national mandatory energy efficiency standards began in 1978, and the program has developed (and in six cases updated) 28 residential and commercial product standards, the investment has proven to be very effective. Over the first 19 years of the program, the U.S. government spent $104 million in developing and implementing these standards. The U.S. government spent an average of $5.5 million annually, with never more than $11.3 million or less than $2.3 million spent in a single year. This corresponded to a range of 2¢ to 12¢ per household per year. The payback on this endeavor has been enormous, with energy benefits well over 100 times the program cost (Koomey et al 1998.).

Experience shows that even a small level of initial investment yields a significant reward. For many developing countries, coming up with that initial investment is difficult. Or even if it is possible, they may lack the technical expertise necessary to complete all the necessary assessments and analyses to develop a proper program with long term viability and benefit. Typically, a combination of financial and technical assistance yields the best results.

2.2 Countries Need Assistance and Access to Technical Support Programs

Even after standards and labeling programs are identified as priority activities and countries make investments in them, to claim success there often remains a recognized need among developing countries for continued financial and technical assistance. In a letter to the Collaborative Labeling and Appliance Standards Program (CLASP), a representative from the Sri Lankan Ceylon Electricity Board (CEB) outlined that even though:

- "CEB has identified appliance energy labeling as an important measure to be implemented jointly with the Sri Lanka Standards Institution in a major drive to assist and encourage customers to select energy efficient appliances," and
- "CEB expects to extend (testing and labeling) to most of the appliances by year 2002,"

there is still a need for follow-on assistance including :

- "additional equipment to strengthen testing capabilities; further training on testing and labeling for testing authorities, suppliers and concept promoters; and....development of large scale publicity programs for labeling."

Similarly, the China Energy Conservation Association expressed that among the issues critical to the successful implementation of energy efficiency standards and labels in China are,

- "Technical assistance services, including policy, analytical, logistical and advocacy support; training education and

compliance programs; advanced measuring and testing methods and equipment," and
- "Convenient access and exchange of the information on standards and labels."

A list of additional countries and institutions that have expressed interest in CLASP's help with labeling and standards programs appears in Table II.[2]

Technical assistance programs are as varied as the countries requesting support. Help is often available through bilateral and multilateral grants and loans to do such things as:
- Assess the potential benefits and costs of standards and labels.
- Establish appropriate legal frameworks for standards and labels.
- Develop test procedures, laboratory services, and labeling schemes.
- Set cost-effective standards, utilizing various analytical methodologies.
- Monitor and report on standards and labels.
- Train government officials, utility company employees, product manufacturers, product distributors and salespeople, architects and designers, environmental activists, and/or consumers in any aspect of the design, development, implementation and use of energy efficiency standards and labels.

The good news is that the benefits of standards and labeling programs are being recognized and actively promoted by a number of institutions. For example, the US President's Council of Advisors on Science and Technology introduced in its 1999 Initiative on Buildings the goal to:

> "Reduce energy use of new buildings in developing and transition economies by 2020 by assisting them to develop efficiency standards, ratings and labeling for building equipment as well as design tools, energy codes, and standards for building shells. Encourage multilateral banks and the Global Environment Facility in support of these measures."

Table II: A Selection of Countries and Institutions Seeking Support for Energy Efficiency Labeling and Standards Programs

Country	Institution(s)
Bahrain	Conservation Department, Ministry of Electricity and Water
Bulgaria	Black Sea Regional Energy Centre Committee of Standardisation and Metrology Center for Energy Efficiency
China	Resources Efficiency and Utilization Department, within SETC China Energy Conservation Association, within SETC China Energy Conservation and Investment Centre Shanghai Energy Conservation Supervision Centre

Egypt	Egyptian Electricity Authority
Ghana	Ministry of Mines and Energy
	Energy Foundation of Ghana
Hungary	International Finance Corporation – Hungary Energy Efficiency Co-financing Program
India	Centre for Environment Education
Indonesia	University of Indonesia
Iran	Ministry of Energy and Sharif University of Technology
Lebanon	Electricite du Lebanon
Malaysia	Department of Electricity and Gas Supply
Mexico	National Commission for Energy Conservation (CONAE)
Poland	National Energy Conservation Agency (NAPE)
	Foundation for Energy Efficiency (FEWE)
	Department of Energy, at the Ministry of Economy
	Network of Energy Cities, Association of Municipalities
Romania	Romanian Energy Policy Association (APER)
	Agency for Energy Conservation,Min.of Industry and Trade
Russia	Ministry of Fuel and Energy
	Center for Energy Efficiency
Saudi Arabia	Energy Research Institute, KACST
Sri Lanka	Ceylon Electricity Board
Thailand	Department of Energy Development and Promotion
	Electrical and Electronics Institute
Tunisia	Agence Nationale Des Energies Renouvelables
Vietnam	Energy Conservation Office
	Vietnam Standard and Consumer Association
Ukraine	State Committee for Energy Conservation
	Agency for Rational Energy Use and Ecology (Arena-Eco)
Yemen	Public Electricity Corporation, Min.of Electricity and Water

In addition, the 1999 United Nations Foundation (UNF) Strategic Discussion on Climate Change states,

> "Within the broad area of the changes required in the energy systems of both developing and developed countries, UNF has chosen two specific programmatic areas which would have a highly leveraged impact on the future development patterns of the developing world: energy efficiency labeling and standards, and community-based rural electrification using sustainable energy technologies."

Several organizations have grant programs providing technical expertise to developing countries specifically for developing energy efficiency standards and labeling programs. The most prominent of these are:

- The United States Agency for International Development (USAID) -- offering training and technical assistance with energy-efficiency standards and labeling programs for most countries, with special emphasis on the Western Hemisphere. USAID has been active in the development of a variety of support tools, such as a Website, a Guidebook and web-based training tools.

- The United Nations Department of Economic and Social Affairs (UN/DESA) -- assisting six Arab countries with energy standards, implementing a refrigerator efficiency project in China, and now offering assistance for all aspects of energy efficiency standards and labeling programs worldwide through a grant from the United Nations Foundation.

- United Nations Economic Commission for Latin America and the Caribbean (UN/ECLAC) - is working with several countries in Latin America to enact legal and regulatory reform for energy standards through a parliamentary approach.

- United Nations Economic and Social Commission for Asia and the Pacific (UN/ESCAP) -has organized workshops in numerous countries in Asia promoting energy standards.

- United Nations Economic Commission for Europe -(UN/ECE) promotes standards under its Energy Efficiency 2000 program and manages some European Commission programs in Eastern Europe

- The Global Environmental Facility (GEF) administered through the World Bank, UNDP and UNEP to provide grants for greenhouse gas mitigation -- For example, GEF has contributed $9.8 million to a $40 million program to improve the efficiency of refrigerators in China, including the development of stringent energy efficiency standards. Other GEF activities include support for standards and labels in Egypt and Tunisia.

- The United Nations Development Program (UNDP) – See GEF entry above.

- The United Nations Environmental Program (UNEP) – See GEF entry above.

- The European Commission's Directorate General for Transport and Energy (DG TREN) sponsors projects to promote energy efficiency programs, including labeling and appliance market transformation, in European Countries outside the European Union. It also has programs to foster collaboration with Latin America and Asia on energy efficiency.

- ADEME, the French Agency for the Environment and Energy Management, collaborates to promote energy efficient appliances in North Africa, the Middle East and Asia.

In addition to grant programs, multilateral banks are increasingly recognizing that energy efficiency standards and labels are cost effective for the implementing government and have been providing loans to fund various aspects of their development. So far we are aware of such loans from:

- The Asian Development Bank (ADB),
- The Interamerican Development Bank (IDB), and
- The World Bank (The International Bank for Reconstruction and Development, IBRD).

Various foundations are also supporting standards and labeling activities. Among these are:

- The United Nations Foundation,
- The Energy Foundation, and
- The Packard Foundation.

Furthermore, there are many other organizations worldwide involved in the various aspects of developing standards and labeling programs. These organizations include manufacturer associations, standards setting organizations, testing laboratories, government agencies, lending institutions, consultants, universities, and public interest advocacy groups.

3 Conclusion

International experience indicates that standards and labeling program benefits are being realized by a handful of countries, most of which are in the developed world. But, the developing world is rapidly recognizing the beneficial effects of this effective policy tool and the implementation of these programs is spreading.

Today, some developing countries are making commitments to standards and labeling programs and assessing their needs to make such programs successful. For each country, including those just beginning, determining its stage in the standards and labels development and/or implementation process and the readiness of its relevant institutions to support such a program will enable it to identify specifically its technical and financial assistance needs. There are a number of financial and technical support institutions that are aware of this need and are involved in realizing the benefits of standards and labels. Organizations like CLASP are working to develop a network for energy professionals and policymakers where they can access and share information, tools and contacts to support their efforts.

Endnotes

[1] The focus of the paper is on standards and labeling approaches for appliances and energy-using equipment. It does not address building codes or standards for building materials.

[2] At the request of the UNF, CLASP and its partner UN/DESA invited national institutions to draft letters of support for the CLASP concept. The countries and institutions cited in this section and listed in Table II responded to this invitation. Their letters of interest are a part of a CLASP/UNDESA UNF Project Document.

4 References

[1] Agra Monenco, Inc. 2000. *DSM Program Evaluation, Conservation Program, Final Report. Volume 5: Impact Evaluation.* Agra Monenco
[2] Duffy, J. 1996. *Energy Labeling, Standards and Building Codes: A Global Survey and Assessment for Developing Countries*, International Institute for Energy Conservation, Washington, D.C.
[3] Du Pont, P., S. R. Na Phuket. 1999. *International Needs Survey: Information on Energy Efficiency Standards and Labeling.* International Institute for Energy Conservation-Asia, Bangkok, Thailand
[4] Greening, L. et al. 1996. *Retrospective Analysis of National Energy Efficiency Standards for Refrigerators.* ACEEE Summer Study. 9: 9.101-9.109
[5] Koomey, J., S. Mahler, C. Webber, and J. McMahon. 1998. *Projected Regional Impacts of Appliance Efficiency Standards for the U.S. Residential Sector.* LBNL-39511, February.
[6] McMahon, J. et al. 1996. *Assessing Federal Appliance and Lighting Performance Standards.* ACEEE Summer Study 9: 9.159-9.165
[7] Rumsey. P. and T. Flanigan. 1995. *Standards and Labeling: The Philippines Residential Air Conditioner Program.* International Institute for Energy Conservation
[8] United Nations Foundation. 1999. *Strategic Discussion on Climate Change.*
[9] U. S. President's Committee of Advisors on Science and Technology, Panel on International Cooperation in Energy Research, Development, Demonstration and Deployment. 1999. *Powerful Partnerships: The Federal Role in International Cooperation on Energy Innovation.* June.
[10] Waide, P., B.Lebot, and M. Hinnells, 1997. *Appliance Energy Standards in Europe.* Energy and Buildings, Elsevier, Vol. 26, No. 1, p.45.
[11] Wiel, S., J. McMahon et al. 2000. *Energy Efficiency Labels and Standards: A Guidebook for Appliances, Equipment and Lighting*, Collaborative Labeling and Appliance Standards Program, Washington, DC. (forthcoming)

The Calculation Method and Potential of Energy Saving in Energy Labeling Program for Indonesia

Rinaldy Dalimi Ph.D.

Electrical Department University of Indonesia

Abstract. Implementation of Energy labeling program is one of the efforts for stimulating the producers of electricity equipment to produce an efficient energy consume, which is compared to other products. As the result, it will create a competition between producers. The competition will be exist if every customer, before buying, asking about the energy label of the electricity equipment which is going to buy, and do not buying it if no energy label on it.

In Indonesia, base on the survey, the need of saving energy by customer is still low. Which is 84 % of responders in the survey saying that they do not consider the energy used before buying the electricity equipment. It is because the price of the electricity is low (government subsidizes the electricity in Indonesia). However, the government of Indonesia is going to reduce the subsidy gradually, started from April 2000. It is a chance for implementing of the energy-labeling program.

If competition between the producers is already exist and the improvement of efficiency of the electricity equipment (because of competition) increasing gradually, the potential of energy consume reduction for Air Condition in Indonesia is 588,000 MWh, for lamp is 741,000 MWh, and for electric fan is 7,109 MWh in 2005.

The government of Indonesia is already preparing a Master Plan for implementation of energy-labeling program. The laboratories, which are accredited internationally, are being prepared. And also about the promotion of labeling program to the people is prepared comprehensively, because it is one of the important things for the success of the program. In term of the requirement for energy labeling, the refrigerator, lamp and electric fan are ready to be labeled in Indonesia.

1   Introduction

To control the energy consumed there are three basic approach could be done. First, at the supply side or energy producers, there are several activities could be done such as improve the efficiency of energy producers, increase the added value of energy primer, etc. Second, at the producers of electricity equipment, implementing the energy standard and energy labeling are the program could be done. And third, at the demand side, Demand Side Management activities is one of best.

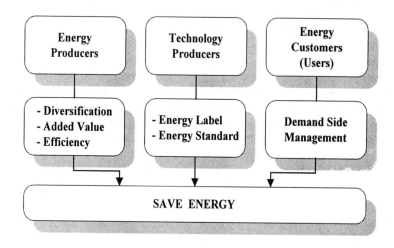

Figure 1: Three strategies for saving energy

Indonesia is preparing the energy labeling program right now. The Master plan of the Energy Labeling program has been made and the regulation for the implementation is being prepared. One of the good thing for the successful of the energy labeling activities in Indonesia is the supporting of producer association However, there is also an obstacle for implementing the labeling program in the near future in Indonesia, because there is not so many testing laboratory available, which is internationally acceptable. That is why, the government of Indonesia is going to implement the program by starting with Refrigerator and lamp because the testing laboratory for both are already available.

Because of the economic growth in Indonesia, as the result of the successful of development, the quality of life is improving and the energy need automatically is increasing. As a matter of fact, there is a close relationship between the energy used per person and his physical quality of life. The problem today, some of the electricity equipment available in the market are not efficient in energy consume. Since the government of Indonesia gives the subsidy for electricity, increasing the energy consume mean increase the total subsidy have to be given to the customers. And because of the subsidy, the price of the electricity is cheaper and the people do not aware about the energy saving. One of a best effort to promote the energy saving program is energy labeling program. The customers will save the energy used automatically if they already used an efficient equipment.

2 Energy Efficiency

The purpose of Energy Labeling programs is to give an information to the customers, before buy an equipment, about how the energy efficiency of the equipment compare to the others. The efficiency information will be more valuable by customers if they aware about the energy saving and environment or if the price of energy is already based on their economic value. It will be described in this section about the methodologies of energy efficiency calculation of Refrigerator, Air Condition, Lamp, and Fan which will be used as a reference for the implementation of energy labeling program in Indonesia.

2.1 Refrigerator

The energy consume calculation for Refrigerator is done in the laboratory with interpolation method which refer to ISO-7371-1995(E) standard. For the Refrigerator, we do not calculate the efficiency but the total energy consume which is compared to the other refrigerator with the same type and size. The formula for the energy consume calculation is :

$$kwh = kwh1 + \frac{(t - t_2)}{(t1 - t_2)}(kwh2 - kwh1)$$

Where :   kWh   :   The energy consumed for 24 hours of Refrigerator at reference temperature.
          kWh1   :   The energy consumed for 24 hours of refrigerator at the higher than reference temperature.
          kWh2   :   The energy consumed for 24 hours of refrigerator at the lower than reference temperature.
          t   :   The reference of temperature.
          t1   :   The higher than reference temperature (the first condition).
          t2   :   The lower than reference temperature (the second condition).

2.2 Air Condition

For the Air Condition energy labeling program, the Efficiency Energy Ratio (EER) is the factor which is going to be compared to the others. EER is the ratio between total cooling capacity (Btu/h) and the electricity needed by the compressor of the Air Condition (watt). The procedure of energy efficiency calculation is based on JIS S151-1994 and ANSI/SHARE 16-1983 standard. And the formula is :

$$EER = \frac{CC\,(\,Btu/h\,)}{P\,(\,Watt\,)}$$

Where   : CC = The cooling capacity
                   P   = The electric consumed

2.3 Fan

The variable is going to be used as a reference for energy labeling of Fan is the ratio between air flow (m3/S) and the electricity consumed by Fan (watt) which is called efficiency of Performance. The test procedure is based on IEC 879 (1986) standard : *Performance and construction of Electric Circulating Fans and Regulators*. The formula is :

$$FPE = \frac{AF\,(\,m^3/s\,)}{P\,(\,Watt\,)}$$

Where   : FPE = the Fan Performance Efficiency
                 AF   = the Air Flow
                 P   = the Electricity consumed

2.4 Lamp

The variable is going to be used as a reference for energy labeling of lamp is Performance Efficiency (PE) of the lamp. The PE is the ratio between light intensity (lumen) and electricity consumed (watt). The test procedure is based on IEC 81-1984, IEC 64-1987, and IEC 921-188 standard. The formula is :

$$LPE = \frac{F\,(\,lumen\,)}{P\,(\,Watt\,)}$$

Where   : LPE = the Lamp Performance Efficiency
                 F   = the Light Intensity of the lamp
                 P   = the electricity consumed

3 Energy Saving Potential

The activities of Energy Labeling program are not directly to produce the saving energy in kWh. However, the activities will encourage the electricity equipment producers to improve the efficiency of their products in term of energy consumed, and changing the attitude of consumers about the energy, where the energy is one of the important factor to consider before they buy an electricity equipment. The main indicator of successful of Energy labeling program is when the people have already asked to the seller how efficient the equipment which is going to be bought compare to the others. The information about efficiency or energy consume is available in the energy label of the product. If everybody has already asked about energy efficiency before buying a product, it will encourage the producers to improve the efficiency performance of their product in order that more efficient than the others. This condition will create the competition between producers to produce an efficient equipment. As the result of the competition, the improvement of efficiency will exist automatically, and the energy saving in the customer side will be existed.

With an assumption that the competition will be existed because of labeling program, we predicted the percentage of the efficiency improvement that can be achieve by each type of electricity equipment available in Indonesia. Base on the prediction, we calculated the total energy save every year, as seen in the table below showing the potential of energy saving until 2010.

Table I: Energy saving projection

Year	Air Condition	Fluorescent Lamp	Incandescent Lamp	Fan
2000	225.801.821	2.067.322	4.273.775	1.743.718
2001	269.329.233	3.078.679	4.586.733	2.010.654
2002	318.375.670	4.426.091	4.899.701	2.318.453
2003	439.973.593	6.055.726	5.212.638	5.346.744
2004	510.851.580	7.993.165	5.525.616	6.165.24

				8
2005	588.238.374	10.263.978	5.838.574	7.109.051
2006	672.131.702	12.893.713	12.303.064	8.197.335
2007	762.536.113	15.907.932	12.928.980	9.452.220
2008	1.289.167.173	19.332.227	13.554.916	14.168.968
2009	1.444.296.802	23.192.169	14.180.832	16.338.014
2010	1.609.186.230	27.513.329	14.806.747	11.185.722

4   Closing Remark

Energy labeling program is going to be important for the global market, it will become one of the factors to be considered in the global competition. Because of that, every country is trying to have and implementing the energy label for their product. However, it must be an effort for the harmonization of the labeling program in order that the label energy from a country can be accepted by others. We hope that the energy label will not become a new tool for some of Industrial Countries to reject the other product come in their market, when the global market is already started.

6   Reference

[1]   *The Master Plan of Energy Labeling Program In Indonesia*, Directorate of Electricity and Energy Development, Jakarta 1998.
[2]   *Testing Procedure of Energy Efficiency for Electricity Household Appliance*, Center for Energy Study University Of Indonesia, Jakarta 1997.

Making it Obvious: Designing Feedback into Energy Consumption

Sarah Darby

Environmental Change Institute, University of Oxford

Abstract. The process of giving feedback on consumption motivates consumers to save energy through reduced waste, yet the body of evidence testifying to this is rarely acted upon in any systematic way. The paper reviews the literature on the effectiveness of three types of feedback to domestic consumers: direct feedback in the home, indirect feedback via billing and 'inadvertent' feedback (a by-product of technical, household or social changes). The lessons learned on the importance of clear, immediate and user-specific information are then applied in a survey of the opportunities for better feedback to consumers in terms of technology, design and location of meters and display panels, energy billing and services such as audits and advice programmes.

The paper concludes that feedback has a significant role to play in raising energy awareness and in bringing about reduced consumption of the order of 10%; and that opportunities exist for designing it into energy-related systems which have yet to be realised.

1 Introduction

While some aspects of energy usage may be highly visible, domestic energy consumption as such is largely hidden from view. This 'invisibility' hampers our ability to learn about how to use energy more intelligently and less wastefully. Evidence from the survey of implementation of the EU directive on labelling of cold appliances in the EU indicates that 'the message about energy saving and the environment has been noted by consumers in every country', but that few actually link the importance of energy saving to their own personal behaviour (Winward et al, 1998). 'Noting' a message is clearly not enough to spur people to action: much work remains to be done to build on a low level of awareness of a need to save energy, by developing peoples' ability to identify what can be done in *specific* terms to improve the situation.

This paper begins an investigation into *the extent to which householders can teach themselves* about energy usage in the way in which they teach themselves about so many other things: by using feedback signals from their own actions and their own consumption.

2 Conceptualising Energy

How do we think of energy? At the level of the individual consumer, in three main ways: as a commodity, a social necessity and an ecological resource (though see Sheldrick and Macgill, 1988, for a fuller account). All of these suggest ways of making consumption more visible, while pointing to shortcomings in policy and practice aimed at carbon reductions.

1. Energy is a *commodity*: much policy is based on this conception. With the liberalisation of utilities, customers have become more aware of fuel price, but most only have fleeting contact with the financial cost of their energy services, when they receive a bill or bank statement or if they change their fuel supplier. Those who are constantly reminded of their usage because they rely on solid or 'packaged' fuel, or because they pay in advance for energy, are in a minority. For the rest, individual metering and prepayment send stronger signals about usage than group metering and payment in arrears (Birka Teknik og Miljo, 1999). For the majority who pay in arrears for their energy, billing can be developed in ways which send more frequent and clearer messages to customers (Kempton, 1995; Wilhite and Ling, 1992).

2. Energy is a *basic human need*: in that sense, it is most noticeable when in short supply. Users of modern energy systems, and especially those living in poverty, need to be able to understand how to control their energy to best effect and to have access to help when it is needed. Feedback and information systems must be as accessible, clear and simple as possible in order to allow for this.

3. Energy is an *ecological resource* – that is, energy use never occurs without side-effects. Production of energy for human use requires mining, tree felling, the growing of fuel crops, gas and oil extraction, the construction of dams, pipelines, power lines and power stations. Some of this production may be highly visible in a localised way, so that there is a vague awareness of the ecological dimension of energy; but electricity and gas, along with carbon dioxide and other waste gases, are largely invisible in consumption. This invisibility comes about in a number of ways: through connection to huge hidden distribution networks; through lack of thought about energy unless it becomes expensive or suddenly scarce; through design for convenience or utility rather than for visibility and learning; and through obscure metering and billing systems.

These conceptions of energy show consumption to have extensive financial, social, ecological and cultural aspects that are inadequately recognised, not least because they are often obscured. It is becoming more and more clear that existing policies aimed at increased efficiency, fuel switching and development of renewables cannot bring about savings in carbon in the timescale necessary to stave off significant climate change. The recent UK study of *Lower Carbon Futures*, for example, concludes that a scenario which does not involve any change to lifestyle, behaviour or standards of service 'will not achieve, by 2020, the reductions in carbon emissions needed to achieve sustainability by 2100. To do so requires behavioural change. Some of this can be encouraged through policy changes, particularly provision of information and feedback to consumers' (Fawcett et al, 2000).

3 How Feedback Works

Feedback is defined as *The modification, adjustment or control of a process or system ... by a result/effect of the process, especially by a difference between a desired and an actual result; information about the result of a process, experiment etc; a response.*

Oxford English Dictionary, 2000

This definition is here applied to the process of learning. Observation of a young child quickly shows how fundamental feedback is as an element in early learning, but we tend to forget that it remains crucial throughout life:

> *We are obliged to act...as intelligently as possible in a world in which...we know very little, in which, even if the experts know more than we do, we have no way of knowing which expert knows the most. In other words, we are obliged to live out our lives thinking, acting, judging on the basis of the most fragmentary and uncertain and temporary information. The point of all this is that this is what very young children are good at doing...The young child is continually building what I like to call a mental model of the world, the universe, and then checking it against reality as it presents itself to him, and then tearing it down and rebuilding it as necessary...We have got to learn...this business of continually comparing our mental model against reality and being willing to check it, modify it, change it, in order to take account of circumstances.* - Holt, 1970

Such an approach to learning helps to explain why environmental information and education do not necessarily lead to behavioural change. Learning is an active process and learning about practical issues is related to 'reality as it presents itself'. Environmental policy aimed at reducing energy use has failed so far to recognise adequately the crucial link between our (generalised) sense of our environment and our (specific) daily needs and actions: there is a need to extend expertise much more widely and to do so by focusing on how people *connect* their lives to the environment (Eden, 1996)

Policymakers have a major contribution to make in providing a 'toolkit' for householders that enables them to learn how to do this. Such a toolkit can be immediate and tangible – as with better direct displays of energy use – but the concept can also be extended to the cultural context (see, for example, Bruner, 1996). Opportunities for learning about energy from the daily usage in homes could connect with learning in the local community, or from interactions with utilities, government and government agencies.

Two general approaches to cutting carbon can be observed. The first begins with identification of carbon reduction targets and aims to meet them in the most efficient way by identifying promising areas for reduction in the hope of persuading or ordering people to implement the necessary actions. The second begins from existing patterns of energy use in their cultural context and looks at needs and aspirations, aiming to identify processes by which people might come to use energy in more environmentally-friendly ways. This paper is concerned more with the second approach: with the processes by which people may learn, by trial and error, to use energy in an ecological fashion.

3.1 Forms of Feedback

The literature on feedback on domestic energy use is limited, but it does supply some pointers as to the approaches most likely to be successful in bringing about energy conservation. A typology, with some examples, provides an outline of what is possible:

A. *Direct feedback: available on demand. Learning by looking or paying.*
  (a) *Direct displays*, such as those tested in Canada and Japan (Dobson and Griffin, 1992; Tanabe, 2000). Customers who have their supply metered in the standard way are unlikely to consult their meter: it will probably be hidden away and difficult to understand. Some more attractive and user-friendly displays of energy usage have now been tested, and the indications are that these do lead to energy savings as well as to increased awareness (eg Tanabe, 2000; Mansouri and Newborough, 1999). An additional benefit is likely to be that better-designed meters will have an appeal because of they will be seen as high quality products: this appears to be the case with high-efficiency cold appliances (Winward et al, ibid).
  (b) *Interactive feedback* via a PC has shown promise and is an obvious candidate for further development (Brandon and Lewis, 1999). Some utilities (eg Scottish and Southern Energy) already offer this service to large business customers.

(c) *Smart meters.* Possibilities include meters operated by smart cards (Birka Teknik & Miljo, 1999) and two-way (automatic) metering (Sidler and Waide, 1999; Kennedy, 1999).

(d) *Trigger devices/consumption limiters.* These are contentious because they can cut the supply of low-income consumers. However, there are possible solutions to this, such as that in use by EdF for providing such customers with help from social services (Ranninger, pers comm.)

(e) *Prepayment meters.* The continued usage of these meters by consumers on low incomes in the UK - in spite of the extra cost - is an indication of the high importance attached to debt avoidance and the value of direct feedback to people with limited resources (Doble, 1999).

(f) *Self-meter-reading.* The review below shows the value of this as part of an effective feedback programme.

(g) *Meter reading with an adviser,* as a tool in energy advice programmes (see LEEP, 1996; Harrigan, 1992).

(h) *Cost plugs* or similar devices on appliances (though they tend to be complicated to operate and can be unreliable).

B. Indirect feedback – raw data processed by the utility and sent out to customers. Learning by reading and reflecting.

(a) *More frequent bills,* based on meter readings rather than estimates (Wilhite and Ling, 1992; Arvola et al, 1994).

(b) *Frequent bills based on readings plus historical feedback* - comparison with the same period of the previous year, weather-adjusted. (Wilhite and Ling, 1995).

(c) *Frequent bills based on readings plus normative feedback* - comparison with similar households. (Kempton and Layne, 1994; Wilhite et al, 1999).

(d) *Frequent bills plus disaggregated feedback.* This is relatively expensive, though popular when tested (Wilhite et al, 1999). The NIALMS and DIACE systems allow for automatic end-use breakdown by pattern recognition (Sidler and Waide, 1999).

(e) *Frequent bills plus offers of audits or discounts on efficiency measures.* Frequent, informative bills can stimulate a demand for audits by raising awareness (see Lord et al, 1996).

(f) *Frequent bills plus detailed annual or quarterly energy reports.* See Wilhite et al (1999) and Kempton (1995).

C. Inadvertent feedback – learning by association

There is little in the way of literature on this, but there are pointers to the potential for such feedback.

(a) *New energy-using equipment* in the home, when a person moves house or when there are changes in the physical fabric of the dwelling, provides an opening for effective 'opportunistic' advice (Green et al, 1998).

(b) With the advent of solar water heaters and photovoltaic arrays, the home can become a site for *generation* as well as consumption of power and it is highly likely that this causes increased observation of energy use and a shift in thinking.

(c) A further possibility for inadvertent feedback is the development of community energy conservation projects, with their potential for *social learning* (see, for example, Sharpe and Watts, 1992).

Two further types are worth noting in passing. They are:

D. *Utility-controlled feedback – learning about the customer*

Utility-controlled feedback is not designed with householders' learning in mind, but it is rapidly being developed and debated with a view to better load management.

E. *Energy audits*

Audits are included here because they provide vital baseline information on the 'energy capital' of a dwelling as well as giving guidance on how to improve it. Audits may be

(a) *undertaken by a surveyor on the client's initiative*

(b) *undertaken as part of a mortgage or other mandatory survey*

(c) *carried out on an informal basis by the consumer* using freely available software such as HESTIA or the UK 'EcoCal'. A series of audits can give a stream of feedback, guiding a motivated consumer towards a target consumption. More formal audits are likely to be infrequent, but can still indicate degrees of progress.

The diagram in Figure 1 shows some of the types on two axes, approximately related to the level of immediacy and the extent to which the energy user is in control of finding and using the information:

If feedback is to promote learning, the discussion above would suggest that immediacy and control of the process by the user would tend to lead to the most effective feedback. What does the literature show?

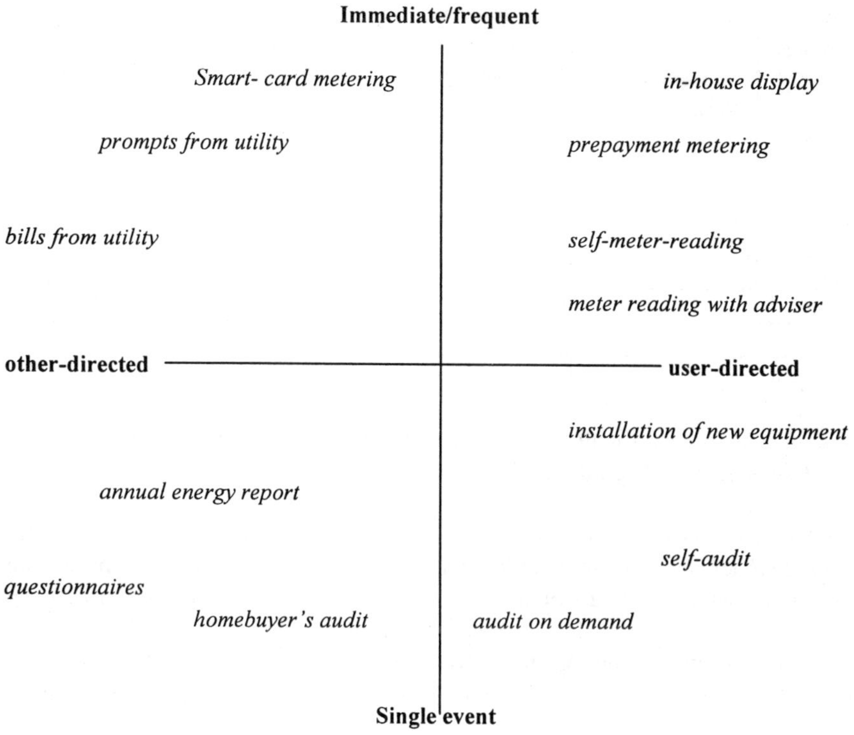

Immediate/frequent

Smart- card metering	*in-house display*
prompts from utility	*prepayment metering*
bills from utility	*self-meter-reading*
	meter reading with adviser

other-directed ——————————————|—————————————— **user-directed**

	installation of new equipment
annual energy report	
	self-audit
questionnaires	
homebuyer's audit	*audit on demand*

Single event

Figure 1: *Feedback in terms of immediacy and control*

3.2 Feedback Effectiveness – a Review

A review of 38 feedback studies carried out over a period of 25 years demonstrated the possibilities of some types of feedback and also some of the issues which affect interpretation of the results. A number of difficulties arise in comparing, and even categorising, these studies: all contain a different mix of elements such as sample size (from three to 2,000), housing type, additional interventions such as insulation or the provision of financial incentives to save, and feedback frequency and duration. The timing of the study itself may also be significant in relation to the energy politics and research paradigms of the period. In spite of these areas of uncertainty, though, some lessons can be learned from the review.

First, feedback has a significant part to play in bringing about energy awareness and conservation. Savings achieved by the 38 projects were as follows:

Table I: Savings demonstrated by the feedback studies

Savings	Direct feedback studies (n=21)	Indirect feedback studies (n=13)	Studies 1987-2000 (n=21)	Studies 1975-2000 (n=38)
20%	3		3	3
20% of peak			1	1
15-19%	1	1	1	3
10-14%	7	6	5	13
5-9%	8		6	9
0-4%	2	3	4	6
unknown		3	1	3

Awareness is more difficult to assess, but an increase in awareness was noted in half of the studies and some continuing or additional effect in 11.

While it is not possible here to go into the detail of each study, it appears that *direct feedback*, alone or in combination with other factors, is the most promising single type, with almost all of the projects involving direct feedback producing savings of 5% or more. The highest savings – in the region of 20% - were achieved by using a table-top interactive cost- and power- display unit; a smart-card meter for prepayment of electricity (coinciding with a change from group to individual metering); and an indicator showing the cumulative cost of operating an electric cooker. In the absence of a special display or a PC display, the feedback was supplied by the reading of standard household meters, sometimes accompanied by the keeping of a chart or diary of energy use. The implication that this meter-reading was a factor in reducing consumption demonstrates how seldom people normally consult their meters (probably hidden away) and/or convert their readings into useful information.

Direct feedback in conjunction with some form of advice or information gave savings in the region of 10% in four programmes aimed at low-income households (with constant or improved levels of comfort), indicating the potential for feedback to be incorporated into advice programmes on a regular basis.

Providing direct financial incentives for consumers to save energy (a method tested during the late 1970s) made little lasting impact: consumption reverted to what it had been once the incentive was removed. Cost signals need to be long-term to have a durable effect.

Where *indirect feedback* is concerned, the range of savings achieved does not go so high, although significant levels are still achievable at relatively low cost (eg Wilhite and Ling, 1995). There was also agreement between most of the studies that interest and awareness levels of consumers were raised as a result of supplying informative bills. One study (Garay and Lindholm, 1995) found no savings at all (but increased customer satisfaction) after providing bills for electricity and water with historical and normative feedback over a period of 18 months. This was an unusual outcome but interesting in that it pointed to at least one possible reason for the lack of change: many of the customers were users of district heating and it could be that they feel less incentive to save than others because of a perception that the heat would be available whether or not they made use of it.

Only three of the studies might be thought of as *inadvertent feedback,* as defined above, but they give an idea of the possibilities for learning using novel technology or situations. The first involved a cable service to over 600 electricity customers which combined energy information to the householder with automatic meter reading, load control by the utility and time-of-use pricing This produced average bill savings (not necessarily energy savings) of 7-10% along with a 2kW peak demand reduction per household (Goldman et al, 1998). The second and third, both unpublished small-scale projects reviewed by Ellis and Gaskell (1978) contained 'trigger' signals which went on when the outside temperature dropped below 68F or when the electricity load went above a specified amount. They achieved a 16% reduction in air-conditioning consumption and a 'moderate' reduction in peak load respectively.

Finally, one community programme involved energy audits for 1,600 households followed by subsidised retrofitting according to customer choice (Sharpe and Watts, ibid). The whole programme was estimated to have achieved a reduction of 20% in peak demand: it could be argued that this was solely due to physical measures, but the strong emphasis on participation and learning suggests a contribution from inadvertent feedback.

In general, there does not seem to be any correlation between the scale of a project and the outcome in terms of reported savings and awareness: the spread of results for the 12 larger-scale projects, with experimental samples of 200-2000, mirrors that for the whole range of studies. Similarly, the best-documented studies show a spread of outcomes which parallels that of the whole range. When the more recent studies are compared with those carried out over the whole 25-year period, the ratio of 'successful' ones (5%+ or 10%+ savings) to the whole is almost exactly the same (although the four most effective projects in terms of savings were all carried out from 1992 onwards).

The implication is that all those studies which demonstrated some effectiveness had enough of a common element (or elements) to succeed; or that they compensated for lack of one element with another. It could be, as a minimal explanation, that *any* intervention helps if it triggers householders into examining their consumption. It could also be that the personal attention of the experimenters motivated the householders into action. However, the documentation of these feedback projects points strongly to other factors at work, of which immediacy or accessibility of feedback data - allowing the householder to be in control - are highly important, accompanied by clear information that is specific to the household in question. Provision of such data is coming well within reach in terms of the technical possibilities for metering, appliance and heating system design. It also requires political determination if it is to be implemented soon.

Feedback is a necessary but not always a sufficient condition for savings and awareness. It should not be treated in isolation: this is also a clear lesson from this review. The range of savings, as well as the accompanying detail, shows the importance of factors such as the condition of housing, personal contact with a trustworthy advisor when needed, and the support from utilities and government which can provide the technical, training and social infrastructure to make learning and change possible.

4   Conclusion

Feedback is an essential element in effective learning: this is as true of domestic energy use as of anything else. A variety of feedback types can be identified and the literature on three – direct, indirect and inadvertent – indicates that they have a significant role to play in raising energy awareness and in bringing about reduced consumption of the order of 10%.

A number of lessons emerge from the literature. Metering displays should be provided for each individual household in a form that is accessible, attractive and clear. Signals which are activated when a given load is exceeded may have potential – though not in isolation, without the means to learn from them - but need testing with great caution, especially where low-income households are concerned. Informative billing, designed and tested on customers before becoming widely available, shows promise as a means of raising awareness. Audits can provide baseline information on each dwelling and are increasingly used to assess the quality of the housing stock. The language of the audit should fit with that of the utility and the householder: there should be a common language for maximum clarity.

Feedback implies monitoring, which can be used at an individual or at a collective level: design for feedback should take this into account, as debate on whether energy-saving initiatives are reaching their goals is often ill-informed. Finally, new technologies are making possible generation of power at a household level, automatic and highly sophisticated metering and more detailed communication between utilities and customers. All these developments hold out the possibility for improved learning and control of energy use, if handled with attention to the principles of immediacy, clarity and specificity.

5 References

[1]   Arvola A, Uutela A and Anttila U (1994) Billing feedback as a means of encouraging conservation of electricity in households: a field experiment in Helsinki. *Energy and the consumer,* Finnish Ministry of Trade and Industry.

[2]   Birka Teknik & Miljo (1999) Att kopa el i livsmedelsaffaren. Stockholm Energi/Familjebostader

[3]   Brandon G & Lewis A (1999) *Reducing Household Energy Consumption: a Qualitative and Quantitative Field Study,* Journal of Environmental Psychology 19, 75-85.

[4]   Bruner J (1996) *The culture of education.* Harvard University Press, Cambridge, Mass.

[5]   Doble, M (1999) Why do poor consumers like prepayment? *Energy Action* issue 76, April 1999.

[6]   Dobson JK and Griffin JDA (1992) *Conservation effect of immediate electricity cost feedback on residential consumption behaviour.* Proceedings, American Council for an Energy-Efficient Economy, 1992, 10.33 – 10.35

[7]   Eden SE (1996) Public participation in environmental policy: considering scientific, counter-scientific and non-scientific contributions. *Public Understand. Sci.* 5, 183-204

[8]   Ellis P and Gaskell G (1978) *A review of social research on the individual energy consumer.* Unpublished manuscript, London School of Economics Dept of Social Psychology.

[9]   Fawcett T, Lane K, Boardman B et al (2000) *Lower Carbon Futures.* Environmental Change Institute, University of Oxford.

[10]   Goldman CA, Kempton W, Eide A, Iyer M, Farber MJ and Scheer RM (1998) Information and telecommunication technologies; the next generation of residential DSM and beyond. *Proceedings, American Council for an Energy-Efficient Economy, 1998,* 2.71-2.82.

[11]   Green, J, Darby S, Maby C and Boardman B(1998) *Advice into Action.* EAGA Charitable Trust, UK.

[12]   Harrigan M (1992) *Evaluating the benefits of comprehensive energy management for low-income payment-troubled customers: final report on the Niagara Mohawk Power Partnerships pilot.* Washington DC: Alliance to save energy.

[13]   Holt J (1970) *The underachieving school.* Penguin

[14]   Kempton, W (1995) *Improving residential customer service through better utility bills.* E-Source strategic memo SM-95-1, August 1995.

[15]   Kempton W and Layne LL (1994) The consumer's energy analysis environment. *Energy Policy* 22 (10), 857-866.

[16]   Kennedy RD (1999) Viewpoint – can an investment in AMR be justified? *Metering International* 3, 1999.

[17]   Lord D et al (1996) *Energy Star billing: innovative billing options for the residential sector.* Proceedings of the ACEEE summer study, 2.137.

[18] LEEP (1996) *BillSavers: securing the savings.* A report on the first two years of the Billsavers project. Lothian and Edinburgh Environmental Partnership.

[19] Mansouri I and Newborough M (1999) *Dynamics of energy use in UK households: end-use monitoring of electric cookers.* Proceedings, European Council for an Energy-Efficient Economy, 1999. Panel III, 08.

[20] Ranninger H, Electricite de France, pers comm., 2000

[21] Sharpe VJ and Watts DR (1992) *Beyond traditional approaches to marketing energy conservation: the Espanola experience.* Proceedings, American Council for an Energy-Efficient Economy, 1992, 10.149-10.156.

[22] Sheldrick B and Macgill S (1988) Local energy conservation initiatives in the UK: their nature and achievements. *Energy Policy,* December 1988.

[23] Sidler O and Waide P (1999) Metering matters! *Appliance efficiency* issue 4 volume 3, 1999

[24] Tanabe K (2000) *Energy conservation results of the survey project on the status of energy saving in the residential sector.* Presentation at IEA meeting on standby electricity, Jan 00, Brussels. The Energy Conservation Center, Japan.

[25] Wilhite H and Ling R (1992) *The person behind the meter; an ethnographic analysis of residential energy consumption in Oslo, Norway.* Proceedings, American Council for an Energy-Efficient Economy, 1992, 10.177-10.185

[26] Wilhite H and Ling R (1995) Measured energy savings from a more informative energy bill. *Energy and buildings* 22 pp145-155.

[27] Wilhite H, Hoivik A and Olsen J-G (1999) *Advances in the use of consumption feedback information in energy billing: the experiences of a Norwegian energy utility.* Proceedings, European Council for an Energy-Efficient Economy, 1999. Panel III, 02.

[28] Winward J, Schiellerup P and Boardman B (1998) *Cool Labels*: the first three years of the European Energy Label. Energy and Environment Programme, Environmental Change Unit, University of Oxford.

Impact on Implementing Demand Side Management in Residential Sector

H.A.Rahman[1], M.S.Majid[1], M.Y.Hassan[1] and K.S.Kannan[2]

[1] Faculty of Electrical Engineering, Universiti Teknologi, Malaysia
[2] Faculty of Mechanical Engineering, Universiti Teknologi Malaysia

Abstract. Residential electricity consumption in Malaysia increased at a rate of 14% per year between 1993 to 1997. In 1998, over 60% of population lived in urban areas. The growth of urban population at a rate of 4% per annum is expected to continue to outpace the overall national population growth of 2.2% per annum. With such rapid growth rate Demand Side Management (DSM) activities must be progressive and development – oriented to enhance national productivity objectives.

This paper describes the potential electrical energy reduction through implementing an energy efficient technology in the market. This paper also presents a case study on the energy usage in the urban housing estate and the reduction in kW and kWh with the proposed electrical fittings.

1 Introduction

Electric power is a major energy source in many countries today and it will continue to play an important role in providing electricity for tomorrow's world. Electricity use in Malaysia is increasing rapidly, growing at an average rate of about 9% annually. In most countries, residential electricity consumption ranges from 20% to 40% of total electricity consumption. This energy is used by a variety of appliances providing water heating, food and space cooling, lighting and other end-uses. Data from National Energy Balance indicated that residential and commercial sectors in Malaysia used about 46.3% of the total electricity consumption in 1996. To cater for this growth, the generation, transmission and distribution facilities have to be upgraded continuously. At the same time, concerted efforts will also be undertaken to promote efficient energy utilization and to discourage non-productive and wasteful patterns of energy consumption by demand management. It will ease environmental impact by burning less fossil fuels and constructing fewer new power projects to satisfy the growing energy demand, thereby lowering the need to raise capital to meet other national priorities.

Environmental protection, as one of the most significant global issues to shape our energy path and economic pattern, but we are not sure what we can do to help reduce the growing environmental damage from increased energy use from all kinds [1]. If electricity can be used more efficiently, a significant contribution could be made to the environmental quality. Efficient use of electricity reduces the number and type of power plants that have to be built to meet electricity demands.

Demand Side Management (DSM) is the conservation of energy through reduced demand, to the benefit of consumers, the environment and utility shareholders. DSM could be used to reduce emission to increase the competitiveness of electricity compare to for example natural gas and oil, to fulfill the utilities' role as public service undertaking; to increase the electricity companies competitiveness in a deregulated framework.

DSM also refers to intervention by suppliers across the meter, to modify the customers' use of energy. This 'demand side' includes the meter and all the equipment on the customers' meter is known as the 'supply side'.

This paper highlights the potential electrical energy reduction through implementing an energy efficient technology in the market and also presents a case study on the energy usage in the urban housing estate.

2   System Description

The Tenaga Nasional Berhad (TNB) is currently undertaking electricity supply in Malaysia in Peninsular Malaysia, the Sabah Electricity Board (SEB) in the State of Sabah and the Sarawak Electricity Supply Corporation (SESCO) in the State of Sarawak.

As of 1996, TNB's fourteen major power stations have a total installed capacity about 7,612 MW and about 4,240 MW of capacity from the Independent Power Producers (IPPs). The TNB is a private company with the Government of Malaysia owing over 70% of its equity while SEB and SESCO are statutory bodies.

2.1 Electricity Generation and Consumption in Asia

Table I shows the electricity consumption in selected countries of Asia in 1992 and the projections for the year 1995 and 2025 [2]. A growth of 4.5% is assumed for the year 20-year period between 2005 and 2025.

Table I: Electric Energy Consumption in Asia (billion kWh)

Country	1992	1995	2000	2005	2025*
Asia	1409.3	1787	2467.45	3278.4	7906.5
Bangladesh	6	7.8	12	18.5	44.7
China	701.6	870	1200	1509.8	3641.2
Hong Kong	26.2	31.5	43	58.6	141.3
India	255.4	318.2	398.5	498.9	1203.3
Indonesia	35.6	51.1	93.3	170.3	410.7
Malaysia	**27.1**	**53.2**	**54.1**	**83.2**	**200.7**
Myanmar	1.8	2	2.4	2.9	6.9
Nepal	0.7	0.9	1.5	2.4	5.7
Pakistan	43	53.6	81.2	122.2	294.6
Philippines	23.2	33.2	60.9	101.7	245.2
Singapore	16.7	20.7	29.6	42.3	102
South Korea	123.9	165.8	226.1	308.3	743.6
Sri Lanka	3	3.4	4.4	5.6	13.4
Taiwan	93.6	124.9	156.4	195.8	472.2
Thailand	51.6	68.9	104.3	158	381.2

*Based on an assumption 4.5% growth rate from 2005 to 2025.

The rapid growth of electricity consumption in Asia has taxed the existing generating capacity and even resulted in severe shortages in Pakistan, Indonesia and the Philippines in the late eighties and early nineties. It has also resulted in additional generation fossil fuel plants that emit global warming and acid rain gases. It also indicated that the Asia's electricity consumption is projected increase from 1409 billion kWh to 3278 billion kWh in 2005.

Table II: Percentage of electricity generation by fuel type for selected Asian countries (1992)[3].

Country	coal	oil	gas	nuclear	Hydro/others
Asia	57.6	13.9	5.1	6.1	17.2
Bangladesh	0	18	73.1	0	8.9
China	74.1	8.1	0.3	0	17.4
Hong Kong	96	4	0	0	0
India	71.7	3.7	1.2	2.1	21.3
Indonesia	29.8	49	3.8	0	17.4
Malaysia	**11.3**	**33.6**	**40.1**	**0**	**15**
Myanmar	0.2	10.8	41.1	0	47.8
Nepal	0	4	0	0	96
Pakistan	0.1	18.3	29.1	0.8	51.7
Philippines	6.7	55.8	0	0	37.5
Singapore	0	100	0	0	0
South Korea	16.9	27	9.3	43.2	3.7
Sri Lanka	0	18.1	0	0	81.9
Taiwan	33.4	21	2.8	34.4	8.5
Thailand	25.9	26.4	40.2	0	7.4

Table II above shows the percentage of electricity generation by fuel type for selected Asian countries. About 75% of the continent's electricity is generated by burning fossil fuels [3].

2.2 Electricity Generation in Malaysia

The electricity generated in 1998 was about 33,235 GWh of which 48.3% from the natural gas, 28.9% from oil, 11.4% from hydro and 11% from coal. Tenaga Nasional Berhad currently serves about 4.67 million customers of which 3,909,911 domestic, 718,232 commercial, 18,689 industrial, 22,406 public lighting and 51 mining. Table III below shows the breakdown of electrical energy consumed in Peninsular Malaysia by sectors for the period 1993 to 1998 [4].

Table III: Electrical Energy Consumed by Category of Consumers (1993-1998)

Year/ Consumers	1993 (GWh)	1994 (GWh)	1995 (GWh)	1996 (GWh)	1997 (GWh)	1998 (GWh)
Domestic	4,454	5,006	5,800	6,655	7,203	8,516
Commercial	6,957	7,892	9,132	10,352	12,070	13,151
Industrial	13,710	15,932	18,414	20,704	24,606	24,447
Public Lighting	178	208	229	255	290	358
Mining	178	93	81	68	76	68
Export	102	70	33	140	40	26
TOTAL	**25,579**	**29,201**	**33,689**	**38,174**	**44,285**	**46,566**

The growth of electricity demand in the country is driven by the industrial sector and also by the growth of Malaysian middle class who to some extent increase the energy usage. For the next few years, with the increase in usage of electrical appliances, the domestic sector will pose a higher growth in electricity consumption as compared with the industrial and commercial sectors.

3   Overview of DSM Options in Malaysia

Demand side management practices, as part of energy efficiency measures, are adopted by a number of utilities. Tenaga Nasional Berhad, has begun to investigate the advantages of such activities. However, Peninsular Malaysia is in a situation of having excess generation capacity of about 50%, which is above the 35% planned reserve margin. Demand side management may not be relevant at present in the context of saving or deferment on generation plant due to surplus of energy. But this excess of energy will be used up by the year 2000, as shown in Table IV, giving DSM measures an opportunity to play a role in the electricity sector of Peninsular Malaysia [5].

702

Table IV: Electricity Demand Projection and Installed capacity for Malaysia (1996-2005)

Year	Electricity Generation Requirement (TWh)	Peak Demand (MW)	Peak Demand + Margin (MW)	Installed Capacity	Surplus/ Deficit (MW)
1996	37.0	7063	9535	10110	575
1997	41.4	7901	10666	11400	734
1998	45.6	8677	11714	12548	834
1999	50.1	9528	12863	13048	185
2000	55.0	10448	14105	13988	-117
2001	60.4	11451	15459	13988	-1471
2002	66.3	12545	16936	13988	-2948
2003	72.6	13727	18532	15588	-2944
2004	79.5	15006	20258	15588	-4670
2005	87.0	15389	22125	15588	-6537

TNB through its subsidiary, Tenaga Nasional Research and Development Sdn. Bhd. (TNRD) have been promoting DSM consciousness and activities. Besides, the energy efficiency programs of the Ministry of Energy Telecommunication and Post Malaysia (METP) is coordinated by a committee which include representatives from government ministries and departments such as the Information Department, Economic Planning Unit, Department of Environment, representatives from local institutions of higher learning, the Federation of Malaysia Manufactures and also representatives from private sector, particularly the oil companies.

The energy efficiency programs were to help modernise the economic sectors, to build up public confidence, to develop an energy efficiency industry and to educate the public the benefits of DSM.

There are 12 DSM measures selected for detailed study under these programs and the are aggregated according to sectors industrial, commercial and residential. Only five demand side management measures were focused for residential sector. The measures are listed below:-

a)   Compact fluorescent lamp,
b)   Low-loss magnetic ballast,
c)   High efficiency air conditioner,
d)   High efficiency refrigerator,
e)   Turbo ventilator.

4 Case Study

The electricity consumptions study were conducted in Taman Sri Putri and Taman Permas Jaya, Johor Bahru. Both of these housing area were selected because they were located in urban area and there are approximately 310 double storey terrace houses in Taman Sri Putri and Taman Permas Jaya. About 90% are medium class occupants. The purpose of this study was to determine the demand characteristic energy consumption, to identify the major electricity consuming equipment and also to propose methods of implementing DSM to the major electricity consuming equipment. A total of thirty samples were collected from Taman Sri Putri and thirty-seven samples from Taman Permas Jaya.

The analyses of the energy consumption for the normal and energy efficient appliances were based on the samples collected from Taman Sri Putri only.

Table V shows the quantity, rated power, duration used and the energy consumption of each electrical appliance. It also calculated the percentage of utilization of each appliance.

Table V: Energy Consumption for normal electrical appliances

Item	Watts (W)	Time (h)	Quantity (Q)	kWh	%U*	AU*	KWh/Q	%AU
Air Conditioner	1000	97	17	97	18.2	5.71	5.71	29.27
Refrigerator	200	624	26	124.8	23.4	24.00	4.8	24.63
Water Heater	1500	20.5	14	30.75	5.76	1.46	2.20	11.27
Microven	1000	8	3	8	1.5	2.67	2.67	13.68
Lighting 1.Fluorescent	40x10	2024	259	81.68	15.3	7.89	0.32	1.62
2.Incandescent	60x5	486	75	19.44	3.64	6.48	0.26	1.33
Computer	200	310.5	55	62.1	11.6	5.65	1.13	5.79
Television	240	192.5	36	46.2	8.66	5.35	1.28	6.58
Radio	200	318	56	63.6	11.9	5.68	1.14	5.83
TOTAL				**533.6**		**19.49**		

* U:Utilization
 AU: Average Utilization

Figure 1 below shows the percentage of Utilization of Energy Consumption in Taman Sri Putri for normal electrical appliances.

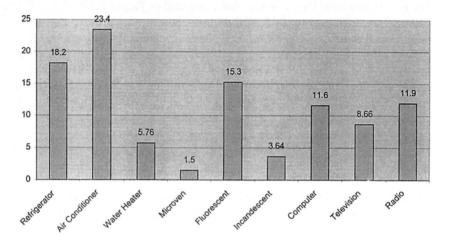

Figure 2 shows the major electricity consuming equipments in Taman Sri Putri.

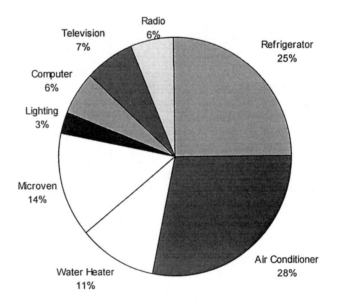

From Figure 1 and Figure 2 the four major electricity consuming appliances are lighting (18.94%), refrigerator (23.39%), air conditioner (18.18%), computer (11.64%) and television (8.66%).

Table VI below shows the quantity, rated power, duration used and the energy consumption of each energy efficient appliance. It also calculated the percentage of utilization of each energy efficient appliance.

Table VI: Analyses of the energy consumption for the efficient appliances if implemented in Taman Sri Putri.

Item	Watt (W)	Time (h)	Quantity (Q)	kWh	%U	AU=H/Q	KWh/Q	%AU
Air Conditioner	746	97	17	72.36	18.02	5.71	4.26	31.83
Refrigerator	100	624	26	62.4	15.54	24.00	2.40	17.95
Water Heater	0	20.5	14	0.00	0.00	1.46	0.00	0.00
Microven	1000	8	3	8.00	1.99	2.67	2.67	19.94
Lighting 1.Fluorescent	36x10	2024	259	73.51	17.29	7.88	0.28	2.00
2.Incandescent	36x5	486	75	17.50	4.36	6.48	0.23	1.74
Computer	200	310.5	55	62.10	15.46	5.65	1.13	8.44
Television	240	192.5	36	46.20	11.50	5.35	1.28	9.60
Radio	200	318	56	63.60	15.84	5.68	1.14	8.49
TOTAL				**405.67**			**13.39**	

Figure 3 below shows the Percentage of Energy Consumption in Taman Sri Putri using the Energy Efficient Appliances.

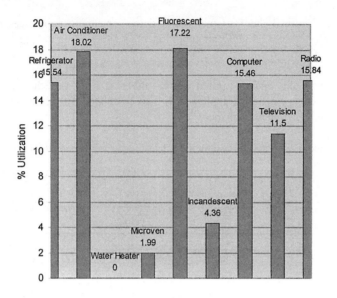

The analysis was performed to compare the energy consumption and savings by using the normal electrical appliances and the energy efficient appliances for one house. This is as shown in Table VII below.

Table VII: Comparison of electricity consumption using the efficient appliances and normal appliances on one house.

	Normal Appliances(kWh)	Energy Efficient Appliances(kWh)	Saving/each(kWh)
Air Conditioner	3.23	2.412	0.818
Refrigerator	4.16	3.224	0.936
Water Heater	1.025	0	1.025
Lighting 1.Fluorescent	2.723	2.45	0.273
2.Incandescent	0.972	0.583	0.389
TOTAL	**12.11**	**8.669**	**3.441**

Total saving/day = Total consumption(without DSM) – Total consumption(with DSM)
$$= 12.11 \text{ kWh} - 8.669 \text{ kWh}$$
$$= 3.441 \text{ kWh}$$

Total saving/month = Total saving/day x numbers of days/month
$$= 3.441 \text{ kWh} \times 30$$
$$= 103.23 \text{ kWh}$$

Total ringgit saved per month = Total savings per month x cost of 1 kWh
$$= 103.23 \text{ kWh} \times \text{RM0.24}$$
$$= \text{RM24.78}$$

Normally the average electricity bill per month for each house is approximately RM70.00

Percentage of saving = [{Total ringgit saved/month}/Electricity Bill] x 100
$$= \{24.78/70\} \times 100$$
$$= \textbf{35.4\%}$$

Figure 4 below is the comparison of average energy consumption for the normal and energy efficient appliances.

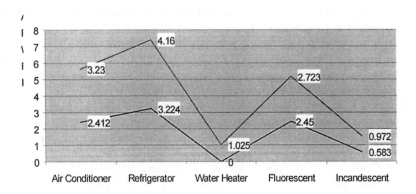

5   Conclusion

The high growth of industrialization in Asian countries has resulted in corresponding increase in their electricity consumption. Malaysia is a developing country with the ambition of becoming a developed nation by the year 2020. The residential sector in Malaysia accounts for 30% of the nation consumption and 35% of the peak load. The rapid growth of electricity demand is due to the increasing consumer demand for electrical appliances such as air conditioners and refrigerators. To promote energy efficiency, the residential sector was seen as a natural starting point for the DSM program, especially to curtail demand in the evening peak hours (18:30 – 21:30).

This paper has provided estimates saving approximately 35% from the impacts of the use of energy efficient appliances in domestic sector. The findings from the survey indicate that majority of the residents are still unaware of the energy efficient appliances. Hence, with the government declared goal of becoming an industrialised nation by 2020, the consumer should use electricity more efficiently and wisely so that a significant contribution could also be made to the environment quality.

6   References

[1]   Saifur Rahman, 'Electric Energy, Environment and DSM', Keynote Paper IEEE International Conference on Electricity Sector Developmeny and Demand Side Management, K.L. Malaysia, 21-22 November 1995.

[2]   Arnulfo de Castro and Saifur Rahman, 'Assessment of Economic Reliability and Environmental Impacts of DSM Technologies',Proceeding IEEE International Conference on Electricity Sector Developmeny and Demand Side Management, K.L. Malaysia, 21-22 November 1995.

[3]   International Energy Agency, Energy Statistics and Balances in Non-OECD Countries: 1991-1992, OECD, Paris, 1994.

[4]   Statistics of Electricity Supply Industry in Malaysia Booklet by the Department of Electricity and Gas Supply Malaysia.

[5]   Maarof Sulaiman and Francis Xavier Jacob, 'Demand Side Management In Integrated Resource Planning of Power Sector In Malaysia', UNDP/ESCAP National Workshop, K.L. Malaysia, 6th May 1996.

[6]   Tee Li Wee, 'Demand Side Management Impacts in Residential Sector', B.Eng Thesis, UTM (1999).

The UK'S Green Electricity Market: Is it Sprouting?

Judith Lipp

Environmental Change Institute, University of Oxford

Abstract. Green electricity (GE) is a generic term for electricity generated from clean, environmentally preferable energy sources. Because of its low-to-zero carbon content, the use of GE is seen as a desirable and important option for the UK's carbon emission reduction strategy. Since 1999 GE has been available to all customers in the UK. The market has had positive beginnings with almost all electricity providers offering, or planning to offer, a green electricity product. Marketing has been launched and consumers are beginning to make the switch to green electricity despite the premium charged. An accreditation scheme guarantees that the green purchases match power entering the grid. While the groundwork has been set for a progressive market to emerge, recent energy policy proposals may undermine the progress made. The Utilities Bill and Climate Change Levy, despite aiming to support renewables, are introducing a number of uncertainties to the market. These uncertainties mean there is some question about where the potential for this market lies. Research carried out by the Environmental Change Institute examines the developments of the UK's GE market; it traces past and present policy and identifies policy options for the promotion of this sector in future.

1 Introduction

Green Electricity (GE) is a generic term for electricity generated from clean, environmentally preferable energy sources such as wind, water, solar, energy-from-waste and energy-from-crops (bio-mass), collectively known as renewable energy. Green electricity is being developed as a customer product in some European countries and is presently available in the UK, the Netherlands, Germany and Sweden. This product distinguishes itself from conventional electricity by giving consumers a choice regarding the environmental impact of their electricity use. Although electricity from renewable energy is produced in many countries, it is usually integrated and sold as conventional electricity (i.e., no distinction is made between 'brown' and 'green' electricity). This paper is interested specifically in green electricity as a product and how it can be encouraged.

Green electricity is largely a phenomenon of the liberalized electricity markets emerging throughout Europe. As the first country to be fully liberalized, the UK's green electricity market has the potential to be the most progressive in Europe, yet at present, development is still slow and uncertain. Progress has been made by way of an accreditation scheme and number of green tariff offers. Customer numbers are low but increasing. In the midst of this progress emerge new policies which are introducing uncertainties about the future growth of this sector. For instance, the newly introduced Utilities Bill will shift the onus for GE growth from the consumer to the electricity suppliers, stipulating renewable energy targets. Although an apparently positive development, the mechanisms suggested to reach these targets could stagnate the market rather than encourage its growth. The Climate Change Levy, which exempts renewable sources, is expected to cause GE demand to rise in the non-domestic sector. Again, this suggests encouragement for the green electricity market, but in reality may frustrate and alienate potential consumers due to shortages in supply.

UK activity in the green electricity and renewable energy sector can serve as an important example for countries considering a green electricity market of their own. This paper thus aims to outline and explore the policies in place in the UK and the effect they have had on the market. The new policy developments are introduced, and additional policy options that could be used to encourage this market are explored.

2   From Renewable Energy to Green Electricity

With privatisation and liberalisation of the UK electricity industry came the diversification of electricity products. Competing for consumers means public electricity suppliers (PESs) have to develop new competitive products with which to retain old, and attract new, customers. Green electricity is one product that has emerged since the liberalisation process began in 1997. Since complete liberalisation of the electricity market in May 1999, every consumer has had the option of signing up for the special GE tariffs offered by most electricity suppliers. But green electricity is a by-product of an earlier policy era; without the generation of energy from renewable sources, green electricity would not be available today.

2.1 A Brief History of Renewable Energy in the UK

The renewable energy industry in the UK was given a policy impetus through the Non Fossil Fuel Obligation (NFFO[1]) in 1990. Under NFFO, electricity providers are obliged to purchase the electricity from renewables generated in their region, thereby creating a market for it (Dukes, 1999). Since inception there have been five NFFO rounds with the last one announced in 1998. Suppliers were subsidized for the additional cost of electricity from renewable sources through the Fossil Fuel Levy, collected from all electricity users. In England and Wales the renewables capacity that became available at the end of the first two NFFO contracts (NFFO1 and 2) was entirely integrated into the competitive electricity system, and formed the major capacity which initiated green electricity tariffs (Mitchell, pers. comm., 2000). Much of this capacity is now sold at premium prices[2]. Presently, renewable generation still supported by NFFO (i.e., NFFO 3, 4 and 5 contracts) cannot be sold as GE, but can be when the contracts end in 2004-2010 (Mitchell, pers. comm., 2000).

The five NFFO orders have between them contracted a total of 3639 MW from renewable energy sources. In 1998 only 706 MW of renewable energy capacity was operational (DTI, 1999), with 318 MW of this from NFFO 1 and 2 contacts (DTI, 1999). Many plants that were commissioned in the 3rd, 4th and 5th NFFO rounds have not yet been built, due in part to planning constraints and planning refusal (EA, 2000). Other non-NFFO green electricity capacity comes from large-scale hydro and bio-fuel generation.

2.2 Green Electricity Accreditation

There are various interpretations of the phrase renewable energy, and consequently which electricity is to be called 'green'. In order to provide clarity and consistency, an accreditation scheme was established. This accreditation, called 'Future Energy', was developed, in consultation, by the Energy Saving Trust and launched in July 1999 (EST, 1999a). 'Future Energy' accreditation covers all non-fossil and non-nuclear sources, but to count as green electricity no more than 50% can come from large scale (greater than 10MW) hydro schemes. The remaining 50% must come from solar,

---

[1] NFFO applies to England and Wales, with similar, but not identical, obligations for Scotland and Northern Ireland made later.
[2] Charging a premium for these may be considered double charging as it has already been subsidised, but it is beyond the scope of this paper to explore the implications of these arrangements.

wind, waves, biomass, biogas, energy from waste, landfill gas or geothermal energy (EST, 1999a).

Backed by the Government, the accreditation scheme aims to raise consumer confidence in suppliers' claims about their green electricity products, thereby stimulating the market for renewable energy. The accrediting body also acts as a regulator, auditing accounts to ensure green electricity contracts are fulfilled. As of March 2000, the EST had accredited 13 of the electricity suppliers offering green electricity products. There has been some objection raised to the inclusion of energy-from-waste as a renewable energy resource, and one supplier has refused accreditation in protest.

The inclusion of energy-from-waste and land-fill gas may pose a problem in the wider EU context. The European Commission, in a recent proposal for a Directive on the promotion of green electricity (CEC, 2000) has excluded energy-from-waste and land-fill gas in its definition of renewable energy. Discrepancies across countries in this regard could hinder inter-European trading of green electricity (or GE certificates); accreditation should be based on standard guidelines developed through the European Commission (Odgaard, 2000).

3 Overview of the UK Market

3.1 Green Electricity Products

There are two broad classifications of green electricity products (also known as green tariffs) currently on the UK market. These are described as 'green source' and 'green fund'. Green source consumers buy electricity from suppliers marketing renewable generation. Although the electrons entering customers' homes and business cannot be guaranteed as green (because all electricity is mixed within the grid), customers are guaranteed that for every unit of electricity they consume, the corresponding amount of renewable generated electricity will enter the network over one year.

Green fund customers, on the other hand, donate money into a fund that supports new renewable capacity or other related initiatives (Lovell, 1999). Green funds are often administered through an independent body established by the supplier, or through an unrelated charity. In some cases the fund will pay for new capacity to be developed by the utility and in others it is invested in new generation by a generator. Green fund contributions are matched by the electricity suppliers in a few cases.

3.2 Electricity Suppliers: Green Electricity Players - Part 1

There are currently 14 electricity suppliers offering a GE product in the UK. A summary of these products is provided in Table I (see end of document). Green source products are offered by eight of the suppliers and four suppliers offer a green fund scheme. Two companies offer both green source and green fund to their customers. All but two companies sell their green product to domestic customers and roughly half are available throughout Great Britain, with the other half offering their products only regionally.

The prices in Table I are based on 3,300 kWh electricity; the average consumption of household lights and appliances each year[3]. The prices given in Table I provide only a rough estimate of the cost of green electricity as GE tariffs vary.

Despite offering GE as a product, few companies are pursuing an intensive marketing strategy to promote their products. Most suppliers view GE as a niche market, which is "still too young and too volatile to set realistic targets" regarding marketing pounds and numbers of customers (Wincott, pers. comm., 1999). Even companies who joined the market with an extensive marketing campaign, such as Unit Energy, have recently slowed the pace because of the introduction of new policies. A representative from Unit Energy has said they slowing their marketing campaign until they have a better understanding of market movement, since domestic consumers may no longer be where the potential for sales lie (Davenport, pers. comm., 2000).

3.3 Consumers: Green Electricity Players - Part 2

At the end of March 2000, the total number of customers to have signed up for a GE tariff was about 15,000 across the UK (EST, pers. comm., 2000). The majority of these are domestic customers, with only about 100 non-domestic users, including government, business and NGOs. This equates to about 0.06% of UK households, and 7223 tonnes of carbon saved.

These numbers appear low when compared with 1999 UK market research indicating "25% of domestic electricity customers, representing 5.7 million households, would be interested in a green electricity tariff, even if this means paying a little more than the lowest prices to ensure their electricity comes from renewable sources" (MORI, 1999).

---

[3] The average for all domestic uses (including domestic space and water heating) is 4,200 kWh (DUKES, 1999), but 3,300 kWh is the figure used by the electricity suppliers.

But there has been progress; since December 1999 the numbers have grown from 10,000 to 15,000. Bearing in mind that having a choice of electricity product and electricity supplier is a new concept for consumers, even this small amount of growth can be viewed with optimism. Combine this with the premium charged for GE and it is little wonder that uptake has been slow. As the Dutch[4] and American[5] markets suggest, there is potential since, respectively about two percent of households have made the switch (Greenprices, 2000; TRENDS, 1999).

3.4 Government: Green Electricity Players - Part 3

Recent events in UK renewable energy policy may mean that domestic customers will not form the main demand pull for green electricity. The Utilities Bill will place the onus on suppliers to achieve a renewable energy target, while the Climate Change Levy is expected to stimulate demand for green electricity through the non-domestic sector.

The Utilities Bill
In response to the need for a new renewable energy policy to replace NFFO, the Government introduced the Utilities Bill in February 2000, which describes a Renewables Obligation and an associated Renewables Scotland Obligation. These 'Obligations' will require all licenced electricity suppliers "to supply a specified proportion of their electricity to their customers from renewable sources of energy. Any additional cost of supplying electricity from renewables will be met by suppliers and may be passed onto their customers" (DTI, 2000). No new levy (to replace the Fossil Fuel Levy) will be charged. A price cap will be imposed to limit the cost to consumers. A fixed price will be established, at which suppliers can buy out their Obligation (in the form of tradable GE certificates), as an alternative to meeting the renewables requirement. The proportion of supplies which must come from renewables is expected to be set at 5% in 2003, and increase to 10% in 2010 (DTI, 2000).

Although in theory this Bill appears to support renewable energy, and consequently green electricity, in practice there are a number of points to consider which may undermine their growth. Because renewable energy is in short supply relative to the

---

[4] In the Netherlands 140,000 households are buying green electricity. GE has been available to all customer since 1999 with the first GE offer made in 1995 (Mikkers, pers. comm., 2000). There are currently 23 GE tariffs available in the Netherlands (Greenprices, 2000).
[5] The percentage for 'America' applies to some States only since not all States offer green tariffs.

targets (currently about 2% to increase to 5% in three years), the most cost effective renewable energy technologies will be supported, effectively killing off developments in costlier but potentially more suitable technologies (e.g., off-shore wind) (Mitchell, 2000). Furthermore, "Cheap renewable power with a premium below the Obligation's buy-out price will be bought up, leaving almost none for marketing through the green tariff", thereby undermining the GE market (Greenergy, 2000). Of course there is still much speculation about the full effect the Bill will have, but these are some of the immediate concerns it raises.

The Climate Change Levy

First announced in the March 1999 Government Budget, the Climate Change Levy (CCL) is an energy tax to be applied to industry, commerce, and the public sector from April 2001. The increase in tax will be offset by a reduction in employers' National Insurance contributions. The use of renewable energy, and energy generated from combined heat and power plants, will be exempt from the Levy (DTI, 2000).

By exempting renewable-sourced electricity from the Levy, the price of green electricity is reduced relative to electricity from other sources. This exemption may potentially create a strong demand pull for green electricity by non-domestic users. The full effect this policy will have on the GE market is difficult to predict and is dependent on the economics for individual users, i.e., the price of GE compared to the cost of paying the CCL, the role of negotiated agreements on energy efficiency, and so forth. But the result may be demand quickly outstripping supply, leading to an increase in price.

4   Achieving Growth in the Domestic Market

The previous sections have shown there is a great deal of uncertainty in the UK's GE market. Although four players are identified in the cycle: the Government; the electricity supply companies; GE generators; and potential customers, what role each will play is still debateable. The onus for achieving renewable energy targets is being placed in the hands of the electricity suppliers. It remains to be seen how suppliers will respond and which customers will have to pay for the additional cost of renewable generation. Electricity suppliers have three main options as a means to meet and pay for their obligation and will probably use all three in some combination:

- Targeting non-domestic users who will be prompted to buy GE through the climate change levy;
- Encouraging domestic demand and charging customers a premium to buy 'pure green' electricity;

- Internalise the additional cost of power from renewables by increasing prices for all electricity users (below the specified price cap) thereby making brown electricity slightly greener.

Despite these uncertainties, the continued growth of GE in the domestic sector is desirable as its demand pull is seen as an important market signal and could help to stimulate development beyond government targets. In the next section means of transforming the domestic GE market are explored.

Market Transformation for Green Electricity

Market transformation (MT) can be described as a strategy to move the market from a point where a particular product has a low or very low market penetration, to a point where it has a very high or completely competitive penetration. While many products are left to the natural forces within the market to make this transition, for some products it may be socially or environmentally desirable to accelerate the less polluting product's uptake. Accelerating the movement from low to high market penetration can be achieved by applying a number of policies including incentive-, regulation- and information-type approaches.

4.1 Information

Information and Education

In order to penetrate any market, consumers, suppliers and other market players need to be aware of the existence of a product. Widespread knowledge of green electricity as a choice for all consumers has not yet been achieved, and consumers need to become informed and educated about green electricity as a product option (Wincott, pers. comm., 1999). Moreover, those who might be aware of green electricity may not necessarily make the connection between their energy use and its environmental consequences and thus see no reason to switch to green. To this end, the case for the creation of a carbon market is made (Fawcett, 2000). Creating a carbon market in which carbon emissions are monitored not only at the national and industry level, but also the household level, could be a powerful way of bringing together green electricity, energy efficiency and fuel-switching policies. Education and awareness complementing this approach are required to help inform consumers and other actors about their carbon impact. The Power Content Label is one means of developing carbon consciousness.

The Power Content Label

Developed in 1998 in California, the Power Content Label (Figure 1) shows consumers the supply mix of their particular supplier in comparison with the average supply mix within the State. It provides the consumer with the means to easily compare the power content of one supplier with that of another – providing transparency, consumer choice and education. The label appears on consumer bills and on all advertisements sent to customers (Davis and Tutt, 1996).

POWER CONTENT LABEL

ENERGY RESOURCES	PRODUCT A* (projected)	1999 CA POWER MIX** (for comparison)
Eligible Renewable	55%	12%
-Biomass & waste	-	2%
-Geothermal	-	5%
-Small hydroelectric	-	3%
-Solar	-	<1%
-Wind	-	2%
Coal	10%	20%
Large Hydroelectric	10%	20%
Natural Gas	16%	31%
Nuclear	8%	16%
Other	<1%	<1%
TOTAL	100%	100%

* 50% of **A** is specifically purchased from indivdual suppliers.

** Percentages are estimated annually by the California Energy Commission based on the electricity sold to California consumers during the previous year.

For specific information about this electricity product, contact (Company Name). For general information about the Power Content Label, contact the California Energy Commission at 1-800-555-7794 or www.energy.ca.gov/consumer

Source: http://www.energy.ca.gov/consumer/power_content_label.html

Figure1: California's Power Content Label

In California, the label was mainly intended to provide transparency about the kinds of products offered, but it can also be used as a valuable tool for educating the consumer and raising awareness about the implications of their energy use. It could be a very useful marketing tool for 'greener' suppliers, providing an important measure of comparison between two companies claiming to offer a similar product. It is an approach that could prove to be effective in the UK market. The label should include the supply mix of all UK suppliers, on average, compared with a particular electricity company's supply mix. One variation on the California label may be to include a measure of carbon intensity of the fuels in use, in this way linking the power content to its environmental (climate change) implications.

The carbon content label would be appropriate for labelling bills of all energy consumers, not only those purchasing a green electricity product. In this way the label is not a policy to promote green electricity specifically, but rather renewable energy more generally. Whether a company was selling GE as a separate product ('pure green') or integrating renewable energy in its supply mix ('greener brown electricity'), would no longer be relevant with this kind of labelling scheme because the label would show the proportion of electricity coming from renewable sources, regardless of how it was sold.

4.2 Incentives

Incentive-type policies can be targeted at any one, or all, of the market players. They are usually in the form of a financial inducement to the consumer to purchase a particular product or to the producer (supplier) to supply a product. Incentives can either be in the form of rebates and subsidies that lower the cost of a product, or in the form of a tax exemption which reduces the burden of purchasing the product, thereby making the purchase of a comparable product more expensive.

In the case of green electricity, most suppliers are charging the consumer a premium for green electricity products, thereby introducing a disincentive to uptake. To help transform the market, the premium needs to be removed and GE seen to be competitive with, or carrying a price advantage over, brown electricity.

4.3 Regulation: Example of Obligated Purchases

Regulatory approaches aim to apply standards and minimum requirements to the producer or supplier of a particular product or service. This approach ensures that a certain standard or level is achieved within a set time frame. In the case of GE, the government has announced it will oblige electricity supply companies to supply a certain percentage of their power from renewable sources under the proposed Utilities Bill. A complementary approach may be to set a carbon target per household, which would encourage not only the purchase of green electricity but also the use of other lower carbon fuels (such as natural gas), and more efficient appliances.

5   Conclusion

Green electricity has taken its first tenuous steps in the liberalised world of electricity supply and demand. Since May 1999, GE has been available to all customers in the UK and consumers can pick and chose their energy supplier and electricity product. Fourteen electricity suppliers are offering some form of green tariff; either green source or green fund, and all but one of these schemes are accredited by the Future

Energy programme. An estimated 15,000 domestic customers have signed up for a GE scheme. For most consumers, options regarding their electricity product are a new and novel concept, consequently, the green electricity market is progressing slowly. Companies view GE as a niche market and are uncertain about pursuing domestic customers due to the introduction of two new energy policies.

There is a need for a market transformation of renewable energy and green electricity. This requires a set of complementary policies that will move green electricity from its niche market position to one where it is integrated into the energy market as a competitive electricity product. Although certain mechanisms have been introduced to this end, much more could, and should, be done. Customers need to become much more aware about the choice of green electricity, its environmental merits, and the implications their electricity use has. A Power Content Label provides one means of achieving this, but needs to be linked to a wider educational process.

Until the price of green electricity and renewable energy is competitive with conventional sources, incentives may need to be provided to ensure its growth. The exemption of renewable energy from the Climate Change Levy provides one such incentive and may lead to a large increase in demand by the non-domestic sector. This growth in demand is encouraging for the industry, but the question of how supply is to keep pace with demand has not yet been adequately addressed. The 5% renewable stipulation by 2003, which may be imposed through the Utilities Bill, is a case in point; it is not clear how these kinds of targets can be met in practice.

These sometimes-conflicting policy developments still need to be properly addressed. A holistic policy approach aimed at transforming and developing the market can be achieved, but requires the integration of various considerations to provide an increasingly greener energy future for the UK

6   References

[1]   CEC (2000). Proposal for a Directive of the European Parliament and of the Council on the Promotion of Electricity from Renewable Energy Sources in the Internal Electricity Market. Commission of European Communities, Brussels.
[2]   Davis, C. and Tutt, T. (1996). The Power Content Label: A Tool For Building A Consumer Driven Market for Renewable Power in ACEEE 1996 Proceedings, Volume 9 - Energy and Environmental Policy.
[3]   DTI (1999). The Energy Report 1999. Department of Trade and Industry.

720

[4] DTI (2000). New and Renewable Energy – Prospects for the 21st Century: Conclusions in Response to the Public Consultation. Department of Trade and Industry.

[5] DUKES (1999). Digest of United Kingdom Energy Statistics – 1999 Edition. Department of Trade and Industry, HMSO London.

[6] EA (2000). The UK Electricity Industry and the Environment 1999. Electricity Association.

[7] ENDS (2000). Renewables industry braces itself for "lean and mean" future in ENDS Report 301, February 2000.

[8] EST (1999a). EST briefing: Renewables provide 'Future Energy'. Energy Saving Trust, October 1999.

[9] EST (1999b). EST briefing: The Case for the Climate Change Levy. Energy Saving Trust, October 1999.

[10] Farhar, B. and Houston, A. (1996). Willingness to Pay for Electricity from Renewable Energy in ACEEE 1996 Proceedings, Volume 9 - Energy and Environmental Policy.

[11] Fawcet, t. et. Al (2000). Lower Carbon Futures for European Households. Environmental Change Institute, University of Oxford.

[12] Fouguet, R. (1998) The United Kingdom demand for renewable electricity in a liberalised market in Energy Policy, Vol 26, No. 4 pp.281-293, March 1998.

[13] Greenergy (2000). Website at http://www.greenergy.com – June 2000

[14] Greenprices (2000). Website on green electricity in Europe http://www.greenprices.nl -June 2000

[15] Lovell, H. (1998). Green Electricity in the UK: a significant new product for the renewable energy industry? MSc dissertation, University of Oxford.

[16] Mitchell, C. (2000). Tradable Green Certificates in The Utilities Journal, May 2000.

[17] MORI (1999) Options in Domestic Energy. MORI Environment Research

[18] Olgaard, O. (2000) The Green Electricity Market in Denmark: Quotas, Certificates, and International Trade. Unpublished report.

[19] TRENDS in RENEWABLE ENERGIES Issues #100-2, and #116   (October 1999 – February 2000)

[20] WHICH on-line (1999). Press release: Electricity Shake Up Leaves Consumers Confused. http://www.which.net/pr/may99/general/confused.html 07/10/99

[21] WHICH (2000). Going Green in WHICH Magazine, February 2000 issue.

[22] Wincott, A. (1999) Personal Communication, December 1999. London Electricity.

Table I: List of UK green electricity suppliers and their products

Company name	Name of Product	Premium	Price/kWh* (includes 5% VAT)	GE Type	Status	Accreditation	Renewable energy generated/ supported	Target Customers
Eastern Electricity	Eco-Power Eco-Power+	5% 10%	6.97 7.13	Fund	Available since November 1997	'Future Energy'	Sun, wind and waves	National / Domestic
London Electricity	No name	5%	Prices tailored to individual customers	Source	Available since Summer 1999	'Future Energy'	Energy from waste	National / Non-domestic
MANWEB (a Scottish Power company)	Green Energy	5%	7.01	Fund	Available since December 1998	'Future Energy'	Wind/small hydro	Regional (MANWEB region) / All customers
Midlands Electricity plc. (MEB) part of National Power	EverGreen	£5/ year donation		Fund	Available since summer 1999	'Future Energy'	All renewable sources	National / Domestic
Northern Electric (GE tariff planned)	Renewable Resources			Source/ Fund	Planned launch spring 2000		Wind/small hydro/landfill gas	

Northern Ireland Electricity	Eco-energy		9.01 (for 10% GE) 9.06 (for 50% GE) 10.15 (for 100% GE)	Source /Fund	Available since fall 1998	'Future Energy'	Mostly wind; PV and small hydro	N. Ireland only / all customers
PowerGen	Green Supply	10%	Prices are tailored to individual commercial users	Source		'Future Energy'	All renewable sources	National / non-domestic
The Renewable Energy Company (STS)	Eco-tricity	No premium charged	Prices are tailored to individual commercial users	Source	Launching GE product for domestic customers in 2000		Hydro/wind/landfill gas/sewage gas/energy from waste	Regional – close to source to avoid distribution costs / Non-domestic
Scottish and Southern Energy (SSE)	RSPB	No premium charged	5.81 to 7.20 (depending on region)	Source		'Future Energy'	Hydro, wind, landfill, bio-gas	National / all customers
SEEBOARD plc	Go green	.5 pence/kWh	10.40	Fund	Available since December 1998	'Future Energy'	Small hydro, wind, solar, waves	SEEBOARD region / Domestic
Southern Electricity	ACORN	5%	6.70	Source		'Future Energy'	Hydro, wind, landfill gas, biofuels, energy-from-waste	Southern electric region / domestic

				Source / Fund	Available	'Future Energy'	Small	Regional / Domestic
SWALEC	Green Energy	Less than 50 p / week	8.10		Available since April 1999	'Future Energy'	Small hydro/tidal/PV/landfill gas	SWALEC region only
SWEB (South Western Electricity Board)	Green Electron	Varies by amount of electricity used	7.64	Source	Available since October 1998	'Future Energy'	Small hydro/wind/landfill gas/ PV	England and Wales / all customers
Unit Energy Ltd. (STS)	Unit [e]	15%	7.36	Source	current	'Future Energy'	Wind/small hydro	England and Wales / all customers
Yorkshire Electricity	Green Electricity	8%	6.43	Source	Available since Summer 1999	'Future Energy'	Wind/biomass	National / all customers

Sources: Company brochures; personal communication; Which 2000; ENDS 1999; FOE website; company websites; EST Future Energy info pack.

Selling a Function Instead of a Product: Renting White Goods Via Functional Service Contracts (FUNSERVE)

Stefan Thomas[1], Christiane Dudda[1], Pelle Petersson[2] and Kai Schuster[3]

[1] Wuppertal Institut für Klima, Umwelt, Energie
[2] New & Future Business, Electrolux Home Products
[3] Energiestiftung Schleswig-Holstein

Abstract. The paper presents first results from the FUNSERVE project, which is co-funded by the European Commission SAVE programme and started in June 1999. The project aims to examine and field test a new concept, which offers customers the services that they need (e.g. refrigeration) instead of an appliance that provides this service. With this concept, the leading European appliance manufacturer Electrolux and the three electric utilities Stadtwerke Leipzig, Stadtwerke Bremen (both Germany) and WIENSTROM (Austria) will jointly offer to customers a package consisting of a very energy-efficient appliance, full maintenance, and perhaps the electricity and water it needs, for a fee collected with the utility bill.

This approach will increase the market share of energy-efficient appliances by overcoming the barrier of a higher initial investment. It can be a successful value-added service for the electric utilities in the liberalised market. And it can reduce waste, since used appliances are expected to be refurbished by Electrolux and rented or sold again.

An intensive customer survey in the cities of Bremen, Leipzig and Vienna yielded promising results. Between 10 and 20 % of the respondents showed a high interest in using such a service, and the proportion was even higher for those who are about to acquire new appliances. Customer surveys and a limited field test in Sweden confirmed this. In Germany and Austria, field-testing of the concept will also be the next step.

1   Introduction

In the European white goods market, successful policy initiatives like the EU labelling scheme, minimum efficiency standards and negotiated agreements, and co-operative procurement, have considerably improved the average energy efficiency of domestic appliances. However, on the market there is still a wide span of the energy consumption required by different appliances serving the same function, making it desirable to increase the market share of the most energy-efficient models (e.g., those with EU label class A). On the other hand, there are a number of barriers for this, most notably the often higher price of energy-efficient models, particularly as the market is increasingly split between high-value, high-efficiency models and low-value, lower-efficiency models. Furthermore, not only energy efficiency but also more efficient use of other resources is needed, so that an extension of the useful life of appliances is a target.

Manufacturers of white goods face a stagnating market, which makes them look for new business opportunities. On the other hand, they face a continuing demand from policy and public to increase the energy efficiency and reduce the resource use of their products. Similar demands are put upon energy and water utilities, who are under public pressure, e.g., to contribute to the abatement of global warming. In the liberalised and competitive European electricity and gas markets, the costs of demand-side management programmes to increase energy efficiency can often not be passed on to customers via the electricity prices as before; this is currently the case in Germany and Austria. Utilities are, therefore, looking for new ways, which are compatible to competition in energy supply, to provide energy efficiency services to their customers.

Against this background, the leading European appliance manufacturer Electrolux, the three electric utilities Stadtwerke Leipzig, Stadtwerke Bremen (both Germany) and WIENSTROM (Austria), and the Wuppertal Institute created the idea to rent to the utilities' customers the services (e.g., refrigeration, cooking, washing dishes and clothes) they need, instead of just selling appliances. This "functional service" would be a package consisting of a very energy-efficient appliance, full maintenance, and perhaps the electricity and water it needs, for a fee collected with the utility bill. After use, the appliance would be collected and refurbished by Electrolux and rented or sold again.

This approach allows several objectives to be addressed simultaneously.
- First, the market share of energy-efficient appliances is increased, thus contributing to increased economic welfare through reduced energy costs, and combating global warming by reducing electricity production and related emissions. The functional service approach considers the total lifetime costs of appliances, which is often lower for energy efficient appliances despite their higher initial cost.

726

- Second, since no initial investment is needed, the functional service approach makes energy-efficient appliances affordable to customers who would otherwise buy cheaper models, which are often inefficient and expensive to run.
- Third, it can thus be a successful value-added service for the electric utilities in the liberalised market, which also allows the utilities to contribute to their environmental targets in a way compatible to competition in energy supply.
- Fourth, it creates new business chances for the manufacturer and the retail trade in a stagnating white goods business, and increases the market share of the high-value, high-efficiency models.
- Finally, it can reduce waste, since used appliances are expected to be refurbished by Electrolux and rented or sold again.

This paper presents first results from the FUNSERVE project. In the next section, the project will be described with a bit more detail. The following two sections will present first results from the German and Austrian partners, and from the development and field-testing in Sweden. First conclusions will wrap up the paper.

2 The FUNSERVE Project

The FUNSERVE project has the aim to develop and field-test the new approach in four EU Member States: Austria, Germany, Sweden and the UK. The project partners are:
- the appliance manufacturer Electrolux AB;
- Stadtwerke Bremen AG;
- Stadtwerke Leipzig GmbH;
- Wiener Stadtwerke WIENSTROM;
- the Wuppertal Institute for Climate Environment Energy (SAVE project Co-ordinator);
- the Environmental Change Institute at the University of Oxford.

The targets of the project are:
- to assess the benefits and costs of the concept for customers, manufacturers, utilities, and society as a whole, and to determine for which appliances and under which conditions the economical and ecological advantages outweigh possible disadvantages;
- to assess the environmental as well as the market potential for the functional service, related to the given or to possibly improved conditions;
- to examine in detail the technical, economical and legal feasibility of this new

approach, depending on the situation, in particular on the degree of liberalisation and the regulatory framework of the electric utility industry, in four European countries chosen as examples;
- to test the market acceptance of such a service through empirical research and a field test;
- to give hints on a possible extension of the new service to other energy end use products and other customer segments.

The project is divided in two main tasks:
- In the first task, the possibilities and chances of the functional service concept were assessed in a concept and feasibility study in all four countries, including empirical research on customers' needs in Austria, Germany and Sweden. This task is now concluded, and the results are the main basis for this paper.
- Based on this, if the feasibility and the market potential looks promising, in the second project task full-scale field tests including both the marketing and the use phases of the service will be held and evaluated by the partners.

The project concentrates on the renting of energy- and water-efficient appliances; the post-use phase (refurbishment and resale) is not the focus, but will be assessed based on existing data.

The FUNSERVE project started in June 1999. The project is co-funded by the European Commission SAVE programme and by Bremer Energie-Konsens GmbH. At the time of writing this paper, in Sweden the concept has been developed using market research, and a small-scale field test in the island of Gotland has been held. In Germany and Austria, the concept has been developed and the empirical research on customers' needs has been performed, but decision on the design of the field tests is pending; more information will be given at the Conference. In the UK, concept development is still ongoing. More detailed information will be found in the project intermediate report to the European Commission, which is soon to be published.

3 First Results from Austria and Germany

3.1 The Concept of FUNSERVE for Austria and Germany

In Austria and Germany, the functional service concept was analysed for a range of most energy- and water-efficient appliances of the AEG brand. This range included almost all major types of white goods – clothes washers and driers, dish-washers, refrigerators, freezers and fridge-freezers, and gas cookers.

A detailed analysis of the existing technical options for measuring electricity and water consumption and for automatically transferring the data to the utility revealed that it is currently still too expensive to include the energy and water consumption into the service, unless a data transmission infrastructure is already in place (see Swedish example below). So the team decided to proceed with the analysis of a renting scheme for the most energy- and water-efficient appliances. This scheme includes

- free delivery and installation, with instructions on the optimal use of the appliance in terms of energy and water consumption;
- free repairs, and a service hotline;
- and removal of the appliance after the agreed contract period.

After the removal, Electrolux will take the used appliance back and refurbish or recycle it.

Some of the Stadtwerke (municipal utilities) are interested to be the supplier of the service, while others only want to be a partner in marketing the service; so the supplier would be either the manufacturer or the retailers. The retailers are important partners for the AEG brand, so it is clear that they should also be partners for the functional service. There are, thus, three possible concepts for organising the service, and for the roles of the partners in the different steps; and sometimes there are alternative options for the steps:

Table I: Roles of the partners in organising the service – three alternatives

Supplier of FUNSERVE	Utility	Manufacturer	Retailer
Marketing	Jointly: utility, manufacturer, retailers	Jointly: utility, manufacturer, retailers	Jointly: utility, manufacturer, retailers
Point of sale	Utility info centre or retailers	Retailers	Retailers
Delivery/ Installation	Retailers / Utility?	Retailers	Retailers
Hotline	Utility	Manufacturer or retailers	Retailers / Utility?
Service	Retailers or manufacturer	Manufacturer or retailers	Retailers
Billing	Utility	Utility as service; manufacturer	Utility as service; retailers
Removal after contract period	Retailers / Utility?	Retailers	Retailers
Refurbishment/ Recycling/ Resale	Manufacturer	Manufacturer	Manufacturer

Electrolux provided appliances prices for wholesale of the new, and for taking back the used appliances, and Electrolux and the utilities provided some other cost data. Based on these and on estimates of the missing data, possible prices for this renting service were calculated. These prices also sought to provide a fair offer to customers compared to buying an appliance, and a fair business to the manufacturer, the retailers, and the utility, after such a service will be fully introduced in the market. However, it proved crucial for an objectively fair offer to the customers that the appliances achieve a certain residual value when sold back to the manufacturer. I.e., the full economic (and ecological) benefits of the concept can only be harvested if really a market for high-quality, manufacturer-refurbished used appliances is established in parallel to FUNSERVE.

3.2 Results of the Customer Surveys

After the concept was developed, it was tested in a number of surveys. The most important survey was based on a written questionnaire, which was sent to 3,000 of their customers by each of the three municipal utilities. This quantitative survey was complemented by focus groups and interviews with both institutional buyers (housing companies, student's homes) and retailers.

The written questionnaire asked for the attractiveness of the offer for a washing machine and a fridge-freezer, using three different combinations of contract duration (3, 5 and 10 years) and price for each of the two appliances. The questionnaire also asked for the motivations of customers why they would use the service, also for other types of appliance, which features would be most interesting, and for some optional features. Furthermore, the questionnaire allowed for differentiation of lifestyle groups and socio-economic factors.

The customer surveys revealed a promising level of interest for this new service. As Figure 1 shows, between 12 and 21 % of the respondents showed a high interest in using the renting service, and in addition more than 20 % may be interested. When looking only at those customers who are about to acquire an appliance, the share of customers who are interested becomes even higher.

It is also interesting that those who are interested in the service consider the utility as the preferred supplier of such a service: Between 60 and 80 % of those who are interested in the service say the utility should be the supplier; 40 % each would accept the manufacturer or the retail trade (multiple answers possible).

Furthermore, depending on the location, most of the interested respondents prefer the 10 years, or the 5 years contract duration, but much less the 3 years duration.

Overall, the survey showed that this renting offer is perceived as something new and different from leasing by the customers. The option to buy the appliance after the contract period, like in leasing schemes, did not receive a high score by the respondents. The interested customers seem to like the freedom from possessing an appliance that the service creates for them.

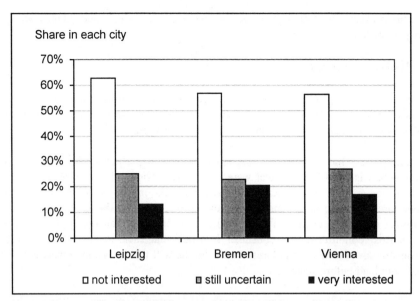

Figure1: Customer interest in the Functional Service in Austria and Germany

4 First Results from Sweden

4.1 Background

In Sweden, the major utility Vattenfall has for the last few years been running a technical project testing intelligent electrical meters. It is among other tasks possible to perform a remote reading of the electricity consumption and get access to various appliances and products connected to the powerline.

Vattenfall and Electrolux have been in discussion about various partnership projects during some years, all related to get the consumers to use energy efficient appliances and get them to use the appliances in a smarter and more energy efficient way.

The technical platform, used in Vattenfall's project in the island of Gotland makes it possible to test an interesting concept of supplying appliances to the households. Instead of selling or renting out the appliances it is possible to get the payment from the consumers to the producer according to how much the appliances are used.

When the project was started, 7.000 smart meters had been installed, mainly around the city of Visby.

4.2 Project Concept

The idea behind the project was to design a business model around selling a function instead of a product, i.e. to base the commercial relationship between the user and supplier of the appliance on how much the appliance is used.

Since the project has its origin in the environmental department of Electrolux, the environmental issues of the project were highlighted. The appliance selected for the test was one of the most energy efficient products in the range of Electrolux. Furthermore one important assumption in the project was that the appliances used in the project should be used in a more energy efficient way than the appliances sold under normal conditions. If a consumer is aware that each washing cycle costs a specific amount of money, it is foreseen that he/she will fill the washing machine with a maximum load before starting it.

The end consumer signs an agreement with AB Electrolux, represented by our local Home-dealer. Initial payment for the consumer is 495 SEK, covering the cost for the installation. The agreement period is not written for a certain time period, instead we set the limit to 1000 "washing cycles". A "washing cycle" is defined as 1 kWh of electricity used. This gives an additional incentive to use the lowest washing temperature possible. For each washing cycle the consumer pays 10 SEK. If the consumer terminates the agreement before 1000 cycles the consumer has to pay for the cost to collect the washing machine (max 500 SEK).
During the duration of the agreement Electrolux owns and services the machine. At the end of the agreement Electrolux is responsible for the used product. It can be scrapped, remanufactured or used as a source of spare parts.
The consumer has the right to get a new machine, after signing a new contract when the agreement ends.

4.3 Project Phases

Potential analysis
650 telephone interviews regarding their view of the Pay-per-Wash-concept.
5 % easy to convince (10 on a scale 1-10)
13 % possible customers on short sight (8-10 on a scale 1-10)
24 % possible customers on longer horizon. (6-10 on a scale 1-10)

Focus group interviews
6 focus groups in Sweden to find out more details about the importance in a Pay-per-Wash-offer, i.e. service level, additional functions etc.

Pricing model test
Investigation of possible pricing models, willingness to pay, value of a washing cycle.

Technical adaptation
Test and preparation of the machines for the test in Gotland. The machines used are standard except for an added energy meter. During the installation an extra connection, compared with a standard installation, has to be placed between the energy meter of the appliance and the electrical meter of the home.

4.4 Involved Parties

AB Electrolux
Owns the machine and signs the agreement with the consumer.

Home-dealer (Local)
Representing Electrolux locally and will provide the service during the agreement period (which will be invoiced to AB Electrolux).

Vattenfall
Owns the infrastructure (=intelligent electrical meters plus communication system) we are using to communicate with the individual washing machines.

GEAB (Gotland Energi AB)
The local utility on Gotland, owned by Vattenfall. Is responsible for invoicing the consumers. Transferring payment and information about washing frequency per consumer to Electrolux.

5   Preliminary Conclusions and Outlook

The customer surveys in Germany, Austria, and Sweden, and the Gotland field test have provided promising perspectives. There seems to be a considerable fraction of domestic customers who are interested in renting an energy- and water-efficient appliance. The retail trade also has shown a high interest in being a partner in such a service. Based on these results, the partners can take a decision on further field tests.

734

These will be needed to prove if really this new approach can provide a fair offer to customers compared to buying an appliance, and a fair business to the manufacturer, the retailers, and the utilities.

Such a service could also yield high environmental benefits. If only 10 % of the appliances market could be converted to a rental market for efficient appliances in some years from now, we estimate that in the EU within 10 years the annual electricity demand could be reduced by ca. 7 TWh.

The savings can be increased, if remote metering, as in the Swedish pilot, becomes available at reasonable prices for more households in Europe, or for implementation in all appliances. Then the energy and water consumption can be included in the price and the bill for the FUNSERVE, making it a real Functional Service, and giving incentives to the customers for optimal use of the appliances, e.g. to use lower wash temperatures for clothes and dish washers, and to use the machines only when they are full.

Furthermore, the economic analysis confirmed that the full economic (and ecological) benefits of the concept can only be harvested, if really a market for high-quality, manufacturer-refurbished used appliances is established in parallel to FUNSERVE. In Sweden, Electrolux has already installed a refurbishment line in one of its factories; in Germany, Austria, and the UK the possibilities for refurbishment still have to be explored.

6 Reference

[1]   Wuppertal Institute et al. 2000: *Selling a Function Instead of a Product: Renting White Goods via Functional Service Contracts (FUNSERVE)*, Project Intermediate Report, Wuppertal (forthcoming)

ENEA Activities on Demand Side Management

Franco Iacovoni and Giuseppe Massini

ENEA

1 Introduction

DSM (Demand Side Management) is based on a series of activities (planning, execution, monitoring and evaluation) designed, promoted and executed by gas and/or electricity utilities, government agencies or private ESCOs (Energy Services Companies) to encourage users to modify the amount and type of energy consumption in order to benefit the companies, the users themselves and society in general.

Thus, the objective of DSM is to modify the trend of energy consumption; in particular, electric energy but without decreasing the services (lighting, heating, air-conditioning, etc.) by acting on the demand.

ENEA's DSM activities mainly involve two areas:

1. the development of methodologies for DSM programme benefit/cost analysis;
2. the promotion and implementation of DSM pilot programmes with electric utility companies.

2 Area No. 1

An initial analysis conducted by ENEA on DSM state of the art has revealed, among other things, the particular importance that DSM programme economic analysis calculation methodologies have for utilities.

Besides allowing the utilities to classify possible activities and choose the most economically effective, these methodologies enable a rationalisation of utility resource planning by including DSM activities among the possible plans that can be undertaken to adjust a firm's capability to face variations in user demand.

A DSM programme entails many complex aspects and parameters such as the system demand curve and its projection over subsequent years, the interested users' demand curve (referring both to the period prior to and subsequent to the application of more efficient systems), the supposed participation of the customer basin and its distribution over the period in which the programme is implemented, as well as incentives, administrative costs, loans and relative interests, costs of marginal production, generation, transmission and distribution, missed income, the possible results of energy source shifts and environmental and social costs as well as other parameters for the most complex programmes.

Therefore, one of ENEA's first activities was to analyse the DSM programming software available on the international market. The only commercially available software of this kind was found in the United States.

The choice fell upon DSManager, a software programme developed by the Electric Power Research Institute (EPRI) to guide American utilities through the analysis of DSM projects. DSManager is commercialised by EPS Solutions Inc., Minneapolis (MN).

DSManager runs the standard tests outlined in the California "Standard Practice Manual for Economic Analysis of Demand Side Management Programs." These include the Utility Test, the Participant Test, the Ratepayer Impact Test, the Societal Test and the Total Resource Test.

The main parameter provided by each test is the benefit/cost ratio between the present value of the monetised benefits that each subject enjoys following the project implementation and the corresponding implementation costs. The Participant Test analyses the benefits that participating users derive from the DSM programme and their expenses. The resulting benefit/cost ratio provides an indication regarding the probable participation of the eligible customers in a given programme. Similarly, the Utility Test provides us with a "quantitative" idea of the convenience that a utility will have in undertaking a given DSM programme. The Ratepayer Impact Test analyses the benefits and costs that are reflected on all of a given utility's customers. The Total Resource Test evaluates the benefits and costs of a programme that is bound to one type of energy resource. It also allows us to compare these results with the other energy resources with which a given utility could satisfy its customers' demands. The Societal Test analyses the benefits and costs of a programme by studying the external costs (such as social/environmental expenses) that a utility may have to take into account. It allows a comparison to be drawn to other energy resources.

The data required to run DSManager can be divided into three main categories:

1) data concerning the utility system which provides an overall picture of the company's technical-economical status. This data includes: the annual load curve and its projections for the entire period under study, marginal costs, generation, transmission and distribution costs, the relative projections for the subsequent years and the discount rate. Other data may be required for particular calculations.

2) data concerning each individual DSM project to be analysed, which includes data on the new technology to be adopted and that concerning the customers participating in the project. The main data is the average life of technology, the cost of machinery and its installation, maintenance costs, global consumption curves for the average customer participating in the programmes, energy demand curves for typical users (only for the technology under study), forecast energy savings or the decreased peak power of the utility based on the use of new technology and the number of customers participating in the programme over the years. Other data may be required for particular calculations.

3) data concerning the rates applied to the customers interested in the DSM programme. DSManager can analyse different types of rates based on consumption blocks, power blocks, mixed blocks, etc..

The results, gathered in three charts (Standard Benefit/Cost Tests), provide the following information for each test mentioned above: benefit/cost ratio, internal rate of return, total gross and net benefits, total costs, current benefits and current costs for each kW decrease in peak demand and for each kWh saved, as well as a breakdown into the various cost and benefit components both for the participating customers and the utility.

Other tables (the Summary Reports) summarise the most important variations concerning the utility and the project participants such as the peak reduction of the load curve, the reduction of generated energy, saved production costs, decreased receipts, etc.

The main problem with the use of DSManager for Italian utilities concerns the fact that the American regulations (targeted by the Utility Test) are not the same as those governing Italian utilities. Therefore, this test cannot be considered valid for Italian utility companies. The Ratepayer Impact Test remains valid as it analyses the benefits and costs that actually apply to Italian utility companies.

Lastly, although the rates inside the programme allow for a wide spectrum of possibilities, these does not apply completely to the particular Italian rating regarding the charges added to the bill that are not listed as VAT and which the utility company does not withhold and must pay or to the *Cassa Conguaglio* (balance fund) or as taxes. Naturally, this must be taken into account in order to maintain the utility's benefit/cost ratio valid. These difficulties required a few devices to be implemented in order to correctly use the software programme.

This model has been successfully applied to various Italian utility companies.

3 Area No. 2

ENEA has been - and will be - running DSM pilot programmes with utility companies. The following is a brief overview.

3.1 Joint Work with ACEA (Rome Utility)

ENEA ran a project with ACEA, a utility company based in Rome, financed by the EU (SAVE II): "Medium-Term Integrated Resource Planning for the ACEA Electricity System." The objective of this project was to set a plan for technological development, medium-term investments and the development and improvement of the ACEA electricity system according to "Integrated Resource Planning" (IRP) criteria and methodology.

The IRP methodology entails a general definition of technical-economical planning measures that use the same criteria to examine and evaluate any plan regarding both supply and demand. Typical plans regarding supply include the creation of new power plants, the improvement of existing plants and the acquisition of energy from third parties, while typical demand plans include load control and efficient technology for electric energy users.

In particular, with regard to demand, many possible plans have been identified for both the residential sector and the public/services sector that can be implemented in the medium term (3-5 years) based on safe, reliable and efficient technology. Following a preliminary evaluation of the feasible plans, projects for the residential sector focused on high-efficiency lamps, solar heaters, refrigerators/freezers, washing machines and other household appliances; for the public/services sector, the stress was placed on air-conditioning/heating systems such as centralised systems with electric compressors, cold storage systems and absorption systems. Moreover, plans were devised to increase the diffusion of cogeneration and district-heating with gas turbines, cogeneration for hospitals and combined cycles with district-heating and air-conditioning systems.

Special attention was paid by the DSM programme to the substitution of one or two incandescent lamps with CFL lamps in the residential use sector. In these cases, economic calculations were carried out with DSManager due to the greater complexity of the hypotheses formulated for customer participation trends during the period under examination.

The analysis lead to the selection of four DSM activities: the substitution of one or two incandescent lamps with CFL lamps in the residential use sector; the optimisation of the existing combined cycle through a connection to a winter heating service (5 million m^3) and a summer air-conditioning service with absorption units (0,4 million m^3); and the implementation of a combined cycle connected to a summer/winter service similar to the one described above.

These DSM activities have been compared to two supply projects: the construction of a new 100 Mwe combined cycle plant and that of a new 5 Mwe gas turbine plant with a power supply capability equal to the peak electrical power decrease possible on account of previous demand activities.

In order to carry out the IRP analysis, two possible scenarios were evaluated. The first hypothesised that the peak annual system load in the period under study (1998-2013) always occurs during the winter; the second scenario hypothesised that the peak annual system load would occur during the winter between 1999 and 2000 and subsequently during the summer. In both cases, the demand activities led to benefit/cost ratios greater than those related to supply. In particular, with regard to lighting, the activities provided the best benefit/cost ratios for both the company and the participants.

3.2 Joint Work with ASM Brescia

The Brescia Municipal Services Company (ASM Brescia SpA) and ENEA jointly devised a pilot programme, partially financed by the EU (SAVE II) as part of a project entitled "Cost Effective DSM Programmes as an Energy Resource for a Medium-sized European Town Utility." The project was implemented between April 1997 and April 1999.

The project was divided into various phases.

During the initial phase, the data necessary for the utility company to reproduce its system in the DSManager calculation model was gathered. This data regarded the electricity demand of all ASM customers over the past years. Based on the resulting trends (and using 1997 as the reference year), forecasts were formulated for the development of demand during the subsequent twelve months. This was done by evaluating the importance played by the growth in population and local industries as well as the changes in energy demand in the various sectors. Furthermore, costs were calculated both for the current marginal production of energy and its probable evolution over the coming years as well as the costs that the company would have to face in order to conform its production, transmission and distribution capability if it were not to meet the demand.

Lastly, external environmental factors were evaluated for the current marginal energy production as well as for the production of postulated new plants built to meet the growing demand.

The second phase focused on the identification of DSM programmes presenting interesting characteristics for the utility company and that could be implemented in a reasonable period of time in relation to the project. Eight DSM programmes were chosen out of all the possible ones that had been identified. Then, for each selected programme, data was gathered in order to assess it economically. In particular, this included: demand curves for the participating customers, specific demand curves for the equipment that would be substituted as well as for the more efficient equipment, rates applied to the participating customers, market interest, costs for the acquisition, installation and maintenance of more efficient equipment, estimates of the number of potentially interested customers and their adhesion to the actual project and the number of probable free-riders and free-drivers.

As the Brescia ASM manages quite an extensive district-heating network, it was interested in developing this particular resource and increasing its use, especially during the summer months, as a substitute for electrical energy which undergoes increasing demand peak during this period. Therefore, some programmes were devised specifically for this situation, while others focused on rationalising domestic energy consumption and improving the electrical energy demand distribution.

Here is a short description of the programmes that were chosen and evaluated, as described above, with DSManager:
1) Replacement of an electric compressor for air-conditioning with an absorption system using district-heating absorption systems in a medium-sized industrial plant.
2) Replacement of traditional washing machines with new models that use the hot water provided by district-heating. This programme addresses domestic customers with district-heating services in order to increase their use.
3) Replacement of traditional dishwashers with new models that use the hot water provided by district-heating.
4) Promote the replacement of traditional refrigerators/freezers with more efficient models. The programme intends to stimulate the diffusion of high efficiency refrigerators by allowing customers to pay the added cost of these new models through instalments on their electricity bills.
5) Promote an awareness raising campaign regarding the replacement of refrigerators, washing machines, dishwashers and traditional lamps with more efficient models.

6) A nocturnal energy consumption programme based on new electrical plugs with timers that will stop the use of washing machines and electric water heaters during certain parts of the day implemented together with differentiated rating. The objective is to shift the operation of certain appliances requiring high energy levels to the night-time, through an awareness raising campaign targeting users. Campaign pamphlets would indicate the type of plugs to purchase, how to regulate the timers and information regarding the different rates applied during the daytime and at night.

7) Installation of new deposits for refrigerated water in conjunction with the application of differentiated rates even for low voltage usage. The objective is to reduce electricity demand during the peak summer hours and create a relative increased usage of electricity during the night-time. This programme will target commercial users that have an elevated electricity consumption for air-conditioning during the summer months and urge them to install deposits for refrigerated water in their conditioning systems. This would be made convenient by very low nocturnal energy rates.

8) Control systems to limit the load of users supplied with medium level voltages and 200-400 kW power. This programme targets industrial and handicraft customers. Its objective is to introduce users to the concept of load management (in particular, for machinery consuming high energy levels) in order to reduce the demand peak on the system.

The economical analysis of these eight programmes was carried out with DSManager for two different scenarios.

The first scenario postulated that in the period under study (1999-2013) the supply of energy would remain abundant and therefore the company would not have to improve its generation systems. Extra energy requirements during peak hours could easily be satisfied through purchases from third parties. Therefore, in this scenario, the costs to increase the energy generation capability would be zero; the marginal production costs during peak hours would be considered equal to the cost of purchasing kWhs from third parties. No plans would be carried out to increase energy transmission and distribution capability.

External environmental factors were also considered in order to run the Societal Test. The cost of environmental damage for the production of electric energy was calculated on the basis of the type of power plant and the fuel used in order to satisfy a medium-high production level energy demand. This, in fact, would be the area affected by DSM programmes.

The second scenario hypothesised a lack of electrical energy on the market during the period under study. In order to satisfy the peak demand for electrical energy the company would be forced to install new gas turbine power plants. The cost for increasing the generative capability and the marginal production costs per kWh have been evaluated on the basis of this type of power plant. As in the first scenario, the second does not entail an increase in energy transmission and distribution capability; thus, these costs are considered to be zero.

The cost of environmental damage in the Societal Test for the production of electric energy was assumed to be the same produced by a turbo gas plant.

The analysis results demonstrated the validity of these programmes for the participating customers for whom the benefit/cost ratio was advantageous. On the other hand, the company would profit from implementing some of the programmes described for the second scenario, since it would be able to reduce the demand and avoid having to construct new power plants.
On the basis of these results, the feasibility studies and the necessary implementation time (on account of contractual obligations), as well as other factors that are not easily quantifiable in an economical analysis, the decision was made to implement programme no. 5 (see above) on a small scale. This programme entails the promotion of an awareness raising campaign to replace refrigerators, washing machines, dishwashers and traditional lamps with more efficient models.
All the customers located in the selected areas received a pamphlet containing essential and easily comprehensible information on the meaning of the household appliance labelling system. The pamphlet emphasised the amount of energy and money that could be saved with each new appliance, as well as the environmental benefits resulting from reduced energy consumption. It also contained practical advice for a more rational use of household appliances.
In order to verify the campaign results, a telephone survey was conducted among the participants who had received the information pamphlet. The aim of this survey was to determine to what extent the information had been assimilated, if the information was considered clear, how much had been effectively understood and accepted. Furthermore, the survey investigated the efficiency level of the participants' household appliances as well as their reactions (factors influencing choice, age of appliance, plans for purchasing new appliances, forecast life length of new appliances, interest in novelties, etc.).
The survey results indicated the effectiveness of the awareness raising campaign and allowed an estimate to be made regarding the participation level and the percentage of free-riders. The final programme cost balance also allowed the utility company to verify the hypotheses postulated regarding administrative costs.
Clearly, the information that has been gathered is extremely useful both in order to define a similar, but more expensive programme targeting the utility company's entire customer basin as well as for other similar projects.

3.3 Joint Work with Turin AEM and Cremona AEM

Project 2.2 "Promotion of DSM Pilot Programmes" is a part of the agreement signed by the Ministry of the Environment and ENEA on November 28, 1998 in order to link the study, research and development activities conducted by ENEA with the government's high priority environmental protection objectives.
The purpose of this project is to promote and back up DSM activities implemented by Italian electricity companies (especially municipal companies) and provide them with the knowledge and know-how that ENEA possesses in the fields of

DSM programme analysis, energy consumption rationalisation, innovative and more efficient technology and the calculation of avoidable environmental costs in order to create a reference point for future DSM activities.

The project entails the implementation of DSM pilot programmes in some electric utilities and a survey of the demand curve of typical customers. Unlike other countries, in fact, in Italy, scant attention has been paid to experimental assessments of the effective typical daily power demand of various users (single household appliances or apartments, shops, offices, etc.). Therefore, the curves representing energy demand over time are either hard to find or non-existent. These graphs are extremely useful for the implementation of DSM activities.

In order to carry out the project, ENEA signed agreements with the Turin Utility Company [Azienda Energetica Metropolitana S.p.A. (AEM) di Torino] and the Cremona Utility Company [Azienda Energetica Municipale S.p.A. (AEM) di Cremona].

ENEA and the Turin AEM are carrying out three joint lighting pilot projects: in the residential sector, through the replacement of incandescent lamps with CFL lamps; in the services sector, through the replacement of the lighting systems in two public offices and a school building; and on a public road of a Turin residential area by replacing the existing lighting system with a more efficient one.

The domestic sector activity will take place through the customers connected to predetermined sub-distribution stations for which the Turin AEM has gathered a large amount of data, including average annual load diagrams, as well as information regarding the households (acquired through recent surveys). The replacement of a number of lamps in each household and the repetition of these measures both in the pertinent sub-distribution stations, as well as in neighbouring ones, both prior to and subsequent to the DSM activity should reveal a remarkable reduction in energy demand.

Lamps will be distributed to the customers following an awareness raising campaign.

In the services sector, the pilot programme will involve edifices that are representative of the general building size used in the tertiary sector. The existing illumination systems will be replaced by electronic fluorescent lamps with an adjustable luminous flux system that will vary according to the level of natural light present in the workplace. Energy consumption measuring campaigns before and after the installation of the new lighting systems will reveal the actual monetary saving that can be reached.

The activity regarding public lighting will take place on a Turin road and will entail the removal of the present lighting system on one side of the road and the replacement of the system on the other side with a sodium lamp system. This will allow a power reduction from 11 to about 7 kVA.

Moreover, a campaign will also be conducted to measure electricity demand curves both in some of the households and buildings involved in the project.

The AEM of Cremona, which distributes both heat and methane, besides electrical energy, and ENEA are implementing a DSM pilot project, among the utility's customers, to replace conventional electric compressor air-conditioning systems with absorption systems powered by district-heating (or methane in the areas in which district-heating networks are not present). This project will entail: - a study of the economic and environmental benefits of absorption plants as opposed to conventional plants; - the identification of potential customers through modern and efficient research strategies and the use of a custom-tailored software programme, as well as the definition of a methodology for this type of research; - the selection of standard projects and their technical-economic evaluation; - the preparation of commercial packages and contracts to be proposed to customers in order to facilitate the diffusion of these new systems; - the creation of a prototype that can be used to gather further data on its operation and verify the project hypotheses, improve the project, demonstrate its technical-economical validity and dissipate any possible doubts that potential clients may have; - gathering and elaborating data on the system and its customers; - publicise the results, even through marketing activities, in order to open the market to this new form of technology and raise the interest of other Italian utility companies that distribute both electric energy and district-heating or methane.

744

Europe's Most Energy Efficient Refrigerator-Freezers – the European Countries Working Together on the Forefront of the Market

Anna Engleryd[1] and Sophie Attali[2]

[1] Swedish National Energy Administration, STEM, Sweden
[2] International Consulting on Energy, ICE, France

Abstract. This paper presents the outcome and the lessons learned from the first round of the Energy+ procurement project of refrigerator-freezers carried out under the SAVE program. The aim of the project, that was preceded by a thorough market study, is to give procurement as a market transformation tool a trial run on a pan-European level. According to the study this instrument could be a means to achieve market transformation on highly international markets with fairly standardised products where national initiatives would not be enough, and could constitute an opportunity for the European Union Countries to work together to fulfil their Kyoto commitments.

The Energy+ project has in a first round brought together nearly 90 retailers, institutional buyers and other organisations from 10 European countries, eager to purchase and promote the most energy efficient refrigerator-freezers on the market. The specifications issued by the project specified an appliance using three-quarters or less of the energy of equivalent appliances meeting the minimum class A requirement for the European energy labelling scheme.

This call was answered by the manufacturers who submitted 2 models complying with this high efficiency criteria available on the European market from February 2000 and 5 complying models to be introduced during the year. In a second round, additional optional criteria are added to the initial specification to lead the manufacturers to even further developments to lower the environmental impact and rise the user utility of their appliances. The results of this second round will be presented at the Domotechnica trade fair in March 2001 when the most advanced units will receive the European Energy + Award.

In September 2000 the Energy+ project gives new entrants a possibility to come on board when an update of the available appliances meeting the specifications is made. The result of the update will be made official at the SAVE conference in Naples.

1 Small Might Be Beautiful... ...but not Strong Enough

Procurement programs of different kinds have been recognised as useful tools to work on the demand side of energy efficiency. Two kinds of procurement programs can be distinguished 1) *technology procurement*, referring to a process with the explicit aim of promoting technological development and used to increase the market availability of products or systems that better correspond to the buyers needs than those existing when the process is initiated (Westling, *IEA DSM Annex III*, 1996), and 2) *co-operative procurement*, referring to a process designed to create significant markets for already existing technologies with good qualities, in this context energy efficient technologies(STEM, 1998).

The idea of procurement is to create a bridge between demand and supply of high quality products and show manufacturers that there is a demand for such products. This is done by bringing a group of purchasers together to identify potential improvements of a product, or stating their interest in existing but poorly available products and issue a specification. Manufacturers are free to choose technical solutions to meet the demanded improvements and send in tenders; the tenders are compared and evaluated, and a winner is selected. The winner is offered certain benefits: in any case publicity and support by powerful and credible institutions, but in some cases also a large initial order, or rebate schemes for its products.

Programs of both kinds have been successfully used at national level in a number of European Countries like for example the Netherlands, Finland and Sweden. However, following the internationalisation of goods production and the emerging internal European market, it is getting more and more difficult to interest manufacturers to take part in these processes and change their production with a buyer group from one country only. This is true for fairly standardised products sold by large companies on international markets. In such situations a one-country buyer group will constitute only a small fraction of the total demand for the product in question. By gathering purchasers from the whole European Union a demand-pull sufficient enough to influence such markets could be created.

Combined forces would further lead to a more significant transformation of the European market in its totality. The current prevailing differences between national markets would be softened contributing to the completion of a single European market for energy efficient products thus providing a basis for a more integrated market transformation approach i.e. easier introduction of stringent standards and labelling schemes. Enabling the introduction or strengthening of such measures would in their turn further contribute to rise the energy efficiency in the European Union.

In the case of the Energy+ refrigerator-freezers procurement, it is expected not only to promote the best products on the market and rise their market share, but also to reinforce the European labelling scheme, ease its revision and complement the EU Council Directive banning the sales of the most inefficient units effective as of September 1999.

Considering the fact that the European countries, in accordance with the Kyoto protocol, are committed to reduce their green house gas emissions, actions at the EU level supported by the European Commission provide a clear political signal and constitute a good means to make sure that both the common European target and national targets are met. Working to transform the refrigerator and freezers market is in this context of great importance considering that the electricity consumed by these appliances in European households is estimated to contribute to some 62 million tonnes of carbon dioxide emissions per year meaning 2% of EU's total CO_2 emissions from manmade sources[i].

2 A Project Preceded by thorough Preparations

The aim of the ongoing Energy+ project is to give procurement projects a trial run at pan-European level. It was preceded by a thorough one year study investigating the possibilities and the potential for such actions on European level. Nine countries[ii] participated in the study and carried together out 36 market studies regarding four different products: electric motors, solar-energy systems for water heating, office lighting systems, and refrigerator-freezers.

The study focused on the process of procurement and identified two main variables which greatly determine the best way to ensure a successful process at international level 1) the similarities in product usage and the level of standardisation of the product; and 2) the international character of the market i.e. presence of international actors. A generalised picture can be drawn as in Figure 1.

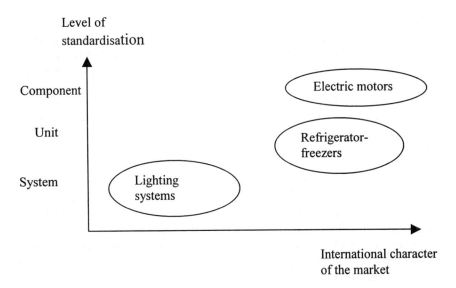

Figure 1: Product type related to international character of the market and level of
standardisation. (STEM, 1998)

In general, components and units are highly standardised and sold on more or less
international markets, while systems are difficult to standardise. Related to the low
level of standardisation is the need to have the system made to fit the individual
end user and his/her specific environment. This means that the geographical
distance between supplier and end-user is important and correspondingly, markets
are national, regional or even local. The systems can however be divided into
components and whole systems showing different degrees of standardisation and
international character of the markets.

The study recommended three organisational forms:
- *Full EU wide procurement*, where a common procedure is developed and
implemented at the EU level by a European process co-ordinator in co-operation
with national co-ordinators; This scenario is suitable for international markets with
more or less standardised products;
- *EU co-ordinated procurement*, where a common procedure is used
concurrently on the individual national markets, is judged suitable for products
with a low level of standardisation or large differences in usage, or on markets
where the countries are very different in terms of policy and regulations;
- *Two-step procurement*, suitable for complex systems that can be broken into
more or less standardised units. This scenario is a combination of the two above: a
full EU wide procurement is carried out for the units or components, combined
with EU co-ordinated national procedures for the entire systems.

The study concluded that procurement is a both feasible and promising instrument for common European actions. Given the fact that the European countries differ in terms of climate, culture (and therefore product usage), market structures and policy aspects present challenges. One of the main barriers to co-operative procurement being implemented at European level was identified as the lack of experience and therefore trust in the process as such. To gain such experience the Energy+ pilot project was started up using the first one of the organisational forms mentioned, full EU wide procurement, on a highly international market with fairly standardised products, the refrigerator-freezer market.

When the European Union introduced its labelling scheme for refrigerators-freezers only a few appliances qualified for the most stringent level "A". A quick evolution of the market has however necessitated the ongoing revision of the scheme. Since it will take a few years for the new requirements to come into force, the Energy+ project has an important role to play in giving visibility to the most efficient products available and help the buyers distinguish them from the "normal" A-rated units.

3 A Project Design Adapted to Specific Products and Actors

In order to adapt the project design and the product to be targeted to the prevailing circumstances, the first step of the Energy + project was to thoroughly investigate the markets in the participating countries. National market characteristics were mapped-out and the actors, their interests and working routines identified.

This investigation gave at hand that the average model that can be sold in most countries (although some differences in national preferences exist) would be a free standing two door fridge/freezer with a total net volume between 250 and 300 litres. Production of highly efficient such units was not considered to imply any technological problem for manufacturers. The main barrier to the large diffusion of already existing efficient units was the high sales price which was more due to branding and marketing policies than expensive technical components.

The European market is dominated by a few large manufacturing companies that are present, in one way or the other, on most national markets. Well over 100 different brands were found in the ten countries studied: appliances from the very same manufacturer are sold under different brands in different countries. This shows that, although cold appliances are fairly standardised units, strong national preferences prevail in what respects total volume, position (top or bottom) and volume of the freezer compartment, and above all brand and marketing arguments. The brand and marketing strategy on different markets is consequently of high importance to the manufacturers, i.e. not every brand can be sold in every country under any marketing argument and in any price class. Even if most manufacturers would be able to participate in an international scale procurement from a production point of view, only the most influential ones have the commercial resources and the distribution network needed to widely deliver the product throughout the Union.

One of the distinctive features of this procurement program is that the most important actor on the buyer-side was identified as being retailers that would not buy the product for their own use, as was the case with the so called institutional buyers[iii] in earlier national experiences. This implied certain issues to be tackled. Retailers tend to prefer a product that attract consumers through for example design or brand recognition, features that may be expensive and of less importance to other types of buyers i.e. institutional ones. The retailers favour top range products permitting them to put a high mark up on the price, something that could counteract the aim of the project. They further showed themselves unwilling to co-operate in a buyers group with their direct competitors.

Retailers generally have yearly agreements with specific manufacturers. The price they pay their supplier is often not set until the number of appliances sold (of all kinds, not only refrigerators-freezers) is known. Bargaining practices between supplier and retailer imply that the price for a cold appliance can be influenced by the number of washing machines of the same brand sold. The relations between retailer and supplier are often of long-term character and new brands will only be included in the range if they are expected to be fairly easy to sell. Retailers would not include the products conforming to the Energy+ specifications in their range if the brand in question is not in line with their business strategy and/or does not fit the national context. Retailers are for these reasons not able to place any firm orders and can not commit themselves to buy a product, no matter how energy efficient and of good quality the unit would be.

Given this situation the project would not be able to place any orders for highly efficient units from manufacturers that would in their turn not be willing, and in some cases not have the possibility, to deliver suitable brands in all of the participating countries. Moreover, organising a bulk purchase would risk to be in conflict with European competition legislation. These market characteristics obstructed the setting up of a buyer group in the traditional sense. To meet these challenges and at the same time maintain the market transforming effect on a European scale, the project was instead designed to push for the best of the best products and the forefront demand side actors, but let the commercial interactions take place outside the scope of Energy+.

4   Energy + Step by Step[iv]

The units targeted correspond to the following specifications, mandatory for units submitted to the project. The units:

- Are 4-star refrigerator-freezers.
- Have a total net volume of between 200 and 300 litres.
- Have an energy efficiency index equal to or below 42% (in accordance with Directive 94/2/EC and the EN 153 test procedure).
- Are available on the European market (EU and Norway).

The project is being carried out in two rounds. For the first round, a list of refrigerator-freezers that meet the Energy+ specifications and are currently available on the market has been compiled and published. Another section of the list covers products to be available by December 2000. The first round list will be updated once in September to include new entrants. At the time of writing the outcome of this update is not yet known, but should be ready for discussion during the presentation of this paper in Naples in September 2000.

Participating retailers and institutional buyers sign a document where they declare their intention to promote and/or purchase appliances according to the Energy + specifications. The names of all the participating retailers, institutional purchasers and supporters are also published in a separate list. The second round is dedicated to Energy+ appliances available in 2001. New lists of participants will be compiled and promoted.

The Energy+ logo is used to publicise qualifying models complying with the high efficiency criteria in catalogues and promotion media. It can also be used by retailers and other demand-side actors to manifest their participation in the project and thereby show that they are forefront actors.

Figure 2: The Energy+ logo (in black and green)

As an additional incentive for innovation and increased energy efficiency, the European Energy+ Award competition was launched in February 2000. The competition is based on the current Energy + specification presented above, and optional requirements of importance to the buyers are added to guide the manufacturers in the development of products having increased energy efficiency and low environmental impact. These optional specifications include even better energy efficiency, refrigerants and foaming agents with low environmental impact, low noise, clear temperature displays, reasonable price, and user friendliness. The competition is open to units in three separate classes one-door models, two-door models and prototypes. The winning products will be presented at a public ceremony to be held at Domotechnica, Germany, in March 2001.

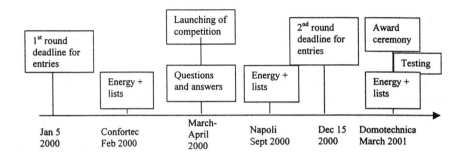

Figure 3: Energy plus step by step

An independent jury composed of internationally recognised experts in the fields of energy efficiency, technological aspects and consumer issues will evaluate and select the winner of the award competition using a pre-defined formula to weight the different criteria. To verify the information in the testing protocols sent in by manufacturers, random testing of the appliances in all classes will take place at an internationally recognised testing laboratory in The Netherlands using standardised testing methods.

5 A First Call Well-Answered

The response to the first round of the project was overwhelming with nearly 90 European Organisations backing up the initiative. On the demand side three groups can be distinguished:

• *Retailers*, (1/2 of the demand side participants), regrouping large international retail chains as well as individual retailers, mail order companies and electric utilities. Altogether this makes more than 15 000 retail outlets spread all over Europe.

• *Institutional buyers*, (about 1/5 of the demand side participants), comprising housing companies, national housing associations, and holiday parks. These buyers together possess more than one million dwellings and come mainly from Scandinavia and Finland.

- *Supporters*, (about 1/3 of the demand side participants) including environmental organisations, energy agencies and other organisations who in their daily work will push for and inform about the Energy+ appliances.

These demand side actors have been actively involved in the process by contributing with their experience and expressing their interest when developing the specifications for both rounds of the project and the award competition. Numerous discussions, interviews and meetings with buyers have been held in each participating country. This is of utmost importance for the project to have a strong demand side support based upon true interests and needs of the buyers and not wishes and guesses of the project organiser or other initiative taker.

The response from the supply side was exceeding the expectations in what respects the technical features of the submitted units (see below). However only two manufacturers, Electrolux and Fagor, were on board the first list. One reason for the relatively low manufacturer participation is in part related to the very short notice given for the first round, meaning that only products already in the development line could be submitted. This is so far the major default of the project expressed by the market actors, and an important lesson for future projects. In order to adapt to market actor responses (whether they be manufacturers or buyers), the process needs to be flexible in terms of schedule and budget. This is true for procurement projects in general, but is even more important when it comes to international scale actions. However, frameworks offering the possibility to financially and politically back up projects of this kind comprising as many as 10 countries (as the SAVE program) can not offer this flexibility; all projects have to stick to pre-determined work programs and firm deadlines.

Another reason for the few manufacturers sending in entries is thought to be a lack of confidence end experience concerning the process and precautionary measures before the outcome and impact of the first round is known. In this case, the realisation under a European Commission program like SAVE, will most likely help to counteract doubts in the seriousness of the action.

5.1 Europe's Most Energy Efficient Refrigerator-Freezer

Electrolux presented two models, a one-door and a two-door refrigerator-freezer, with energy efficiency indexes of respectively 38% and 36%. This means thus that they only use 38 and 36 per cent of the energy used by an average European cold appliance of comparable size and type. The two-door appliance, for which the interest is the greatest among the Energy+ buyers, has a total net volume of 288 litres of which a frozen food compartment of 95 litres. It uses only 219 kWh/year making it the most energy efficient refrigerator-freezers of its type in Europe and meaning that it largely surpasses the requirements of the project. To reach this energy efficiency level the unit is equipped with a frequency controlled compressor and the door is insulated with vacuum panels.

In the second category of the first round, qualifying appliances expected to be on the market before December 2000, four new units from Electrolux are found as well as one unit from Fagor.

6 Perspectives for the Future

In September 2000, we are now approaching the end of the first round of the project and some lessons can already be learnt :

Good timing is of utmost importance and it is sometimes difficult to harmonise the agendas of different actors from different countries: time given for manufacturers to participate, time to retailers to decide upon their yearly or biyearly product range, difference in time perception between public and private participants, suitable time and event to announce participating organisations and availability of Energy+ products on the market, etc.

Patience is needed in order to effectively evaluate the true market transformation (although some results are already visible) : the first Energy+ appliances are, as mentioned above, highly satisfactory from a technological and energy efficiency point of view. At the moment they are high range products that are quite expensive, something that to a certain extent hampers their large diffusion on the market and thus a thorough market transformation. It is however often observed that modified and technically advanced units are first introduced as "top of the range" products, but that the innovations gradually penetrate the whole range making them standard features. Increased interest from the buyers side will, in combination with the perspective of more manufacturers offering Energy+ products, most likely drive down prices and bring up sales volumes.

Communication and marketing, regarding the project itself as well as the products, are central elements of co-operative procurement, especially when it involves retailer participation. The Energy+ steering group has actively been working to develop several communication tools to be used by the participants across Europe : a logo (which might be used for other appliances in the future to help consumers identify good existing products and give an identity to European procurement projects), a web site, a newsletter called "the Energy+ Bulletin" and a CD Rom providing communication information for retailers and their sales staff. These measures provide a good support for manufacturers whose contribution to the success of the project is that they push for their own Energy+ products.

To bring more actors on board the project and reach a large market diffusion resulting from larger volumes purchased and offered, activities to reinforce and secure the demand is now ongoing all over Europe. Press releases have been massively sent out to specialised press as well as to daily papers which have resulted in a large number of articles. The project also got radio and TV attention in Sweden, the home country of the manufacturer of the first Energy+ appliances. Information on these Energy+ appliances are also presented at different trade fairs and exhibitions by the organisations, often energy agencies, behind the project. Buyers and wholesalers meetings have been organised in some countries, while buyers in other countries are kept informed by close phone contacts. Rebate schemes are planned and training courses for salesman are underway.

Being managed at European level and targeting national markets, the Energy+ project holds a peculiar and critical position, making the pilot project interesting and complex at the same time:
• the pros of being a European scale project is clearly visible : the European Commission is undertaking actions in line with the European commitments taken in Kyoto; a larger number of interested buyers can be reached thanks to the participation of several countries; a consistent signal is given to manufacturers showing a general desired evolution towards increased energy efficiency and also in what respects European policy (the Energy+ projects fits well into the European agenda in-between the minimum efficiency standard Directive and the re-evaluation of the Energy label criteria);
• some difficulties related to the European-scale of the project can also be observed : even though a few large manufacturers supply all European countries, the demand for refrigerator-freeezers remains very much linked to national preferences in terms of brands (due to concentration movements in the industry during the last 20 years, appliances from the same manufacturer are sold under different brands in different countries); agendas and thus ideal timing of the different project phases vary between countries; national differences exist regarding procurement practices and the perception of such actions.

As any European-wide action, the Energy+ project experiences what could be referred to as a "subsidiarity effect" where the guidelines and the impulsion are common and set at European level, but where the implementation is the responsibility of each member state (in this context each participating country). Neither the European Union nor the European market for refrigerator-freezers were built in one day. We must allow the two rounds of the Energy+ project to be completed and give some time to observe market dissemination before any final conclusions can be drawn.

7 References

[1] Engleryd, A. *"technology procurement as a policy instrument"*, NUTEK, Sweden, 1995.

[2] NUTEK et al, *"Procurement for Market Transformation for Energy Efficient Products. A study under the SAVE programme"*, Sweden, 1998.

[3] Westling,H. *"Co-operative procurement – Market Acceptance for Innovative Energy-Efficient Technologies"*, annex II of the IEA DSM Implementing agreement, NUTEK, Sweden, 1996.

[4] Energy+ Steering group, *"International Co-operation on Energy Efficient Cold Appliances Procurement – ICECAP feasibility study"*, intermediate report for the European Commission, SAVE contract XVII/4.1031/Z/98-273.

[5] Energy+ Steering group, *"Who will provide Europe's most efficient refrigerator-freezer?"*, information brochure for manufacturers.

---

[i] This is estimated considering that most electricity consumed in European households is generated from fossil fuel combustion.

[ii] Austria – EVA, the Austrian Energy Agency; Finland – Motiva, Energy Information Center for Energy Efficiency and Renewable Energy Sources; France – Ademe, French Agency for the Environment and Energy Management; Germany – Wuppertal Institut with support from UBA, the Federal Environment Agency; Italy – Ministry of the Environment and Politecnico di Milano; The Netherlands – Novem, the Netherlands Agency for Energy and the Environment; Portugal – CCE, Center for the Conservation of Energy; Sweden – STEM, the Swedish National Energy Administration; United Kingdom – BRE, Building Research Establishment Ltd. In the ongoing pilot project Norway has joined the initial group of nine countries and are represented by the Norwegian Water Resources and Energy Directorate and NEE, Norwegian Energy Efficiency and Energy Management Inc. BRE that represented the U.K has in the pilot project been replaced by the ECI, the Environmental Change Institute of Oxford University.

[iii] Institutional buyers are buyers buying the product in question for their own use, for example large housing companies equipping their flats with white goods (Scandinavia and Finland), holiday homes, homes for elderly people etc. These buyers are in a position to place firm orders of products fulfilling their specified needs and requirements.

[iv] For a more detailed description of the Energy + project, visit the project internet site: www.energy-plus.org

Domestic Appliances as Virtual Power Stations

Nico Beute

Cape Technikon

Abstract. Implementing demand-side management can be seen as an alternative to erecting power stations. This also applies to the domestic sector. This paper will evaluate domestic sector demand-side management options as applied to appliances.

The paper will investigate the use of various domestic appliances, and how their use can be controlled by a utility. Water heaters are appliances that can easily be controlled, so the effect of controlling water heaters is investigated as an example of appliances that can easily be controlled. Centralised control for water heaters is investigated because many installations have been commissioned and many are in operation. The value of saving a kVA generating capacity is not so easy to calculate because it depends heavily on the condition of the national grid. If there is spare generating capacity, the value is very little, but if centralised control of water heaters avoids the building of a power station, then the value of centralised control is exceeds the cost of the power station . Similar calculations can be done for transmission and distribution increasing the value of kVA's saved.

A variety of domestic appliances are considered in the same way. The principal of the virtual power station is also demonstrated by a practical application at a local university.

1   Introduction

The International Energy Agency (IEA) concluded in a recent workshop that projections of increasing emissions of greenhouse gases (GHGs) confront us with the enormously complex issue of global climate change. In order to stabilise GHG concentrations in the atmosphere at any level, significant technological changes, regulatory and institutional changes and cultural changes will be required (IEA 1997). A strategy to ensure responsible use of energy is also of essential importance to South Africa. Wise and effective use of energy depends largely on the time of day the energy is used. Electrical grid energy is very scarce at certain times of the day, but there are also times of the day when electrical grid energy is freely available. Hence load shifting is a strategically important component of wise and effective use of electrical grid energy.

Demand management may require customers to take one of the following actions:

- Allow the supply authority to install a device which will control an appliance automatically and is not adjustable. If the customer sees this control as assisting him, he will be receptive to it. But if the customer thinks that this device is going to behave in a way that will take control out of his hands and that the control is inconveniencing him, he will not want the device. An example of such a device is the intelligent thermostat. Customers could respond in either way as described above.

- Use appliances differently to what the consumer would normally do. He could for example, use his washing machine at times other than times of peak demand, or switch lights off in unused rooms during times of peak demand. In this case the consumer may be more willing to do it, but he may forget about it or it may require too much effort from him to do it. In such a case automatic control will be to the advantage of the customer as well as the supplier, but the customer may not be prepared to relinquish control of his appliance.

- Allow the supply authority to install something in his house which will control an appliance without him being able to change the control, such as the control of water heaters, where the supply authority switches the water heater without any input from the consumer.

The paper will discuss some appliances and how they may be controlled to achieve a better load factor, leading to a strategy to apply demand management involving various end-uses of domestic consumers.

2 The Principle of Virtual Power Stations

When a supply authority experiences an increase in electricity demand, it may be necessary to bring another generator or power station on line. This may be for a short time or the base load may increase and additional power may be needed for a longer period. Reducing the demand may also provide for increased capacity. Switching of some selected domestic loads may provide this reduction in demand,. Some components of the domestic load or some appliances are suitable for this. Using the domestic load in this way may be very effective if the shortage of supply is expected to last for only a few hours.

3 Appliances that Could Be Targeted for the Virtual Power Station

The following are some activities that can be rescheduled to times other than peak load times without inconvenience to the customer: (Beute 1999) (see Table I)

3.1 Hot Water

Devices which convert electrical energy into other forms of energy are most suitable for moving the load away from peak demand because energy is not necessarily needed while the product is used. A good example is the water heater: hot water can be used at a time other than the time when the water is heated.

3.1.1 Possible Ways of Controlling the Hot Water Load

Centralised Control of Water Heaters
South Africa has extensive experience in centralised control of domestic water heaters. Many local authorities have used various types of control for many years.

Intelligent Thermostat
To improve the efficiency of a hot water storage system, just sufficient water should be heated immediately before it is used so that the temperature of the water is as low as possible when hot water is not needed and heat loss is reduced to a minimum. A storage water heater should also be used to store energy to reduce energy use during peak demand time. For this purpose water should be heated just before peak demand time if hot water is needed during or immediately after peak demand time. If the electricity is normally disconnected during peak demand time, and the water tends to be used up while the heater is disconnected then it would be better to heat the water above the normal operating temperature before the time of disconnection to ensure that sufficient energy is stored to avoid water getting too cold during switch off time. An intelligent thermostat has been developed which senses and predicts the time of use of hot water and regulates the temperature accordingly to improve efficiency and storage capacity of the storage water heater.

Intelligent Thermostat with Centralised Control
The intelligent thermostat on its own is not dynamic enough and cannot respond to unforeseen circumstances in the supply network. To give the intelligent thermostat an added dimension, an intelligent thermostat that can be controlled by a central control desk needs to be developed. This will improve the effectiveness of the intelligent thermostat and it will give control to the supply authority or the generator of electricity.

This type of control can also be introduced where there already is system of control for the domestic water heaters. The added intelligence will in most cases not only give additional reduction of the peak load, but also less discomfort to the customer.

3.2 Pool Pump

Pool pumps do not need to run 24 hours per day, so it seems obvious that they should not be switched on during peak load times. Pool pumps are normally controlled by means of a time switch, so all that needs to be done is to convince the consumer to set the timer accordingly. This may not be as easy as it seems, as the consumer must respond to requests made to him.

3.3 Refrigeration

Freezers and refrigerators have a large thermal capacity and can be left disconnected from electricity for a few hours without the temperature in the freezer rising noticeably. The appliances could be connected to a supply which could be controlled by the supply authority to be energised only during the off peak periods.

3.4 Washing Appliances

Washing appliances (laundry and dish washing) can be used any time during the day or night and it is sometimes hard to justify them being used during peak load times, but convincing customers of this is a challenge. These appliances normally use hot water, and this is an additional reason for not using these appliances during the peak period. These appliances could also be connected to a supply, which can be controlled by the supply authority.

3.5 Space Heating and Air Conditioning

Houses have a large thermal capacity, so if precautions are taken to have the house thermally insulated, houses can be pre-heated in winter (or pre-cooled in summer) during off peak times so that only minimal space heating (or air conditioning) is used during peak time. Additional thermal storage can be provided in water tanks or in the structure of the building. Load shifting opportunities for space heating in South Africa are not as great as in countries with colder climates, but investigation is needed to identify opportunities and niche applications for South Africa.

4. Factors Influencing the Implementation of Residential Demand-Side Measures

4.1 National or Localised Control

In some cases the local supply authority experiences overload on its local distribution network and in such cases the control of water heaters is based on the loading of the distribution network and not on financial considerations based on tariffs.

There is often also a difference between the time of peak load of the local supply authority and that of the national load. This means that the local supply authority may reduce the hot water load at a time which is not the national peak load, so while the control does reduce the electricity account of the local authority, it does not reduce the peak national load.

4.2 Customer Satisfaction

Installing control equipment in people's households has been relatively easy in the past. However, people are becoming more aware of their rights and will not allow control of their equipment unless they can get benefit from what is installed. Equipment, which has been installed for demand management, is easily blamed for any irregularities of electrical equipment.

4.3 Voluntary or Compulsory Participation of Customers

Consumers want to be able to make their own decisions, and to decide what gadgets they will allow to be installed in their house. For these issues to be addressed it is important to consult extensively with the consumer before any attempt is made to install control equipment in their home and to explain clearly what the value of the equipment is to the consumer. Each demand management option must be evaluated according to the value to the consumer and the consumer's perception of the value to himself and to the environment.

5 Possible ways of Using Water Heaters as a Virtual Power Station

A summary of a selection of load shifting options is listed in Table I. A discussion of some of the possible options follows.

Many ways of controlling the hot water load have been listed and it is clear that the hot water load is most suitable as a virtual power station. This is especially so because the time of use of the hot water need not coincide with the time when the water is heated, or when the electrical energy is used to heat the water. If the hot water load is to be used as a virtual power station, the supply authority needs to control the hot water load. Items 3.1.1.1, (centralised control) and item 3.1.1.3, (intelligent thermostat with centralised control) are suitable methods of controlling the hot water load.

Knowledge of the national load on a long-term as well as short-term basis and also on an immediate basis is essential for the effective control of the virtual power station. The load acting as a virtual power station must be controlled on line by the national or regional control centre. In order to predict the medium and long term load, the South African utility Eskom is implementing a national Geographic Information System (GIS) to help manage the national grid. Network planning is supported by a system called TIPS (The Integrated Planning Solution), that runs operates in a GIS environment and allows a planner to view and interrogate several spatial data sets at any point in the country. This system involves comprehensive data sources with load profiles that give present as well as expected future load (Jones 1999). The load consists of various data sets such as customer data, billing data, census data and network data. This information would be very useful for the virtual power station concept because it gives information of the size and position of the virtual power station.

6   A Pilot Project Using the Hot Water Load as a Virtual Power Station

The University of Pretoria (UP) has implemented a Virtual Power Station (VPS) using the hot water load of their student residences in collaboration with the City Council of Pretoria (CCD). Real-time energy consumption, data logging capability and hot water load control switches allow for real-time control. The control algorithms use the data, available in 1-minute intervals, to determine the switching schedules of the hot water cylinders.

Emphasis is placed on controlling the hot water cylinders for two main reasons [4]:

- The hot water load is a shiftable load, which implies that it is able to store energy.
- Control of the hot water load is invisible to the University community.

The flow of information can be described with the help of Figure 1.

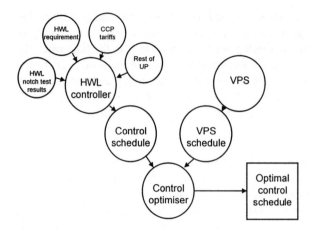

Figure 1: Flow of Information to Determine the optimal Switching Schedule

The control room has the capability, with radio controllable switches, to control the 2.5MW installed hot water load of the hostels. The notch test results can be interpreted to give the required hot water load. The load of the 'rest of the University', which is the total load minus the hot water load, can be determined. The controllable part of the hot water load, according to the hot water load models (Delport GJ 2000) is used to minimise the cost to the City Council of Pretoria (CCP), according to the tariff. This will produce a switching schedule that will give the biggest saving without making the control visible to the hostel students.

The University joined Eskom's virtual power station (VPS). Eskom will give a request to the University to shift load according to their load profile, which shape is dominated by the domestic sector. This means a morning and afternoon peak. The VPS will then pay an amount per kWh shifted on their request.

The postulated solution included the following:
- creating advanced measuring and control hardware, and
- creating a dynamic pricing mechanism.

The result will be that a large amount of under-utilised customer-side (demand-side) load shifting capacity in the current base of installed residential hot water load control systems in South Africa can be more optimally utilised to the benefit of the national grid operator as well as its customers.

7 Financial Considerations of Load Shifting

The cost load shifting initiatives is often difficult to determine. In the case of installing and maintaining centralised control for water heaters it is known because many installations have been commissioned and many are in operation. The cost in South Africa is in the order of $ 50 per water heater which converts to about $80 per kVA saved. The value of saving a kVA generating capacity is not so easy to calculate because it depends heavily on the condition of the national grid. If there is spare generating capacity, the value is very little, but if centralised control of water heaters avoids the building of a power station, then the value of centralised control is the cost of the power station which is in the order of $120 per kVA. Similar calculations can be done for transmission and distribution increasing the value of kVA's saved

8 Conclusion

Technically, parts of the domestic load can be shifted, and virtual power stations can be implemented, bringing benefits to customers in terms of a more cost-effective supply of electricity. But one must remember that the challenge in bringing advanced techniques to the consumer even when cost is not an obstacle is marketing these innovative ideas. Scientists and engineers too often believe that the technical superiority of what they have developed will be sufficient to ensure market success - this is a simplistic notion, which overlooks economic, financial and sociological reality (Laponche 1997). Demonstration and effective marketing is needed to introduce and implement the options mentioned in this article.

9 References

[1] Beute N: *"Initiatives to shift the domestic load to improve the load factor"*, Domestic use of energy conference Cape Technikon, Cape Town, 30 March to 1 April 1999

[2] Delport G.J.: *Hot Water Load Control Models Used on Campus. The 7th Domestic Use of Energy Conference*, Cape Town, 18 April to 19 April 2000.

[3] Delport J & Chuen-Hsiung Chen, *"Hot Water Load Control on campus as part of a Virtual Power Station"* Domestic Use of Energy Conference, Cape Technikon, Cape Town, 18-19 April 2000.

[4] International Energy Agency(IEA), *"Electric technologies: Bridge to the 21srt century and a sustainable future",* IEA workshop , Paris, 15-16 September 1997

[5]   Jones LR et al: *"Improving forecasting practice GEO based load forecasting"*, Domestic use of energy conference Cape Technikon, Cape Town, 30 March to 1 April 1999

[6]   Laponche, B at al: "Energy Efficiency for sustainable world", International Conceil Energie, Paris, 1997

10 Bibliography

Eto J, Vine, E Shown R & Sonneblick R;   *"The total cost and measured performance of utility sponsored energy efficiency programmes";* Energy Journal Vol 17 no 1 pp 31-51; 1996

Forlee C;   *"Residential load research: the Eskom way"*   Domestic use of electrical energy conference Cape Technikon, Cape Town,   1-2 April 1996

International Energy Agency,   *"Implementing agreement on demand-side management technologies and programmes 1994 Annual report";*   Swedish national board for industrial and technical development, Department of energy efficiency, Stockholm,   March 1995.

Joskow PL, and Marron DB,   *"What does a Megawatt really cost?   Evidence from utility conservation programmes"*   Energy Journal Vol 13 No 4 pp 41-74, 1992.

Lane I,   *"RDSM Stakeholder Value Forum",*   Workshop held in Johannesburg, 3-4 September 1997.

Laquatra, J & Peters SKC,   *"Determinants of homeowner's response to energy observation in non metropolitan areas";*   Energy Vol. 14 No 7 pp397-408, 1989.

UNIPEDE;   *"Integrated Resource Planning and Demand Side Management in Europe: Present status and potential role."*   Economics and Tariffs Study Committee,   60.04 TAROPT   1994. Vine E;   *"International DSM and DSM Program evaluation: an INDEEP assessment";*   presented at the European Council for an Energy Efficient Economy;   Mandelieu, France;   June 6-10, 1995.

Table I: Options enabling the shifting of domestic loads

Option Criteria	1.1 Central HW control	1.3 Intelligent Thermostat with central HW control	2 Reschedule Pool Pumps
Packaging Description/Definition	Customer's HWC's are remotely controlled by supplier using for example ripple or radio control. Switched off during peak periods and emergencies. Inconvenience to customers must be minimised.	An intelligent thermostat is installed to control customer's HWC's. Thermostat detects usage patterns. It optimises control of individual HWC's, by pre-heating to a higher temperature before peak times. Influenced by signals from a central control, it disconnects load during peak times where appropriate.	Ensure that customers set the timers of their pool pumps so that pool pumps do not operate during peak periods. In some cases more reliable time switches are needed to replace existing timers.
Training required	Customers must be informed so that they understand the reason for HW control.	Customers must be informed so that they understand the reason for HW control	Customers must be informed so that they understand the reason for HW control
Appropriate tariffs	TOU or fixed reduction of account	TOU or fixed reduction of account	TOU or fixed reduction of account
R/kW/year	Existing: R 53 New R 38	Existing: R 40 New R 30	Existing: R 26 New R 4
Total kW potential	Existing: 272 MW New 226 MW	Existing: 555 MW New 429 MW	Existing: 111 MW New 57 MW
Mutual exclusivity	No other hot water load control is feasible for customers with this control.	No other hot water load control is feasible for customers with this control.	Must not be duplicated in Home Automation

Economic potential/			
Cost benefits	Costs are considerably less than comparable generating costs	Costs are considerably less than comparable generating costs	Considerable benefits are possible with very little effort, but it requires co-operation from customers
Participant	Participants have not always been given their share of the financial benefits in the past	TOU tariffs could be used to pass benefits on to customers	TOU tariffs could be used to pass benefits on to customers
Distributor	Required control algorithm for distributor may differ from what the bulk supplier requires	Required control algorithm for distributor may differ from what the bulk supplier requires	Difficult to determine whether or not customers respond to request to reschedule
Bulk supplier	Great potential for bulk supplier because the load can be controlled when required with very little discomfort to customer.	Although not remotely controllable, it provides large reduction in peak load	Will reduce peak load considerably if it is monitored to ensure that timing is set correctly.
Technical potential Packaging			
linkage	Links well with TOU tariff	Links well with TOU tariffs	Links well with TOU tariffs
viability	Proven technology. Results are predictable	This intelligent thermostat needs to be developed.	Simple technology
Customer acceptance			
Satisfaction	Some resistance from customers possible due to excessive control.	If correctly implemented, it should not affect the customer	Some customers may want their pool pumps to operate during peak time
Market willingness	Some resistance	Must be tested	Must be tested.
Market potential	About one third op the potential market already has this technology installed	Control of presently controlled water heaters can be adapted. Also suitable for others.	Swimming pools are very popular in SA
Market segmentation	Suitable for upper and middle class	Suitable for upper and middle class	Suitable for upper class
Niche applications			
Targetebility	Most homes in middle to upper class	Most homes in middle to upper class	Target areas where swimming pool is common

Future potential			
Uncertainty	Technology and customer acceptance well known,	Technology needs to be tested	Customer acceptance and present setting of timers need to be tested
too many assumptions	More data are needed to get an accurate estimate of the MW impact	Long term application is likely to be better than normally controlled water heaters	Statistical data of pool pumps and timer setting is required
Degradation	Customer may interfere with apparatus	Regular checking is advisable	Timers may not be stable over time -
Reliability	Reliable technology	Technology is likely to be reliable	Power failure may cause timing error
Availability	Technology is available	Some development work is needed still.	Simple to apply
Maintainability	Customers sometimes de-energise equipment	Access to customer premises is necessary	Routine check needed
Flexibility (risk)	Very good	Very good	Good
System level controllability	Controllable by supplier - could also be centrally controlled	Controllable by supplier - could also be centrally controlled	Not available

Transforming Multifamily Properties to Efficient Appliances and Lighting Via Centralized and Negotiated Procurements[1]

JW Currie and GB Parker

Pacific Northwest National Laboratory

Abstract. This paper presents a novel approach to implementing energy efficiency in California's multifamily sector – a traditionally "underserved" market segment for energy efficiency programs – through centralized or negotiated procurement of new and emerging efficient ENERGY STAR® lighting and appliances. The approach, which relies on reaching the multifamily segment through local/regional apartment associations, is being implemented through market transformation programs at Southern California Edison (SCE) and Pacific Gas & Electric (PG&E), the two largest utilities in California.

The objective is to permanently change the lighting and appliance purchasing behavior of multifamily owners/operators, as well as their tenants, and to eventually expand the procurement program to include other underserved residential buyers groups such as senior communities. Given the success to date, we suggest that other utilities, or market transformation organizations, consider adopting, or testing out, our multifamily market transformation program design.

1   Introduction

Multifamily (MF) households account for a significant portion of the United States residential sector. Approximately 27% of U.S. households reside, as renters, in MF properties (>2 units) with 2/3 of these renter households residing in properties having more than 5 rental units (U.S. Bureau of Census March 1997).

MF renters are relatively mobile with more than 1/3 relocating in any given year (Goodman 1999). Furthermore, the majority of immigrant households reside in MF properties. In California, for example, 14% of all apartment residents have moved into the country during the past 10 years (Goodman 1999). In addition, the median annual income of MF households is less than that of homeowners (Goodman 1999). Finally, the median energy use for a MF unit is significantly less than that for a single-family house (Energy Information Agency 1997).

MF properties are unique in that two different types of energy customers are represented— owners/operators and tenants. Both types of customers are usually located on the same property and within the same building. Owners/operators and tenants are usually assigned different electric and gas tariffs. Many utilities assign commercial tariffs to owners/operators and have then tried to implement standard commercial energy efficiency programs but with little success.

Also, owners/operators typically purchase some equipment for which they have little responsibility for paying the energy bill. For example, owners/operators typically purchase refrigerators, dishwashers, wall/window air conditioners, and lighting fixtures for apartment units but, typically, do not pay the apartment electric bills. This creates significantly different incentives for owners/operators and tenants to purchase energy-efficient equipment.

Taken together, the above characteristics are the primary reasons why energy efficiency programs that have been designed for single-family households and small commercial businesses have not proven successful in MF properties. Thus, the term "underserved" is used to characterize the state of energy efficiency program design and effectiveness relative to the MF sector.

The task now facing U.S. utilities is to design and implement effective energy efficiency programs for MF properties and, in addition, to transform the MF markets for the end-use measures targeted. Because the MF sector has been historically underserved, this is a formidable challenge indeed. The effort requires combining the appropriate, and in some cases, new, technologies with novel implementation approaches.

2 PNNL Multifamily Technology Research

Research at the Pacific Northwest National Laboratory (PNNL), related to transforming the MF sector, has focused on both new and emerging technologies coupled with a novel and innovative implementation approach. The MF technologies considered to date are lighting, refrigerators, dishwashers, wall/window air conditioners, and coin-operated clothes washers. The primary technology focus has been to develop screw-base subcompact fluorescent lamps (sub-CFLs), and field verification of efficient refrigerator and coin-operated clothes washer cost and performance.

2.1 Sub-CFLs

CFLs have been available for well over a decade. Electric utilities have spent millions of dollars subsidizing the manufacture, distribution, and purchase of CFLs. However, the acceptance and use of this technology in the residential sector is abysmal. The most recent U.S. Department of Energy residential lighting survey finds that less than 50% of U.S. households are aware of the technology, less than 9% of the households use the technology, and, incredibly, *"less than 1% of all lights used 15 minutes or more per day are compact fluorescent"* (Energy Information Administration September 1996).

PNNL determined that the primary barriers to accepting this technology were 1) first cost and 2) length. CFLs were too expensive and too long to fit into many fixtures. In response to these problems, PNNL set a goal in 1997 to partner with manufacturers to bring to market, a screw-base CFL that was no more than 5 inches long, $5 in price, and equivalent in light output to a 100W incandescent bulb. These goals are close to being achieved with the availability of a 15W lamp that is 4.56 inches in length, with a delivered cost of less than $5 in lots of one thousand.

2.2 Refrigerators

PNNL has been involved in deploying and monitoring existing and new, high-efficiency refrigerator stock in the New York City Housing Authority (NYCHA) multifamily public housing since 1996. PNNL designed and implemented a durable six-sensor metering protocol to collect detailed time-series data on ambient and compartment temperatures, compartment door-opening activities, and power usage. Metering and demographic data were, also, collected and analyzed for the same representative sample of apartments. Annual savings from installing the highly efficient refrigerators were significant, averaging nearly 550 kWh savings per refrigerator (Pratt September 1998), making these refrigerators highly cost-effective. These findings led PNNL staff to conclude that the efficient refrigerators being installed by NYCHA were excellent candidates to use in market transformation promotion programs.

2.3 Clothes Washers

PNNL recently completed field verification of the performance and cost of high-efficiency, family-size clothes washers in a military barracks (Parker 2000). Six conventional washers were compared with 6 new high-performance washers from each of 4 manufacturers. Each of the 30 washers was metered in real-time for hot water use and temperature, cold-water use and temperature, machine energy use, and the number of cycles completed.

The total average water savings of the high-performance washers, compared to the conventional washers, was 38,780 gallons/year/machine. The machine energy savings was 140 kWh/year/machine and the hot water energy savings was 8.1×10^6 Btu/year/machine. These findings led PNNL researchers to conclude that the high-performance clothes washers are an excellent technology to promote through a variety of market transformation programs.

3 Multifamily Market Transformation Program Design

3.1 California Multifamily Characteristics

We were asked by SCE and PG&E to assist them in designing and implementing new approaches to transform the markets for MF electric end-use equipment. Collectively SCE and PG&E serve more than 8 million households, over 2 million of which are in MF properties (ADM Associates 2000).

Tenant electric bills are relatively small, averaging $30-$40 per month (Currie 1998). Owner/operator bills for MF property common areas average about $100 per month for properties with 2 or more units and $125 per month for properties have 5 or more apartment units. Given that owners/operators purchase the electricity using equipment for which they pay the bill, that their electric bill is significantly higher than tenant bills, and that they purchase much of the electricity using equipment for which the tenant pays the bill, we concluded that the owner/operator must be the first and primary focus for our market transformation (MT) effort.

Focusing first on the owner/operator has additional advantages. For properties having 5 or more units, the owner/operator usually has a permanent presence in the form of an onsite manager, and this person is almost always available. Thus, the cost to contact and interact with the owner/operator is much less than with tenants. Also, owners/operators have an established and maintained line of communication to the tenants. If a successful MT program could be implemented with a significant percentage of forward-thinking owners/operators, this group may be able to be incentivized as *de facto* agents for the utility in assisting with, and promoting, MT activities for the tenants. The first critical step is a successful MT program with owners/operators.

3.2 Buyer Attitudes

An effective MT program design should account for the purchasing attitudes of the targeted buyers since we are attempting to permanently change their purchasing behavior. We relied on survey data we collected at trade shows, as well as published data, to help in this regard (ADM Associates 2000, Opinion Dynamics Corporation 2000, Currie 1998.)

From the surveys referenced above, the following buyer attitudes, regarding appliance and lighting purchases, were identified and were critical in designing our program.

- Low first cost is the overwhelming purchase criterion.
- Direct toll-free or internet-based purchase is only a short-term (trial) option.
- Maintenance of traditional purchase and distribution channels is critically important.
- Significance of ENERGY STAR label is not understood.
- Appliance is purchased when an existing appliance fails; there is little on-site warehousing of spare appliance inventory.
- Apartment associations, of which many owners and operators are members, are the most credible sources of information.

It was clear that an effective MT program should mesh well with how MF owners/operators like to conduct business and that working with apartment associations would be important.

3.3 Program Design Elements

Given our findings above, we concluded that the most cost-effective approach to reaching MF owners/operators and, ultimately, tenants is to develop a collaborative working relationship with apartment associations and lighting and appliance manufacturers. Our approach is shown in Figure 1.

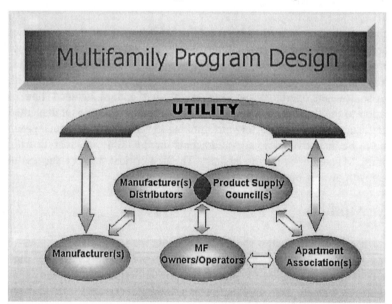

Figure 1: Multifamily Market Transformation Program Design

Our program design is intended to cost-effectively reach MF owners/operators while building credible and sustainable product and information delivery channels. We have had success following the steps described below and believe they can be implemented by other utilities. The key elements of the program design are:

- Build formal relationship with apartment associations
- Build relationship with appliance manufacturers
- Identify specific low first-cost ENERGY STAR appliance or lighting
- Work closely with product supply councils
- Negotiate lowest possible first cost
- Promote purchase of specific appliance or lamp through distributor.

3.4 Build Relationship with Apartment Associations

It is crucial to build formal working relationships with the apartment associations. We have purchased memberships in associations on behalf of the utilities we are working with. We have had several meetings with association officers to explain our objectives and to jointly map out a strategy. The collaboration resulted in the utilities promoting the MT program at association trade shows, attending monthly association meetings, purchasing advertising space in association journals, and publishing technical articles in association journals that explain the benefits of ENERGY STAR lighting and appliances.

3.5 Build Relationships with Appliance Manufacturers

It is important to develop and maintain relationships with appliance manufacturers at the national, regional, and local levels. We provided our contacts with estimates of the market potential for the appliances and screw-base CFLs that are MT technology targets of this program. We met with several manufacturers to emphasize the importance of promoting specific ENERGY STAR appliances having the lowest possible first cost. Finally, it is critical to secure manufacturer support prior to attempting to negotiate prices with distributors. This is because manufacturers may choose to structure special prices for distributors to pursue the MF market, and they will likely need to consider the impact on their local retailers.

3.6 Identify Specific Low First-Cost ENERGY STAR Appliance or Lighting

The importance of having the lowest possible first cost ENERGY STAR appliance cannot be overstated. MF owners/operators are extremely sensitive to first cost—more so than another other market segment we have ever dealt with. The primary reason is the "split incentive" issue. For refrigerators, dishwashers, and wall/window AC, the MF owners/operators buy the appliance but do not pay for the electricity to operate the appliance. As such, the owner/operator will almost always buy the lowest cost, reliable appliance available. We have observed, for example, that just a few dollars difference on the cost of a refrigerator can shift the outcome between many sales and almost no sales.

3.7 Work Closely with Product Supply Councils

All apartment associations have "product supply councils" comprised of members who are focused on providing goods and services to other members of the association; i.e., MF owners/operators. Included in the product supply council are lighting and appliance distributors. The distributors speak at luncheons, attend monthly meetings, advertise in the association journal, and purchase booth space at trade shows. It is critical that distributors are chosen from the product supply councils to receive apartment association support for the MT program.

3.8 Negotiate Lowest Possible First Cost

Once the distributor(s) has been identified, one or more meetings are required to arrive at the carryout price. It is important to describe the extent of advertising and promotional activity that the utility will engage in to promote both the specific appliance and the distributor as the place to purchase it. Furthermore, the distributor and manufacturer should be presented with an analysis showing the size of expected annual sales and the increased "foot traffic" that the utility promotion will generate. With this approach, we have been able to reach carryout prices on selected ENERGY STAR appliances that are only 2 - 5% above the price that manufacturers charged the distributors – without any utility buy-down or incentives.

3.9 Promote Purchase of Specific Appliance or Lamp Through Distributor

After the "deal" is structured, the utility needs to aggressively follow through on its end of the bargain to promote the specific appliance or lighting product. The promotion includes several complementary activities. These include advertisements, flyers, and technical articles placed in apartment association monthly journals and/or mailed to association members. Finally, the utility should purchase booth space at association trade shows and promote the utility's ENERGY STAR program and the specific appliances and lighting products in the program. This includes handing out flyers, sample products, and contests or drawings for ENERGY STAR appliance giveaways.

4 Results

We currently are promoting several ENERGY STAR products for SCE and PG&E; 10 sub-CFLs, 3 refrigerators, and one dishwasher. In addition, we are evaluating 3 different brands of high-performance coin-operated clothes washers. We initially worked directly with 5 apartment associations representing over 10,000 owners/operators and over 100,000 tenants and are expanding our efforts to work with an additional 14 associations representing nearly 30,000 owners/operators and over 250,000 tenants.

4.1 Sub-CFLs

We are promoting ENERGY STAR sub-CFLs (http://www.pnl.gov/cfl) with both SCE and PG&E. There are over one million incandescent lighting sockets in MF exterior and common areas served by SCE and PG&E (ADM Associates 2000). These lamps are on at least 12 hours per day and with electricity costs of 10-12 cents/kWh, sub-CFLs are extremely cost-effective for the owner/operator.

The major activity is with PG&E. We are purchasing 10,000 sub-CFLs outright for giveaway to condition the market. PG&E will then "buy down" the sub-CFLs by $3 per bulb on direct purchase from the suppliers by owners/operators until a maximum of 100,000 bulbs have been sold. At that time, our MT sub-CFL activity will be folded into the statewide California ENERGY STAR retail lighting program.

4.2 Refrigerators

We are promoting 3 ENERGY STAR refrigerators, all of which are at least 30% more efficient than the current U.S. government standard. Owners/operators own over 1 million standard refrigerators across the two California utilities. We were not able to document any owner/operator purchases of ENERGY STAR refrigerators prior to implementing our program with SCE. In our initial program with 5 associations, we have been promoting a 15 ft^3 refrigerator. Distributors report sales to us and their data indicate that this refrigerator is now capturing over 25% of sales to owners/operators. A key reason is that we were able to negotiate a price that was competitive with other standard 15 ft^3 models.
We have not been as successful with an 18.5 ft^3 model. In 6 months of promotion, there were no reported sales because our negotiated price, which is only 2.5% above the distributor cost, is still over $50 above other standard models of similar size.

In July 2000, California implemented a $100 rebate for all refrigerators that are at least 30% more efficient than the federal standard. At that time owners/operators began purchasing the 18.5 ft^3 refrigerator.

4.3 Dishwashers

We estimate that MF owners/operators, served by PG&E and SCE, own at least 1.5 million standard dishwashers (ADM Associates 2000) with roughly 200,000 being replaced annually. We recently initiated a promotion of an ENERGY STAR dishwasher program for SCE's owners/operators. The distributor is selling the dishwasher at a very competitive price and at a 6% markup over his cost. Early indications are that this promotion will result in many sales.

4.4 Coin-Operated Clothes Washers

There are approximately 200,000 standard coin-operated clothes washers owned by owners/operators across the two utilities (ADM Associates 2000). We are initiating a field verification and MT program in a MF senior citizen community that has more than 1000 coin-operated clothes washers. We will first meter the existing washer stock to determine a baseline and then substitute 4 high-performance washers from each of 3 different manufacturers to demonstrate the performance and monetary savings. Included in the field verification will be survey data from consumers on the acceptability of the high-performance washers. Next, we will promote the results through the news and TV media in the community. Finally, we will disseminate the results to our network of apartment associations via articles in the association journals and presentations at monthly meetings. We will then negotiate with one or more of the manufacturers and their distributors for very low promotion prices and we will jointly advertise with the distributors and promote the washers.

5 Summary & Conclusions

The primary objective of our efforts has been to permanently change the purchase behavior of MF owners/operators for appliances and lighting – and to do so without relying upon direct utility financial incentives such as rebates. We have devised a cost-effective approach that utilizes formal working relationships with apartment associations and equipment manufacturers. Finally, we have achieved some successes – measured as significant and/or first time sales of ENERGY STAR appliances to owners/operators who have never before purchased energy efficient equipment. We have, also, determined that owners/operators are exceptionally sensitive to first cost. In some cases, prices negotiated with manufacturers and distributors may not be low enough to induce sales. In these instances, direct financial incentives are required if the target market segment is going to be penetrated.

Our program has been in existence for about two years. Much of that time has been spent designing and testing our approach of reaching owners/operators through the apartment associations. We believe that we have demonstrated the value and cost-effectiveness of this paradigm – that this approach can permanently impact the purchasing behavior of owners/operators – and the concept can now be transferred to other utilities and MT organizations. However, the challenge remains to reach the tenants.

In calendar year 2000, we plan to test approaches for reaching the tenants. Our current plan is to begin working with some of the more forward thinking owners/operators we have encountered. We have ideas on how to help them promote specific ENERGY STAR appliances and these include fliers targeted for tenants, free workshops, and a fully metered "green apartment building" demonstration to show the monetary savings that are produced by purchasing energy and water efficient equipment.

6   References

[1]   ADM Associates, TecMRKT Works. May 2000. *Statewide Survey of Multi-Family Common Area Building Owners Market Final Report Volume 1: Apartment Complexes.* Prepared for Southern California Edison Company, Rosemead, California.

[2]   Currie, JW, GB Parker, DB Elliott. November 1998. *Private Multifamily Market Transformation Via Centralized and Volume Procurement: Initial Program Design.* Prepared by Battelle for Southern California Edison, Rosemead, CA.

[3]   Energy Information Agency. 1997. *1997 Residential Energy Consumption Survey.* Table CE1-4C. Total Energy Consumption in U.S. Households by Type of Housing Unit.

[4]   Energy Information Administration. September 1999. *Residential Lighting Use and Potential Savings.* DOE/EIA-0555(96)/2. U.S. Department of Energy. Washington, DC.

[5]   Goodman, J. 1999. "The Changing Demography of Multifamily Rental Housing." *Housing Policy Debate.* Volume 10 Issue 1.

[6]   Opinion Dynamics Corporation. February 11, 2000. *Multifamily Baseline Study Final Research Report.* MR-99-38. Prepared for Pacific Gas and Electric Company, San Francisco, California.

[7]   Parker, G and G Sullivan. September 2000. "High Performance Clothes Washer In-Situ Demonstration in a Multi-Housing Multi-User Environment." *In Proceedings of 2nd International Conference on Energy*

Efficiency in Household Appliances and Lighting. September 27-29, 2000. Naples, Italy.

[8] Pratt, RG and JD Miller. September 1998. *The New York Power Authority's Energy-Efficient Refrigerator Program for the New York City Housing Authority – 1997 Savings Evaluation.* PNNL-11990. Pacific Northwest National Laboratory, Richland, Washington

[9] U.S. Bureau of Census. 1997. *Current Population Survey 1997.* Machine readable data file. Conducted by the Bureau of Census for the U.S. Bureau of Labor Statistics. Washington, District of Columbia.

Endnotes

[1] Portions of this research are funded by California Utility Customers and administered by Southern California Edison and Pacific Gas & Electric, under the auspices of the California Public Utilities Commission.

[®] ENERGY STAR is a registered trademark of the U.S. Environmental Protection Agency that has been licensed to the U.S. Department of Energy.

The Idea Network and Appliance Efficiency

Nils Borg[1] and Boudewijn Huenges Wajer[2]

[1] Borg & Co AB, Sweden
[2] Novem, The Netherlands

Abstract. Following close discussions between the European Commission and a number of interested European energy agencies and other parties, NOVEM (the Netherlands), took the lead in establishing IDEA (The International Network for Domestic Energy-Efficient Appliances) in 1997 and in launching the free *Appliance Efficiency* newsletter in partnership with ADEME (France) and STEM (Sweden). More recently, the Danish Energy Agency, and Swiss Federal Office of Energy have joined IDEA and contributed to its costs. *Appliance Efficiency*, the Newsletter of IDEA has a world wide circulation to 2400 key policy makers, manufacturers, programme managers, efficiency advocates, consultants, journalists and academics. *Appliance Efficiency* has established itself as a leading instrument for gathering information and contacts addressing domestic appliance energy efficiency. The newsletter has publicised the results and recommendations of all the major European Commission SAVE programme studies concerned with domestic appliances, has publicised articles on various national and EU energy policies and legislative initiatives, and has been highly instrumental in informing key actors in appliance efficiency area programme about relevant actions. The IDEA network has also overseen the development of a web site which carries electronic copies of *Appliance Efficiency* and copies of related SAVE studies and provides forums for discussion about appliance efficiency technology, policies and programmes. The IDEA network also produces and maintains an extensive database of key individuals and institutions concerned with appliance energy efficiency matters in Europe and the rest of the world. This database provides a very useful resource for the organisation of conferences, workshops, mailshots and other dissemination activities.

1 The Need for Two-Way Communication

The initiative to start a network for domestic appliances began in early 1997. The initiative was taken by Novem, Netherlands Agency for Energy and the Environment, who identified two major needs:
Cost-effective dissemination of information: Large-scale dissemination of information is not always cost effective in the way it has previously been done. By creating a network where readers actively state their interest to receive a high-

quality newsletter, information can be disseminated much more cost effectively than in normal bulk mailings where a large number of address lists are utilised. The web site will complement and increase the effectiveness of the printed newsletter .

Two-way communication: Cost effective dissemination will only solve part of the problem: With the assumption that a network of active and professional readers also will be a major source for the information that is disseminated, Novem envisaged a two-way network where the major dissemination tool–the *Appliance Efficiency* newsletter–is so attractive that it can attract professionals and key actors to state their interest for being included in the readership database. Thus, information flows in two directions.

The discussions were later continued in close contact with the European Commission's SAVE programme, and an informal workshop was held in Brussels in the summer of 1997. The network was given the name IDEA, the International Network for Domestic Energy Efficient Appliances. Once IDEA was started, it was encouraged by the Commission, and during 1998 and 1999, SAVE also supported the network financially. From the onset, The French Agency for the Management of Energy and the Environment (Ademe) and the Swedish Energy Administration (STEM) supported IDEA financially. Later, The Danish Energy Agency (DEA) and the Swiss Federal Office of Energy (SFOE) joined the network's board and decided to support it financially.

The design of IDEA was largely modelled upon IAEEL, the International Association for Energy Efficieny Lighting. IAEEL had been started in 1992, and has since then issued a newsletter of energy-efficient lighting, as well as a web site. IAEEL has developed a database of more than 5000 people who are active readers of the IAEEL newsletter. (For more information about iaeel, see: www.iaeel.org.)

2   Overview of Idea and its Activities

IDEA is a network supported by five national European energy agencies which aims at complementing other networks and initiatives. Although regional energy agencies as well as agencies from outside Europe are welcome to support IDEA, the network's core has not yet been extended beyond the five agencies mentioned above.

The network carries out 4 primary dissemination tasks as follows:
- Internet module (IDEA web site)
- Newsletter research, contents and production
- Database with addresses, newsletter distribution
- Operating agent and secretariat

2.1 Internet Module (IDEA Website)

A provisional website for the IDEA network has already been developed with priming funding from SAVE and IDEA, see *www.idea-link.org*. The web site has the following primary goals:
a) to maintain a library of previous *Appliance Efficiency* editions on the web
b) to add fresh web editions of *Appliance Efficiency* throughout the period
c) to create a depository for electronic copies of European Commission SAVE studies, which concern appliances
d) to create a depository for electronic copies of other important studies, which concern domestic appliances
e) to initiate web forums to discuss appliance energy efficiency developments
f) to provide links to other web sites concerned with SAVE and related Community programmes, national energy agencies and the EnR and OPET networks, appliance industry and market news, energy efficiency agencies, international programme sites, environmental links, etc.

2.2 IDEA Newsletter: Appliance Efficiency (Developing Contents)

The 16 page newsletter is issued 3-4 times a year (3 issues 16-page issues are planned for the year 2000), written in English, is independent and of high-quality. It is a journalistic product and not a publication that reflects the official view of IDEA or its members. It reflects the activities in the area with a positive attitude towards initiatives for energy savings actions.

Target groups
The newsletter is not aimed at consumers but at professionals with an interest in appliances and/or energy issues. These are, for example, policy makers working at all levels in local or national/regional energy agencies and governments, people from consumer organisations, electric utilities, Non Governmental Organisations, appliance manufacturers, the press (trade press journalists and specialised writers in the general press), test institutes, research organisations and universities, and professional buyers such as housing organisations.

Subscriptions are free of charge, but an active readership communication where database information is gathered from the readers along the lines of the IAEEL newsletter (International Association for Energy Efficient Lighting) are applied. This is not only a free publication that disseminates information from one source to selected target groups, but each reader in the address database is a source of information. Thus readers are welcome from all over the world, although European readers dominate the readership stock.

Themes

Appliance Efficiency provides a broad and highly informed coverage of: national and multinational policies and programmes concerning appliance energy efficiency; major market developments and efficiency trends (it often gathers source data that is useful for providing an up to the minute snapshot of appliance energy efficiency trends in the world); new high efficiency products and technologies; key end-uses such as cold appliances, wet appliances, consumer electronics, space heating and cooling, cooking, and water heating; research activities, meetings and events, and international product standard developments.

Since its inception it has published major articles on SAVE product studies addressing: refrigerators, clothes-dryers, standby power consumption, air conditioners, storage water heaters, TVs, end-use metering, and most recently ovens. This coverage has enabled a much wider awareness of these studies and related issues and implications to be achieved among people who can have a major impact on residential energy efficiency developments. In addition many other activities that have been funded through or supported by the SAVE programme have been publicised and promoted in the newsletter.

Appliance Efficiency has also given extensive coverage to policy and programme developments at the European level as well as the wider global community. This has considerably heightened awareness of parallel activities taking place in many major economies of the world including: the USA, Japan, Central and Eastern Europe, Australia, China, Brazil, India, Thailand, South Africa, Korea, Iran, etc. Through publicising these activities and their protagonists *Appliance Efficiency* has encouraged good ideas to be adopted more rapidly and has facilitated important programme co-ordination efforts. This has helped to promote successful European programmes such as the European energy labelling scheme, which has now been adopted or imitated in many parts of Central and Eastern Europe, Iran, Mexico and Brazil and is currently being considered by Argentina, Tunisia, Algeria, Columbia, and China. It has also bought very relevant news of international best practice in appliance efficiency, such as Japan's Top Runner programme and Australia's minimum energy efficiency standards programmes to the attention of European policy makers and programme managers.

Its readership of 2400 people covers 63% in Europe and 37% from approximately 50 countries in the rest of the world shows the large interest there is in this subject and in European-backed energy efficiency initiatives.

The newsletter is flexible to explain or discuss new trends and problems. Since the appliance market is global, the newsletter covers the international scene but the primary focus is in Europe.

Review and quality assurance

The organisation in charge of the newsletter (Novem) and the main editor make sure that there are competent and independent technical reviewers that can support the newsletter editor and their writers. A group of technical advisors, which also includes the editorial board of IDEA act as advisors and, if appropriate, as reviewers. Novem manages the whole production process and has been appointed by the IDEA board to secure quality and information flow.

To date (summer 2000) there have been a total of 9 editions of *Appliance Efficiency* since the first in 1997; however, due to irregular funding it has only been possible to manage one year where the full quota of four issues was produced (1999). SAVE funding was not secured for the calendar year 2000 and as a result it will only be possible to produce 3 copies of the newsletter in this year.

2.3 Database of IDEA Network, and Distribution of Newsletter

The address database is an important asset of the appliance network. The database containing the contact details and efficiency interests of more than 3000 *Appliance Efficiency* readers and *potential-readers* is a unique resource that can be used not only to target and refine the newsletter readership but is also very valuable in aiding the organisation of conferences, sending out communiqués and other forms of targeted dissemination. The database itself is also used as a resource when collecting information for newsletter articles and other research purposes. It can help to reduce costs for several projects by core members and others where targeted communication with relevant market actors is important. The database has already been made available for use in organising the SAVE conference on Energy Efficiency in Household Appliances and Lighting in Naples, as well as workshops on: standby power, installed appliances, and lighting, as well as smaller national workshops in the involved countries.

Core member organisations of IDEA can have pass-word-protected access to the database via the internet. The password protection is important to ensure that information on individuals is kept confidential and secured from commercial abuse.

Maintenance of the database and communication with newsletter readers

The maintenance of the database is to be done on varying levels. The addresses of potentially interested readers are supplied and entered from numerous sources. To ensure that the recipients of *Appliance Efficiency* are active and interested in the subject area reply forms are sent out when people receive the newsletter. If no reply is received at the third attempt the subscriber is removed from the mailing list after some time, unless there is a good reason to believe that the person concerned is an important and active user of the newsletter. By constantly renewing and revitalising the database in this manner and by using the large resources of the IDEA network members, it is possible to ensure that the readership remains highly targeted and relevant to the subject area. Thereby the newsletter and network have maximum impact for minimum cost. This method has been proved successful. Apart from adding the contact network, these reply forms provide extremely valuable feedback on reader interest and appreciation for the newsletter and help the editors and board to target the content. The reply frequency is indeed very high: Of 2400 readers, more than 1400 have actively stated their interest. (A large share of readers are always newly entered into the database, so there will always be a number of persons who cannot have replied at any given time.) When discussing the response rate with experts in market communication, the IDEA network has been told that the response rate is to be seen as extremely good (well above 50 percent), since commercial campaigns rarely yield a response rate above 5-10 percent.

Communication with Newsletter distributors, conferences etc.
Updated address lists, including selections from the database, can be easily distributed to the mail company that distributes the newsletter, to member organisations of the IDEA for other purposes, and for the organisers of appliance conferences on request.

2.4 Operating Agent, Secretariat, Co-ordination and Travel

This secretariat role includes maintaining a helpdesk and a permanent IDEA network email address with an information response function. The secretariat is also responsible for launching ancillary actions to support and grow the IDEA network, such as promoting membership, canvassing views on the direction and goals of the network, and encouraging events and actions through the network. The goal of the secretariat is for the network and newsletter to attract enough supporting members to become self-sufficient within a few years. The secretariat is also responsible for network co-ordination activities including staging meetings of IDEA members to discuss network-planning issues.

Overall organisational structure of the network
The organisation has started as an informal European initiative (former activities see below) based on the following entities:
- Core members have a place in a general board of the network. Governmental/public organisations that has put up the minimum contribution of money (financial and in-kind contributions) constitute the board of IDEA. Currently these core members are: ADEME, DEA, Novem, SFOE and STEM.
- The editorial board act as counsellors for the editors and other persons who are disseminating information on behalf of the network. The editorial board will consist of representatives from the general board and/or other persons selected by the members of the IDEA board to represent them. Via the

operating agent the board members can give comments and input about/to the information to be published.

- The editorial team is responsible for writing the newsletter, the content of the internet site and other information to be published by the network. Editors and technical advisors to the editor are to be appointed by the general board.

Operating agent

The tasks of the operating agent (prime lobbyist for the network) includes:

- Preparing board meetings and follow-up on decisions
- Co-ordinate information flow between editorial board and editing team
- Co-ordination of all tasks which are added to the network and at co-ordination with conferences
- Maintaining a secretariat (which deals with practical issues of the network such as preparing board meetings, keeping track of finance, etc.). The secretariat facilitates the network and all organisational aspects (e.g. general printing and mailing). The secretariat will direct all interested people to the right people within the network which amongst others facilities the general board.

IDEA Secretariat (Office Manager)

The Operating agent makes use of an IDEA Secretariat (Office Manager) who:

- assists the Operating Agent in the tasks mentioned before
- make financial reports
- mans a helpdesk acting as a contact to the outside world supplying general information on IDEA and "connecting" people. Automated internet functions are created to release the burden of the secretariat
- does acquisition of subscribers (by collecting address files and free publicity in magazines) and by making a judgement of what addresses to include.
- prepares subscriber evaluation mailings
- co-operates with conference organisers (e.g. Naples conference)

3   Expected Impact

The continuation and extension of the IDEA network and newsletter is expected to continue to serve the informational needs of the energy efficiency community working with household appliances. It will greatly assist dissemination of the launch and impact of programme actions. It enables actors in various energy programmes and beyond to be familiar with each other's work and to contact each other. It raises the profile of the EU's appliance energy performance programmes and enhances the prestige of the SAVE programme, and the same is true for other programmes covered by the newsletter. Thus, the funding organisations see a very important role for IDEA in the multiplier effect that IDEA has in the appliance arena. By sharing experience passed through the constructive filter of the network and newsletter it encourages best practice among various energy programme participants and strengthens their contributions to the funders' appliance efficiency efforts. These facts can be summarised as:

- Information on new study reports and other relevant information from the European Commission, national energy agencies etc. from around the world and research institutes can be disseminated efficiently through the tools of the IDEA network.
- European and national policy making on matters of energy efficiency can be more rapidly implemented because EU officials and representatives from member states have immediate access to relevant information on household appliances.
- International conferences on energy efficient domestic appliances can be supported by the network organisation and thereby assuring high quality of the content. Moreover the important target groups are more easily and more cheaply reached.
- Research and development of energy efficient concepts of household appliances or technology can benefit from acceleration and improved access of information by engineers and scientists.
- Programmes and schemes such as labelling, efficiency standards, procurement etc. can be promoted via the international conference and the network and thus implemented more quickly.

3.1 The Network's Complementarity with Other Actions

The network is international in its nature, clearly reflecting the global appliance market. Each national agency is too small to start such an activity and very few would be allowed to do so with so many people outside that given country benefiting from the activity.

The network's ability to influence the global manufacturers is drastically increasing by running a multi-country project. An initiative launched by one nation will not be taken so seriously and therefor will not take place.

National activities focusing on information dissemination such as the international conference in Naples can, under certain conditions, make use of a selection of the address database. Recruitment of speakers and chairmen for conferences can be facilitated by the network organisation.

Maximising the Improvements, Minimising the Investments. A Case History from a Manufacturer

Chino Mozzon, Massimo Cazzaniga and Matteo Rossignani

CANDY Elettrodomestici

This paper describes the latest improvement of energy consumption on refrigerators.
The new energy limits and the market requirements have forced a technical revision of the past products. The approach of Candy group was based on a selection of the best technological options with the aim to achieve the best performance with the lowest investments costs.
The Candy experience offers an experimental comparison among the majors technologies and an overall analysis of the economic impact on the final product cost.

1   Introduction

The latest Candy strategy has been focused on a massive energy improvement of the whole product range. The market trend of these years has been driven by the new legislation that have banned energy classes above C. This situation have forced manufacturers to find technical solutions compatible with the new requirements and to keep the product cost within acceptable levels.
Most of the companies have introduced in their range more performing models before the due date of Sept. 99.
During the last years the average energy consumption has been improved of about 25% . The Figure 1 shows the energy classes distribution during the last 5 years for the Candy models.

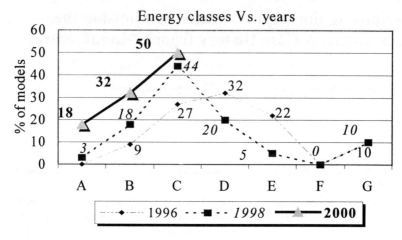

Figure 1

This significant energy improvement has been achieved by using more efficient components and developing new products. Of course this new performance have had a cost impact on the   final product cost/price. The Figure 2 shows the refrigerator cost during the last years (Gfk data)

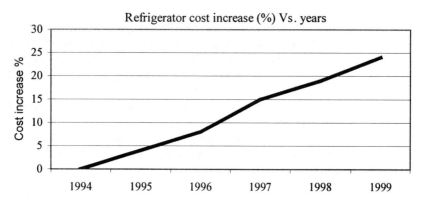

Figure 2

From the above mentioned market scenario, the product cost and the investments have been considered as key item for the development of the new A class product range.

Due to the performance required, the feasibility study for this range has taken into consideration several technological options and, after an experimental evaluation and costs/investments analysis, the most convenient have been chosen.

To achieve the goal, Candy has approached the product revision through an intensive experimental program concerning the following areas : components, insulation and temperature control.

The following paragraphs is a brief summary of Candy experience encountered during the last years. Of course, the enclosed results should be considered as the average outcome of measurements of the different product families.

1.1 Components

Three are the components with a direct impact on the product performance, i.e. compressor, evaporator and condenser. In most of the cases these are not competitive advantages because most of the refrigerator manufacturers depend on the supplier know-how. For this reason the decision to get the maximum performance from components can be considered as a conservative approach waiting for a better understanding of the market direction and to avoid risky investments.

1.2 Compressor

This is the easiest change, a drop-in solution completely delegated to the suppliers. At the same time it is the most expensive one, especially if compressors with a COP higher than 1.6 are required.

The Figure 3 shows the average compressor delta cost in relation with its efficiency and with the result of energy saving measured on refrigerator.

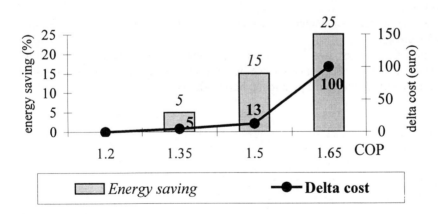

Figure 3

1.3 Heat Exchangers

Both the heat exchangers, evaporator and condenser, are among the most common elements to be changed for improving performance of refrigerators. The experimental data of energy measurements are pretty well correlated with the component size. The obvious limit is the overall dimension of these components in relation to the product size.

1.4 Evaporator

The changes of this component are strongly linked with the product configuration, i.e. double door, freezer, etc. As a general rule, the bigger are heat exchangers and the better is the performance. In reality the approach should be tailored on the different product families and sizes.

There are several possibilities to increase the heat exchange function of the fridge evaporators and once again the final combination is a compromise between the ideal solution and the process/product constrains, either for the supplier or the fridge manufacturer.

The most common fridge evaporator type is the steel pipe foamed inside the refrigerator. A few company have roll bond evaporator type, either foamed or inside the fridge cell.

The thermodynamic should suggest a roll bond type evaporator placed inside the compartment. This solution maximises the heat exchange but introduces other issues like the aesthetic appearance, the safety conditions for R600a applications and the overall cost that seems to shift the interest of most of the producers to the foamed solutions.

In addition the use of roll bond inside the compartment requires investments for new liners.

The relation between evaporator size and energy saving on refrigerators is shown in Figure 4.

Today, any further size increase is impossible having reached the maximum surface covering. The only chance for further improvement is the roll bond solution with all the above mentioned implications.

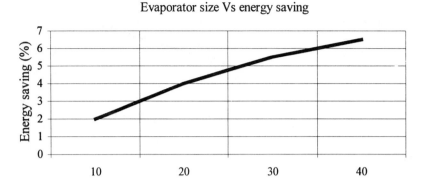

Evaporator size Vs energy saving

Figure 4

1.5 Condenser

Like the evaporator, the condenser dimension can be enlarged to obtain an improvement of the product performance. The correlation between size and energy saving on the product is similar to the one reported in Figure 4.

1.6 Insulation. Foam

The effect of a thicker insulation is easily predictable but there are drawbacks.
First of all any thickness change has an impact on production tools and requires significant investments. In addition, the final product is penalised by a lower net volume.
The insulation effect on performance is shown in Figure 5 a) b), where wall thickness and energy improvement are compared either for a larder or for a freezer (same size).

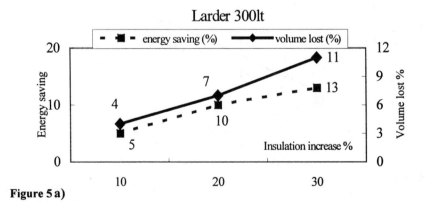

Larder 300lt

Figure 5 a)

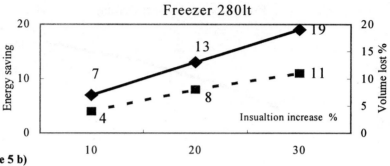

Figure 5 b)

The material cost of this solution is essentially the new foam quantity and if compared with components seems to be more advantageous (see Figure 6).

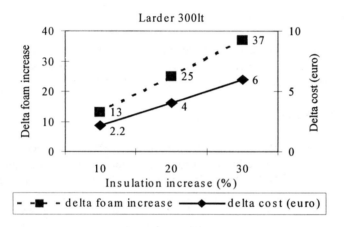

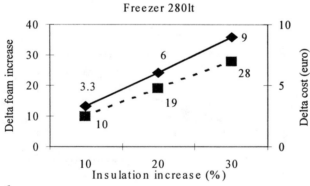

Figure 6

In reality the new product costs should consider the depreciation allowance determined by the investments. This explains why insulation changes are obtained by simple tool modifications, limiting the thickness variation to the most critical areas of the cabinets.

Then, the typical manufacturers approach is focus on the rear insulation, sometime combined with the inner door thickness. The Figure 7 shows the result of 10 mm insulation increase at the rear of the cabinet and the overall effect when is combined with 10mm extra door insulation. This kind of solution has a very little impact on the net volume of refrigerator.

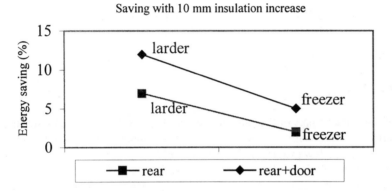

Figure 7

1.7 Insulation. Vacuum Panels

Insulation means also new materials like the Vacuum Panels and a new production process management.

This is a controversial item because the experimental tests of many companies has shown great difference in the final energy saving. But even if the energy benefit of this technology does not get a general consensus, the cost is universally deemed inapplicable for mass production of standard refrigerator.

Figure 8 shows the energy saving measured on 250lt freezer with different product configurations and the cost estimation for each solution. The surface covered by panels was about 70% of the total area.

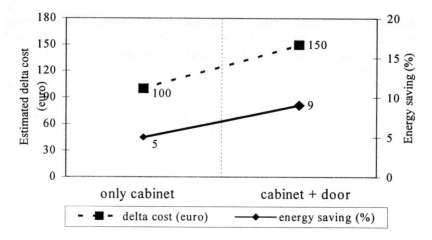

Figure 8

1.8 Gasket

One of the weakest part of the cabinet is the sealing, especially around the freezer compartment. The sealing is based on a well-established technology and it is managed by just a few providers all over the world.

More than material or supplier progress, the sealing improvement can be achieved by redesigning the door closing system, in particular reducing the distance between the cabinet and the door. This approach requires big investments because involves new hinges and new door assembly.

1.9 Temperature Control

Temperature control represents a potential area of improvement. Its success depend on the effectiveness of electronic control that should minimises the on-off losses by reducing the number of defrost cycles. Applying this rule the product becomes cooler and runs for shorter periods. The result of this basic principle depends on the product configuration. As a matter of fact, double doors (i.e. 80% of the market) needs to be re-balanced because the shorter running time affects freezer temperatures. This explains why the real application of this technology is limited to just a few models, in particular larders recently developed.

2 Conclusion

The massive energy improvement reached during the latest years have already taken the advantage of most of the options described above. Most of the refrigerator companies have shared or followed the Candy initiative to maximise the improvements limiting investments and controlling final product costs.

Most of the current A class models have already introduced these options, therefore the next improvement should be concentrated on thickener insulation products. The remaining technical options based on the present know-how does not allow to forecast a level of improvements similar to the one achieved in the recent years and the new energy limits should contemplate this situation.

Strict limits of the next energy classes could affect the overall sales due to the excessive product costs with the consequent extension of the use of old refrigerators poorer performance.

In addition, an adequate consumer education could help the energy saving by following a proper use of the refrigerators.

The Danish Electricity Saving Trust: the Campaign in Autumn 1999 – A Review of Experience

Peter Karbo

Elsparefonden

Abstract: Three activities were carried on simultaneously during 1999:

- a large government campaign, which was also supported by retailers, utilities and manufacturers, urging consumers to 'Buy A';
- subsidies for A rated appliances for a limited period;
- a web site of A rated appliances, their availability and prices was launched and widely publicised.

Together these have had a significant impact on sales and prices of A rated appliances in Denmark and are expected to continue to influence the market in future.

The paper will present the approach, final results and analysis

1   Rebate Programme

1.1 Background

Refrigerators are turned on for 24 hours a day, so when you buy a new refrigerator you are also buying part of your electricity bill for the next 10-12 years. The introduction of energy labelling in Denmark in the middle of the 1990s had a significant effect on the sale of B and C rated refrigerators. The market share of B and C rated refrigerators rose from 20% in 1994 to 40% in 1997, almost wiping out the next lowest energy classes. B and C rated refrigerators conquered a large part of the market and the competition focus in the trade switched over to these products, rationalising production and logistics and making it possible to lower prices. The basic selection in the shops came mainly to consist of these products and it was thus B and C rated refrigerators that were sold. A positive development had started. Between 1994 and 1997 the prices of C and D rated products fell by about 20% as a result of the increased focus. The products had been brought into the "market place" and into competition. The excess price of these products had been reduced and for consumers the value of the electricity saving corresponded to a pay-back time of 3 years – a time scale that Danish consumers accept. The 80-20 rule also applies in this trade: the 20% most sold products represent 80% of the turnover.

On the other hand the same rise did not take place in the sale of A rated refrigerators, which only rose from a market share of 4% in 1994 to 9-10% in 1998. The Danish Electricity Saving Trust therefore decided to do something for these products and the desire was expressed to "shift from 1st to 2nd gear – by bringing A rated products into the market place".

1.2 Aim

The aim of the campaign was to strengthen the market share of A rated appliances by bringing more focus to bear on energy-correct products. This was a double aim: The Danish Electricity Saving Trust wanted to influence consumers to ask for more energy-correct products and to encourage the manufacturers to offer more A rated appliances.

1.3 The Objective of the Campaign

The Danish Electricity Saving Trust estimated the sales of A rated appliances in a normal 11 week period during the autumn at approximately 2,500 out of total sales of approximately 74-83,000 appliances. The Trust wanted to boost this figure to 19,300 by means of a rewbate programme.

The Trust's expectations of the impact of the campaign on electricity saving and CO_2 reduction can be seen in Table I.

Table I: Electricity saving and CO_2 reduction

	Number of A rated appliances	Electricity saving MWh)	CO_2 reduction (tons*)	Subsidy (DKK)
Com. fridge-freezers	8,000	8,350	6,582	4,000,000
Refrigerators	8,000	4,800	3,784	4,000.000
Upright freezers	3,000	3,000	2,,365	1,500,000
Tumble dryers	300	930	733	300,000
Total	19,300	17,080	13,463	9,800,00

*   Calculated on an appliance lifetime of 12 years, CO_2 factor $= 0.78825$ kg/kWh on average per year up to 2011

1.4 Description of the Campaign

Between 20.9.99 and 4.12.99 The Danish Electricity Saving Trust paid a direct subsidy to the consumer who bought an A rated appliance.

2 door combined fridge-freezer (DKK 500), Refrigerator with/without an internal frozen food compartment and chiller (DKK 500), Upright freezer (DKK500) and Tumble dryer (DKK 1,000).

The subsidy budget for the campaign was DKK 10.5 million and a reserve pool of DKK 5 million was set aside should the campaign prove to be more successful than first expected. An additional approximately DKK 1.5 million was set aside for running the campaign, marketing and control.

1.5 Before the Campaign

The campaign was co-ordinated with the electricity utilities, which were also planning an A rated appliance campaign and an external project leader was appointed.

To achieve the greatest possible effect, the subsidy was to be paid out in the purchase situation and all retailers who wished to participate in the campaign had to be contacted. This proved to be a bigger job than first assumed as there were more small and fewer large retailers who were part of a chain than had been thought. This ended with a contract being signed with almost 100 retailers.

In the contract, the retail chains committed themselves to the following, inter alia:

* to exhibit a reasonable selection of A-rated appliances during the campaign period if they had a showroom
* to make use of the in-store and trade materials of The Danish Electricity Saving Trust during the subsidy period itself
* to inform the retailers in the chain of the approaching campaign and the subsidy scheme
* to inform the customers about the campaign and to provide advice about the advantages of the A rated appliances during the sales situation
* to inform about the price before the subsidy, the amount of the subsidy and the price after the subsidy when advertising and displaying special offer signs
* to pay the subsidy directly to the customer
* to write an invoice that included the following: brand name and model ID, gross price and subsidy, the full name, address and telephone number of the customer, date of invoice and delivery address if these differed from the customer's address. If the product was part of a larger purchase, the product was to be separately noted with brand name and model ID, price before subsidy, the subsidy and the sales price
* to submit the last payment request before 20 December corresponding to valid invoices
* to pay back the subsidy for sales that were cancelled and any unjustified subsidy which was revealed in subsequent control
* to keep separate records with all mandatory information concerning subsidised appliances and cancelled sales for 5 years in the form of copies of invoices or electronic records – and to place the records at the disposal of The Danish Electricity Saving Trust or its representatives and the National Auditor Office of Denmark, and for analytical purposes

* to place retail outlets at disposal for unannounced spot checks during the campaign period to supervise special offer signs, the number of A rated appliances on display, and the use of in-store material and mandatory energy labelling
* to submit a report by 20 December 1999 at the latest of the number of appliances sold with details of brand name and model ID within the four categories mentioned.

It was also a requirement that retailers were not to raise the price of A rated appliances during the campaign period so that they did not "take the top" off the subsidy scheme. On the contrary, retailers were urged to lower the prices during the campaign period in order to ensure that it was successful.

In return retailers were given permission, on the part of The Danish Electricity Saving Trust, to pay out the subsidy and this make use of the name and logo of the Trust in their marketing. The material had to be approved by the project leader before an advertisement was placed..

The retailers were extremely positive about the campaign, a fact witnessed by the great number who joined it.

1.6 During the Campaign

Before the campaign kicked off, various types of marketing materials were prepared, aimed at the retailers, the consumers and the press.

A TV spot was shown in the days up to the campaign and then regularly during it, and the total number of contacts was estimated at 6.3 million.

There was a radio spot on DR P3 and The voice, and the estimated number of contacts was 5.2 million.

Apart from the centrally-controlled advertisements placed by The Danish Electricity Saving Trust, there was enormous media pressure by retailers and manufacturers who placed advertisements in national and local newspapers, flyers etc. The retail outlets were involved to a high degree and Grey Promotion dealt with many phone calls and inquiries, approval of marketing material etc.

Unannounced visits were paid to the shops to check whether they were living up to the requirements – for example concerning energy labelling – that they had committed themselves to contractually. If they were not doing so, they got 14 days to correct their mistakes and if this did not take place they were excluded from the campaign.

Analyses were conducted and random samples were taken to measure the energy consumption of the A rated refrigerators for the Danish Energy Agency, and one model had to be withdrawn from the campaign when it proved unable to live up to the A classification.

1.7 After the Campaign

A market research bureau keyed in and checked all the invoices and the number of invoices lacking information about the date of sale, the number of the article or the like was approximately 5%.

1.8 The Result

The main results:
* Focus was created on the A rated appliances during the campaign period, and this focus has subsequently been maintained
* Approximately 35,000 A rated appliances were sold during the campaign period
* The number of A rated appliances on offer has tripled
* The electricity saving on the basis of the campaign has been estimated as 333,182 MWh (333 million kWh) including the calculated impact over the following 3 years, see Table IV
* The corresponding CO_2 reduction is estimated at 266,572 tons
* Due to the sharpened focus, the prices of A rated appliances have fallen by about 15%
* CO_2 saving per DKK 1 subsidy: 1.20 kg per DKK 1
* Socio-Economic shadow price:- DKK 138 per ton CO_2

Table II: Sales realised during the campaign period

	Budget	Sales realised	Variation (%)
Com. fridge-freezers	8,000	14,420	80.3
Refrigerators	8,000	15,437	93.0
Upright freezers	3,000	4,345	44.8
Tumble dryers	300	692	130.7
Total	19,300	34,894	80.8

Table III: Electricity saving and CO_2 reduction based on the realised sales during the 11 weeks campaign

	Number of A rated appliances	Electricity saving (MWh)	CO_2 reduction (tons*)	Subsidy (DKK.)
Com. fridge-freezers	14,420	32,556	25,662	7,210,000
Refrigerators	15,437	17,355	13,680	7,718,500
Upright freezers	4,345	10,147	7,998	2,172,500
Tumble dryers	692	1,436	1,132	692,000
Total	34,894	61,495	48,743	17,793,000

* Calculated on an appliance lifetime of 12 years, CO_2 factor = 0.78825 kg/kWh on average per year up to 2011

It general it may be said that not only has the campaign had an effect in the short term, but that it would also appear to have had more permanent significance for the sale of A rated models. This is supported by figures from ELDA: White Goods Statistics for the first quarter of 2000, where there is clear fall in the development in the average efficiency index (cf. Figure 1).

Figur 1: Udvikling i gennemsnitlige effektivitetsindeks

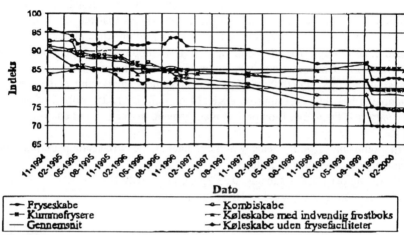

Kilde: ELDA

The figure shows the different types of refrigerators and freezers since 1994.

It is the assessment of The Danish Electricity Saving Trust that the effect of the campaign is a general shift from 15% to 20% in the market share of A rated appliances. The impact of such an increase on electricity saving and CO_2 reduction can be seen in Table IV.

Table IV: If the general market shift towards A rated appliances from 15 to 25%; the effect is calculated on 3 years' sales:

	Number of A rated appliances	Electricity saving (MWh*)	CO_2 reduction (tons*)
Com. fridge-freezers	33,938	141,523	111,556
Refrigerators	39,439	62,408	49,193
Upright freezers	11,272	68,032	53,626
Tumble dryers**	2,250	4,725	3,724
Total	86,899	276,687	218,099

*   Calculated on an appliance lifetime of 12 years, CO_2 factor $= 0.78825$ kg/kWh on average per year up to 2011.

** Based on a 1% market share (750) for 3 years, 500 kg clothes per year assumed, A and C rated appliances' consumption set at 0.35 and 0.7 kWh/kg, respectively, cf. ELDA

The great success has also resulted in the campaign budget being overrun. The original budget was DKK 10.5 million in subsidy ands a reserve pool of DKK 5 million was set aside, but the subsidy actually paid out amounted to DKK 18.5 million. The Danish Electricity Saving Trust expects that the final subsidy paid put will amount to in the region of the DKK 18 million when all the invoices have been gone through. An additional approx. DKK 800,000 was spent on marketing, DKK 250,000 on the mandatory announcement of the campaign in the national press, and DKK 950,000 on project management, marketing and control. The campaign has thus cost approx. DKK 20 million, DKK 18 million of which was subsidy, and DKK 2 million project management, marketing and control.

Where the analysis only deals with the impact on the purchasing pattern during the campaign, the result is as follows:
CO_2 saving per DKK 1 subsidy: 0.23 kg CO_2 per DKK 1
Socio-economic shadow price: DKK 311 per ton CO_2

Based on the general shift in in purchase patter as en effect og the campaign, the overall result during *and* after the campaign is:
CO_2 saving per DKK 1 subsidy: 1.20 kg CO_2 per DKK 1.
Socio-economic shadow price: -DKK 138 per ton CO_2

Comments on the economic evaluations:
The CO_2 saving per DKK 1 of subsidy is on a level with the *most* cost-effective measures, cf. the evaluation of the subsidy schemes of the Danish Energy Agency, February 2000.
Similarly, analyses show that the economic impact of the joint campaign on purchasing patterns has been extremely inexpensive as the economic shadow price is negative, i.e. that this has an economic benefit in itself.
As appears from the above-mentioned report from the Danish Energy Agency, by far the greatest number of energy policy measures have a positive shadow price, i.e. saving CO_2 represents an economic expense for these measures.

2 www:hvidevarepriser.dk

2.1 Background

The background for www.hvidevarepriser.dk was the same as for the subsidy scheme: the desire to focus more on the sales price of A rated appliances.

2.2 The Aim of the Campaign

The campaign target was that every 5th purchase of white goods should be based on information on the site, i.e. every week approximately 800 consumers should visit the home page.
In addition 40% of the retailers should report to the system.
The target group was private households with their own white goods and with internet access. Buyers in companies, consumer organisations and public institutions were also part of the target group.

2.3 Description of the Campaign

An internet bureau was asked to develop a home page which was to contain the following elements:
*	A product table with information about A rated appliances
*	A table of retailers with information about white goods retail outlets
*	A price table with information about the prices charged by the individual shops for the appliances in the product table
The consumers were to get
*	An overview of A rated products in different product categories
*	Information about the individual A rated products
*	The possibility of conducting a search for retailers and offers within a geographically limited area
*	The possibility of searching for information and sending it electronically via e-mail
*	The possibility of printing out the overview
*	Similar information from the National Energy Information Centre if the consumer did not have internet access.

2.4 Before the Campaign

There was some criticism and the home page was boycotted by the large-scale chains, who did not wish to take part. For this reason The Danish Electricity Saving Trust contacted the independent retailers and initially succeeded in getting 45 independent retailers to join the campaign.

2.5 During the Campaign

Because of the attention created by the home page, there was a great deal of attention from the media when www.hvidevarepriser.dk was launched and it also received mentioned on the news programmes on national TV.

A TV spot that included mention of the subsidy campaign was shown on national TV during the first five weeks of the campaign and advertisements were placed in most of the large newspapers. Advertising banners were also placed at internet portals.

A total of approximately DKK 2.5 million was spent on marketing www.hvidevarepriser.dk.

2.6 Results

www.hvidevarepriser.dk got off to a flying start when the number of users is examined. From the time it was launched on 13.10.99 to the end of the subsidy campaign on 4.12.99 there were more than 62,000 visitors, meaning a daily average of just under 1,200.

2.7 What was Successful/less Successful?

To sum up the good and less good sides of the campaign in a few words: the consumers were happy and the retailers unhappy.

According to the statistics, consumers have made frequent use of the home page and The Danish Electricity Saving Trust has received several hundred 'fan mails' from enthusiastic users. The retailers, however, have been more critical, which the Trust takes in its stride.

There is a lot of guesswork involved in such a campaign. For example, one retailer has estimated that less than 1% of retailers had joined in www.hvidevarepriser.dk but 10% of the subsidy we paid out during the campaign went to those retailers. And it looks as if their price level is well under the prices charged by the five big chains – this is their sore point and also probably the cause of their dissatisfaction. They all claim to be the cheapest – but this is not possible at the same time.

Besides, their prices are obviously not good enough to stand up to comparison. Their well-designed images about price guarantees are precisely based on the fact that consumers don't really have a chance to check them. The consumer would have to happen to pass a shop where precisely the same product is on display at a lower price or see it advertised shortly after buying his or her own appliance. The consumer very rarely registers it and can be bothered going back and proving it. If he goes back the guarantee is honoured without any problems, but how often does this happen in practice? This is the way all the five big chains for years have been able to look as if they were the cheapest – without anyone being able to get a clear picture of the situation. Who can get a clear picture of a market with over

800 different products? How many products are shown in the flyer or on display at a time? I do understand their resistance. but that's the way it is on the market. You have to adapt to new situations all the time – if you want to stay in the market – even if it would be a lot easier if things remained as they were.

3 Conclusion

3.1 The Rebate Programme

It is probably too early to say anything definitive about the extent of the campaign's impact on electricity consumption because this is dependent on whether a successful effort has been made to accelerate development towards more energy-correct products. There are, however, several indications that the campaign has had an effect that extends beyond the campaign period. It is a realistic guess that the market share of A rated appliances has risen from approximately 15% before the campaign to approximately 25% after the campaign, and the price level has fallen by 10-15%. If it is to happen, this means that CO_2 saving per DKK 1 of subsidy is 1.2 kg CO_2 per DKK 1 of subsidy and the economic shadow price is –DKK 138 per tons of CO_2. This also means that the campaign is on a level with the most cost-effective measures, cf. the evaluation of the subsidy schemes of the Danish Energy Agency, February 2000. The gear *has* shifted from 1st to 2nd.

3.2 www.hvidevarepriser.dk

It was an objective that about 800 consumers a week should visit the site. It is a proven fact that there were more than 62,000 visits to the site a week during the first seven weeks – about 1,200 a day on average.

The consumers have really accepted the site and even though the element of novelty has disappeared, at the middle of May 2000 there are still 300-500 visitors a day.

On the other hand the number of retailers who join the system have fallen from about 50 at the outset to about 30.

It has, thus, been possible to get into contact with the consumers but more problematic in relation to the retailers. The future of the home page probably depends on whether it can retain the retailers and maybe attract some new ones. Because there can be no doubt that the consumers appreciate the site.

In the autumn The Danish Electricity Saving Trust will produce an English version of the Danish www.hvidevarepriser.dk to serve as inspiration for other countries.

The Sky is the Limit!
Or Why Can More Efficient Appliances Not Decrease the Electricity Consumption of Dutch Households

Harm Jeeninga and Martine A. Uyterlinde

ECN Policy Studies

Abstract. In The Netherlands, the only energy carriers playing a significant role in households are natural gas and electricity. Natural gas is mainly used for space heating, hot water production and cooking, whereas electricity is used for appliances and lighting. Prognoses indicate that the total residential electricity consumption will rise continuously. In this paper, the impact of policy instruments on the development of electricity consumption in the Netherlands is investigated. First, a brief overview of policy efforts over the last two decades is given. It appears that the main focus of these policy instruments is on restricting the natural gas consumption rather than electricity consumption. Next, recent policy initiatives are discussed. Special attention is given to the effects of energy rebates granted on A-labelled appliances. The subsidy effectiveness as well as the total impact on CO_2 emissions are determined by means of model calculations. Finally, some remarks are made with respect to the feasibility and unfeasibility for limiting the growth of electricity consumption. Part of the factors responsible for the growth of the residential electricity consumption, such as disposable income, composition of population and life style factors, are beyond the reach of policy makers. However, the energy labelling system offers some good opportunities on condition that it is upgraded regularly. Nevertheless, even at best it is unlikely that the labelling of appliances solely will be able to bring about a turnover in the growth of residential electricity consumption.

1   Introduction

In The Netherlands, the total gas consumption per household has been decreasing since 1980 continuously, even though the decrease tends to flatten out. The electricity consumption per household has decreased continually by 1.5% per year within the period 1980 – 1988. However, since then, it rises rapidly by on average 1.9% per year. Obviously, it is important for policy makers to understand the factors underlying the upward trend in electricity consumption, in order to be able to direct policy efforts in such a direction that the increase can be turned into a stabilisation or even a decrease. A number of factors could explain this trend:

* Policy intensity and macro-economic circumstances;

- The level of energy prices as an incentive to save energy;
- The development of the penetration rate of electrical appliances. This is closely related to a number of lifestyle trends;
- The development of the efficiency and hours of usage and performance of electrical appliances.

Discouraging or even banning the ownership of certain types of appliances is not a politically favoured option for decreasing residential energy consumption. Moreover, a decrease in the hours of usage often involves a change in the daily practise. This is not easy to accomplish since this involves a structural change in behaviour (Antonides, 1998). Therefore, most policy instruments aim at decreasing the specific energy consumption by means of increasing the share of energy efficient appliances. In this paper, the factors explaining the upward trends are investigated and current policy efforts, like the energy rebates on very efficient appliances, are discussed.

2   Policy Efforts Since 1980

Looking back over the years since 1980, a number of shifts have occurred regarding the different factors influencing residential energy consumption. Three periods can be distinguished. The first period, 1980-1986, can be characterised by high fuel prices, economic recession, and active conservation policy. In this period, one of the main policy instruments was the National Insulation Plan. This Plan subsidised various insulation measures in existing dwellings as well as efficient condensing boilers, and was highly cost effective. It received a wide support, mainly due to the rising fuel prices in this period.

The second period, 1986-1991, is characterised by low fuel prices, economic recovery, and less attention paid to energy conservation. In 1990, the utilities started their Environmental Action Plans (MAP), as a result of the aim of the government to have target groups in the energy sector take responsibility for energy conservation. The plans included subsidies for insulation in dwellings and energy efficient lighting.

The third period, 1991-present, is characterised by low fuel prices, economic growth, and increasing effort in conservation policy, however now the main driving force is the climate problem. The Environmental Action Plans still play an important role, but there are also governmental initiatives, such as the Energy Performance Standard for new buildings. As a result of the changing role of utilities in the context of market liberalisation, the Environmental Action Plans will not be continued after 2000.

During the different periods, a shift in actors can also be observed: from government to utilities (MAP), and back to government, including a more important role for the EU (e.g. labelling). There is also a trend towards more market-based instruments, under influence of the ongoing market liberalisation. Policy measures have concentrated to a great extent on achieving natural gas savings. Apart from stimulating energy efficient lighting, measures aiming at reducing electricity consumption are from a very recent date.

Energy prices and expenditures
Between 1980 and 1986, energy prices were high (Jeeninga, 1997). The electricity price reached the highest level of 28,7 cent/kWh in 1985, then decreased to 22,9 cent/kWh in 1986, on which level it stabilised for a number of years. Only recently, since 1996 the electricity price has been rising again to 25,8 cent / kWh in 1998, as a result of the newly introduced energy tax. However, when the share of total expenditures on energy use in the disposable income is calculated, it turns out that in the first half of the eighties, expenditures on energy as share of the disposable income were twice as high as after 1985. Despite the rapid increase in electricity consumption, the ratio between total expenditures on energy use and disposable income remained almost constant. In the eighties, people were intrinsically motivated to change their behaviour to achieve energy savings. The economic recession added a sense of urgency to the need to save energy. After 1986, when prices decreased, the energy costs were no longer an incentive for energy efficient behaviour. Also in the current situation of substantial economic growth, the impact of financial incentives (energy tax, discount on A-labelled appliances) is limited.

3   Recent Policy Initiatives

In June 1999, the *Action Programme Energy Conservation 1999-2002* was published. It aims at increasing the energy efficiency improvement from 1.6% to 2% annually. The Action Programme recognises that the increase of residential electricity consumption is one of the most difficult issues to be solved, partly because consumer behaviour plays an important role. It aims at an efficiency improvement of appliances of on average 1.8% annually in the period 1995-2010. Core-instruments for the residential sector are levies such as the Energy Regulatory Tax (REB), the Energy Performance Advice (EPA) for existing buildings and the Energy Performance Standard (EPN) for new dwellings.

Specifically for appliances, the following instruments are deployed.
* The Energy Rebates ('Energiepremies') programme for households has started in January 2000. Buyers of efficient (A-class) appliances receive a financial incentive. Subsidies are also available for insulation measures. The

subsidies are paid from the revenues of the Regulatory Energy Tax (most of the revenues however are recycled back through income tax reductions).

- On EU level, there are negotiated agreements with manufacturers of televisions and VCRs to reduce the stand-by energy consumption of those appliances, and with manufacturers of washing machines to improve their energy efficiency. The covenant on televisions and VCRs seems to be effective, as research shows that the stand-by consumption of these appliances has gradually diminished in the past four years (Ministry of Economic Affairs, 1999). Agreements on decoders, battery chargers, loaders and feeders are in preparation. Energy labelling, another initiative on EU level, is described in the next section.

The government intends to counterbalance the undesired environmental effects of the expected price decrease in a liberalised market will be counterbalanced by an increase in the energy tax (REB). However, consumers are hardly aware of the existence and height of this tax. The REB will be raised in three steps: 1999, 2000 en 2001, and it is anticipated that this will generate 3,4 billion Dutch guilders every year. The extra gains will be returned to taxpayers through a reduction of the income tax and the tax on wages.

4 Effects of Energy Rebates on the Purchase of Efficient Appliances

4.1 Introduction

Basically, the development of the energy efficiency of household appliances is determined by the following factors:
- the availability of energy efficient technology
- the additional investment and total investment
- the energy saving and the pay back period
- purchasing behaviour

However, these factors are mutually interconnected. The application of energy labels on household appliances intends to change the purchase behaviour. However, the change in consumer preferences may result in a change in the line of products as shown in shops as well in a change in the manufacturing strategy. The existence of energy labels offers also an opportunity for manufacturers to distinguish themselves in a positive sense from other manufacturers and enhances the aspect of energy efficiency in the design process of appliances. In practice, there are some indications that this last process indeed occurred. Even before the energy label was attached to specific household appliances, manufacturers adapted their line of products and enhanced their efforts in developing energy efficient appliances. In the remainder of this chapter, some of the aspects related to increase in the energy efficiency of household appliances are discussed in more detail.

4.2 Purchase and Availability of Cold Appliances

Figure 1 shows the sales figures for fridge/freezers by label class (Waide, 1998). Comparable sales figures are available for independent refrigerators and freezers (Boonekamp, 2000). At first sight, the results are very favourable. Labelling seems to be effective, because the share in sales of the more efficient models is increasing continuously. This observation is supported by the fact that for all types of cold appliances, refrigerators without freezers, fridge/freezers as well as independent freezers, show the same tendency towards efficient models.

The shift in the share of sales of cold appliances towards more efficient appliances can be explained by:
- an increase in the availability of efficient appliances,
- changes in purchase behaviour as a result of particularly the application of energy labels, and
- other factors such as the total investment, additional investment and energy prices.

Since appliance efficiency is continuously improving, the number of appliances available in the more efficient classes will increase every year, implying that the sales figures for efficient appliances will tend to increase anyhow, compared to those for less efficient models.

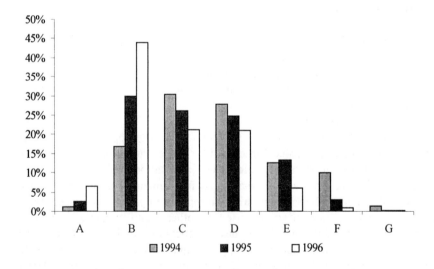

Figure 1: Purchase of new fridge/freezers per energy class in 1994, 1995 and 1996.

By comparing the changes in availability of efficient cold appliances and the development of the share of efficient appliances in total sales, a rough indication is obtained to what extend the sales per label class is driven by the availability of efficient appliances. For all cold appliances with energy label A, the availability exceeds the sales share. In other words, the sales share of A-labelled cold appliances is not determined by the availability solely. Other factors such as the change in purchasing behaviour resulting from the application of energy labels, the cost-effectiveness of the investment and the consumer preferences with respect to the purchase of appliances of a specific brand, are likely to play a role as well. However, it should be noted that the availability[1] as defined in this report is not necessarily equal to the average line of products as shown in shops.

The strong increase in sales share of A-labelled cold appliances might well be (partly) explained by some autonomous trends. On the other hand, it is the only and best evidence available at the moment. However, two comments have to be made. Although it was not compulsory to apply energy labels on cold appliances before 1996, this was already done by part of the retailers. Moreover, there are reasons to believe that the autonomous efficiency improvement of cold appliances in the early 90's was almost zero, since manufacturers were forced to change the coolant of the appliances[2]. The observations done in this section stress the need for monitoring the effectiveness of policy measures such as labelling.

4.3 Availability of Washing Machines, Clothes Dryers and Dishwashers

In 1997, the availability of A-labelled washing machines is marginal at 1%. Most models belong to energy class B (65%). A limited number of models belonged to energy class D or higher (5%). In 1999 however, about 45% of the washing machines belonged to energy class A (EnergieNed 1999). The availability of clothes dryers of energy class A is negligible. No data are available with respect to the availability of dishwashers. However, based on the number of appliances per energy class shown at international exhibitions, it is estimated that the availability of A-labelled dishwashers amounts to 20% at maximum (AE 2000).

4.4 Additional Investments for Very Efficient Appliances

Based on (EnergieNed, 1999), the additional investments for efficient cold and wet appliances are determined. By doing this, the additional investment is corrected for several side effects. For e.g. cold appliances, a correction was made for the (1) type of appliance (single door fridge with and without small internal freezing compartment, two door fridge/freezer, chest freezers and upright freezers), (2) size and (3) the brand name. These corrections appeared to have a significant influence on the additional investments. Without these corrections, the additional investment of an A-label refrigerator in comparison to a C-labelled refrigerator[3] amounts to about 210 Euro. The correction for the brand name specifically appears to have a large influence on the additional investment. When correcting for the brand name, the total investment drops by 50 Euro. This effect also applies to wet appliances.

The additional investments[3] of A-labelled refrigerators, fridge/freezers and freezers amount to respectively 160 Euro, 195 Euro and 155 Euro. The additional investments for washing machines, clothes dryers and dishwashers were estimated at respectively 140 Euro, 550 Euro and 140 Euro. It should be noted that the (statistical) uncertainty in the additional investment is of the same magnitude as the additional investment itself.

4.5 Effectiveness of a Subsidy on A-Labelled Appliances

As stated in section 3, the Dutch government has introduced energy rebates for A-labelled appliances. Recently, ECN was asked by the Ministry of Finance to evaluate the effectiveness of these energy rebates. In general, the effectiveness of a subsidy is restricted due to several factors, in particular as a result of the limited role of pay back time in the purchase behaviour and due to free rider effects. Important aspects are the total investment rather than the additional investment, purchase habits, preferences regarding the brand name, comfort level and easiness to use and (the expected) quality of the product.

In order to enhance the transparency and therefore the effectiveness of the policy measure, the height of the rebates for all appliances, except the clothes dryer, was set at 45 Euro. For A-rated clothes dryers as well as gas fired clothes dryers, the subsidy amounts to 200 Euro. The effects of the energy rebates on CO_2-emissions were determined by means of the ECN-model SAVE-Households (Boonekamp, 1994). Calculations were performed for the Global Competition scenario (CPB,1997), as scenario with a high growth of GDP (3.5% per year). As a result of the energy rebates, the cost effectiveness of A-labelled appliances improves and therefore the penetration rate increases. It is assumed that the energy rebates will not effect the total penetration rate of appliances which have not reached total market saturation yet (e.g. the dishwasher and clothes dryer). Besides that, it is assumed that, as a result of the transparency of the policy instrument and the way the energy rebates are used as sale offers by retailers, that for all A-rated appliances sold, an energy rebate will be granted. Finally, it is assumed that within the period 2000 – 2010, no changes will be made in the current energy label system.

As a result of the granting of the energy rebates, the share of A-labelled appliances increases. In Table I, the effect on CO_2 emissions as well as the total subsidy granted is given. The total CO_2 reduction amounts to 0,15 Mton in 2005 and 0,25 Mton in 2010. For comparison, the total CO_2 emissions for domestic appliances in the Netherlands is expected to increase to approximately 9,5 Mton in 2005 and 11,7 Mton in 2010 (ECN, 1998). A more extensive description of the results of the model calculations, including penetration rates per appliance type and energy class, can be found in (Boonekamp, 2000).

Table I: CO_2 reduction and total subsidy granted (Million Euro per year) as a result of the energy rebates on A-rated appliances (Mton per year).

	Total CO_2 reduction		Total subsidy	
	2005	2010	2005	2010
Cold appliances	0,11	0,16	51	59
Washing machine	0,02	0,04	11	15
Clothes dryers	0,00	0,00	1	2
Dishwasher	0,02	0,04	5	9
Total	0,15	0,25	68	85

The majority of the energy rebates (about 70%) is allocated to cold appliances. This is mainly due to the large free rider effects. In the reference scenario, the penetration rate of A-labelled cold appliances in the total stock increases from 14% in 2000 to 43% in 2005 and 65% in 2010. When an energy rebate is applied, the share of A-labelled cold appliances increases to 56% in 2005 and 85% in 2010. For the clothes dryer, the granting of energy rebates has no effect on total CO_2 emissions. This can be explained by the high additional investment and the limited availability of A-labelled clothes dryers.

In order to obtain an indication of the effectiveness of the energy rebates, the so-called cumulative subsidy effectiveness is calculated. The subsidy is more ore less regarded as an investment that has to be amortised over the lifespan of the energy conservation option. Therefore, the yearly subsidy is multiplied with an annuity factor. The effectiveness of the energy rebates is basically determined by two factors: (1) free rider effect and (2) the ratio between the subsidy and energy savings.

For cold appliances, the washing machine, the cumulative subsidy effectiveness amounts to over 300 Euro per ton CO_2 in 2005, see Table II. For the dishwasher, a value of 165 Euro per ton CO_2 is found in 2005. This implies that, in comparison to the dishwasher, twice as much subsidies are needed for cold appliances and washing machines in order to save 1 ton of CO_2. For cold appliances, the poor subsidy effectiveness is due to the large share of free riders. For washing machines, this is explained by the relative low energy savings[4] in comparison to the size of the energy rebate. Mainly as a result of an increase in efficiency of electricity generation, the amount of subsidy needed in order to reduce 1 ton CO_2 increases in time.

Table II: Cumulative subsidy efficiency (Euro per ton CO_2) in 2005 and 2010.

	2005	2010
cold appliances	305	441
washing machine	323	423
clothes dryer	273	432
dishwasher	164	214
average	291	395

5 Discussion and Conclusions

Over the last decade, the residential electricity consumption in The Netherlands has increased rapidly. This rise poses a difficult task on the government, since the growth in electricity consumption is to a large extent due to growth and composition of population and growth of GDP, factors that can not be influenced by policy makers, even if they would want to. In a period of economic growth as The Netherlands is currently experiencing, the influence of energy prices is limited. Expressed as a share of disposable income, the energy costs do not impose a heavy burden on household budgets. Therefore the impact of financial incentives such as the energy tax is limited, in particular given the fact that most customers are hardly aware of the existence and height of the tax. Proper feedback on their energy consumption and a clearer presentation of the energy bill could improve on this situation.

The purchase and ownership of appliances is closely related to lifestyle trends such as social recognition, individualisation and scarcity of spare time (Jeeninga, 1998). Energy labels can and do influence purchasing decisions. Offering subsidies (the 'Energy Rebates') on the most efficient models can stimulate the choice of an efficient appliance. However, subsidising energy efficient equipment may also have some adverse effects. For example, a subsidy on clothes dryers or dishwashers could accelerate the increase in penetration rate of these appliances. Moreover, this subsidy might suggest that the purchase of very efficient types of these appliances has an environmental benefit (the rebate serves as governmental quality mark, increase of social acceptance). However, this is not the case, since the use of efficient appliances is only less harmful to the environment instead of beneficial.

Due to free riders, subsidising appliances that already have a high penetration rate is expensive. In principle, subsidising very efficient refrigerators might be a good option, since the penetration rate is almost saturated and the potential for energy saving is sufficiently large. However, model calculations show that under current conditions free rider effects are strong in case a subsidy is granted on A-labelled cold appliances. When looking at the Dutch situation, a fast revision of the classification of cold appliances is urgently needed in order to maintain a sufficient rate of energy efficiency improvement. From a technical point of view, this is beyond doubt feasible on the short term, see for example (EP 2000). This yields to a lesser extent for washing machines.

In case of limited availability of number of models within energy class A, brands with relative high initial expenses are overrepresented. In this case, the additional investment is quite small in comparison to the total investment. However, total expenditures are high in comparison to identical but less efficient appliances. As a result of the high initial investment, only customers with a relative high disposable income are able to afford the purchase of an A-rated model. Under these circumstances, the subsidy on A-rated appliances is mainly collect by people with a high disposable income. In case even the less expensive brand names are able to fabricate an A-rated model, the additional investment drops and free rider effects become dominant. This imposes a serious and hard to overcome problem with respect to the effectiveness and desirability of energy rebates on very efficient appliances. This problem becomes even more important when the justification imposed by the rebate as a quality mark for the purchase of non-indispensable appliances such as dishwashers and clothes dryers is taken into account.

By changing the rebate system in a way that it only reaches specific target groups, the effectiveness as well as total CO_2 reduction decreases and transaction costs will increase. However, the energy rebates are currently financed by means of a general levy (the REB) on residential energy consumption paid by all energy consumers. This levy is supposed to be neutral to the consumer budget and returned to the sector by means of a decrease in income tax and energy rebates on investments in energy efficiency. For social reasons, the situation in which all income classes profit from the energy rebates is preferable over a situation in which only high-income customers receive the energy rebates, even if this involves large free rider effects.

Prognoses indicate that the total electricity consumption related to the use of domestic appliances will increase continuously. The share of cold and wet appliances in total residential electricity consumption decreases from 35% in 1995 to 33% in 2010, 31% in 2020 and 29% in 2030. In order to restrict the growth of electricity consumption, it is desirable to extent the labelling system. One of the main merits of the energy labelling system is the pressure it imposes upon the appliance manufacturers to increase the efforts with respect to designing and manufacturing of energy efficient appliances. However, a timely upgrade of the labelling scheme is necessary. In this case, the manufacturers have the ability to discern themselves from other manufacturers. Only under these conditions, the development of very efficient appliances is remunerative.

6 References

[1] Antonides G.A., W.F. van Raaij. *Stuurbaarheid van milieu-relevant consumentengedrag.* Erasmus University, Rotterdam, the Netherlands, in Dutch 1998.

[2] AE. *Issue 1, volume 4, 2000.* Appliance Efficiency, Newsletter of IDEA, the International Network for Domestic Energy-Efficiënt Appliances, page 10, 1999.

[3] Boonekamp P.G.M. *SAVE-Module Huishoudens. De modellering van energieverbruiksontwikkelingen.* ECN, Netherlands Energy Research Foundation, report nr. ECN--C-94-045, Petten, the Netherlands, in Dutch, 1994.

[4] Boonekamp P.G.M. and H. Jeeninga. *Gedrag en Huishoudelijk Elektriciteitsverbruik. Kwalitatieve en kwantitatieve analyse 1980 – 1997.* Netherlands Energy Research Foundation (ECN), report no. ECN-C-99-057, Petten, the Netherlands, in Dutch, 1999.

[5] Boonekamp P.G.M., H. Jeeninga and H. Heinink. *Effectiviteit Energiepremies. Analyse voor het Huishoudelijk Verbruik tot 2010.* Netherlands Energy Research Foundation (ECN), report no. ECN-C-2000-062, Petten, the Netherlands, in Dutch, 2000.

[6] CPB. *Economie en fysieke omgeving; beleidsopgaven en oplossingsrichtingen 1995-2020.* Netherlands Bureau for Economic Policy Analysis (CPB), The Hague, in Dutch,1997.

[7] ECN. *Nationale Energie Verkenningen 1995 - 2020.* Netherlands Energy Research Foundation (ECN), Petten, the Netherlands, in Dutch, 1998.

[8] EnergieNed. *Basisonderzoek Elektriciteitsverbruik Kleinverbruikers.* Arnhem, the Netherlands, in Dutch, several years.

[9] EnergieNed. *EnergieWijzers, diverse jaren.* EnergieNed, Arnhem, 1999.

[10] EP. *Energy-plus.* http:\\\www.energy-plus.org\, april 2000.

[11] Jeeninga H. *Analyse Energieverbruik sector Huishoudens 1982 - 1996, Achtergronddocument bij het rapport Monitoring Energieverbruik en Beleid Nederland.* ECN, Netherlands Energy Research Foundation, report nr. ECN--I-97-051, Petten, the Netherlands, in Dutch, 1997.

[12] Jeeninga H., *Domestic Appliances and Life Style. Consequences for domestic electricity consumption in 2010,* Report prepared for Novem, Sittard, the Netherlands, 1998.

[13] Ministry of Economic Affairs, *Action Programme Energy Conservation 1999-2002,* Den Haag, June 1999.

[14] Waide P., *Monitoring of energy efficiency trends of European domestic refrigeration appliances: final report.* Manchester, United Kingdom, 1998.

[15] Winward J., P. Schiellerup, B. Boardman, *Cool labels, The first three years of the European Energy Label,*Oxford, Environmental Change Unit, University of Oxford, 1998.

Footnotes

[1] Availability is defined as the number of models available for sale in the Netherlands.

[2] As a result of the ban on the use of CFK's

[3] In this analysis, C-labelled appliances are regarded as the reference technology.

[4] The average energy consumption of washing machines in the Netherlands amounts to about 50% of the average energy consumption of cold appliances

Criteria Definition and Recognition Mechanisms for Accelerated Innovation

Hans Westling

Promandat AB, Stockholm, Sweden

Abstract. Lessons learned from the International Energy Agency six-year project "Co-operative Procurement", which has just ended, show the large importance of criteria definition and creation of mechanisms for recognition of future-oriented solutions. The objectives were to establish a co-operative demand-pull procedure to bring more energy-efficient products to the marketplace and to test and draw lessons from pilot projects. Within this collaborative project, the participating eight countries have, with support also from the European Commission, defined an international procedure and possible areas for joint actions and fulfilled pilot procurement and promotion projects.

In general, theoretical research about innovation instruments on the demand side, better communication, building up of networks in fragmented areas and the creation of powerful buyer groups have been stressed as important components for successful technology procurement projects. It is, however, very time-consuming to build up new buyer groups in areas with a large degree of fragmentation. In the "Co-operative Procurement" project, it has been found that the creation of other recognition mechanisms, such as Awards with high prestige, can have a large influence on the suppliers´ decisions to concentrate development efforts in the actual area.

The first Class A clothes drier with 50 % energy reduction, industrial motors with 20-40% reduction of losses and a copier with reduction by 70-75% down to 30% are promising results. The involved project managers and experts identified 60 lessons learned from the pilot projects and these were compared with findings made by an external evaluator at a workshop.

Understanding the market as a whole and all different performance criteria, rules for procurement, competitions and the right mix of different support mechanisms, prestigious awards and an international forum, on a formal or informal basis, to enable exchange of ideas are important aspects for accelerated innovation.

1 Annex III and the DSM Agreement within the International Energy Agency

Annex III "Co-operative Procurement" is one of nine different Tasks within the International Energy Agency (IEA) Demand-Side Management (DSM) Agreement. Starting in 1993-94, seventeen member countries in IEA have in co-operation with the European Union and the World Bank carried out various studies and specific projects (Westling, 2000).

2 Objectives

The objectives of Annex III have been to establish a process for activities in international co-operation based on the demands of the market. The process is intended to facilitate for more energy-efficient and environmentally adapted solutions to be developed, introduced and spread on the market (Westling, 1996). Areas, which are suitable for specific procurement activities but which have not been introduced on the market, have also been identified. Finally, experiences from the process have been compiled.

The following countries and organisations have taken part in Annex III: Denmark, Danish Energy Agency; Finland, Motiva; Korea, KEMCO; The Netherlands, NOVEM; Spain, ENHER and ADAE; Sweden, Swedish National Energy Administration – STEM (earlier NUTEK); United Kingdom, DETR, BRE and ETSU; United States, U.S. Department of Energy (DOE) and the Environmental Protection Agency (EPA); and the Commission of the European Union, DG XVII, Energy. Operating Agent for Annex III has been Hans Westling, Promandat AB, acting on behalf of the Swedish National Energy Administration.

3 Need for a New Process

IEA and the European Commission have stated in their programmes that it is urgent to follow up with concrete actions the international climate agreements made in Rio, Kyoto and Buenos Aires. The member countries of the organisations have agreed to contribute, in various ways, to reducing the risks of climate changes, among other things by facilitating the development and diffusion of more efficient energy solutions, which will lead to less emission of greenhouse gases. A proper market transformation will also be accelerated where more efficient solutions will have an increasing market share.

Use of mandatory regulations, as well as large rebate programmes, are expected to encounter increasing difficulties and opposition and, in may cases, lack of funding. The present trends are towards a desire for more individual choices, deregulation and privatisation, which will diminish the possibilities for government interventions using regulations and other traditional methods. Joint procurement activities with expressed innovative purposes can offer good alternatives for governments, buyers and users, but also for manufacturers, to bring about reliable solutions that are quicker accepted on the market.

4   Definitions and Theoretical Framework

Technology procurement is an entire acquisition process with the expressed purpose of stimulating innovation. Technology procurement is a steering mechanism working on the demand side – *market pull* (Edquist, 1990). The creation of new networks, across time and trade borders, has proven to be of particular importance for innovations (Håkansson, 1987 and Teubal, 1991). The importance of long-term work in *interactions* between buyer and manufacturer has also been pointed out (Lundvall, 1988). It is particularly essential to bring about collaboration between future-oriented, influential buyers and users, so called *lead users* (von Hippel, 1986), or, as sometimes expressed in the United States, *anchor buyers*. Generally, requirements are made up in the form of *performance requirements*. The requirements are often expressed on two levels, *mandatory* requirements, which must be met, and *desired* requirements, which are evaluated as positive in procurements and competitions. A complete performance specification, with requirements in various areas – e.g. in addition to energy when working with more energy-efficient products, also other requirements, such as reduction of noise and water consumption, etc. – will facilitate the diffusion of new solutions.

Co-operative procurement includes also, in addition to technology procurement, a joint procurement, which aims at increasing the use of the best products. Internationally, the terms *volume* or *bulk purchasing* are also used.

The market transformation is very dependent on various other supportive measures (rebate- and subsidy programmes, information, labelling, competition elements with awards etc.). Different steps in the preparations are shown in Figure 1.

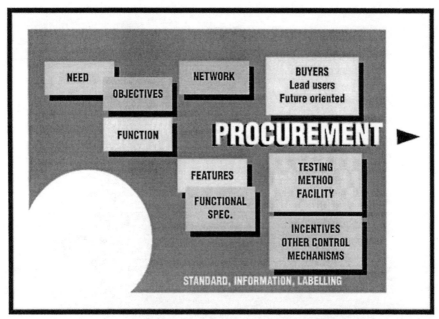

Figure 1: Some important preparatory steps in technology procurement (Westling, 2000)

5 Some Historic Examples

Technology procurement is not a new process introduced in connection with energy efficiency. The method of technology procurement has been used earlier, above all in order to develop new national systems, e.g. in the transportation, power supply and telecommunications areas. A well-known historic example is the competition at Rainhill in Great Britain in 1829, which was something of a breakthrough for the new technology with steam-operated locomotives in the railway area. The company who wanted to build a railway between Manchester and Liverpool announced a competition. A number of performance requirements were laid down as regards pulling capacity, speed, various safety demands and also an economic target of £500. The arranging company promised then to buy at least five "locomotives" from the winner. So far, individual railways had used steam engines on railway carriages, above all for transports of material, but there had been no breakthrough for railways with transportation of both goods and people in regular traffic. Several interesting solutions entered the competition. The winner was Robert Stephenson with his locomotive "The Rocket" (Figure 2), who was given the guaranteed orders and many successive ones.

Figure 2: Robert Stephenson's winning locomotive "The Rocket" (photo from the Science Museum, London)

6 Results – Process and Projects

The experts and project management of the Annex III work have developed a preliminary process, which has been documented in the report "Co-operative Procurement. Market Acceptance for Innovative Energy-Efficient Technologies" (Westling, 1996). The report gives a broad background with previous results of measures which have similarities with technology procurement and which have been taken on the market side in several countries. It also contains analyses of results from projects where the process has been used. Similar methods have been used on a broad scale in weapon programmes and space industry. In the energy field, some 20-30 projects have been carried through in NUTEK's and STEM's energy-efficiency programme (Nilsson, 1992). In the United States, the Department of Energy, the Environmental Protection Agency and the Consortium for Energy Efficiency have carried out a number of projects. The very extensive and expensive programme "Super Efficient Refrigerator Program", for example, has attended much attention. Similar projects have followed in the areas of washing machines and refrigerators, particularly for apartment blocks.

In connection with a workshop in Paris in 1994, the Annex III group pointed out three possible areas for joint procurements. The areas were *wet appliances, lighting* and *copiers*. In total seven areas have since been subjected to joint analyses. During the preparatory stages, the process has more developed towards a combination of procurement and promotion activities.

Various alternatives have been identified to give recognition to new, successful solutions, not necessarily by guaranteed large-volume purchasing only. An Award, "the IEA DSM Award of Excellence", has been introduced, which has been given to three projects. The first Award was granted to a *drier* for household laundry where the energy consumption has been reduced down to 50%. This drier is now being introduced on the European market as the first "*Class A drier*" with a heat pump (Figure 3).

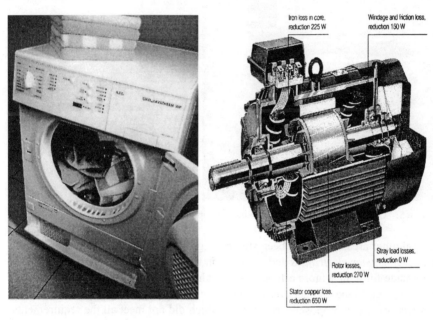

Figure 3: The winning AEG drier with heat pump – the first "Class A drier" in Europe.

Figure 4: Longer lifetime through reduced losses – M2BA280, one of the winning ABB motors.

The second award has been granted within the area of *Hi-Efficiency Motors* for two motors of different sizes from ABB, where the losses have been reduced between 20 and 40 % (Figure 4). In this area, calculations also prove that very short payback periods can be achieved when the motors are used in process industry, down to between one and three years. The third area is "*Copier of the Future*" (Figure 5) where copying equipment from RICOH shows that it is possible to reduce the energy consumption down to 25%. This is close to meeting the *"factor 4"* goal introduced in an international context (i.e. only one fourth of the original consumption).

Figure 5: "Copier of the Future" – the winning copier from RICOH.

Several other areas have also been studied, e.g. *vending machines* for hot and cold drinks, *consumer electronics* and a project in the *lighting* field, where preparations were made for a 30% more efficient bulb with considerably greater possibilities of use compared to the compact fluorescent tubes used at that time. In the latter case, however, only one proposal was received which did not meet all the requirements. Now preparations are going on for *traffic lights with LED* (light emitting diodes) solutions on a large scale, which could possibly also include control systems including transformers. At a seminar in the autumn of 1999, it was illustrated that with the combined results of more efficient lamps with highly efficient light emitting diodes, with refined control systems and more efficient transformer solutions, it would be within reach to reduce the use of energy to *one tenth*, i.e. reach in total *factor 10* (1/10 of the original consumption and consequently also of the environmental load). The area does not represent a very large use of energy, but the purpose of the projects has also been to gather experience and use pioneering projects to inspire efforts in more areas.

7 Conclusions

At the Annex III international workshop "Lessons Learned" which was held in London in 1999, the project managers and experts presented experiences from the pilot projects. About 60 different "lessons learned" have been identified in the field. An evaluation report by an external evaluator and results from a number of U.S. projects, particularly through the Consortium for Energy Efficiency, were presented (Annex III London Proceedings, 1999).

Increased interest has been shown lately for using procurement initiatives for innovation and market transformation. A number of EU initiatives and IEA studies are examples in this area. Furthermore, experiences from several large procurement projects in various countries have been analysed in a comprehensive study (Edquist & Hommen & Tsipouri, 2000).

Two of the Annex III projects have been specially pointed out as successful examples of IEA projects in connection with the IEA 25th anniversary in May 1999 (International Energy Agency, 1999).

Some of the most important lessons learned are:

- To stress the importance of understanding the market as a whole and the underlying conditions and to consider a product with all different performance criteria.
- To stress the importance of creating buyer groups and preparing performance specifications in functional terms.
- To combine competitions and procurements with different support mechanisms, such as labelling, bulk purchasing, minimum standards, information and rebate programmes (Figure 6).
- To formulate clarifications of national and international rules for procurement in order to facilitate innovative efforts. The rules for public procurement within the World Trade Organisation and the European Union are drawn up for normal cases – purchase of already developed and introduced products and systems – and are not bearing development projects in mind. Experiences show that it is possible to carry out technology procurement according to the present rules within the World Trade Organisation and the European Union if it is strictly observed that the planned project must be made public and the competitive aspect and objectivity can be maintained. However, in connection with planned changes it is desirable to clarify some particulars to facilitate the accomplishment of technology procurement and development projects.
- To stimulate innovations and more efficient solutions by using different processes which have similarities with technology procurement (performance specifications in functional terms and competitions or purchasing). Joint

development activities have been studied in an EU study (SNEA, 1998) and a specific project for more energy efficient refrigerators/freezers through international joint procurement has started (Energy+, 2000).

- To change the role of governments from being only a big buyer to also playing an *intermediary role* as stressed in a couple of EU studies (EC DG XII, 1998).
- To create an international forum – initiated by interested countries and IEA – to gather experiences of technology procurements and similar competitions with development purposes.
- To strengthen the prestige of Awards such as the "IEA DSM Award of Excellence".

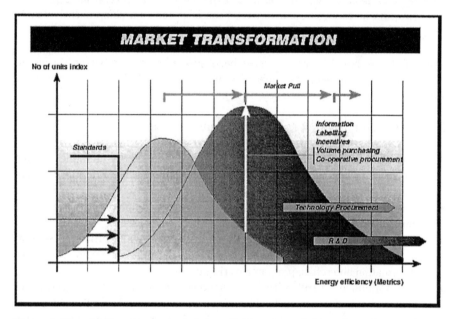

Figure 6: Use of different instruments is essential to get a market transformation.
(In Westling, 2000, adoption after John Millhone, DOE, USA)

8 References

[1] Annex III London Proceedings (1999): *Proceedings from the Annex III International Workshop 'Accelerate Innovation and Market Transformation of Energy-Efficient Products, London, 24-25 February 1999*, compiled by Westling, H, Promandat AB, Stockholm.
[2] Annex III homepage: http://www.stem.se/IEAprocure

[3] EC DG XII (1998): Report in the EC DG XII TSER Programme by Lundvall, B.Å. & Borrás, S, *The Globalizing Learning Economy: Implications for Innovation Policy*. Luxembourg.

[4] Edquist, C (1990): Technology Policy. In *Tema teknik och social förändring*. /In Memorandum Theme technology and social change/. Linköping.

[5] Edquist, C & Hommen, L & Tsipouri, L (2000): Public Technology Procurement and Innovation. In *Economics of Science, Technology and Innovation*, Vol. 16. Boston (Kluwer Academic Press).

[6] Energy + (2000): Brochures and competition documents for "Energy+ Second Round, 2000", ADEME/ICE, Paris.

[7] Hippel, E von (1986): Lead users: A Source of Novel Product Concepts. *Management Science*, Vol. 32, No. 7, July 1986. Cambridge, Mass., USA. (The Institute of Management Sciences).

[8] Håkansson, H (ed) (1987): *Industrial Technological Development. A Network Approach*. Worchester (Billing & Sons Limited).

[9] International Energy Agency (1999): *International Collaboration in Energy Technology. A Sampling of Success Stories*. OECD/IEA, Paris.

[10] Lundvall, B.Å (1988): *Innovation as an Interactive Process: From User-Producer Interaction to the National System of Innovation*. In Dosi, G et al: Technical Change and Economic Theory.

[11] Nilsson, H (1992): Market Transformation by technology procurement and demonstration. In *Proceedings of the ACEEE 1992 Summer Study*, Washington D.C.

[12] SNEA, Swedish National Energy Administration (1998): *Procurement for Market Transformation for Energy-Efficient Products. A study under the SAVE-programme*. ER15:1998. Eskilstuna.

[13] Teubal, M & Yinnon, T & Zuscovitch, E (1991): Networks and Market Creation. In *Research Policy*, no. 20, 1991.

[14] Westling, H (1996): *Co-operative Procurement. Market Acceptance for Innovative Energy-Efficient Technologies*. B1996:3. NUTEK/IEA. Stockholm.

[15] Westling, H (2000): *IEA DSM Annex III Co-operative Procurement of Innovative Technologies for Demand-Side Management - Final Management Report*. Stockholm.

Major Appliance Energy Efficiency Trends in the European Union

Paul Waide

PW Consulting (UK)

Abstract. Energy labelling for cold appliances[1] has been in place in Europe since 1995, while mandatory minimum energy performance standards (MEPS) have been implemented since September 1999. Labelling for clothes-washers and washer-dryers[2] has been implemented since 1996. This paper presents the results of two European Commission-sponsored studies (ADEME 1998 & 2000) to monitor the impact of these policies on the efficiency trends of cold and wet appliance sales across the EU. Monitoring of the efficiency trends was achieved through merging national cold and wet appliance sales databases, gathered commercially by market research agencies, with comprehensive technical databases containing relevant energy and market data for individual models. The combined databases, the largest of their kind ever assembled, contain almost 60 000 records for cold appliances alone and cover 11 countries, that account for ~95% of all EU cold and wet appliance sales. The cold appliance panel coverage averages 83% of sales for the national markets included and therefore encompasses 78% of all cold appliance sales in the EU. Results are reported for the period 1994–97 for cold appliances and 1996–97 for wet appliances; however, 1998 data are being added under a new project and non-sales-weighted data are available for some countries for 1999.

The main finding is that the energy efficiency of cold appliances sold in 1997 improved by an average of ~12% compared to pre-policy levels circa 1992, and early indications from non-sales-weighted data show that the average efficiency of the EU market following the implementation of MEES in late 1999 is likely to have improved by ~30% compared to pre-policy levels. By 2010 it is estimated that this will have avoided 212 TWh of electricity consumption, ~27 billion Euro in electricity bills of and 104 M-tonnes of CO_2 emissions. These findings indicate the great success of the two policies and vindicate their initiation.

For clothes-washers the results indicate a very pronounced peak in the sales of products in energy class B and the almost complete eradication of E, F and G product sales. This also confirms the success of the labelling policy and of the voluntary agreement negotiated between the European major appliance manufacturers' association, CECED, and the European Commission.

---

[1] Refrigerators, freezers and their combinations
[2] Clothes-washers, washer-dryers, dish-washers and clothes-dryers are collectively known as 'wet appliances'

1 Cold Appliance Sales in the EU

Sales of cold appliances within the EU have grown from just under 17 million units per year in 1994 to approximately 18.7 million units in 1999. The market is relatively saturated and sales over the long term sales are projected to be reasonably constant with a gradual upward trend in response to the increase in the number of households and rising average living standards. A small dip occurred in 1996 that is likely to be linked to changes in general macroeconomic conditions and consumer confidence levels. Since that time annual sales in the EU have grown by 15%, with the UK market, where sales have increased by a staggering 47%, being the primary driver, Table I and Figure 1.

Table I: Sales of cold appliances in the EU from 1994 to 1999 (1000 units)[1]

Type & year[2]	Country or region															
	AUS	BEL	GER	SPA	FRA	UK	ITA	NL	SWE	DEN	FIN	IRE	GRE	POR	LUX	EU
Freezers																
1994	122	102	1164	246	787	708	381	190	121	94	68	28	33	75	8	4126
1995	131	115	1142	240	749	722	390	192	122	94	68	28	33	75	8	4108
1996	129	119	1089	249	703	710	371	181	107	82	71	28	36	71	8	3954
1997	123	166	1084	242	741	862	411	191	135	92	73	31	36	73	8	4268
1998	124	157	1104	275	721	899	420	203	137	95	74	34	36	79	8	4366
1999	126	159	1083	286	753	985	424	203	153	104	79	37	36	89	8	4525
Refrigerators and refrigerator-freezers																
1994	233	299	3705	1328	2057	1711	1638	573	262	237	155	110	305	236	15	12862
1995	245	216	3600	1316	2054	1773	1676	593	261	236	155	110	304	235	15	12789
1996	248	333	3367	1231	1953	1762	1611	586	236	209	161	115	310	243	15	12380
1997	245	402	3269	1298	1962	2246	1589	631	285	228	176	122	310	265	15	13043
1998	262	405	3234	1415	2109	2264	1688	679	288	226	181	131	310	292	15	13499
1999	271	409	3197	1463	2242	2645	1796	698	296	233	184	143	310	303	15	14205
All cold appliances																
1994	356	401	4869	1574	2844	2418	2019	763	383	330	223	138	338	310	23	16988
1995	376	448	4742	1556	2803	2495	2065	785	383	330	223	138	337	310	23	17015
1996	376	452	4457	1480	2656	2471	1982	768	343	291	232	143	346	314	23	16334
1997	368	568	4353	1540	2703	3108	2000	822	420	320	249	153	346	338	23	17311
1998	386	562	4338	1690	2830	3163	2108	882	425	321	255	165	346	371	23	17865
1999	397	568	4280	1749	2995	3630	2220	901	449	337	263	180	346	392	23	18730

[1] Source: GfK Marketing Services. 1999 values are provisional. Data for Luxemburg and Greece for 1997 to 1999 were not supplied by GfK and are therefore assumed to be constant with 1996 sales.

[2] Freezers includes both upright & chest types, Refrigerators and refrigerator-freezers includes all categories in the labelling Directive except 8 and 9, which are freezers.

830

In value terms the EU cold appliance market was worth about 7.9 billion Euro in 1998, but this is comfortably exceeded by the ~14.3 billion Euro paid in electricity bills to power cold appliances each year.

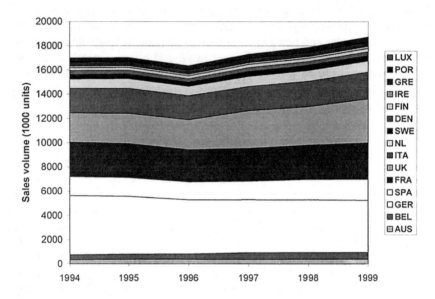

Figure 1: Sales of cold appliances in the EU from 1994 to 1999[1]

[1] Source: GfK Marketing Services. 1999 values are provisional. Data for Luxemburg and Greece for 1997 to 1999 were not supplied by GfK and are assumed to be constant with 1996 sales.

2 Market Developments in Cold Appliance Energy Efficiency

Analysis of the matched database gives the evolution of EU cold appliance sales by energy label class from 1994 to 1997 as shown in Figure 2. There is a steady shift to higher efficiency classes over the three year period, such that by 1997 class A appliances accounted for over 6% of total sales and class A to D appliances ~79% of sales. By comparison with the distribution of models by class in the GEA database[3] this represents a significant efficiency improvement

---

[3] A database of models offered for sale in Denmark, France, GB, Germany, Italy, Portugal, Spain from 1990-1992 assembled for the *Study on Energy Efficiency Standards for Domestic Refrigeration Appliances*, the Group for Efficient Appliances for DGXVII of the European Commission, March 1993

831

although the GEA data is not sales-weighted and hence provides a less reliable portrait of the appliance market.

The analysis of the matched database indicates that the average energy efficiency index of cold appliances sold in the EU in 1997 was 87.9%, some 12.1% lower (more efficient) than the reference levels used in the energy labelling and minimum energy performance standards Directives. The average energy efficiency index of EU cold appliances sold in 1996 was 90.7%, in 1995 was 92.5% and in 1994 it was 92.6% (Table II). However, these figures are based on analysis of the cold appliances for which it was possible to identify the parameters needed to evaluate their energy efficiency index. The proportion of EU appliance sales in the database where the energy efficiency index is known is 94.7% in 1997, 96.7% in 1996, 95.7% in 1995 and 89.2% in 1994. A full analysis shows that the appliances with unknown indices tend to be lower efficiency models and therefore it was necessary to apply a compensating correction methodology for unknown efficiency appliance sales to increase the reliability of the time series analyses. Applying this methodology results in 'corrected' estimates of the EU cold appliance sales-weighted annual average energy efficiency index of 89.9% in 1997, 91.8% in 1996, 93.9% in 1995 and 96.1% in 1994. These corrected values may be a more accurate representation of the actual efficiency progression from 1994 to 1996.

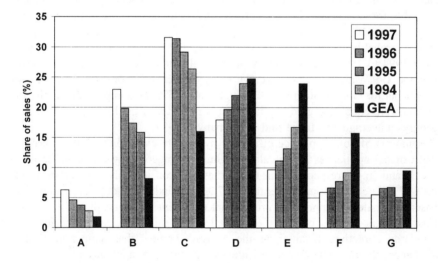

Figure 2: EU cold appliance sales share by energy label class for 1994 to 1997 (also showing the distribution of GEA models by label class)

Table II: Cold appliance sales-weighted annual average energy efficiency indices by Member State for 1994 to 1997 (%)[1]

	EU	Aus	Bel	Den	Fra	GB	Ger	Ita	Nl	Por	Spa	Swe
1997[5]	87.9	81.5	87.7	87.9	91.7	99.5	75.4	94.8	80.6	NA	91.5	86.8
1996	90.7	85.0	93.4	89.8	97.4	101.7	77.8	96.9	84.1	101.2	92.9	87.4
1995	92.5	87.2	95.2	91.5	100.3	103.1	80.0	99.3	87.7	-	94.2	90.1
1994	92.6	88.1	97.0	-	102.2	96.0	83.3	101.1	91.5	-	89.6	-
GEA	100.2	-	-	94.0	103.9	107.5	95.4	107.5	99.3	125.5	101.7	-

[1]the efficiency index is determined according to the Commission Directive (94/2/EC). The data does not include sales of cold appliances with unknown efficiency indices and thus the cited EU average values are the sales-weighted values based on those countries where data is available. The 1994 and 1997 data for France, GB and Germany is for branded models only and does not include appliances sold under a private label. By contrast the 1995 and 1996 values for these countries includes private label sales. The GEA data is not sales-weighted and includes appliances available for sale on the national markets indicated between 1990 and 1992. Danish data is not sales-weighted.

A post hoc treatment of the GEA database used to define the reference energy efficiency indices for the energy labelling and minimum energy performance standards Directives reveals that the average energy efficiency of cold appliances available for sale on the EU market between 1990 and 1992 was more probably 102.2% than 100% as previously thought. Thus, the average efficiency of the EU market probably increased by 12.3% from 1990/2 to 1997 and by at least 6.2% from 1994 to 1997. The efficiency of national markets varies considerably, such that the 1997 sales-weighted average energy efficiency index for Germany was 75.4% but for Great Britain was 32% higher at 99.5%. This implies that if all other factors were equal that cold appliances sold in Britain would consume 32% more energy on average than those sold in Germany in 1997.

An analysis of cold appliance efficiency by climate class found that sub-tropical and tropical appliances had a sales-weighted average energy efficiency index of 71.1% and 76.4% respectively in 1997. These are considerably more efficient than the 1997 EU cold appliance average of 87.9% and indicate that there is no market-based rationale to support the preferential treatment of sub-tropical and tropical class appliances as is currently embodied in the minimum energy performance standards Directive. If anything the analysis implies that temperate N class appliances should be treated more leniently than the other appliance classes.

The sales-weighted distribution of the 1997 EU cold appliance energy efficiency index, shown in Figure 3, illustrates a clear influence of the energy label on the market in that very large peaks in sales were found to occur for energy efficiency indices coincident with the minimum requirement of the higher efficiency classes A, B C and D. These peaks that developed from 1994 onwards, which is the period when the label structure was first known[4], seem to be partially a result of manufacturers preferentially positioning products such that they just pass the threshold into a higher energy labelling class. Non-sales data for later years suggests that this trend has been increasing and that the majority of models available on the market in 1999 were specifically designed to meet a given label class. The transition from a roughly random Gaussian efficiency distribution in pre-labelling times to a highly targeted distribution within just a few years is strong confirmation of the positive impact of the categorical energy label design used in the EU.

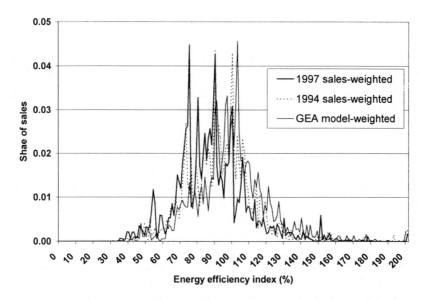

Figure 3: Comparison of the sales-weighted and model-weighted distribution of all EU cold appliances by efficiency index in 1997

Analysis of the distribution of sales by efficiency for each national market enabled the fraction of appliance sales in 1997 satisfying the efficiency thresholds specified in the minimum energy performance standards legislation to be determined. The values ranged from as low as 28.0% of sales on the British market up to 83.9% of sales on the German market. Overall 59.6% of the cold appliances sold in the EU in 1997 would have satisfied the efficiency thresholds due to be enforced in September 1999, which is up from 53.4% in 1996. This finding indicates that the EU cold appliance market in 1997 was part way toward

---

[4] The label requirements were published on January 1st 1994 and the first Member States to implement the label did so in January 1995. Some Member States did not implement the label until 1998.

meeting the September 1999 thresholds but still required some significant progress to achieve full compliance.

Cold appliance energy consumption

Sales-weighted average EU cold appliance energy consumption was 398.2 kWh/year in 1997, 406.1 kWh/year in 1996, 410.9 kWh/year in 1995 and 409.5 kWh/year in 1994 (Table III).

Table III: Cold appliance sales-weighted annual average energy consumption for 1994 to 1997 (kWh/year)[1]

	EU	Aus	Bel	Den	Fra	GB	Ger	Ita	Nl	Por	Spa	Swe
1997	398.2	321.6	406.5	400.7	425.4	423.0	299.9	458.2	356.0	-	499.9	429.2
1996	406.1	333.0	427.0	406.7	445.5	441.0	313.5	465.2	368.8	488.9	511.5	422.5
1995	410.9	340.0	433.1	413.5	455.7	443.6	317.5	483.1	380.6	-	526.8	423.1
1994	409.5	343.1	440.2	-	458.7	411.6	328.8	485.6	398.1	-	516.6	-
GEA	449.8	-	-	415.0	473.4	468.7	403.5	529.1	429.0	642.7	538.1	-

[1]The data does not include sales of cold appliances with unknown energy consumption, while the cited EU average values are the sales-weighted values based on those countries where data is available. The 1994 and 1997 data for France, GB and Germany is for branded models only and does not include appliances sold under private label. By contrast the 1995 and 1996 values for these countries includes private label sales. The GEA data is not sales-weighted and includes appliances available for sale on the markets indicated between 1990 and 1992. Danish data is not sales-weighted.

These figures are evaluated from the sales of appliances in the database with a known energy consumption and as with the energy efficiency index this value decreases from about 98.5% in 1997 to about 90% in 1994. It is probable that the appliances for which the energy consumption was unknown consume proportionately more than those for which the data is known. Thus it is difficult to be confident of the sales-weighted average values for 1994.

The seemingly less rapid rate of reduction in sales-weighted annual average energy consumption compared to the efficiency index between 1994 and 1997 is partly explained by a corresponding increase in their sales-weighted average volume. Between 1994 and 1996 the average adjusted volume of the cold appliances sold in the EU increased by 10.4 litres from 284.3 litres to 294.7 litres. In 1997 the sales-weighted average adjusted volume was identical to that in 1996. Almost all the increase from 1994 to 1996 was attributable to an increase in the sales-weighted average volume of fresh food storage space as frozen food volume scarcely changed. As a result the sales-weighted average share of total cold appliance adjusted volume accounted for by frozen food storage fell from 53% to 46%.

The sales-weighted average price of cold appliances increased by 5.4% from 408.7 ECU in 1994 to 430.8 ECU in 1996 and then fell by 2.7% to 419.2 ECU in

1997. The price per litre of adjusted volume grew by less than 1% over the same period. For the European market as a whole and within most Member States there appears to be a significant positive correlation between average price and average efficiency such that an average A class appliance was 116 ECU more expensive than the average of all appliances sold in the EU and an average B class appliance was 18 ECU more expensive in 1997; however, these efficiency related price differences had declined sharply from 1996 levels, which probably reflects a lowering novelty of higher efficiency products as their presence in the market increases.

3   Market Trends Since 1997

A new SAVE sponsored project is underway to repeat the market analysis described above for the year 1998; however, there are already strong indications that the pace of improvement in cold appliance energy efficiency has increased over the intervening years. Minimum energy efficiency standards regulations came into effect from September 1999 and have excluded the majority of D, E, F and G appliances from sale. A comprehensive survey of all cold appliances being offered for sale on the German market at the Domotechnica trade fair held at the beginning of 1999 revealed a very strong movement toward higher efficiency products and has also offered further concrete evidence of the impact of the energy labelling thresholds on the products offered for sale. The range of cold appliances offered for sale in Germany at the beginning of 1999 had an average energy efficiency index of 70.7%. Some 28% were class A appliances and 43% class B. Interestingly, class D, E, F and G appliances only accounted for some 7.5% of the models which indicates that manufacturers had already mostly phased out appliances that would be prohibited from sale from September 1999, see Figure 4. This fact, combined with the growing tendency for manufacturer's to position products at higher efficiency levels associated with class A and B appliances shows the important combined influence of the energy labelling and minimum energy efficiency standards Directives. If the few appliances on sale in the German market at the beginning of 1999 that did not satisfy the September minimum efficiency standards requirements are removed from the database the average efficiency index of the permitted appliances is improved to 66.9%.

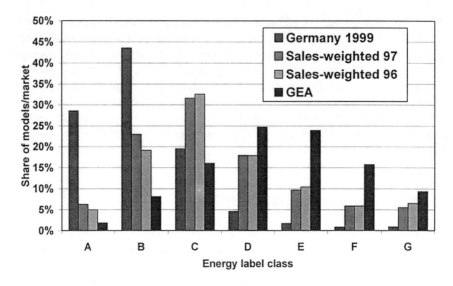

Figure 4: Evolution of market share by efficiency class from 1991 to 1999 (EU and Germany)

For a variety of reasons it seems likely that the cold appliances offered for sale in the EU as a whole immediately after the implementation of the minimum energy efficiency standards Directive in September 1999 will have an average efficiency index of about 72%, which represents a 30% efficiency improvement from pre-regulation levels, Figure 5. It is necessarily speculative to imagine how the energy efficiency of the European cold appliance market would have progressed without the stimulus provided by the two European Directives; however, the average market efficiency had been static or had even shown a slight deterioration in the years immediately preceding the two Directives. Furthermore, from the development of tell tale characteristics in the product offer energy efficiency distribution it seems certain that the two Directives have had an appreciable impact on the market and most likely that the majority of the measured efficiency improvements are attributable to their influence. The actual sales-weighted market trends will become apparent when the full sales data has been acquired and matched to the technical databases as has already been done for the years 1994 to 1997. However, it can be tentatively concluded that as a result of the efficiency improvements that are largely attributable to the two Directives the cumulative energy consumption of cold appliances sold in Europe between 1991 and 2010 is likely to be ~16% lower than would otherwise have been the case and 21% lower by 2020 (some 212 and 528 TWh respectively), Figure 6. These savings estimates are made by comparison with a simple static efficiency scenario where it is assumed that without the EU policy intervention that cold appliance efficiency levels would have been frozen at 1991 levels. Although this is unlikely to have been the case the scenario compensates for this over pessimistic assumption by conservatively assuming that the energy efficiency of the cold appliance market remains static after 1999 (i.e. that after 1999 labelling and minimum efficiency standards have no further impact on the market). Under the same projection annual energy savings are forecast to reach 8.5 TWh/year by 2000, 26 TWh/year by 2010 and 35 TWh/year by 2020. This last figure is equivalent to an annual saving in electricity demand of 1.7% of all electricity consumption in the EU in 1995. Annual CO_2 savings are forecast to reach 4.2 Mega tons CO_2/year by 2000, 12.6 Mega tons CO_2/year by 2010 and 17.2 Mega tons CO_2/year by 2020. Cumulative CO_2 savings of 104 Mega-tons are forecast for 2010, Figure 6.

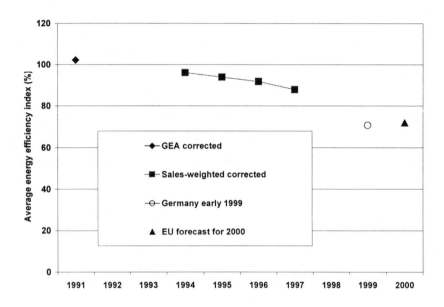

Figure 5: The average energy efficiency index of the EU and German cold appliance market from 1991 to 2000 (forecast value for the EU in 2000)

If average electricity prices in the EU remain at 0.13 Euro/kWh these efficiency improvements will translate into customer energy bill savings worth 4.6 billion Euro per year by 2020 (~33 Euro per household per year or 495 Euro over a typical cold appliance lifetime of 15 years). With an average of about 1.8 cold appliances per household across the EU the reduction in the typical household's electricity bill as a result of lower running costs is about 275 Euro per appliance. Given that the average B class cold appliance was 18 ECU (or 18%) more expensive than an average cold appliance sold in Europe in 1997, but used 83.4 kWh/year less electricity then the simple payback period associated with buying an appliance having an energy efficiency index of 69.5% as opposed to 87.9% was 1.8 years. For the remaining average 13.2 years of product life the typical

838

consumer makes net savings worth 143 Euro equal to about 33% of the cost of the appliance. Even for the relatively expensive A class appliances the average payback was 5.5 years and in this case the average consumer would make net savings worth 40% of the cost of the appliance over a 15 year period. In practice the incremental cost of a class A appliance is likely to continue to diminish as their market share increases.

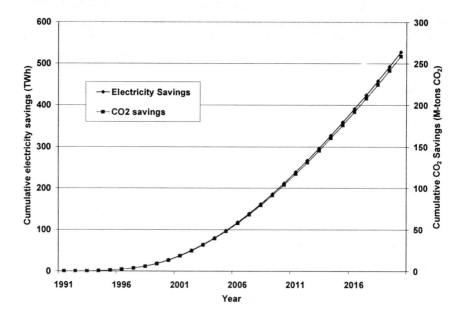

Figure 6: Forecast cumulative electricity and CO_2 savings resulting from the current EU energy labelling and minimum energy efficiency standards regulations.

4 Wet Appliance Sales in the EU

Sales of wet appliances (clothes-washers and washer-dryers) within the EU declined from just over 12 million units per year in 1997 to approximately 11.2 million units in 1999, Table IV. As with cold appliances the market is relatively saturated and sales over the long term are projected to be reasonably constant with a gradual upward trend in response to the increase in the number of households and rising average living standards.

Table IV: Sales of clothes-washers and washer-dryers in the EU from 1997 to 1999 (1000 units)[1,2]

Year	Country or region															
	AUS	BEL	GER	SPA	FRA	UK	ITA	NL	SWE	DEN	FIN	IRE	GRE	POR	LUX	EU
1997	217	272	2550	1412	2090	2294	1560	553	181	166	138	109	232	239	NA	12013
1998	222	271	2540	1326	2020	2126	1450	516	170	171	148	116	239	269	NA	11584
1999	217	272	2570	1200	1920	1974	1390	484	158	187	162	125	243	274	NA	11176

[1] Source: GfK Marketing Services. 1999 values are provisional. [2] Includes both clothes-washers and washer-dryers.

5 Market Developments in Wet Appliance Energy Efficiency

Clothes washers

Analysis of the matched database gives the evolution of EU wet appliance sales by energy label class from 1996 to 1997 is shown in Figure 7. There is a steady shift to higher efficiency classes over the three year period, such that by 1997 class A appliances accounted for over 6% of total sales and class A to D appliances ~79% of sales. By comparison with the distribution of models by class in the GEA database[5] this represents a significant efficiency improvement although the GEA data is not sales-weighted and hence provides a less reliable portrait of the appliance market.

---

[5] A database of models offered for sale in Denmark, France, GB, Germany, Italy, Portugal, Spain from 1990-1992 assembled for the *Study on Energy Efficiency Standards for Domestic Refrigeration Appliances*, the Group for Efficient Appliances for DGXVII of the European Commission, March 1993

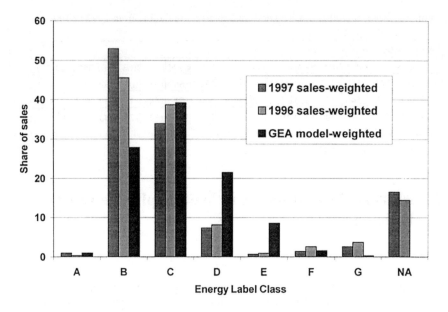

Figure 7: EU clothes washer sales share by energy label class for 1996 to 1997
(also showing the distribution of GEA models by label class)[1]

[1] GEA data is mostly tested under the old EN60456 procedure and is converted from a 90oC to 60oC energy consumption result using a simple correction factor.

The sales-weighted average specific energy consumption for a 60°C cotton wash cycle fell from 0.250 kWh/kg in 1996 to 0.243 kWh/kg in 1997. This compares to an average model value of 0.26 kWh/cycle reported in the 1995 GEA study that was used to define the energy label structure. However, it should be remembered that the data used in the GEA study was tested under the old EN60456 test procedure and that the 60°C wash cycle data was mostly derived from 90°C wash cycle results using a simple correction factor. This fact makes it difficult to make a direct comparison between the two data sets. The sales-weighted average water consumption of clothes-washers for a 5 kg cotton cycle according to the EN60456 test fell by 5.7% from 74.7 litres in 1996 to 70.5 litres in 1997, which will have been a major determinant driving the 2.8% improvement in energy efficiency.

The clothes-washer energy label also ranks spin-drying performance in a scale from A to G and washing performance also on an A to G scale. Figure 8 shows the distribution of sales by spin drying class in 1996 and 1997 for the EU. The results indicate a slight shift toward the higher labelling classes which is correlated with a 3.1% increase in the sales-weighted average spin speed from 841.5 rpm in 1996 to 867.7 rpm in 1997. These EU average figures mask a significant national difference in the sales-weighted average spin speed between northern and southern Europe as indicated in Table V.

Table V: Sales-weighted average clothes-washer spin-speed in 1997 (rpm)

Year	Country or region									
	AUS	BEL	FRA	UK	GER	ITA	NL	SPA	SWE	EU
1997	1022	1087	737	1018	1085	544	1149	602	1055	868

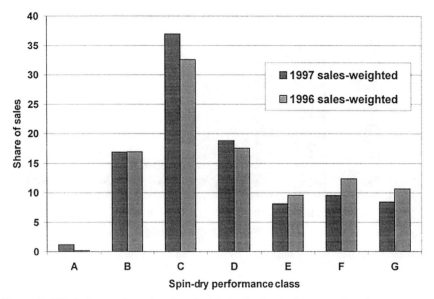

Figure 8: EU clothes washer sales share by label spin-dry performance class for
1996 to 1997

The advent of the clothes-washer energy label appears to be focusing at least as much attention on to the wash performance as to the energy efficiency. Sales of A, B and C class appliances increased from 62% to 71% from 1996 to 1997, Figure 9.

Washer dryers

Sales of washer dryers were about 650 000 appliances across the EU in 1997, which is equivalent to about 5.8% of clothes washer sales. The quality of the 1996 database was significantly poorer than the 1997 database in terms of the number of appliances for which it was possible to establish their energy label class and as a result the energy class is only known for some 45% of sales in 1996 as compared to some 63% of sales in 1997. In both cases the coverage is significantly poorer than for clothes-washers or cold appliances. The distributions by energy label class are shown in Figure 10. The apparent fall in the share of A class sales and the increase in the share of F and G class sales from 1996 to 1997 is almost certainly just an artefact caused by the increase in market coverage and probably does not reflect a real trend.

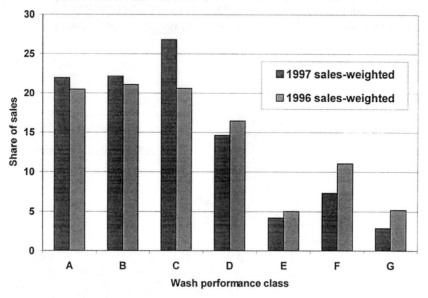

Figure 9: EU clothes washer sales share by label wash performance class for 1996 to 1997

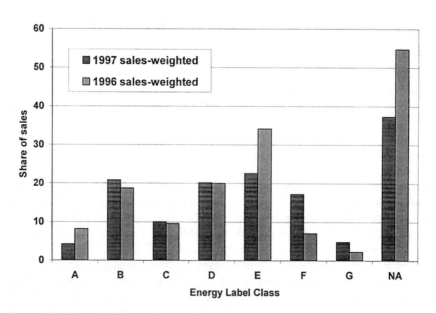

Figure 10: EU washer-dryer sales share by energy label class for 1996 to 1997

The sales coverage of the energy consumption per complete wash-dry cycle measured according to EN 50229 was also only 45% in 1996 and 63% in 1997, which may explain why the sales-weighted average value for the EU apparently rose from 4.69 kWh/cycle in 1996 to 4.84 kWh/cycle in 1997.

6   New Challenges Ahead

Successful as the cold appliance energy efficiency policies have been it is important to realise how much more energy could have been saved had they been implemented earlier, more forcefully and with greater support. The energy labelling programme has suffered from a number of weaknesses (very late implementation in some Member States, piecemeal and poor enforcement, insufficient promotion and consumer awareness and insufficient efforts to engage retailers are just some of them) that have reduced its effectiveness and slowed the rate of market transformation. Passage of the minimum energy efficiency standards Directive was also significantly delayed. This lack of decisiveness in implementing appliance efficiency policy has undoubtedly delayed the impact of the policies which could have made a stronger impact toward the Kyoto protocol greenhouse gas targets than will now be the case. Nonetheless, there are clear signs of an almost classical combined market transformation effect taking place through which a combination of minimum energy efficiency standards and an effective categorical energy label are raising the energy efficiency of the market. With the revision of both the energy labelling and minimum efficiency thresholds for cold appliances and energy labelling for washing machines currently under consideration there will soon be an opportunity for policy makers to strengthen both schemes and to build on their initial success.

7   References and Bibliography

[1]   ADEME (1998) *Monitoring of energy-efficiency trends of European domestic refrigeration appliances: final report*, PW Consulting and ADEME for DG-XVII of the Commission of the European Communities, SAVE contract No. XVII/4.1031/D/97-021.
[2]   ADEME (2000) *Monitoring of energy efficiency trends of refrigerators, freezers, washing machines and washer-dryers sold in the EU*, PW Consulting and ADEME for DG-TREN of the Commission of the European Communities, SAVE contract No. XVII/4.1031/Z/98-251.
[3]   Commission Directive 94/2/EC of 21.1.94: Energy labelling of household electric refrigerators, freezers, and their combinations OJ No L 45/1-19.
[4]   Commission Directive 95/12/EC of 23.5.95: Energy labelling of household washing machines, OJ No L 136/1-21.

[5]   Commission Directive 96/60/EC of 19.9.96: Energy labelling of household washing machines, OJ No L 266/1-21.

[6]   Council and European Parliament Directive 96/57/EC of 3.9.96: Energy efficiency requirements for household electric refrigerators, freezers and combinations thereof, OJ No L 236/36-43.

[7]   ECU (1998) *Cool Labels: The first Three Years of the European Energy Label*, Environmental Change Unit, University of Oxford, September.

[8]   GEA (1993) *Study on energy efficiency standards for domestic refrigeration appliances*, Group for Efficient Appliances for DG-XVII of the Commission of the European Communities, March.

Creating a Carbon Market

Brenda Boardman

Environmental Change Institute, University of Oxford

Abstract. The emphasis of policy needs to change from greater energy efficiency to energy conservation and carbon saving. This reflects the progress that has been made in identifying the opportunities for improved energy efficiency and the accompanying failure to reduce total energy consumption. The focus on carbon dioxide would align energy policies with climate change targets more precisely.

The debate increasingly needs to incorporate the role of fuel-switching at the household level and the contribution that can make in achieving carbon savings. This paper will examine the issues involved in these debates and specifically:
- the labelling of gas and electric appliances;
- the market transformation of systems, rather than products;
- the interplay with supply-side policies (extent of the gas network);
- household carbon targets, both for suppliers and for consumers;
- promotion of household-level renewable energy supply (e.g., solar thermal).

The widespread understanding of appliance energy use developed in Europe must be placed in this wider context as a contribution towards future sustainability.

1   The Kyoto Context

There is now no serious doubt that anthropogenic greenhouse gas emissions are leading to increased global warming. Since the 1970s, the world has warmed by about 0.15°C per decade, and 1998 was the warmest year on record. The European Union (EU) and national governments have signed up to legally-binding greenhouse gas reduction targets. At Kyoto, the member states of the EU agreed jointly to undertake an 8% reduction of six key greenhouse gases from 1990 to 2008-2012.

However, the policy measures planned and implemented so far will not be sufficient to meet these targets; further action is needed. The European Commission has published an Action Plan on energy efficiency, which sets out principles, but does not suggest detailed policy actions needed to meet the Kyoto targets. These details are a primary concern of the French Presidency, during the last six months of 2000. The Action Plan states that 'there is a pressing need ... to promote energy efficiency more actively'. There are opportunities to reduce energy intensity by 1% pa, below the reference case, whereas at present no reductions are being achieved. The EU deficit by 2010 is likely to be 118MtC from the domestic sector.

There are policies other than energy conservation that will contribute to the Kyoto targets (e.g., increased use of renewable energy, combined heat and power). However, reduced domestic demand has an important contribution to make both to Kyoto targets – which only the UK, Luxembourg and Germany are likely to meet – and to savings beyond Kyoto. The substantial benefits that can be achieved through more efficient equipment (Fawcett *et al.*, 2000) will make an important contribution, but it is necessary to go beyond energy efficiency to both energy conservation and to carbon reductions and to address policy on fuel switching to less carbon-intensive fuels. This will move the policy debate from a focus on energy to a focus on carbon dioxide.

2 A Carbon Focus

Fuel switching has been occurring on the supply side, for a variety of reasons, for several years. The British 'dash for gas' has resulted in substantial carbon reductions as a result of gas-fired generating capacity displacing coal-fired stations. Demand-side fuel switching occurs naturally, in both directions: from gas to electric ovens; from electric to gas water heating. There is the potential to enhance lower carbon choices by both expanding the gas network and by more intensive use of gas in households on the network: more use of gas for water heating, cooking and tumble dryers.

The policy challenge for fuel switching is not the same as that for energy efficiency. The pricing of fuel already generally works in favour of gas use across the EU. The challenge, on the demand side, is to persuade those consumers who have not connected to the gas network to do so, and those who are connected to use gas for as many functions within the house as possible. This is also in the interest of the gas industry. Policy considerations include the way in which the decision to change fuels is made, as well as the additional costs, within the house, of providing a gas supply to the new equipment. To effect accelerated network expansion, policy boundaries extend to the choices of and regulations imposed upon the gas utility.

Once policy moves beyond efficient use of electricity to gas use and fuel switching a broader approach is required if the maximum carbon reductions are to be achieved. The policy realms are schematically portrayed in Figure 1: the inner circle is the appliance, the second extends to the water heating system, the third is the whole house (infrastructure and energy demand), and the outer circle includes national fuel choices (availability, supply system and policies on renewable energy). The breadth of the policy agenda is part of the debate. Product-level policies (efficient appliances) are no longer sufficient. It is also necessary to consider the interactions between individual components in the whole system (e.g., central heating) which has labelling and training implications. The way the consumer chooses between competing fuels and who advises her defines the opportunities for intervention and education (Fawcett *et al.*, 2000). National policies to encourage fuel choices provide the overall framework.

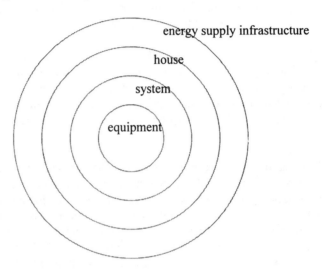

Figure 1: Schematic representation of policy realms

The main focus in this paper is on fuel switching from electricity to natural gas, although the opportunities provided by renewables are of increasing importance, both household-based options like solar thermal and collective schemes through green electricity (Fawcett *et al.*, 2000 – Appendix G). There is concern that by encouraging a switch to natural gas people will be locked into a carbon-based fuel, as opposed to concentrating on providing renewable alternatives. However, given the very slow progress towards generation of renewable electricity in many EU countries, gas is likely to remain the lower carbon fuel, compared with electricity, for at least 10-20 years. Switching to natural gas is thus a valuable component of a lower carbon policy in many countries and should facilitate adoption of renewable options.

3   Fuel Switching Potential

Electricity consumption across the EU is still rising, primarily due to an increase in household numbers and increasing ownership and use of appliances and lighting.

In the short term, major potential carbon reductions arise from the more efficient use of electric lights, refrigeration and consumer electronics appliances. The main emphasis continues to be on the more efficient use of electricity, partly because this is the more polluting fuel in most the EU, but also because product-level policy is the easiest to implement. The powerful approach of market transformation strategies, if supported by strong EU commitments, can improve the efficiency of products with certainty and speed. Guaranteed savings are available but depend on the Commission, Council and Parliament to be more active, with encouragement from member states.

Efficiency improvements to gas boilers do reduce gas consumption though they do not provide large reductions in carbon by 2010, when compared to the potential reduction by electric appliances and lighting. This is mainly because, by 2010, the natural rate of turnover means that much less than half of all, older and less efficient, boilers will have been replaced. Boilers last longer than fridges or light bulbs.

Piped domestic gas is becoming the norm, with perhaps 70% of all EU households connected by 2020. Therefore, household-level fuel switching from electric to gas is relevant on the national level and increasingly so for EU policy. Fuel switching, to reduce carbon emissions, should be directed primarily towards water heating. The opportunities from saving from cooking and tumble dryers will decrease significantly if the projected improvements in efficiency of their electric counterparts under Kyoto+ are reached (Fawcett et al., 2000). The potential carbon reduction in carbon is significant, about 3MtC (Table I); space heating savings would be additional.

Table I: Potential effects of fuel-switching across EU

	RC, 1998 electricity (TWh)	RC, 2010 electricity (TWh)	Kyoto+ – RC, 2010 electricity TWh)	Kyoto+ – RC, 2010 gas (TWh)	Kyoto+ – RC, 2010 Net carbon reduction (MtC)
Cooking	51	45	15	17.3	0.75
Tumble dryers	9	10	3	3.5	0.15
Water heating	87	82	41	47.2	2.06
FS-TOTAL	147	137	59	68	3

RC = reference case; Kyoto + = efficiency and fuel-switching scenario

The estimated potential carbon reductions from fuel-switching is from consumers choosing gas rather than electric. However, there will be some reduction in carbon emissions if consumers switch from bottled LPG to piped natural gas (over 20% reduction in carbon emissions per delivered kWh). LPG water heating and cooking is common in some countries where there is no piped network. However, not all this potential may be realised since there is some evidence (Fawcett *et al.*, 2000, Appendix P) that switching from LPG to natural gas increases consumption, especially for water heating where there is always access to gas; with bottled LPG there is a limit due to running out of LPG supplies.

Where district heating exists, there is the potential to reduce carbon emissions using heat-fed appliances (Fawcett *et al.*, 2000). There is an extensive domestic network in Denmark that could use this technology. Attaching the heat-fed appliances to a central water heating system could also reduce emissions though the cost-effectiveness is less certain at present. Increased production could reduce costs sufficiently, to be viable when connected to a household's central water heating system.

4   Energy Supply Infrastructure

The gas debate opens up a further set of issues in relation to the remit of the different policy makers. Natural gas is not equally available throughout the EU: 1% of households use natural gas in Sweden, up to 97% in the Netherlands (Table II). The opportunities for switching between fuels therefore varies and means that European-wide policies on comparative fuels may not be applicable.

Table II: Households with natural gas and carbon content of electricity

Country	% with gas (1997-99)	Electricity kg C/kWh (1995)
Austria	37	0.06
Belgium	55	0.09
Denmark	12	0.22
Finland	2	0.09
France	41	0.02
Germany	42	0.16
Greece	0.2	0.24
Ireland	27	0.20
Italy	70	0.14
Netherlands	97	0.17
Portugal	3	0.16
Spain	26	0.12
Sweden	1	0.01
UK	82	0.14
EU – 15	48	0.11

As the carbon emissions per unit of electricity varies across Europe, the benefits of, and reasons for, fuel switching are nationally specific. Policies on fuel switching have to be the responsibility of the member state rather than the European Commission at present. Half of all European households have natural gas by 2000 and this is expected to grow to about 70% by 2020. The carbon content of gas is 0.05 kg C/kWh, so provided gas is used at an efficiency of 65% or more, there are carbon savings from the more extensive use of gas in all European countries, except Austria, France and Sweden. Even with these countries, if increased gas use displaces low-carbon electricity that can be sold elsewhere in Europe, for instance through the inter-connector to the UK, then there are net carbon reductions.

At state level, however, there are several reasons for a reluctance to promote one fuel over another. The priorities vary in different member states, but include:

- the problems of favouring one nationalised industry over another and of favouring one private company over another;
- the problems of agreeing the units of measurement (e.g., price, primary energy, delivered energy, useful energy and, now, carbon dioxide emissions) as this affects the comparison;
- changing conditions over time, for instance, the carbon content of electricity has dropped by 30% in the last seven years in the UK;
- the need to retain the goodwill of both gas and electricity industries where they fund energy savings organisations, as is the case with the UK Energy Saving Trust;
- the concerns of those industries that manufacture products solely for one fuel, including utilities that sell only one fuel.

Some of these barriers will be reduced as, increasingly, utilities sell both fuels to domestic customers and the retail outlets are energy centres, rather than gas or electricity showrooms. The liberalised energy market in Europe will have the same effect. Conversely, common standards on frequency of bills, information provided, metering units and other administrative measures will by required by increased liberalisation and could result in a greater demand for carbon information. The trade in green electricity is one example of where consumers are going to require consistent information across the utilities in the EU.

The reluctance to promote fuel switching to natural gas does not exist in all EU countries. For example, the Danish government has set up the Electricity Saving Trust with the express objective of reducing electricity demand, particularly through fuel switching (www.ens.dk). However, in the UK fuel switching is not, and has not (Boardman, 1991), been encouraged either specifically by government or by other energy agencies, such as the electricity and gas regulator, OFGEM (Office of gas and electricity markets) or the EST. Nor has fuel switching been promoted by the gas utilities, even though they would benefit most.

5 Labels for Gas and Electrical Appliances

The recent and ongoing studies on appliances that use both gas and electricity, for instance ovens (Kasanen, 1999), space and water heating, and the question of fuel switching raises the issue of how these two fuels are to be treated under an energy-labelling scheme. This is one clear area for EU policy. The key problem with comparative labelling for gas and electric appliances is that their relative rankings change depending on the unit of measurement chosen (Table III). For instance, a gas appliance generally uses more delivered energy but less primary energy than an electric appliance. And, as explained above, a gas appliance will generally emit less carbon dioxide than an equivalent electric appliance. This will also vary between member states, depending on the carbon intensity of electricity, and will vary through time as fuel sources for electricity production change.

Table III: Comparative measures for gas and electric appliances

Measurement	Gas	Electricity
delivered energy consumption	More	less
primary energy consumption	Less	more
carbon dioxide emissions*	Less	more
running costs	Less	more

*See Table II

The EU Energy Label, found on most of the major electrical appliances and light bulbs, contains two key pieces of information:
- the relative ranking of an appliance – the A-G scale;
- the absolute energy consumption of an appliance under the appropriate test procedure.

5.1 Options for A-G Scale

The advantages and disadvantages of various options for the A-G scale on an energy label are set out (Table IV).

Table IV: Options for A-G scale on gas and electric Energy Labels

Options for A-G scale	Advantages	Disadvantages
a common scale based on primary energy or carbon	• indicates true environmental cost of each fuel to consumer • may help in fuel switching away from electricity	• all gas appliances bunched together at top end of scale, all electric at bottom • problems of changing ratio of primary to delivered over time, plus different ratios in each member state • scale does not match energy figures

a common scale based on delivered energy	• scale will match energy consumption figures	• misleading to consumer in terms of environmental impact of different fuels • all electric appliances bunched together at top end of scale, all gas at bottom • may encourage fuel switching towards electricity
separate scales based on delivered energy	• scale will match energy consumption figures for separate fuels • makes use of full range of A-G scale for both electric and gas appliances • neutral on fuel switching, thus should be more acceptable than other options to manufacturers	• does not help (and may even hinder) the consumer make environmental fuel switching choice • need to design distinctive gas Energy Label to minimise confusion

From this analysis, a good option is separate scales based on delivered energy. This option offers useful information to the consumer, avoids very considerable practical problems involved in using primary energy as the basis for labelling, and should be more acceptable to industry than other options. In addition, it will fulfil the primary aim of the labelling directive in enabling the consumer to choose a more efficient appliance. However, under this scheme, energy labelling would not be an instrument of fuel switching policy.

If it is thought critical that energy labels contribute to fuel switching, then the option to choose would be that of using a common scale based on primary energy or carbon.

This is an interim recommendation, as the whole subject of fuel comparisons needs to be addressed by the EU, to include household-level renewables. The Energy Label may have to become a carbon label.

5.2 Options for Reported Absolute Energy Use

For gas, the simplest and best option is to use the same principles as applied to the measurement of electricity consumption. The values should be:
• based on delivered energy
• reported in units used by the consumer. These vary across Europe (m^3, kWh) which complicates the issue.

6 Supporting Fuel Switching

One of the key fuel-switching policies which national government must undertake is to accelerate the expansion of the gas network. Two alternative policy routes are suggested to meet this goal: firstly, the Government or Regulator could require the company in charge of the gas network to extend the gas network as an environmental objective. Alternatively, it may be possible to alter the pricing formula and place greater emphasis on the number of customers, rather than on the amount of gas sold.

The use of gas could also be consolidated by requiring that all new homes within the gas network have gas space and water heating installed with additional connection points for other gas appliances. Based on existing patterns of adoption for the UK, it is assumed that if the gas network is expanded, little in the way of incentives will be needed to persuade people to connect for space and water heating. Utilities in Portugal have developed successful strategies to persuade consumers to switch from LPG to natural gas. Incentives may be necessary in order to persuade them to also connect for gas hobs, ovens and tumble dryers; simply offering a financial incentive may not be sufficient to encourage or enable consumers to choose gas. The cost of installing gas pipes and the choice of fuel for cooking are important parameters for consumers.

Member states will have to promote the debate about fuel switching at a national level. As well as general advice and information, this could include, for instance, in the UK putting the Energy Saving Trust's energy efficiency logo on gas appliances, rather than electrical ones (e.g., gas hobs, not radiant electric hobs). Similarly, existing investment funds, such as those supporting housing improvements, could be channelled towards efficient gas appliances, rather than electrical ones, where there is a fuel choice.

7 Creating a Carbon Market

Developing a carbon market for domestic consumers is, at least initially, about awareness and education. The majority of private individuals in the UK do not realise that their activities result in carbon dioxide releases, or that these are the primary cause of climate change. There are other policy initiatives occurring that will create this awareness, for instance, the EU is requiring carbon dioxide emission levels to be shown on the energy labels for cars, mandatory from January 2001. A wider approach to education on emissions is timely and appropriate, and would be doubly effective if occurring in two household-related energy use sectors concurrently.

Creating a carbon market could be a powerful way of bringing together energy efficiency, renewable energy and fuel switching policies. Focussing on carbon reductions could help governments clarify the links between these sometimes disparate policy areas, and enable considerable carbon emission reductions to be made regardless of EU (in)action. In addition, informing consumers (and other actors) about their carbon impact empowers them to choose lower carbon futures for themselves.

Creating a carbon market involves actors other than simply consumers. The proposal suggested here is called average utility carbon per household, or AUCH. National governments would set sector targets for carbon reductions, and based on this would give each utility a reducing cap for carbon emissions. Initial allocation of emissions permits would be based on the number of customers, with separate allocations for gas and electricity use. From the government point of view, the great advantage of this scheme is that by setting carbon reduction targets so directly, it can be sure they will be achieved. The Electricity Regulator in Northern Ireland is already piloting schemes to highlight the carbon impact of supply decisions.

The utility will achieve lower average household carbon emissions through the investment in both lower carbon technologies (including renewable energy) and in reducing demand per household. The utility can encourage fuel switching by their customers, particularly if it is a dual-fuel utility, but this is not the only option available. The extent to which higher energy services will be available to consumers will depend upon improvements in energy efficiency and the use of fuel switching options. One of the interesting implications of AUCH is that the energy companies would have a strong interest in ensuring that appliances become more efficient, as this helps to reduce the annual carbon emissions per household. Discussions on industry agreements in Brussels would no longer be a debate between government and the appropriate trade association. The utilities would be alongside the government representatives trying to persuade the manufacturing industries to be more ambitious in their efficiency targets.

There are various ways in which information could be provided to the householder, but the simplest way would be for fuel bills to include carbon dioxide emissions, and for this to be compared to the average level for similar households. Research would be required as to whether the information should be on the normal quarterly bill, whether there is an annual total, how to combine different fuels and what form of comparison would be most appropriate. Informative bills have been shown to produce useful energy savings. This method of information provision would not work for bottled LPG, so an alternative mechanism would need to be found.

Having provided information on consumption and emissions to consumers via energy bills, further information could be provided via a carbon audit of the home, so that all properties were graded in terms of carbon emissions. This would ensure that household emission levels had an importance and status and can be built into other policy initiatives. The eventual aim would be to have a formal ranking system, such as existing household energy audit systems, so that properties and households could be graded from highest to lowest levels of carbon emissions. As householders become carbon aware, they can make across-fuel decisions on an environmental basis.

Increased information to consumers and AUCH should be combined, so that both householders and utilities are moving towards the common goal of reducing carbon emissions. There is a strong logic in involving both energy users and suppliers in the climate change targets, through using the same method of measurement – carbon. If the duty is placed on only one of the players, then the actions of the other could offset any progress made in carbon conservation.

8 Discussion

Market transformation is the current philosophy underlying much of EU and member state's energy efficiency policies on domestic appliances. This approach has a great deal to offer in delivering more efficient equipment to consumers, and protects lower income consumers from the disadvantages of an energy or carbon tax. Fuel switching raises policy issues which extend the scope of market transformation policy beyond energy efficient products to carbon efficiency.

With the introduction of a competitive fuel – gas – into the assessments of carbon savings, policy can no longer focus solely on the equipment: account has to be taken of the system, the household and even the availability of the fuel. This will be equally true when other fuel switching options are considered, for instance household-level renewables such as solar thermal and photovoltaics, and small-scale combined heat and power.

Labels for appliances where there are competing fuels is a problematic policy area but still an EU concern. With the present diversity between member states, it is recommended that the labels for gas and electric appliances are separate and based on delivered energy. This will increase the importance of national policies on fuel priorities if appropriate switching is to occur. As renewable energy products become more prevalent, the issue of labels will have to be revisited: how would gas, electric and solar water heaters be compared? The EU energy label may have to become the EU carbon label; this would fit with policies on car labels.

The variations across Europe in gas ownership and the carbon content of electricity mean that fuel-switching policy is country specific at present. This puts the onus onto the member states to develop their own policies on fuel-switching.

The main policy focus should be on educating consumers on the carbon impact of their activities (for instance, through informative energy bills) and on achieving a greater coherence between energy supply and energy demand decisions. These policies work together in developing a carbon market.

The policy agenda is becoming much broader, and requires greater integration between national and EU policies. The method of regulating the utilities may be as important as agreements with appliance manufacturers. A policy agenda that encompasses everything from individual products to European policies on energy liberalisation is particularly challenging. But the efficiency of a combined cycle gas turbine producing electricity is about 50%, whereas burning the same gas in a domestic boiler would achieve efficiencies of 75% or more quite easily. There are real carbon reductions available from a focus on household fuel switching and these would make an important contribution to the Kyoto targets and to improved sustainability after 2010.

9 References

[1] Boardman, B. (1991) *Fuel Poverty*. Bellhaven Press, London.
[2] Fawcett, T., K. Lane, B. Boardman *et al.* (2000) *Lower Carbon Futures*. Environmental Change Institute, University of Oxford.
[3] Griffin, H. and T. Fawcett. (2000) *Country Pictures*, supporting document for *Lower Carbon Futures*. Environmental Change Institute, University of Oxford.
[4] Kasanen, P. (1999) *EU domestic ovens, SAVE study*. TTS, Finland.

Acknowledgements

This work is based on the results from a SAVE and UK DETR-funded project, *Lower Carbon Futures*. Further information can be obtained from Fawcett *et al.*, 2000 (the main report); our web site, where several appendices are also posted, www.eci.ox.ac.uk; and Griffin and Fawcett, 2000, the *Country pictures*. The DETR contribution is part of the Market Transformation Programme. I am indebted to both these funding agencies for enabling this research to be undertaken. I have relied heavily on contributions from the other members of the Energy and Environment Programme. My thanks to you all.